Contaminants of Emerging Concern in the Environment: An Analytical Framework

Contaminants of Emerging Concern in the Environment

An Analytical Framework

Edited by

Núria Fontanals Torroja
Analytical and Organic Chemistry Department,
Universitat Rovira i Virgili, Tarragona, Spain

Rosa Montes Goyanes
Department of Analytical Chemistry, Nutrition and Food Science &
Aquatic One Health Research Center (ARCUS), Universidade de
Santiago de Compostela, Spain

Noelia Salgueiro-González
Environmental Health Sciences Department, Istituto di Ricerche
Farmacologiche Mario Negri—IRCCS, Milan, Italy

ELSEVIER

Elsevier
Radarweg 29, PO Box 211, 1000 AE Amsterdam, Netherlands
125 London Wall, London EC2Y 5AS, United Kingdom
50 Hampshire Street, 5th Floor, Cambridge, MA 02139, United States

ISBN: 978-0-443-33821-2

For information on all Elsevier publications visit our website at https://www.elsevier.com/books-and-journals

Publisher: Peter Llewellyn
Acquisitions Editor: Yashila Girdhar
Editorial Project Manager: Abdus Salam Mazumder
Production Project Manager: Fahmida Sultana
Cover Designer: Bakyalakshmi S

Typeset by Integra Software Services

Contents

Part 2
Sampling and sample preparation techniques for environmental samples

9 Suspect and nontarget screening in environmental analysis: Principles, current directions and real-world applications

Alberto Celma, Lidia Belova, Rubén Gil-Solsona, and Adrià Sunyer-Caldú

10 Portable techniques and sensors for environmental analysis

Saidy Cristina Ayala-Durán, Matheus Schiavon Kronka, Hernán Dario Rojas-Mantilla, Maria Valnice Boldrin Zanoni, and Paulo Clairmont Feitosa de Lima Gomes

Part 4
Quality assurance, quality control, and validation in environmental analysis

Contributors

Iker Alvarez-Mora Faculty of Science and Technology, Department of Analytical Chemistry, University of the Basque Country (UPV/EHU) Leioa; Research Centre for Experimental Marine Biology and Biotechnology (Plentzia Marine Station; PiE-UPV/EHU), University of the Basque Country (UPV/EHU), Plentzia, Basque Country, Spain

Luan Valdemiro Alves De Oliveira Food Science and Technology Department, Federal University of Santa Catarina (UFSC), Florianópolis, Brazil

Juan F. Ayala-Cabrera Faculty of Science and Technology, Department of Analytical Chemistry, University of the Basque Country (UPV/EHU), Leioa; Research Centre for Experimental Marine Biology and Biotechnology (Plentzia Marine Station; PiE-UPV/EHU), University of the Basque Country (UPV/EHU), Plentzia, Basque Country, Spain

Saidy Cristina Ayala-Durán Department of Analytical, Physical Chemistry and Inorganic Chemistry, Institute of Chemistry, National Institute of Alternative Technologies for Detection, Toxicological Assessment and Removal of Micropollutants and Radioactive—INCT-DATREM, São Paulo State University (UNESP), São Paulo, Brazil

Rodrigo Barcellos Hoff Federal Laboratory of Animal and Plant Health and Inspection, Ministry of Agriculture and Livestock, São José, Brazil

Lidia Belova Toxicological Center, University of Antwerp, Wilrijk; Toxicological Centre, University of Antwerp, Antwerp, Belgium

Maria Valnice Boldrin Zanoni Department of Analytical, Physical Chemistry and Inorganic Chemistry, Institute of Chemistry, National Institute of Alternative Technologies for Detection, Toxicological Assessment and Removal of Micropollutants and Radioactive—INCT-DATREM, São Paulo State University (UNESP), São Paulo, Brazil

Tim Boogaerts Toxicological Center, University of Antwerp, Wilrijk; Exposome Center of Excellence, University of Antwerp, Antwerp, Belgium

Magda Caban Faculty of Chemistry, Department of Environmental Analysis, University of Gdansk, Gdansk, Poland

Gabriela Castro Department of Analytical Chemistry, Nutrition and Food Chemistry, Aquatic One Health Research Center (ARCUS), Universidade de Santiago de Compostela, Santiago de Compostela, Spain

Rafael Cela Mestrelab Research Center (CIM), Santiago de Compostela, Spain

Alberto Celma Department of Aquatic Sciences and Assessment, Swedish University of Agricultural Sciences, Uppsala, Sweden; Environmental and Public Health Analytical Chemistry (Q-AMS), Research Institute fos Pesticides and Water (IUPA), University Jaume I, Castellón de la Plana, Castellón, Spain

Luke Chimuka Department of Chemistry, University of the Witwatersrand, Johannesburg, FL, South Africa

Adrian Covaci Toxicological Center, University of Antwerp, Wilrijk; Exposome Center of Excellence, University of Antwerp, Antwerp, Belgium

Luana de Souza Futigami Food Science and Technology Department, Federal University of Santa Catarina (UFSC), Florianópolis; Federal Laboratory of Animal and Plant Health and Inspection—LFDA/RS, Ministry of Agriculture and Livestock, São José, Brazil

Fatima den Ouden Toxicological Center, University of Antwerp, Wilrijk; Exposome Center of Excellence, University of Antwerp, Antwerp, Belgium

Milena Dutra Pierezan Food Science and Technology Department, Federal University of Santa Catarina (UFSC), Florianópolis, Brazil

Paulo Clairmont Feitosa de Lima Gomes Department of Analytical, Physical Chemistry and Inorganic Chemistry, Institute of Chemistry, National Institute of Alternative Technologies for Detection, Toxicological Assessment and Removal of Micropollutants and Radioactive—INCT-DATREM, São Paulo State University (UNESP), São Paulo, Brazil

Núria Fontanals Department of Analytical Chemistry and Organic Chemistry, University Rovira i Virgili, Tarragona, Spain

Rubén Gil-Solsona Institute of Environmental Assessment and Water Research (IDAEA-CSIC), Barcelona, Spain

Belén González-Gaya Faculty of Science and Technology, Department of Analytical Chemistry, University of the Basque Country (UPV/EHU), Leioa, Basque Country, Spain; Research Centre for Experimental Marine Biology and Biotechnology (Plentzia Marine Station; PiE-UPV/EHU), University of the Basque Country (UPV/EHU), Plentzia, Basque Country, Spain; IKERBASQUE, Basque Foundation for Science, Bilbao, Spain

Irina Guerra-Martín Laboratorio de Materiales para Análisis Químico (MAT4LL), Departamento de Química, Facultad de Ciencias; Unidad de Investigación de Bioanalítica y Medio Ambiente, Instituto Universitario de Enfermedades Tropicales y Salud Pública de Canarias (IUETSPC), Universidad de La Laguna, San Cristóbal de La Laguna, Spain

Celine Gys Toxicological Center, University of Antwerp, Wilrijk; Exposome Center of Excellence, University of Antwerp, Antwerp, Belgium

Mireia Irazola Faculty of Science and Technology, Department of Analytical Chemistry, University of the Basque Country (UPV/EHU), Leioa; Research Centre for Experimental Marine Biology and Biotechnology (Plentzia Marine Station; PiE-UPV/EHU), University of the Basque Country (UPV/EHU), Plentzia, Basque Country, Spain

Cristian Rafael Kleemann Food Science and Technology Department, Federal University of Santa Catarina (UFSC), Florianópolis; Federal Laboratory of Animal and Plant Health and Inspection—LFDA/RS, Ministry of Agriculture and Livestock, São José, Brazil

Hanna Lis Faculty of Chemistry, Department of Environmental Analysis, University of Gdansk, Gdansk, Poland

Naroa Lopez-Herguedas Faculty of Science and Technology, Department of Analytical Chemistry, University of the Basque Country (UPV/EHU), Leioa; Research Centre for Experimental Marine Biology and Biotechnology (Plentzia Marine Station; PiE-UPV/EHU), University of the Basque Country (UPV/EHU), Plentzia, Basque Country, Spain; Department of Cell Toxicology, Helmholtz Centre for Environmental Research—UFZ, Leipzig, Germany

Lawrence Mzukisi Madikizela College of Science, Engineering and Technology, Institute for Nanotechnology and Water Sustainability, University of South Africa, Florida Science Campus, Roodepoort, South Africa

Norberto Medina-Rodríguez Laboratorio de Materiales para Análisis Químico (MAT4LL), Departamento de Química, Facultad de Ciencias; Unidad de Investigación de Bioanalítica y Medio Ambiente, Instituto Universitario de Enfermedades Tropicales y Salud Pública de Canarias (IUETSPC), Universidad de La Laguna, San Cristóbal de La Laguna, Spain

Rosa Montes Goyanes Department of Analytical Chemistry, Nutrition and Food Science & Aquatic One Health Research Center (ARCUS), Universidade de Santiago de Compostela, Spain

Carles Moreu-Romero Faculty of Science and Technology, Department of Analytical Chemistry, University of the Basque Country (UPV/EHU), Leioa; Research Centre for Experimental Marine Biology and Biotechnology (Plentzia Marine Station; PiE-UPV/EHU), University of the Basque Country (UPV/EHU), Plentzia, Basque Country, Spain

Mikel Musatadi Faculty of Science and Technology, Department of Analytical Chemistry, University of the Basque Country (UPV/EHU), Leioa; Research Centre for Experimental Marine Biology and Biotechnology (Plentzia Marine Station; PiE-UPV/EHU), University of the Basque Country (UPV/EHU), Plentzia, Basque Country, Spain; Toxicological Centre, University of Antwerp, Wilrijk, Belgium

Somandla Ncube Department of Chemistry, Durban University of Technology, Durban, South Africa

Monika Paszkiewicz Faculty of Chemistry, Department of Environmental Analysis, University of Gdansk, Gdansk, Poland

Leticia Pérez-Mayán Mestrelab Research Center (CIM), Santiago de Compostela, Spain

Verónica Pino Laboratorio de Materiales para Análisis Químico (MAT4LL), Departamento de Química, Facultad de Ciencias; Unidad de Investigación de Bioanalítica y Medio Ambiente, Instituto Universitario de Enfermedades Tropicales y Salud Pública de Canarias (IUETSPC) Universidad de La Laguna, San Cristóbal de La Laguna, Spain; Centro de Investigación Biomédica en Red de Enfermedades Infecciosas (Ciberinfec), Instituto de Salud Carlos III, Madrid, Spain

Maarten Roggeman Toxicological Center, University of Antwerp, Wilrijk; Exposome Center of Excellence, University of Antwerp, Antwerp, Belgium

Hernán Dario Rojas-Mantilla Department of Analytical, Physical Chemistry and Inorganic Chemistry, Institute of Chemistry, National Institute of Alternative Technologies for Detection, Toxicological Assessment and Removal of Micropollutants and Radioactive—INCT-DATREM, São Paulo State University (UNESP), São Paulo, Brazil

Noelia Salgueiro-González Department of Environmental Health Sciences, Istituto di Ricerche Farmacologiche Mario Negri—IRCCS, Milan, Italy

Matheus Schiavon Kronka Department of Analytical, Physical Chemistry and Inorganic Chemistry, Institute of Chemistry, National Institute of Alternative Technologies for Detection, Toxicological Assessment and Removal of Micropollutants and Radioactive—INCT-DATREM, São Paulo State University (UNESP), São Paulo, Brazil

Benigno José Sieira Mestrelab Research Center (CIM), Santiago de Compostela, Spain

Laura Solé-Domènech Department of Analytical Chemistry and Organic Chemistry, Rovira i Virgili University, Tarragona, Spain

Adrià Sunyer-Caldú Science for Life Laboratory, Department of Environmental Science (ACES, Exposure & Effects), Stockholm University, Stockholm, Sweden

María José Trujillo-Rodríguez Laboratorio de Materiales para Análisis Químico (MAT4LL), Departamento de Química, Facultad de Ciencias; Unidad de Investigación de Bioanalítica y Medio Ambiente, Instituto Universitario de Enfermedades Tropicales y Salud Pública de Canarias (IUETSPC), Universidad de La Laguna, San Cristóbal de La Laguna, Spain

Carolina Turnes Pasini Deolindo Food Science, Federal University of Santa Catarina (UFSC), Florianópolis; Federal Laboratory of Animal and Plant Health and Inspection—LFDA/RS, Ministry of Agriculture and Livestock, São José, Brazil

Laura Vallecillos Department of Analytical Chemistry and Organic Chemistry, Rovira i Virgili University, Tarragona, Spain

Alexander van Nuijs Toxicological Center, University of Antwerp, Wilrijk; Exposome Center of Excellence, University of Antwerp, Antwerp, Belgium

Acknowledgments

The editors want to express their sincere gratitude to all the authors who generously contributed their time, expertise, and dedication to the preparation of each chapter included in this book. We deeply appreciate their commitment and understanding throughout the entire process. This book would not have been possible without their valuable contributions. Thank you all.

Núria, Noelia, and Rosa

Introduction

Contaminants of emerging concern (CECs) are increasingly detected across various environmental compartments, often at trace concentrations in the low microgram or nanogram per liter/kilogram/cubic meter range depending on the type of sample. These substances encompass diverse chemical families, arise from multiple sources, and exhibit a wide spectrum of physicochemical properties. In response, highly sensitive and selective analytical methods have been developed to identify and quantify different classes of CECs in a broad array of environmental samples.

This book provides a comprehensive overview of the analytical methods developed from 2010 to the present for the determination of different families of CECs including pharmaceuticals and personal care products, illicit drugs, pesticides, industrial chemicals, and persistent, mobile, and toxic chemicals, among others. Their physicochemical properties, molecular structures, relevant legislation, and environmental occurrence are discussed in Part 1.

The following chapters present detailed information on the different steps of the analytical workflow—from sampling to final determination—organized according to the type of environmental sample (solid, liquid, or gaseous). Part 2 focuses on sampling and sample preparation techniques and is subdivided into: sampling and extraction of atmospheric samples (Chapter 2); sampling of water samples (Chapter 3); extraction of water samples (Chapter 4); miniaturized and sustainable extraction techniques for water samples (Chapter 5); and sampling and extraction of solid samples (Chapter 6).

Part 3 covers the determination of CECs in environmental samples, including chromatographic techniques—such as gas chromatography (Chapter 7) and liquid chromatography (Chapter 8)—along with suspect and nontarget screening approaches (Chapter 9) and portable analytical techniques and sensors (Chapter 10). Finally, Part 4 addresses quality assurance, quality control, and method validation.

Each chapter describes the analytical methods currently in use, enabling readers to identify the techniques most commonly applied in the field. In addition, cutting-edge technologies and future trends are summarized. All of them are supported with tables to provide a concise overview of the state of the art for each specific part of the analytical process. Furthermore, selected case studies are included to illustrate the practical applicability of these methods in the determination of CECs.

This book offers a comprehensive overview of the various analytical techniques actually used to determine CECs across different types of environmental samples. It provides essential, up-to-date information on the state of the art in CEC determination. The editors hope that this book will serve as a valuable resource for researchers, industry professionals, students, policymakers, and interested audiences seeking to understand the approaches used to monitor CECs in the environment.

We would like to acknowledge the authors for their invaluable contributions. Each contributor to this book is a skilled professional and recognized expert in environmental trace analysis, particularly within the specific part of the analytical method discussed in their respective chapters. This book would not have been possible without their dedication and expertise. We also extend our grateful thanks to our readers for their interest and engagement. We hope you find this book informative and enjoyable.

Rosa Montes Goyanes
Noelia Salgueiro-González
Núria Fontanals

Part 1

Contaminants of emerging concern

Chapter 1

Contaminants of emerging concern

Núria Fontanals[a], Noelia Salgueiro-González[b], and Rosa Montes Goyanes[c]
[a]*Department of Analytical Chemistry and Organic Chemistry, University Rovira i Virgili, Tarragona, Spain*
[b]*Department of Environmental Health Sciences, Istituto di Ricerche Farmacologiche Mario Negri—IRCCS, Milan, Italy*
[c]*Department of Analytical Chemistry, Nutrition and Food Science & Aquatic One Health Research Center (ARCUS), Universidade de Santiago de Compostela, Spain*

1 Industrial chemicals

Continuous industrial and commercial development has resulted in the production of vast quantities of chemical compounds used in industry for several applications such as flame retardants, plasticizers, additives, and lubricants, among others. These compounds, produced in quantities exceeding 1000 tons per year in at least one member nation, are classified as high production volume chemicals (HPVCs) in countries belonging to the Organization for Economic Cooperation and Development (OECD). Due to their widespread and intensive use, and the fact that most are not chemically bound to their substrates, these compounds are readily released into the environment and are detected across all environmental compartments.

This section primarily focuses on several widely used industrial chemical classified as contaminants of emerging concern (CECs) including perfluoroalkyl substances (PFAS), phthalate esters (PAEs), organophosphate esters (OPEs), bisphenols (BPs), benzothiazoles (BTHs), benzotriazoles (BTRs), and benzenesulfonamides (BSAs), among others.

1.1 Structure and properties

1.1.1 Per- and polyfluoroalkyl substances

Per- and polyfluoroalkyl substances (PFAS) encompass a wide range of families, each containing various homologous compounds and their isomers. Among the most extensively studied are the perfluoroalkyl acids (PFAAs), which include perfluorocarboxylic acids (PFCAs) and perfluorosulfonic acids (PFSAs) (Table 1.1).

Contaminants of Emerging Concern in the Environment. https://doi.org/10.1016/B978-0-443-33821-2.00012-8

TABLE 1.1 Examples of some industrial chemical and their abbreviation organized by families of compounds.

PFAS	Acronym	Main structures
Perfluorooctanoic acid	PFOA	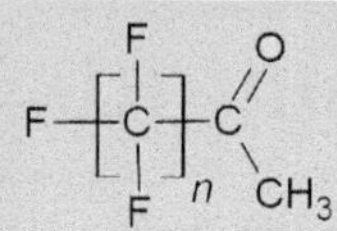
Perfluorooctane sulfonic acid	PFOS	Perfluorocarboxylic acids (PFCAs)
Perfluoralkyl iodides	PFAIs	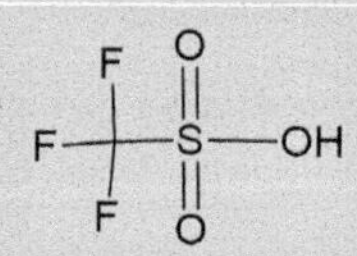
Perfluoroether carboxylic acid; trade name: GenX	PFECAs	
Fluorotelomer olefins	FTOs	
Fluorotelomer acrylates	FTAs	Perfluorosulfonic acids (PFSAs)
Fluorotelomer alcohols	FTOHs	
Perfluorooctane sulfonamides	FOSAs	
Perfluoralkane sulfonamidoethyl acrylates	FASACs	
PAEs		
Butyl benzyl phthalate	BBzP	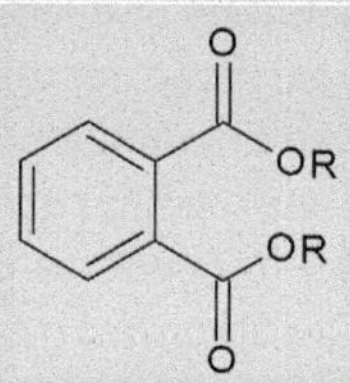
Di(2-ethylhexyl) phthalate	DEHP	
Diethyl phthalate	DEP	
Diisobutyl phthalate	DiBP	
Dimethyl phthalate	DMP	
Di-n-butyl phthalate	DnBP	
Di-n-octyl phthalate	DnOP	
OPEs		
Tris(2-chloroethyl) phosphate	TCEP	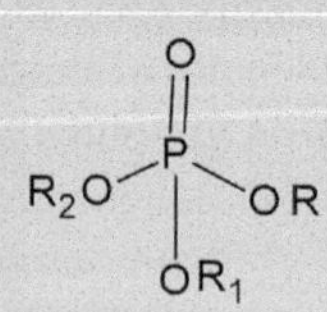
Tris(1-chloro-2-propyl) phosphate	TClPP	
Tris(2-butoxyethyl) phosphate	TBEP	
BPs		
Bisphenol A	BPA	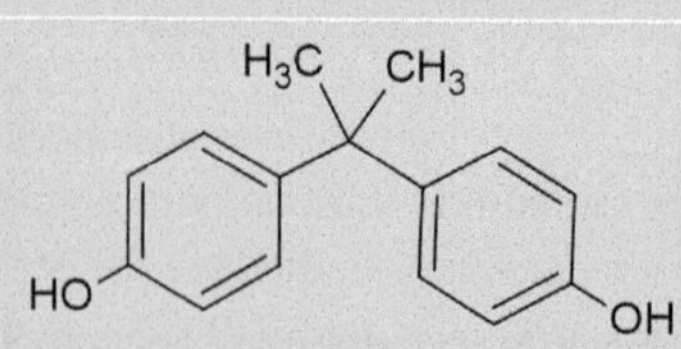
Bisphenol S	BPS	
Bisphenol F	BPF	
Bisphenol AF	BPAF	BPA

PFAS	Acronym	Main structures
BSAs		
Benzenosulfonamide	BSA	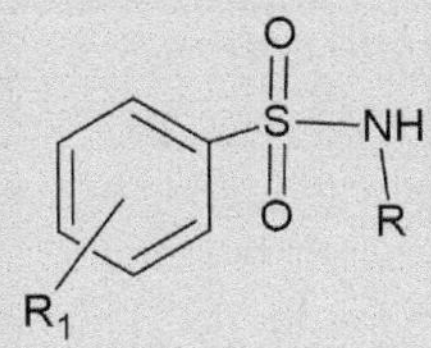
o-toluenesulfonamide	o-TSA	
p-toluenesulfonamide	p-TSA	
BTRs		
Benzotriazole	BTR	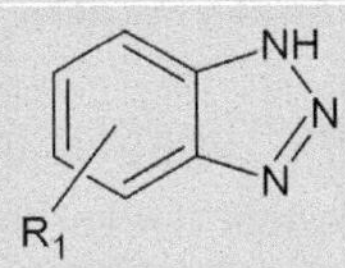
5-chloro-1H-benzotriazole	ClBTR	
4-methyl-1H-benzotriazole (4-tolyltriazole)	4TTR	
5-methyl-1H-benzotriazole	5TTR	
BTHs		
Benzothiazole	BTH	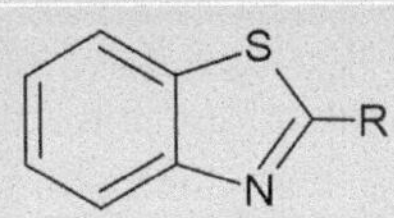
Benzothiazole-2-sulfonic acid	SO_3BTH	
2-methylthiobenzothiazole	MeSBTH	
2-aminobenzothiazole	NH_2BTH	
2-hydroxybenzothiazole	OHBTH	

Source: https://pubchem.ncbi.nlm.nih.gov/: created by the authors.

PFAAs and their anions are frequently referred to as "long-chain" (PFCAs with eight or more carbon atoms, and PFSAs with six or more) or "short-chain," based on the length of their perfluorinated carbon chains. Table 1.1 lists some of the most popular PFAS and the abbreviation used in this chapter. Notable examples include perfluorooctanoic acid (PFOA) and perfluorooctane sulfonic acid (PFOS). Other families are those based on perfluoroalkyl iodides or fluorotelomer olefins, among others (Buck et al., 2011). The common properties of these molecules are that they feature hydrophobic tails and hydrophilic heads, typically containing carboxylic or sulfonic groups. Their remarkable stability arises from the strong carbon-fluorine (C–F) bond—the strongest covalent bond in organic chemistry—due to fluorine's high electronegativity and small size. Fluorine atoms also provide effective shielding and possess nonbonding electron pairs, contributing to PFAS's exceptional thermal and chemical resilience. As a result, PFAS are nonflammable, thermodynamically stable, and highly resistant to degradation by heat, light, chemicals, or biological processes. Owing to these properties and their lipophobicity, PFAS compounds and polymers have been extensively used since the 1950s in industrial applications such as surfactants, water-repellent sprays, lubricants, adhesives, coatings, and paints. In addition, they have been used in a number of commercial applications as nonstick cookware, food packaging, insecticides, and waterproof and fire-resistant fabrics (Glüge et al., 2020; Young et al., 2007).

1.1.2 *Phthalate esters*

Phthalate esters (PAEs) are alkyl-substituted derivatives of phthalic acid (Table 1.1) characterized by varying log K_{ow} values and degrees of volatility. Depending on the alkyl chain length, their log K_{ow} values range from 1.65 for dimethyl phthalate (DMP) to 12.06 for ditridecyl phthalate (DTDP). PAEs are primarily used as plasticizers that are not chemically bonded to materials, serving a key role in numerous consumer products and industrial applications (Net et al., 2015).

1.1.3 *Organophosphate esters*

Organophosphate esters (OPEs) have largely replaced brominated compounds as new-generation flame retardants, and some also serve as plasticizers. Among the most well-known OPEs are tris(2-chloroethyl) phosphate (TCEP) and tris(1-chloro-2-propyl)phosphate (TClPP). Due to their dual roles in preventing fire ignition and altering plastic properties, OPEs are found across a wide range of applications, including industrial products, electronics, textiles, and furniture. Structurally, these compounds feature a phosphate core bonded to alkyl or halogenated groups (Table 1.1). Their physicochemical properties vary significantly, and in generally, solubility decreases when molecular mass increases. When hydrolysis half-life are similar, the OPEs with lower molecular mass are more likely to be found in the aquatic environment than those with higher masses, which is confirmed by their logK_{ow} values (ranging from ca. −9.8 to 10.6) (van der Veen and de Boer, 2012).

1.1.4 *Bisphenols*

Bisphenols (BPs) are synthetic chemicals with two hydroxyphenyl functional groups bonded to a specific skeleton (Table 1.1). Bisphenol A (BPA) was early introduced in the mid-20th century in plastic manufacturing; however, its toxic effect was soon noticed, leading to the emergence of some alternatives such as bisphenol F (BPF) or bisphenol S (BPS). BPs are highly used in plastic production due to their low vapor pressure and low water solubility. Their general low to moderate solubility implies that they will prone to be sorbed to organic matter and surfaces. Their pK_a values indicate they predominantly exist in the neutral form in the environment, with exception of BPS. Due to the presence of the sulfone group, BPS is partially present in its anionic form, making it more water soluble and increasing its acidic character (Czarny-Krzymińska et al., 2023).

1.1.5 *Benzenesulfonamides, benzotriazoles, and benzothiazoles*

Benzenesulfonamides (BSAs) are alkyl-substituted benzene rings bearing an N-substituted sulfonamide group (Table 1.1), and they are used both as plasticizers and as intermediates in the synthesis of pesticides, pharmaceuticals, and artificial sweeteners. Benzotriazoles (BTRs) and benzothiazoles (BTHs) are heterocyclic compounds, with BTRs containing a five-membered ring with three nitrogen atoms while BTHs featuring a 1,3-thiazole ring fused to a benzene ring (Table 1.1). BTH derivatives include 2-substituted forms such as 2-hydroxybenzothiazole (OHBTH), 2-aminobenzothiazole (NH_2BTH), benzothiazole-2-sulfonic acid (SO_3BTH), and 2-methylthiobenzothiazole (MeSBTH), with log K_{ow} values ranging from—0.99 (SO_3BTH) to 3.22 (MeSBTH). These compounds are commonly used as vulcanization accelerators, fungicides, and corrosion inhibitors (Liao et al., 2018). In contrast, BTRs are highly polar compounds with low log K_{ow} values, often used as anticorrosion agents in antifreeze fluids, detergents, and other corrosion-sensitive products. They also serve as antifogging and antifungal agents. BTR derivatives include mono- and polymethylated compounds such as 4- and 5-methyl-1H-benzotriazole (4TTR and 5TTR), 5,6-dimethyl-1H-benzotriazole (XTR), and halogenated variants such as 5-chlorobenzotriazole (ClBTR). In addition, BTRs can be bonded to phenolic groups to form benzotriazole UV stabilizers (BUVs), also known as Tinuvins, which are primarily used to protect dairy packaging and other materials from UV degradation (Shi et al., 2019).

1.2 Legislation

Among the industrial chemicals, PFAS stand out as the most restricted due to their persistent, bioaccumulative, and toxic nature, making them a major focus of environmental regulation. Within the European Union, PFOS has been restricted for over a decade under the Persistent Organic Pollutants (POPs) Regulation (EU) 2019/1021 (European Union, 2019). Both PFOS and PFOA are listed under the Stockholm Convention, which includes 152 signatory countries, as substances that should be eliminated or banned from production and use (United Nations Environment Programme, 2010). The most recent EU legislation addressing PFAS is Regulation (EU) 2024/2462 (European Commission, 2024), which amends Annex XVII of the REACH Regulation (EC) No. 1907/2006 (European Union, 2006). This amendment restricts the use of undecafluorohexanoic acid (PFHxA), its salts, and related compounds due to their extreme persistence, mobility in aquatic environments, and potential risks to human health and ecosystems. PFAS have also been included in the EU Drinking Water Directive (EU) 2020/2184 (European Union, 2020), which sets maximum allowable concentrations of 0.1 μg/L for the sum of 20 specified PFAS of concern and 0.5 μg/L for total PFAS. Similarly, in the United States, the Environmental Protection Agency (EPA) finalized its National Primary Drinking Water Regulation for six PFAS (i.e., PFOA, PFOS, PFHxS, PFNA, HFPO-DA, and PFBS) in April 2024 (Environmental Protection Agency, 2024). Furthermore, the Environmental Quality Standards (EQS) for PFOS, as defined by Directive 2013/39/EU, establish an annual average (AA) EQS of 6.5×10^{-4} μg/L for inland surface waters and 1.3×10^{-4} μg/L for other surface waters. The biota EQS for PFOS is set at 9.1 μg/kg wet weight (European Union, 2013).

In the case of BPs, BPA has been listed in the REACH Regulation (European Union, 2006) as substances of very high concern (SVHC) due to its endocrine-disrupting properties and reproductive toxicity. Consequently, BPA has been subject to use restrictions in countries worldwide. Other BPs, such as BPS and BPF, are currently under evaluation to be included in the SVHC list due to similar toxicity concerns. In case of drinking water, the most recent directive sets a limit value of 2.5 μg/L of BPA (European Union, 2020).

Regarding the other industrial chemicals, few are regulated in environmental samples and some of them have been banned or restricted in some products. Among them PAEs are the most regulated. For instance, butyl benzyl phthalate (BBzP), di-n-butylphathalate (DnBP), and di(2-ethylhexyl)phthalate (DEHP) have been added to the REACH Regulation list as SVHC (European Union, 2006). Six PAEs, namely DnBP, BBzP, DEHP, dimethyl phthalate (DMP), diethyl phthalate (DEP), and di-n-octyl phthalate (DnOP) are recognized as priority pollutants in water by regulatory agencies in the EU, USA, and China. In addition, the EU has proposed a guideline value of 1.3 μg/L for DEHP in inland surface waters (European Union, 2013). Environmental quality guidelines in Australia and New Zealand have also established trigger values for DMP, DEP, and BBzP at 5100, 1300, and 64.6 μg/L, respectively (Net et al., 2015). The World Health Organization (WHO) has set a guideline value of 8 μg/L for DEHP in drinking water (World Health Organization (WHO), 2008). National drinking water standards vary, with recommended maximum concentrations of 6 μg/L in the USA, 9 μg/L in Australia, 100 μg/L in Japan, and 10 μg/L in New Zealand (Net et al., 2015).

In the case of OPEs, BTRs, and BTHs, although several compounds can pose an important risk to human health and to the environment, only some of them, such as tris(2-chloroethyl)phosphate (TCEP) have been included in the REACH list of SVHC (European Union, 2006). This chemical has been also included in the US EPA's Fifth Contaminant Candidate List (CCL5) (Environmental Protection Agency, 2022). However, as the use of these industrial chemicals continues to expand from food-contact and consumer products into the broader environment, it is expected that more stringent regulations will be introduced, particularly concerning their presence in surface waters, wastewater effluents, and industrial discharges.

1.3 Occurrence in the environment

Industrial chemicals have been found in many environmental compartments. Fig. 1.1 shows the number of publications found in Scopus using the query string TITLE-ABS-KEY(*compound name) AND TITLE-ABS-KEY(*contaminant) AND PUBYEAR > 2010 for the most representative compounds included as examples of each category in Table 1.1. Among them, PFASs are the ones that have been ubiquitously reported, even in remote regions as Artic (Wong et al., 2018). For example, PFAS were early detected at pg/L concentration in Arctic

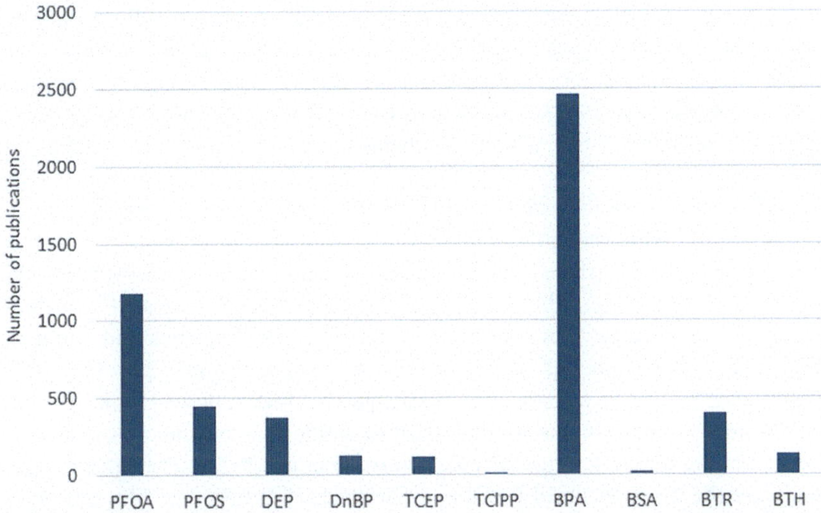

FIG. 1.1 Number of publications indexed in Scopus including some of the most relevant industrial chemicals considered in Table 1.1 using the search string "compound name" AND "contaminant" from 2010. *(Source: Scopus database accessed on 2025/11/25;. created by authors.)*

snow in studies conducted from 2006 to 2024 (Hartz et al., 2024; Young et al., 2007). Moreover, the formation of some PFAS has been linked to the radical-mediated atmospheric degradation of precursor compounds. In addition, the more volatile and neutral PFAS (nPFAS)—including FTOHs, perfluorooctane sulfonamides (FOSAs), perfluorooctane sulfonamidoethanols (FOSEs), perfluoroalkane sulfonamidoethyl acrylates (FASACs), FTOs, and FTAs—are primarily transported through the atmosphere. In contrast, PFCAs and PFSAs are ionizable and therefore more water-soluble than nPFAS. While long-chain PFAS persist in the environment for extended periods, emerging short-chain PFAS are also now increasingly detected due to their limited degradability (Janousek et al., 2019). Accordingly, PFAS in levels from nondetected to low ng/L (levels that are critical since even exposure at such levels may involve adverse effects) occurred in aqueous environmental samples in matrices as wastewater (Gebbink and van Leeuwen, 2020), groundwater (Zhang et al., 2025), surface (Baabish et al., 2021) as well as drinking water (Zhu et al., 2024). In this matrix, Kaboré et al. (2018) revised the occurrence of PFASs in drinking water collected from various countries worldwide, and in all places three PFAS (PFCAs, PFSAs, and perfluoroalkyl acid precursors) were frequently detected with concentrations ranging from below detection limits to 39 ng/L. Regarding to PFAS in air samples, a recent review (Zhang et al., 2025) revealed the detection of significant C_8 species, PFOA and PFOS as well as emerging PFAS, at levels spanning a broad range of pg/m^3. Nevertheless, PFAS tend to be strongly adsorbed in solid matrices with levels ranging from low ng/g in sediments (Guo et al., 2024) to hundreds ng/g in sewage sludge (Behnami et al., 2024). Moreover, due to their properties and high mobility in the environment, PFASs largely impacts on marine ecosystems through bioaccumulation and biomagnification (Lukić Bilela et al., 2023). In this sense, recently 29 perch samples were analyzed where the total PFAS concentration ranged from 0.27 to 3.5 ng/g, being the most prominent PFOS, which was present in all samples (0.27 to 3.5 ng/g) (Zacs et al., 2025).

PAEs, with DEHP followed by DnBP as the most important congeners, have been reported in different environmental samples within levels of concentration of the low μg/L in surface waters or ng/g in sediment samples (Sokołowski et al., 2024). In addition, PAEs, as plasticizers, have been suggested as microplastic tracers because they can be released from microplastics and occur in marine ecosystems, including seawater, sediments and biota. The most frequently detected PAEs in biota are DEHP, DnBP, DiBP followed by DEP, with concentration in muscle tissue samples from edible fish, crustaceans, and mollusks reaching the ng/g range (Hidalgo-Serrano et al., 2022). The same congeners were the most occurred PAEs in air samples. For instance, Maceira et al. (2020) determined these congeners in PM_{10} samples at concentrations ranging from 4708 to 97,690 pg/m^3 for DEHP and 407 to 1310 pg/m^3 for DEP in the sampling point near to the harbor.

In the case of OPEs, total concentrations typically range from 1 to 10 μg/L in aqueous matrices, with TBEP and TClPP being the dominant compounds, likely due to their higher solubility compared to other OPEs (Wang et al., 2020a). In addition, OPEs have been reported to be at one order of magnitude higher in surface waters than in drinking water (Wang et al., 2020a). In a recent study conducted in Colombia, 20 OPEs were determined in sediments and fish. The average concentrations of Σ18OPEs in sediments ranged from 2.80 to 115 ng/g dw, while Σ20OPE concentrations in fish ranged from 212 to 1968 ng/g lipid weight (lw) (Olivero-Verbel et al., 2025). As for air samples, OPE concentrations in indoor air have been found to be 10 to 1000 times higher than in outdoor air. In outdoor environments, their total concentration is typically only a few ng/m^3, and it is widely agreed that OPEs are predominantly associated with the particle phase rather than the gas phase (Wang et al., 2020a).

Studies on BPs have reported the ubiquitous presence and distribution of these compounds in the different environmental compartments, and, for instance, they have been detected for years in aquatic environments at ng/L levels (Liang et al., 2024). In the case of surface water, the levels were at least 10 times higher with respect to the levels found in drinking water (up to 50 ng/L) (Mishra et al., 2023). In a recent review, Czarny-Krzymińska et al. (2023) compiled the concentration of BPA and its substitutes in aquatic environment including samples as wastewater, surface water, groundwater, seawater, sewage sludge, and sediments for different sampling points around the world. The concentration in aqueous samples for BPA were in the range from nondetected up to 1–20 μg/L corresponding the highest levels to effluent sewage samples, meanwhile the concentration of the substitutes (including in most of the instances BPS, BPF, or BPAF) were lower. Interestingly, the studies conducted in Asia (especially China) reported higher occurrence of BPA and related compounds as this region dominates the bisphenol market in terms of production and consumption. In the case of solid samples in China, the levels of BPA were at concentrations 0.1 to 200 ng/g when the sediments were in contact with surface water, but they increased (10 to 1000 ng/g) for sediments in contact with wastewater (Czarny-Krzymińska et al., 2023). BPA has been also reported at even higher levels in landfill leachate from Bosnia and Herzegovina (1.20–1.70 mg/L) and Republic of Serbia (0.6–3.6 mg/L). Unfortunately, other BPs were not evaluated (Narevski et al., 2021). Moreover, due to the ubiquitous presence of BPA, but also the emerging BPs, they were also found in marine species. For instance, a recent study (Castellani et al., 2025) that included 12 sampling points along coastline in central Italy have studied the presence of 8 BPs and 7 PAEs in mussel samples at concentration up to 4.3 ng/g for $\sum$8 BPs, except two sampling points where the concentration raised to 219 and 56 ng/g; in the case of $\sum$7 PAEs the concentrations were from 57 to 107 ng/g.

From the family of BTRs, BTHs, and BSAs, BTR has been reported at concentrations up to 700 μg/L in influent sewage samples, while other compounds such as 4/5TTR isomers and p-TSA at concentrations ranging from 10 to 100 μg/L (Liao et al., 2018). Notably, BTR have been also detected at high concentrations in effluent sewage samples, indicating both low removal efficiency in sewage treatment plants and the possibility that BTR is a transformation product of other benzotriazoles. As a result, BTR and related compounds are frequently released into environmental matrices. Both BTRs and BTHs have been classified as persistent and mobile chemicals. One recent study reported concentrations of four BTRs and five BTHs in drinking and in bottled water, being all the BTHs (except one) detected in all samples analyzed (Wang et al., 2023). Among the different BTHs, BTH is the most abundant in indoor air (Wan et al., 2016). The occurrence of BTRs in fish and seafood at low ng/g concentrations (i.e., 0.5 to 70 ng/g) has been reported in different locations of the world (Kim et al., 2023).

2 Pharmaceuticals and personal care products (PPCPs)

Pharmaceuticals and personal care products (PPCPs) are usually considered as a compact and priority class within CECs, which in turn involve several chemical classes with very different properties, action, and structure. Pharmaceuticals, in simple terms, are therapeutic drugs primarily used to prevent or treat human and animal diseases. In contrast, personal care products (PCPs) are primarily used to enhance the quality of our daily lives (Boxall et al., 2012).

The US Food and Drug Administration (FDA) defines a pharmaceutical as any substance intended for use in the diagnosis, cure, mitigation, treatment, or prevention of disease in humans or animals. This definition also includes substances (other than food) that are intended to affect the structure or any function of the body (Food and Drug Administration, 2017). The World Health

Organization (WHO) have designed the anatomical, therapeutical, and chemical (ATC) classification system to organize pharmaceutical classes, system that has been adopted worldwide (World Health Organization (WHO), 2025). Although useful, this system can be somehow intricate since one substance can be classified with different codes in the first levels. On the contrary, for PCPs there is not such a systematic classification, except in the case they are used as food additives, where the E numbers are assigned (e.g., methylparaben E208). Although the FDA has actually >20,000 approved pharmaceuticals <5% of these substances have been investigated deeply as CECs (Food and Drug Administration (FDA), 2025). In this chapter most studied PPCPs classes, according to CECs published research will be presented (Chakraborty et al., 2023).

2.1 Structure and properties

As the term PPCPs encompasses a wide range of subclasses, a selection of some representative molecules of the main classes are included in Table 1.2 to illustrate some of the most studied CECs.

2.1.1 *Antibiotics*

Antibiotics, classified under the ATC code J01, are pharmaceuticals that act killing or inhibiting the growth of pathogens such as bacteria, viruses, fungi, and parasites. They are divided in classes based on their chemical structure or mode of action; some remarkable categories are macrolides, sulfonamides, beta-lactams, tetracyclines, and quinolones. A major environmental concern is the release of antibiotics through different sources, such as urban runoff, hospitals, and wastewater treatment plant (WWTP) effluents, to the environment thus contributing to the nowadays global threat known as antimicrobial resistance (AMR) (Larsson and Flach, 2022).

2.1.2 *Analgesics*

Classified as ATC code N02, some frequently reported analgesics in the environment are acetaminophen and tramadol, as main examples of nonopioid and opioid analgesics, respectively. Acetaminophen has a core aromatic (benzene) ring substituted in para orientation by two groups: a hydroxyl and an acetamide, and presents high degree of conjugation. On the contrary, tramadol is a more complex molecule, 2-(dimethyl amino)-methyl)-1-(3′-methoxyphenyl) cyclohexanolhydrochloride, a 4-phenyl-piperidine analogue of the opioid drug codeine (Subedi et al., 2019). In some pain relief therapies both acetaminophen and tramadol are prescribed together.

2.1.3 *Nonsteroidal antiinflammatory drugs*

The nonsteroidal antiinflammatory drugs (NSAIDs) are a heterogeneous group of pharmaceuticals (ATC M01) that act as analgesics, antipyretics, and antiinflammatories including first synthetized NSAID, acetylsalicylic acid (mostly

TABLE 1.2 Environmentally relevant examples of pharmaceuticals from different ATC categories, structure and identifiers.

ATC category/ code	Subclass	Name	Formula	Structure	PubChem CID
Antibiotics (J01)					
	Macrolides	Azythromycin	$C_{38}H_{72}N_2O_{12}$	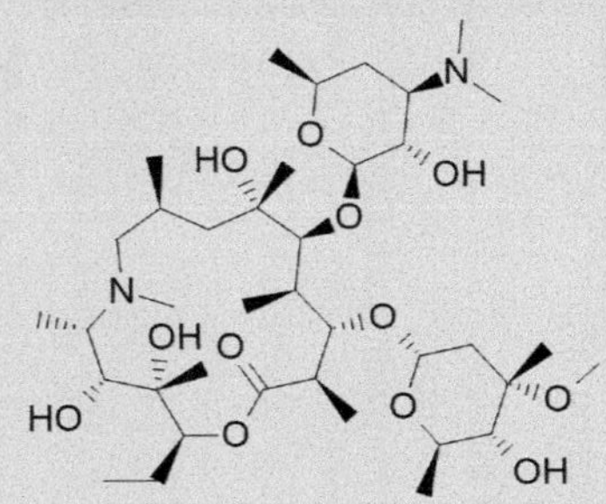	53477736
	B-lactams	Amoxicillin	$C_{16}H_{19}N_3O_5S$	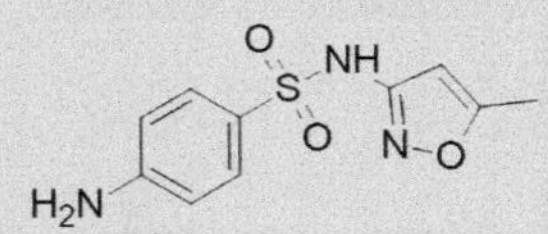	33613
	Sulfonamides	Sulfamethox-azole	$C_{10}H_{11}N_3O_3S$	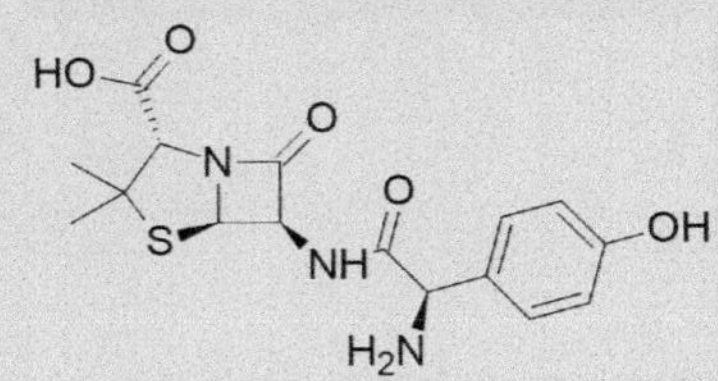	5329

Continued

TABLE 1.2 Environmentally relevant examples of pharmaceuticals from different ATC categories, structure and identifiers—cont'd

ATC category/ code	Subclass	Name	Formula	Structure	PubChem CID
	Tetracyclines	Tetracycline	$C_{22}H_{24}N_2O_8$	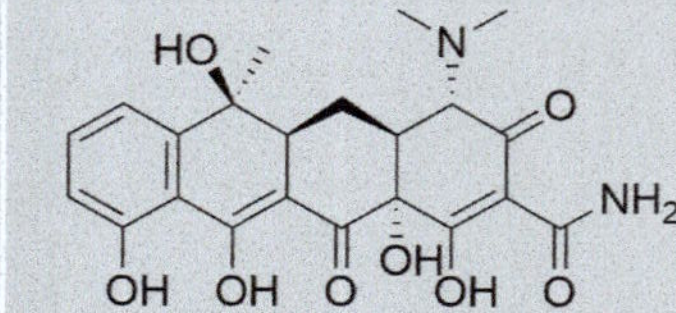	54675776
	Fluoroqui-nolones	Ciprofloxacin	C17H18FN3O3	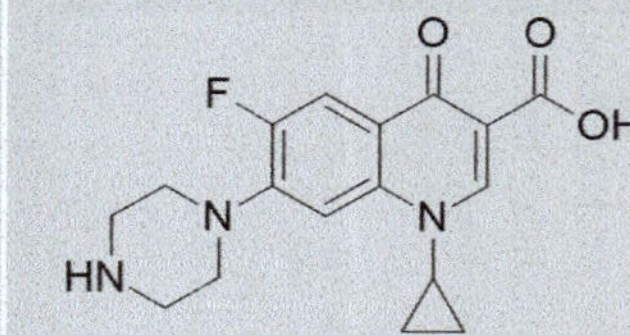	2764
Analgesics (N02)					
	Opioids	Tramadol	$C_{16}H_{25}NO_2$	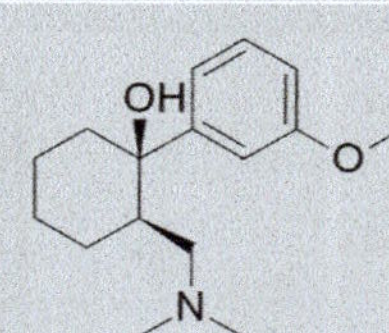	33741
	Nonopioids	Acetamin-ophen	$C_8H_9NO_2$	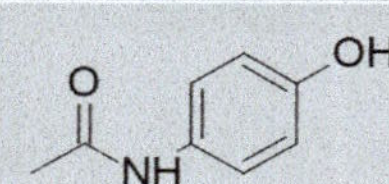	1983

NSAIDs (N01)					
		Diclofenac	$C_{14}H_{11}Cl_2NO_2$	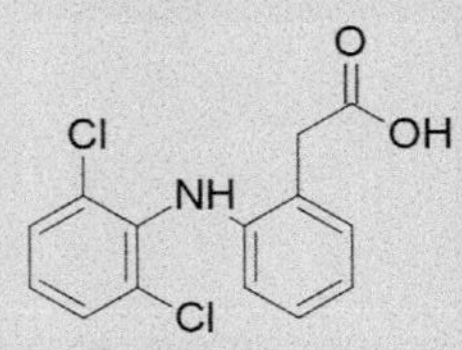	3033
Antiepileptic (N03)					
	Iminostilbene	Carba-mazepine	$C_{15}H_{12}N_2O$	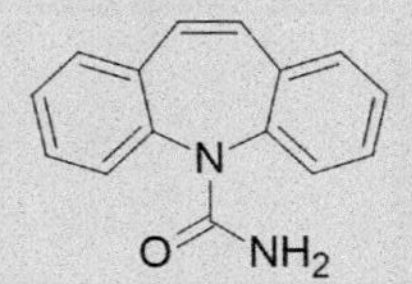	2554
	GABA analogue	Gabapentin	$C_9H_{17}NO_2$	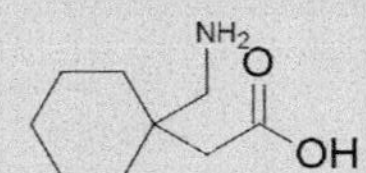	3446
b-blockers (C07)					
		Metoprolol	$C_{15}H_{25}NO_3$	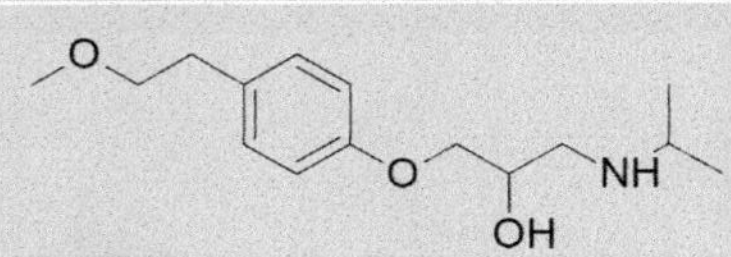	4171

Continued

TABLE 1.2 Environmentally relevant examples of pharmaceuticals from different ATC categories, structure and identifiers—cont'd

ATC category/ code	Subclass	Name	Formula	Structure	PubChem CID
Antidepressants (N06)					
	Phenylamine	Sertraline	$C_{17}H_{17}Cl_2N$	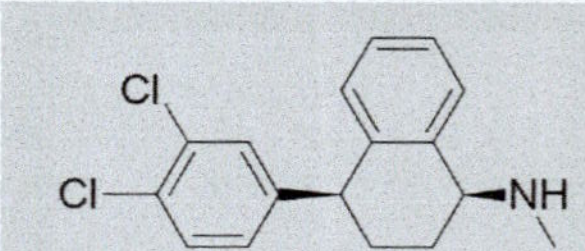	68617
	Aryloxypropa-nolamine	Venlafaxine	$C_{17}H_{27}NO_2$	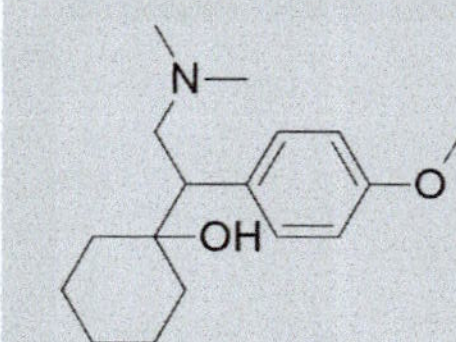	5656

Source: https://pubchem.ncbi.nlm.nih.gov/: created by the authors.

commercialized as Aspirin), and other structurally different groups (coxibs, fenamates, etc.) sharing all of them the same main function as inhibitors of cyclooxygenase (COX) enzymes (COX-1 and COX-2). The aryl- and heteroarylpropionic acids (or profens) and the acetic acid derivative diclofenac are the most studied from the environmental point of view (Lin et al., 2023).

2.1.4 *Antiepileptic/anticonvulsants*

Antiepileptic drugs (ATC N03) such as carbamazepine (iminostilbene derivative) or gabapentin (gamma-aminobutyric acid analogue) are widely utilized against the chronic pain, blocking respectively sodium and calcium channels at cellular neuronal level, and constitute relevant examples, especially in the case of carbamazepine, of ubiquitous contaminants in the aquatic environment (He et al., 2024; Moermond and Smit, 2016).

2.1.5 *β-blockers*

This group of pharmaceuticals (ATC C07) shares a propanolamine structure and acts blocking the effect of different hormones, such as adrenaline, thus reducing hearth rate, muscle contraction, etc. Most known examples are propranolol or metoprolol (Moermond and Smit, 2016).

2.1.6 *Antidepressants*

As explained for previous groups, antidepressants (ATC N06) encompass different categories, being the serotonin inhibitors, such as sertraline (tetralin-based phenylamine) and venlafaxine (aryloxypropanolamine) two examples of well-studied molecules from the environmental point of view (Castillo-Zacarías et al., 2021).

2.1.7 *Preservatives*

Parabens are by far the most studied preservatives class since they present both antimicrobial and antifungal properties. They are esters of parahydroxybenzoic acid presenting different length of the alkyl chain, giving rise to their classification into short-chain (e.g., methylparaben) and long-chain (e.g., butyl paraben) congeners (Haman et al., 2015).

2.1.8 *Bactericides/disinfectants*

Triclosan (5-chloro-2-[2,4-dichlorophenoxy] phenol) is the most studied bactericide from an ecological point of view (Lee et al., 2024), since its structure confers high lipophilicity, which, together with high toxicity, results in particular concern for the health of aquatic organisms.

2.1.9 *UV filters (benzophenones)*

As described for the previous groups, there are different subclasses of UV filters, including benzophenone, benzotriazole, or salicylate derivatives, among others. Oxybenzone or benzophenone-3 (BP-3), 4-methylbenzylidene (4-MBC) or octocrylene are some examples of well-known CECs. A common feature of these compounds is the presence of an aromatic moiety with a side chain, showing different degrees of unsaturation. Their water solubility is variable and can be used as indicator of their possible distribution in the different environmental compartments (Ramos et al., 2015).

2.2 Legislation

As regards the risk of PPCPs in the environment from the regulatory point of view, in some countries such as United States, Canada, Japan and the European Union, environmental risk assessment (ERA) of such chemicals is mandatory; although in other many countries these regulations are still absent or generic as PPCPs are considered in the same level of regulation that any other produced chemical. In the European Union and Canada, ERA is defined as a ratio between predicted environmental concentration (PEC) and predicted no-effect concentration (PNEC), whereas in the United States it is defined as an assessment factor (AF), calculated as the experimental chronic toxicity divided by the maximum expected environmental concentration and thus reflecting an inverse equivalent of the PEC:PNEC ratio (Oldenkamp et al., 2024). Surprisingly, in most countries an unfavorable ERA result not necessarily implies the substance refuse for the market. In the specific case of water contamination, the European Union Water Framework Directive (EU-WFD) (European Union, 2000) and other analogue laws evaluate risk quotients (RQ) as the ratio of measured environmental concentration (MEC)/environmental quality standard (EQS); thus there is an obligation to establish legally binding EQS values for the substances of greatest concern. Some examples of the PPCPs that have been included over the years in the European Priority List or the different Watch Lists (WL) under the WFD for their environmental monitoring are sulfamethoxazole, diclofenac, azithromycin, etc. and the most recent fifth WL includes tetracycline or propranolol among others (European Commission, 2025).

2.3 Occurrence in the environment

Over the last fifteen years research on environmental occurrence and removal strategies for PPCPs has been growing significantly. Fig. 1.2 shows the number of publications found in Scopus using the query string TITLE-ABS-KEY(*compound name) AND TITLE-ABS-KEY(*contaminant) AND PUBYEAR > 2010 for those compounds included as examples of each category in Table 1.2. As can be observed, those one more deeply investigated are those ones included in the different WL over the years, such as diclofenac, sulfamethoxazole, or tetracycline.

The inefficient processes of wastewater treatment and the consequent release of PPCPs into surface water constitute the main entry of PPCPs into the environment (Osuoha et al., 2023). Pharmaceuticals are excreted in variable degree after

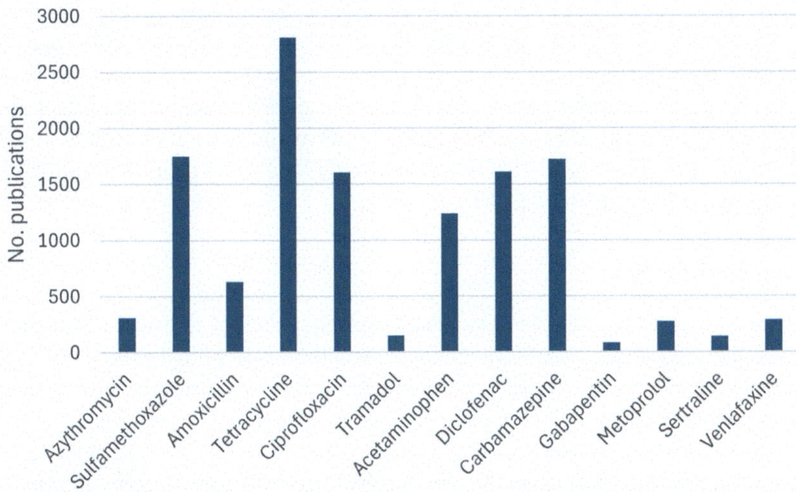

FIG. 1.2 Number of publications indexed in Scopus including some of the most relevant pharmaceuticals considered in Table 1.1 using the search string "compound name" AND "contaminant" from 2010. *(Source: Scopus database accessed on 2025/07/07; created by authors.)*

consumption both as unaltered principle and as conjugates, and PCPs (contained mainly in cosmetics) stay on the skin for a while and are removed through washing or showering entering the sewer system. Most polar compounds will remain mainly dissolved in the water, and if in addition the treatment applied is inefficient, will be directly release to marine or freshwater receiving media, eventually finishing in aquatic organism and/or drinking water and entering the food chain. On the contrary, the most lipophilic ones can be absorbed in sludge and, if this sludge is then used in agriculture as fertilizer also can reach the food chain (Wu et al., 2015). As regards of wastewater, carbamazepine, acetaminophen, and diclofenac are examples of the most frequently detected pharmaceuticals at concentration even higher than 1 mg/L whereas benzophenone and methylparaben stand out within the PCPs group (Reyes et al., 2021). The concentrations reported in surface water (or even drinking water) highly depend on the effectiveness of water treatment processes being some of the most recalcitrant compounds DEET or benzophenone (Pai et al., 2020), but in some cases also high concentrations are reported for compounds which present high removal efficiencies due to the high loads entering the WWTP, as in the case of the antidiabetic metformin (Zheng et al., 2024). The levels of PPCPs reported in sludge are much higher than found in soils and sediments, especially for the most lipophilic compounds (Huber et al., 2016), and there is vast literature on the occurrence of different PPCPs classes. For example, Wang et al. (2018) reported that more than half of the antibiotics loads entering WWTPs remained; there in a monitoring performed in 7 plants in China. Within PCPs triclosan is one of the most reported in biosolids with concentrations over 1 mg/Kg in some cases (Brown et al., 2019).

Finally, as bioindicator of aquatic contamination, freshwater and marine biota has been also investigated for the presence of PPCPs. Antibiotics and NSAIDs have been reported in several species worldwide at the ng/g level, being the highest levels reported in some studies conducted in Asia which reported levels in the μg/g level (Ebele et al., 2017). UV filters, such as oxybenzone (benzophenone-3) and octocrylene, have been also reported at high levels (ng/g) in fish, mollusks and crustaceans (Peng et al., 2017).

3 (Il)licit drugs

The social, economic, and human health consequences of licit and illicit drug use, such as addiction, behavioral changes, and threats to public safety, are well documented. However, growing attention is now being directed toward their environmental impact, driven by the increasing global use, diversification, and persistence of these substances. As a result, illicit drugs (IDs) and their legal alternatives, known as new psychoactive substances (NPS), represent an important class of CECs. Following human consumption and excretion of parent compounds or metabolites, as well as direct discharge from clandestine manufacturing or distribution activities, these substances enter wastewater treatment systems. Conventional treatment processes often achieve only partial removal, allowing these compounds to persist in surface waters, sediments, and even drinking water sources. Their widespread occurrence raises significant concerns regarding ecological disruption and potential long-term risks to human health.

This section primarily focuses on the most prevalent and environmentally significant IDs, including cocaine and methamphetamines. Particular attention is paid to NPS, an evolving group that has rapidly gained prominence due to the increasing prevalence and environmental impact of compounds such as ketamine and synthetic cathinones.

3.1 Structure and properties

3.1.1 Illicit drugs (IDs)

IDs are normally classified according to the United Nations Office on Drugs and Crime (UNODC), which categorizes substances based on their primary effects on the central nervous system (United Nations Office on Drugs and Crime (UNODC), 2022). These categories include stimulants such as cocaine (COC), its primary metabolite benzoylecgonine (BE); amphetamine (AMPH), methamphetamine (METH), and 3-4 methylenedioxymethamphetamine (MDMA), which enhance alertness and energy; depressants such as heroin (HER), which slow down brain function and have sedative effects; hallucinogens such as lysergic acid diethylamide (LSD), which alter perception and mood; and cannabis, which exhibits both depressant and mild hallucinogenic properties.

Table 1.3 summarizes the most relevant IDs from an environmental perspective, highlighting key chemical properties that significantly influence

TABLE 1.3 Chemical properties of environmentally relevant illicit drugs (IDs) and new psychoactive substances (NPS).

Compound	Abbreviation	Class	Chemical formula	Chemical structure	Log Kow	Water solubility	PubChem CID
IDs							
Cocaine	COC	Stimulant	$C_{17}H_{21}NO_4$	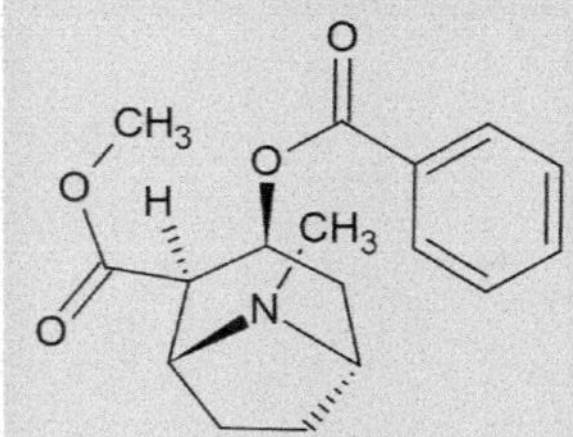	2.3	Moderate	446220
Benzoylec-gonine	BE	Stimulant	$C_{16}H_{19}NO_4$	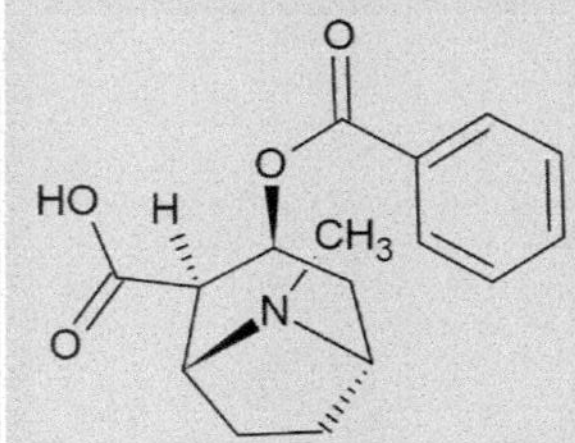	0.6	High	448223
Amphetamine	AMPH	Stimulant	$C_9H_{13}N$	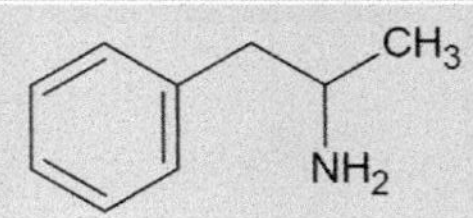	1.8	Moderate	3007

Continued

TABLE 1.3 Chemical properties of environmentally relevant illicit drugs (IDs) and new psychoactive substances (NPS)—cont'd

Compound	Abbreviation	Class	Chemical formula	Chemical structure	Log Kow	Water solubility	PubChem CID
Metham-phetamine	METH	Stimulant	$C_{10}H_{15}N$	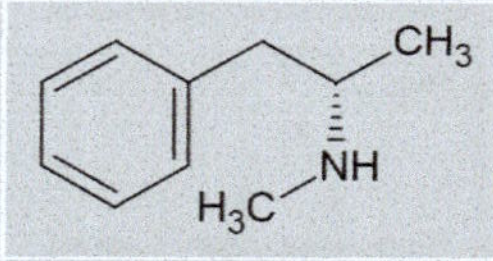	2.0	High	10836
Ecstasy	MDMA	Stimulant	$C_{11}H_{15}NO_2$	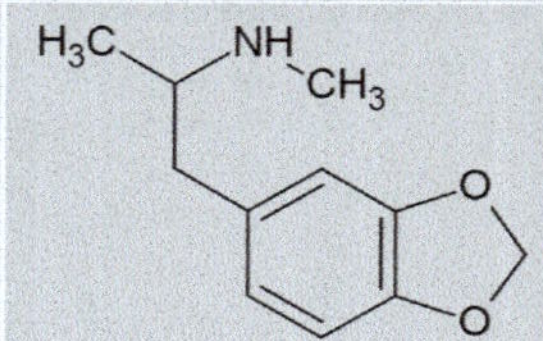	1.7	High	1615
Lysergic acid diethylamide	LSD	Hallucinogen	$C_{20}H_{25}N_{34}O$	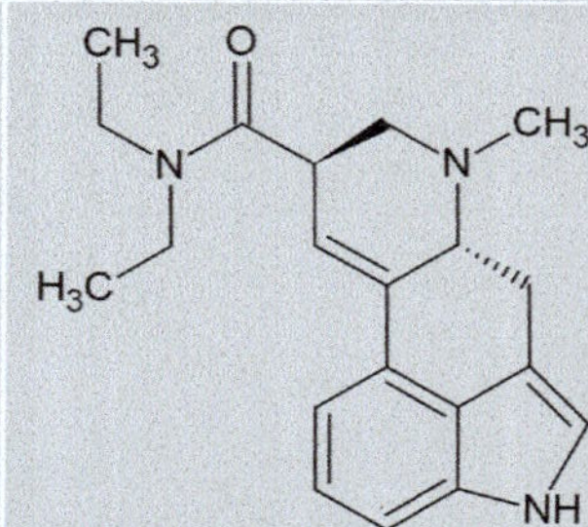	2.3	Low-moderate	5761

11-nor-9-carboxy-Δ9-tetrahydro-cannabinol	THC-COOH	Cannabis	$C_{21}H_{28}O_4$	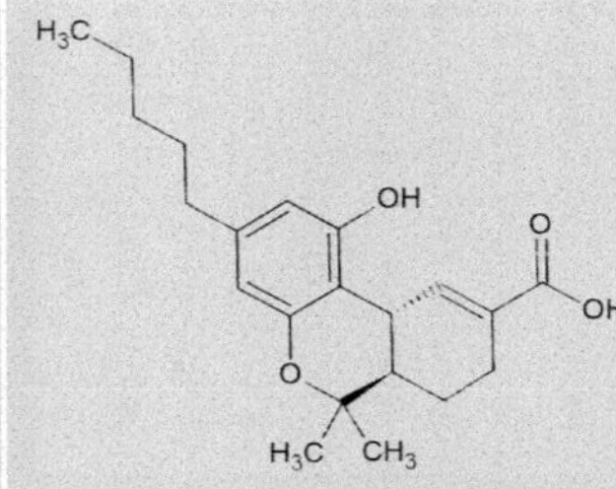	5.1	Low	108207
NPS							
Ketamine	KET	Hallucinogen/ Dissociative	$C_{13}H_{16}ClNO$	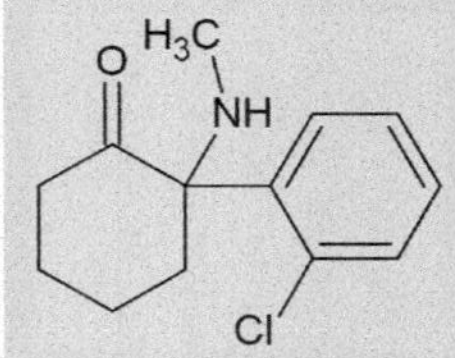	2.2	High	3821

Source: https://pubchem.ncbi.nlm.nih.gov/: created by the authors.

their environmental fate and transport. For example, COC is a plant-derived alkaloid with moderate water solubility and is rapidly metabolized into BE, a more hydrophilic and environmentally persistent compound frequently used as a marker of COC consumption in wastewater analysis (Zuccato et al., 2008). In contrast, synthetic amines such as AMPH, METH, and MDMA possess basic functional groups and exhibit high chemical stability, making them resistant to biodegradation and persistent in aquatic environments (Boles and Wells, 2010). Cannabis-derived metabolites, particularly 11-Nor-9-carboxy-Δ^9-tetrahydrocannabinol (THC-COOH), demonstrate greater water solubility and environmental stability compared to the parent compound, leading to their more frequent detection in wastewater and surface waters (How and Gamal El-din, 2021). HER is a semisynthetic opioid with low water solubility that is rapidly metabolized in the human body into 6-monoacetylmorphine and subsequently into morphine, a natural opiate that is more hydrophilic and environmentally persistent, and frequently detected in treated wastewater effluents and surface waters (Campos-Mañas et al., 2018). Hallucinogens such as lysergic compounds, structurally complex ergoline derivatives, are typically present at very low concentrations in the environment due to their extensive metabolism in the human body, although they may still pose ecotoxicological risks because of their high pharmacological potency, strong sorption capacity, and low biodegradability (Postigo et al., 2010).

3.1.2 *New psychoactive substances (NPS)*

NPS are defined by the UNODC as "substances of abuse, either in a pure form or a preparation, that are not controlled by the 1961 Single Convention on Narcotic Drugs or the 1971 Convention on Psychotropic Substances, but which may pose a public health threat" (United Nations Office on Drugs and Crime (UNODC), 2022). Emerging in the mid-2000s, NPS were initially developed as legal alternatives to traditional IDs by chemically modifying controlled substances to produce similar psychoactive effects while circumventing existing legislation (European Monitoring Centre for Drugs and Drug Addiction and Europol, 2024). According to the latest UNODC Early Warning Advisory (EWA), >1386 individual NPS have been reported by over 150 countries, reflecting the rapid diversification and global spread of this drug class (United Nations Office on Drugs and Crime (UNODC), 2025).

The major categories of NPS include synthetic cathinones, synthetic cannabinoids, phenethylamines, tryptamines, and synthetic opioids, characterized by distinct chemical structures that influence their environmental behavior and fate. Synthetic cathinones (e.g., mephedrone, methylone) contain polar functional groups such as ketones and amines, which confer high water solubility and chemical stability, promoting their persistence and frequent detection in wastewater and surface waters (Fontanals et al., 2017). In contrast, synthetic cannabinoids (e.g., JWH-018, AM-2201) possess highly lipophilic structures with complex aromatic and heterocyclic rings, resulting in strong adsorption onto organic matter. Their rapid metabolism in humans limits the environmental presence of parent compounds, shifting environmental monitoring focus toward their metabolites as key indicators (Fan et al., 2022). Phenethylamines (e.g., 2C-B) and tryptamines (e.g., 5-MeO-DMT), which are often structurally related to natural neurotransmitters, vary widely in polarity and biodegradability but are generally detected at lower concentrations due to the limited environmental stability (Atasoy Aydin et al., 2023). Special attention is paid to ketamine (KET) (Table 1.3), a dissociative anesthetic widely used in human and veterinary medicine and increasingly popular as a recreational drug. Due to the rise of structurally related analogues in illicit markets, KET is often considered alongside NPS. Pharmacologically, KET acts as an N-methyl-D-aspartate receptor antagonist, producing dissociative, hallucinogenic, and analgesic effects. Chemically, it is a cyclohexanone derivative with moderate water solubility and high environmental stability, factors that contribute to its persistence in aquatic environments (Wang et al., 2020b).

3.2 Legislation

IDs are not explicitly addressed by environmental regulations; however, international drug-control conventions provide indirect environmental protection. The UN Single Convention on Narcotic Drugs (1961) (United Nations, 1961) and the Convention on Psychotropic Substances (1971) (United Nations, 1971) impose strict controls on the production, distribution, and disposal of substances such as COC, HER, LSD, and MDMA. While primarily focused on preventing misuse and trafficking, these measures also contribute to reduce environmental pollution by promoting the secure handling and elimination of controlled substances.

At the European level, growing attention has been directed toward the environmental footprint of IDs. The revision of the Urban Wastewater Treatment Directive 2024/3019 (European Union, 2024) introduces new monitoring and treatment obligations for selected CECs in wastewater. Notably, the directive recognizes that IDs and their metabolites can persist through WWTPs and enter into the aquatic environments, where they can pose potential ecological risks. By 2030, Member States are expected to monitor both the presence and removal efficiency of these compounds, including IDs, reflecting their growing significance as CECs.

The regulation of NPS presents further legislative and environmental challenges. Since 2014, several NPS of high concern, such as KET, mephedrone, and NBOMe compounds, have been progressively scheduled under international drug control conventions. Nevertheless, these compounds are frequently still regarded as NPS due to their rapid structural diversification, which enables them to evade existing legal frameworks. This chemical adaptability allows newly synthesized analogues to circulate on the market before regulatory mechanisms can be updated. Consequently, NPS remain a persistent issue not only from public health and legal perspectives but also from an environmental standpoint. Their unregulated production, usage, and disposal raise concerns regarding potential long-term ecological effects, particularly as most NPS are not routinely included in environmental monitoring programs or risk assessment protocols.

3.3 Occurrence in the environment

IDs have been consistently reported across multiple environmental compartments due to their widespread recreational use and subsequent excretion in both parent and metabolite forms (Yedhu Krishnan et al., 2023; Pal et al., 2013; Zergui and Nkontcheu Kenko, 2024). Wastewater has been the most studied compartment, followed by surface water and soils, sediments and sludge. Fig. 1.3 shows the number of publications found in Scopus using the query string TITLE-ABS-KEY(*compound name) AND TITLE-ABS-KEY(*contaminant) AND PUBYEAR > 2010 for the main IDs and NPS.

In urban wastewater, BE and COC are the most studied compounds, at concentrations typically ranging from 10 to 1000 ng/L, depending on consumption patterns and temporal factors (Huizer et al., 2021). Removal efficiencies in conventional WWTPs vary considerably between compounds: while some, such as AMPH and MDMA, are partially degraded through biological processes, others such as THC-COOH exhibit greater persistence due to their hydrophobicity or chemical stability (Yadav et al., 2017). Although advanced treatment

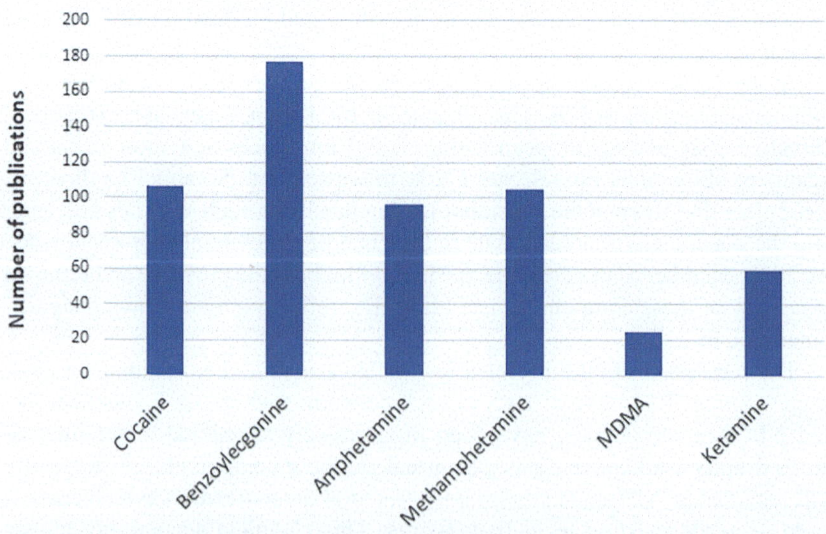

FIG. 1.3 Number of publications indexed in Scopus from 2010 onwards related to the most frequently illicit drugs and NPS investigated in environmental studies. *(Source: Scopus database, accessed on 2025/11/10; created by the authors.)*

technologies, including ozonation, activated carbon adsorption and membrane filtration, shown higher removal efficiencies (Yuan et al., 2024), these technologies are not yet routinely applied in most municipal WWTPs. Consequently, treated effluents continue to serve as a major pathway for the release of IDs into aquatic environments. In surface water, concentrations of traditional IDs typically range from 0.1 to 100 ng/L, with BE as the compound most commonly found (Zhang et al., 2024). Trace levels of IDs (up to 3 ng/L) have also been reported in tap water and groundwater, indicating their environmental persistence and the potential for long-range transport and infiltration into drinking water (Pal et al., 2013).

Despite their generally polar nature, IDs are consistently observed in solid environmental matrices such as sludge, sediments, and suspended particulate matter. Among these, COC and its metabolite BE are most frequently reported, with concentrations averaging 10–13 ng/g in sludge and 1–3 ng/g in sediments (Santana-Viera et al., 2023). THC, a more hydrophobic compound, are usually less frequently detected but can reach substantially higher levels due to strong adsorption to organic matter. For example, THC concentrations in sediments from major rivers in Spain were found to be 6 to 50 times higher than those of COC and BE (López-García et al., 2021).

IDs also bioaccumulate in aquatic organisms, raising concerns for trophic transfer and ecological effects. In mussels from Santos Bay, Brazil, COC concentrations raised 4.58 μg/Kg ww, with field-measured bioaccumulation factors of 130–1454 L/Kg, indicating significant uptake from both water and sediments (Fontes et al., 2021). Ecotoxicological studies corroborate that IDs can induce adverse effects even at environmentally relevant concentrations. For instance, mussels exposed to BE (20 ng/L), COC (500 ng/L), or their combination exhibited oxidative stress, with mixed exposures producing stronger effects than single compounds, suggesting additive or synergistic interactions (De Felice and Parolini, 2020).

Beyond aquatic systems, IDs have also been detected in the atmosphere of urban areas (Cecinato et al., 2016). For instance, COC and BE were found in over 93% of PM_{10} samples collected in Barcelona, revealing an aerosolized transport pathway for these substances in densely populated environments (Mastroianni et al., 2015). This adds another dimension to human exposure, particularly in metropolitan areas with high population densities and recreational drug use.

Environmental research on the occurrence and behavior of NPS is still limited, primarily due to their more recent emergence, sporadic consumption patterns, and generally lower usage prevalence. Consequently, NPS are typically detected at lower concentrations than traditional IDs. Nonetheless, advances in analytical techniques and expanded monitoring efforts have led to more frequent detection of these compounds in various environmental matrices. In urban wastewater systems, KET, 3-methylmethcathinone (3-MMC), and methcathinone (METC) have been quantified at concentrations ranging from 0.35 to 122 ng/L across Europe (Salgueiro-Gonzalez et al., 2024). Temporal monitoring frequently reveals distinct concentration peaks during weekends and major public events, indicative of episodic increases in recreational drug consumption (Bade et al., 2023). The reported removal efficiencies of NPS in the WWTPs are highly variable and compound-dependent, influenced by physicochemical properties such as hydrophobicity, polarity, and volatility. Overall, data on the removal of NPSs remain scarce, and systematic evaluations across different treatment technologies and environmental conditions are still lacking. While some compounds, such as benzylpiperazine (BZP), show high removal efficiencies (>95%), others, including 3,4-methylenedioxypyrovalerone (MDPV), display poor removal (6.8%–21.2%), indicating a high potential for release into receiving waters (Gao et al., 2017). KET is particularly persistent, sometimes showing low or even negative removal efficiencies, likely due to its environmental stability and the formation of chelated complexes that hinder microbial or chemical degradation (Zhao et al., 2023). In fact, KET was found in 91% of surface water samples (68 in total) in China at concentrations up to 1.7 ng/L (Chen et al., 2025). Regarding drinking water, KET, mephedrone (4-MMC), and methylone (METL) were detected in samples collected across the United Kingdom, at concentrations up to 50 ng/L (Peng et al., 2019). Similarly, reported KET concentrations reaching 15 ng/L in treated drinking water in Canada (Rodayan et al., 2016).

Research on the occurrence of NPS in solid environmental matrices and airborne samples remains scarce and is largely limited to KET. In solid compartments, KET has been detected in sewage sludge and benthic sediments associated with WWTPs in China, with concentrations ranging from 0.1 to 500 ng/g, reflecting both the physicochemical properties of the compound and proximity to point-source discharges (Chen et al., 2025). In the atmosphere, WWTP have been identified as sources of aerosolized KET, with concentrations in air measured up to 15 ng/m^3, suggesting a potential pathway for human and environmental exposure (Lin et al., 2020). Evidence from aquatic organisms further indicates the bioaccumulation potential of KET. Adult medaka (*Oryzias latipes*) exposed to environmentally relevant concentrations (0.05–0.5 μg/L) accumulated KET in tissues at levels ranging from 0.55 to 113 ng/g wet weight, corresponding to bioconcentration factors of 1.07–10.94 (Wang et al., 2020b). Overall, these findings highlight that our understanding of NPS occurrence, environmental fate, and bioaccumulation is overwhelmingly centered on KET, emphasizing the need for systematic studies on a broader spectrum of NPS to assess their persistence, ecological transfer, and potential environmental risks.

4 Pesticides

The widespread use of synthetic chemicals in agriculture has led to the release of significant quantities of pesticides into the environment. Pesticides, including insecticides, herbicides, and fungicides, are applied globally to protect crops, control pests and improve agricultural productivity. According to the Food and

Agriculture Organization (FAO), global pesticide consumption was estimated at approximately 3.73 million tons in 2023, with herbicides accounting for roughly 50%, insecticides 22%, fungicides 22%, and other pesticides around 6% (Food and Agricultural Organization (FAO), 2023). Due to their extensive application, chemical stability and mobility, pesticides are frequently detected in various environmental compartments, including surface waters, groundwater, soils, sediments, and even in the atmosphere.

This section primarily focuses on several widely used pesticides that are increasingly being recognized as CECs, such as neonicotinoids (e.g., imidacloprid, clothianidin), pyrethroids (e.g., permethrin), triazole fungicides (e.g., tebuconazole) or triazines (e.g., atrazine). These compounds are of growing environmental concern due to their persistence, potential to bioaccumulate, and adverse effects on nontarget organisms, including pollinators, aquatic invertebrates, and vertebrates. Moreover, they can affect human health by entering in the food chain. Their continued presence in the environment highlights the need for improved monitoring, regulatory measures, and the development of more sustainable pest management strategies.

4.1 Structure and properties

Pesticides are broadly classified into major families according to their target organisms and chemical structures, with each group characterized by distinct modes of action, environmental behavior, and toxicological properties. The main categories (i.e., insecticides, herbicides, and fungicides) can be further subdivided in several subclasses. Table 1.4 summarizes the key representative compounds within the major subclasses of these three primary groups.

4.1.1 Insecticides

Insecticides have played an important role in agriculture and public health through the control of crop pests and disease vectors. Major classes include organophosphates (OPs), carbamates, pyrethroids, and neonicotinoids (Table 1.4). OPs and carbamates inhibit acetylcholinesterase, causing neurotoxic effects in target pests and potentially in nontarget species (Sidhu et al., 2019). These classes largely replaced banned organochlorine pesticides due to the latter's persistence and ecological toxicity. OPs are esters of phosphoric acid, containing aliphatic, phenyl, or heterocyclic groups, which determine their biological activity and degradation pathways (Mali et al., 2023). Some examples are chlorpyrifos, malathion, and diazinon. Carbamates, or N-methyl carbamates, share a similar mechanism of action but are derived from carbamic acid, such as carbaryl and carbofuran (Singh et al., 2025). Both groups are generally degraded via hydrolysis, microbial activity, and photolysis, though certain compounds, such as chlorpyrifos, can persist in soils and aquatic systems (Mali et al., 2023). Pyrethroids are synthetic analogues of natural pyrethrins, extracted from *Chrysanthemum cinerariaefolium*, which act on voltage-gated

TABLE 1.4 Environmentally relevant examples of pesticides divided by categories.

Class	Subclass	Name	Chemical formula	Chemical structure	Pubchem CID
Insecticides	Organophos-phates	Chlorpyrifos	$C_9H_{11}Cl_3NO_3PS$	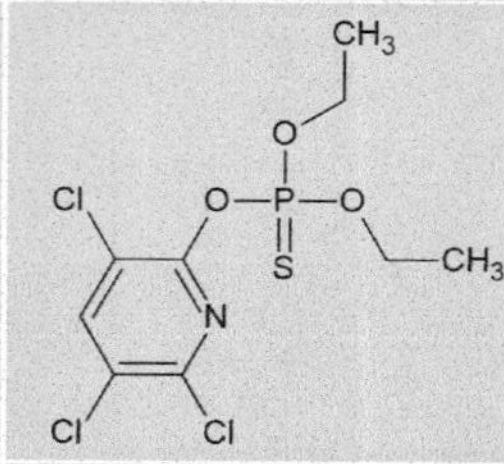	2730
	Carbamates	Carbofuran	$C_{12}H_{15}NO_3$	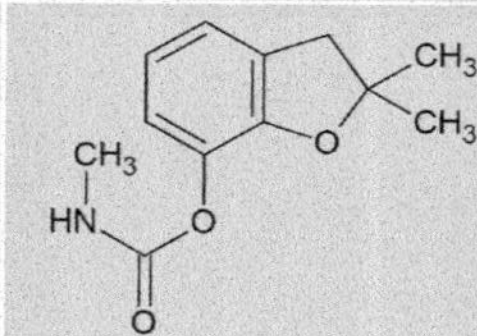	2566
	Pyrethroids	Permethrin	$C_{21}H_{20}Cl_2O_3$	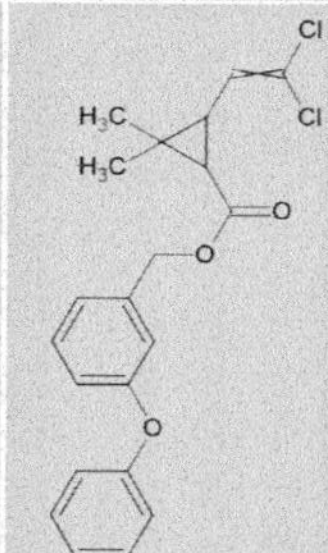	40326
	Neonicotinoids	Imidacloprid	$C_9H_{10}ClN_5O_2$	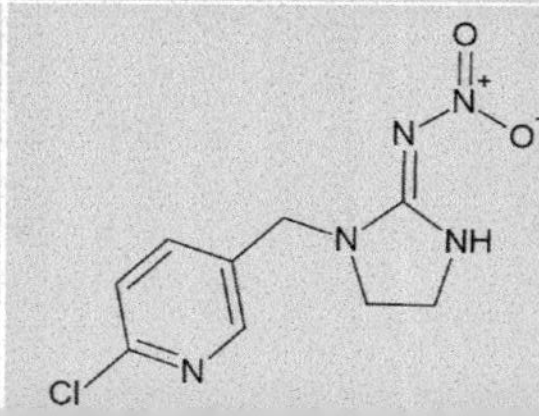	86287518

Herbicides	Triazines	Atrazine	$C_8H_{14}ClN_5$	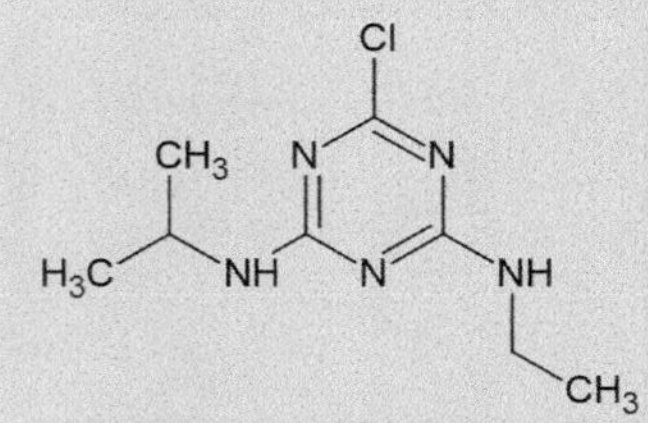	2256
	Phenoxy acids	2,4-dichloro-phenoxyacetic acid (2,4-D)	$C_8H_6Cl_2O_3$	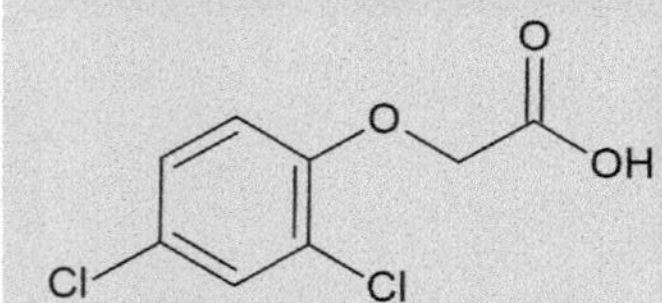	1486
	Glyphosate-based herbicides	Glyphosate	$C_3H_8NO_5P$	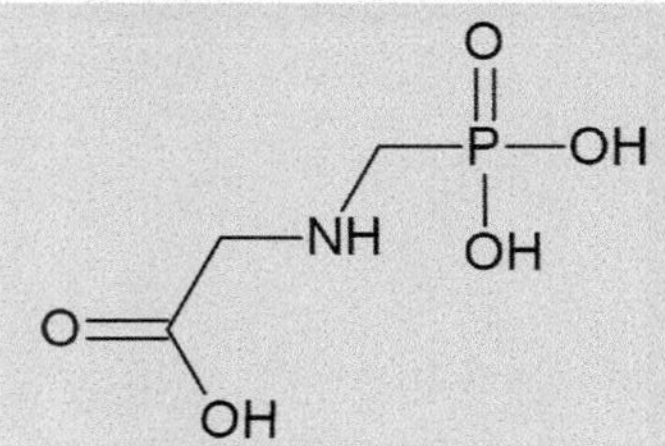	3496

Continued

TABLE 1.4 Environmentally relevant examples of pesticides divided by categories—cont'd

Class	Subclass	Name	Chemical formula	Chemical structure	Pubchem CID
Fungicides	Triazoles	Tebuconazole	$C_{16}H_{22}ClN_3O$	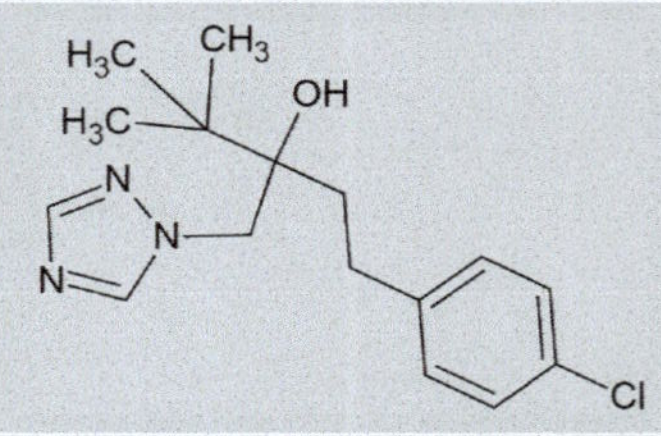	86102
	Strobilurins	Azoxystrobin	$C_{22}H_{17}N_3O_5$	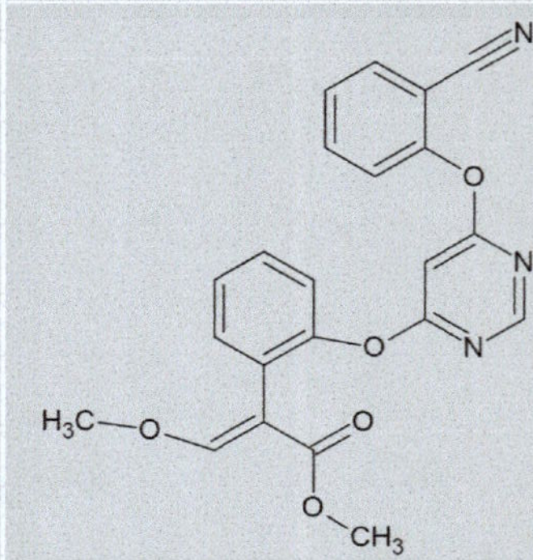	3034285

Source: https://pubchem.ncbi.nlm.nih.gov/: created by the authors.

sodium channels in insect nerve membranes, causing hyperexcitation, paralysis, and eventual death (Tang et al., 2018). These hydrophobic compounds are characterized by low vapor pressure, low water solubility, and high log K_{ow}, promoting strong adsorption to organic matter (Li et al., 2017). Pyrethroids are structurally classified into Type I (non-cyano, or "urban" pyrethroids), such as permethrin, and Type II (α-cyano, or "agricultural" pyrethroids), including cypermethrin, differing in uses, toxicity and environmental persistence (Shafer et al., 2005). Neonicotinoids are broad-spectrum systemic insecticides that act as agonists of nicotinic acetylcholine receptors (nAChRs) in the insect central nervous system, causing paralysis and death (Jeschke and Nauen, 2008). Their novel mode of action avoids cross-resistance with older insecticide classes such as pyrethroids, OPs and carbamates (Elbert et al., 2008). These insecticides are highly water-soluble and persistent, properties that enhance systemic plant protection but also raise concerns for pollinators and aquatic organisms due to mobility and potential bioaccumulation (Morrissey et al., 2015). Some examples are imidacloprid, acetamiprid, clothianidin, and thiamethoxam.

4.1.2 Herbicides

Herbicides are chemical agents that control unwanted vegetation competing with crops for nutrients, water, and sunlight, revolutionizing weed management and significantly improving agricultural efficiency. They are commonly categorized by chemical structure with major classes including triazines, phenoxy acids, and glyphosate-based herbicides (Table 1.4). Triazine herbicides, such as atrazine and simazine, inhibit photosystem II electron transport, blocking photosynthesis and leading to plant deaths. They are 1,3,5-triazine derivatives with three nitrogen atoms in the ring and various alkyl or chloro substituents influencing their reactivity and persistence (Yao et al., 2023). Triazines are weak bases with low water solubility and moderate stability, degrading mainly through microbial activity, and are frequently detected in surface and groundwater due to their mobility and resistance to degradation (Fenoll et al., 2014). Phenoxy acid herbicides control weed growth by artificially elevating phytohormone levels (Mei et al., 2016). Their molecular structures are mainly aryloxy- and aryloxyphenoxy- carboxylic acids, such as 2,4-dichlorophenoxyacetic acid (2,4-D) and 2-methyl-4-chlorophenoxyacetic acid (MCPA). These compounds are highly polar, stable, and water-soluble, which allows herbicides remaining in soil to be transported into rivers, lakes, and groundwater through precipitation and natural drainage processes (Ren et al., 2025). Glyphosate-based herbicides (GBHs) are broad-spectrum, nonselective herbicides that inhibits 5-enolpyruvylshikimate-3-phosphate synthase, blocking aromatic amino acid synthesis (Gandhi et al., 2021). GBHs are phosphonomethyl glycine derivatives with high water solubility and strong soil adsorption, often formulated with surfactants. These derivates are mainly degraded microbially, have low leaching but can reach surface waters via runoff, and may persist in soils and sediments, posing ecological and health risks (Madani and Carpenter, 2022).

4.1 3 Fungicides

Fungicides are chemicals developed to control fungal pathogens that compromise crop productivity and quality. They encompass diverse chemical classes with distinct modes of action, physicochemical properties, and environmental behaviors, including triazoles and strobilurins (Table 1.4). Triazoles, such as tebuconazole, inhibit the cytochrome P450-dependent enzyme C14-demethylase, blocking ergosterol synthesis and disrupting fungal cell membranes (Ore et al., 2023). Structurally, triazoles contain a five-membered ring with two carbon and three nitrogen atoms, conferring strong antifungal activity. However, their moderate hydrophobicity and variable persistence can lead to residue accumulation in soils and crops, particularly of slowly degradable enantiomers, posing potential ecological and human health risks through environmental and dietary exposure (Shuang et al., 2021). On its behalf, strobilurins are derived from the natural products strobilurin A and B and share a common methoxyacrylate pharmacophore responsible for their antifungal activity (Selim and Khalil, 2021). Compounds such as azoxystrobin and trifloxystrobin are moderately to highly hydrophobic with low water solubility, favoring adsorption to soils and sediments but allowing persistence and potential bioaccumulation (Leite et al., 2024).

4.2 Legislation

Numerous studies have documented the ecotoxicological effects of pesticides on aquatic and terrestrial organisms, highlighting risks to biodiversity and ecosystem stability. Contemporary regulatory frameworks address these risks across environmental compartments, food safety, and human health through an integrated, multilevel approach.

Within the EU, pesticide active substances are regulated under Regulation (EC) No. 1107/2009, which defines approval criteria based on key environmental fate parameters such as degradation half-life (DT_{50}), mobility (log K_{ow}), and bioaccumulation (BCF) (European Commission, 2009). Compounds showing high persistence, mobility, or toxicity—such as neonicotinoids (imidacloprid, clothianidin, thiamethoxam)—have faced restrictions under Commission Implementing Regulation (EU) 2018/783 owing to their impacts on pollinators (European Commission, 2018). Complementary to this, the REACH Regulation (EC) No. identifies hazardous pesticide-related substances with PBT/vPvB, carcinogenic, or endocrine-disrupting properties (European Union, 2006). Examples include trifluralin and atrazine, which are listed as Substances of Very High Concern (SVHCs) and subject to further control under Annex XIV (Authorization) or Annex XVII (Restriction).

Protection of water resources is achieved through the Water Framework Directive (2000/60/EC) (European Union, 2000) and Directive 2013/39/EU, which designate certain pesticides as priority substances and establish Environmental Quality Standards (EQS), as 0.6 μg/L for atrazine and 0.03 μg/L for chlorpyrifos in inland waters (European Union, 2013). The newly revised Urban Wastewater Treatment Directive EU 2024/3019 strengthens pollution control by mandating advanced (quaternary) treatment in large facilities to remove micropollutants such as pesticides, aligning with the Zero Pollution Action Plan (European Union, 2024).

Human exposure is managed through drinking water and food safety standards. The EU Drinking Water Directive 2020/2184 limits individual pesticide concentrations to 0.10 μg/L and total pesticides to 0.50 μg/L, reflecting the precautionary principle (European Union, 2020). Similar limits are established internationally: the US EPA sets a Maximum Contaminant Level (MCL) of 3 μg/L for atrazine (USEPA, 2024), while the WHO provides guidance values such as 0.9 mg/L for glyphosate (World Health Organization (WHO), 2008). National standards, including those from Health Canada and Australia's NHMRC, further refine limits for compounds such as 2,4-D and simazine to prevent chronic and endocrine-related effects (Health Canada, 2022; National Health and Medical Research Council, 2011). Dietary exposure is addressed by the European Food Safety Authority (EFSA), which set maximum residue levels (MRLs) in food to ensure safety for consumers, including consideration of mixture toxicity and transformation products (Carrasco Cabrera and Medina Pastor, 2022).

4.3 Occurrence in the environment

Pesticides are extensively used in agriculture and urban environments to control pests, but over 95% of these chemicals spread beyond their intended targets, contaminating air, water, and soil (Syafrudin et al., 2021). Their persistence, mobility, and ability to accumulate in organisms raise significant environmental and health concerns. Fig. 1.4 shows the number of publications found in Scopus using the query string TITLE-ABS-KEY(*compound name) AND TITLE-ABS-KEY(*contaminant) AND PUBYEAR > 2010 for the most environmentally relevant pesticides.

The presence of pesticides in wastewater provide an early indicator of chemical inputs from urban, agricultural, and industrial sources. Influent concentrations at WWTPs vary with season and usage, with herbicides (e.g., atrazine, 2,4-D), fungicides (e.g., tebuconazole, azoxystrobin), and insecticides (e.g., neonicotinoids, chlorpyrifos) commonly detected in the tens to hundreds of ng/L range (Wang et al., 2022). Removal in conventional WWTPs is compound-dependent: hydrophilic pesticides and neonicotinoids often show low or negative removal, while hydrophobic compounds sorb to sludge. Advanced treatments, including ozonation, activated carbon, electrochemical oxidation, and membrane filtration, could achieve >90% removal for some compounds (Disley et al., 2025).

Pesticide contamination of surface and groundwaters varies with land use, with agricultural and urban areas showing higher pesticide occurrence

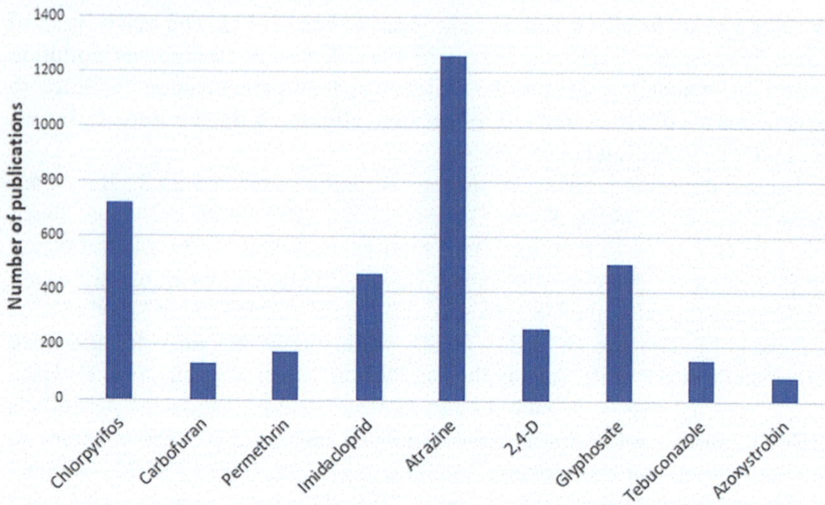

FIG. 1.4 Number of publications indexed in Scopus from 2010 onwards related to the most environmentally relevant pesticides. *(Source: Scopus database, accessed on 2025/11/10; created by the authors.)*

compared to undeveloped regions. Among insecticides, chlorpyrifos is frequently detected, despite restrictions in some countries with reported concentrations up to 1500 ng/L in Bangladesh (Sumon et al., 2018). Tebuconazole and carbendazim are the most commonly reported fungicides, with levels 3200 ng/L in the Ebro River, Spain, reflecting frequent applications in vineyards and nearby crops, as well as their persistence due to low water solubility (Ccanccapa et al., 2016). Among herbicides, atrazine is one of the most commonly detected, with concentrations up to 1000 ng/L in surface waters from China, primarily from runoff from corn and soybean fields (Peng et al., 2017). Elevated levels in specific studies are often explained by intensive agricultural use, irrigation practices, seasonal rainfall patterns, and the chemical stability of the compounds, which allows them to accumulate in aquatic systems (de Souza et al., 2020). Pesticides are also detected in remote environments, highlighting their persistence and mobility in the environment. Chlorpyrifos (3.70 ng/L) and terbuthylazine (1.38 ng/L) have been found in the Alpine glaciers (Italy), where they accumulate in ice cores and are released in meltwater, reflecting both historical contamination and ongoing deposition from agricultural activities in the surrounding lowland regions (Rizzi et al., 2019).

Sediments serve as both reservoirs and secondary sources of pesticides, making them a key but often overlooked factor in long-term aquatic contamination. Across Europe, studies have documented extensive sediment contamination by a wide range of pesticides, including both currently used chemicals and legacy compounds. For example, glyphosate and its main breakdown product,

AMPA, are commonly found in sediments, with detection frequencies ranging from 61% to 76%. In certain cases, such as in sediments from the Czech Republic, total pesticide levels have been reported as high as 600–1200 μg/kg (Khurshid et al., 2024).

Pyrethroids strongly adsorb to sediments and organic matter, resulting in sediment concentrations that are often 10–50 times higher than in overlying waters, creating persistent sources of exposure for benthic organisms. Among aquatic invertebrates, insects such as mayflies (*Ephemeroptera*), caddis flies (*Trichoptera*), and midges (*Diptera*) are particularly sensitive, showing adverse effects on survival, growth, emergence, mobility, and reproduction at concentrations as low as 0.01–0.1 μg/L (Kumar et al., 2023). In Lake Ziway, Ethiopia, the detection of organophosphate and carbamate pesticides in fish further highlights ecological and human health risks. Diazinon and chlorpyrifos were found in all fish species at concentrations ranging from 2.46 to 12.55 μg/kg and 0.42 to 5.88 μg/kg, respectively, indicating bioaccumulation and the potential for neurotoxic effects in fish, as well as exposure risks for human consumers (Tarekegn et al., 2025).

Pesticides such as the neonicotinoids imidacloprid, acetamiprid, thiamethoxam, and clothianidin have been also detected in air samples (Brüggemann et al., 2024). In a study conducted in China, all $PM_{2.5}$ samples contained at least two of these compounds, with imidacloprid showing the highest concentrations, ranging from 4.33 to 1.13×10^2 pg/m^3 (Zhou et al., 2020).

Last but not least, pesticide biotransformation products warrant careful consideration. These metabolites, formed through chemical, microbial, or enzymatic processes, can be more toxic, persistent, or bioavailable than their parent compounds. For example, the herbicide atrazine degrades into desethylatrazine and deisopropylatrazine, which may disrupt endocrine systems in aquatic organisms, while organophosphate insecticides can form oxon derivatives that are stronger acetylcholinesterase inhibitors than the original molecules (Basapuram et al., 2025). Ignoring these transformation products in monitoring programs risks underestimating the true ecological and human health impacts of pesticide contamination.

5 Persistent, mobile, and toxic (PMT) and very persistent very mobile (VPVM) chemicals

5.1 Definition and properties

Some chemical substances, due to their high polarity and resistance to degradation, can travel through the water cycle and even reach underground aquifers. These compounds can pass through both natural and artificial filters, such as soil, WWTPs, etc. (Reemtsma et al., 2016). This group of compounds began to gain attention around 2015, following the work of several researchers who started referring to them using the acronyms PMT (Persistent, Mobile, and Toxic substance), vPvM (very Persistent and very Mobile substance), or PMOC (Persistent and Mobile Organic Compound). These terms are now used to describe a class of CECs capable of circulating through the water cycle and reaching drinking water, posing a potential risk to both human health and aquatic ecosystems (Hale et al., 2020).

The terms PMT and vPvM were developed as derivatives of the well-known PBT (Persistent, Bioaccumulative, and Toxic) and vPvB (very Persistent and very Bioaccumulative) substances, which were already regulated under REACH (Registration, Evaluation, Authorisation and Restriction of Chemicals) due to their classification as substances of very high concern (SVHC). The persistence and toxicity criteria for classifying substances as PMT/vPvM are aligned with those used for PBT/vPvB substances, degradation half-lives (>60 or >40 days) for persistency and different evidences, such as no-observed effect concentration (NOEC) lower than 0.01 mg/L, for toxicity (European Commission, 2023). The property that determines a substance's mobility is mainly polarity, which depends on its molecular structure and the primary parameter used as an approximate indicator of compound mobility is the log K_{oc}. A substance shall be considered to fulfil the mobility criterion (M) when the log K_{oc} is <3 (2 for very Mobile vM) (European Commission, 2023).

5.2 Legislation

A regulatory gap persists for these substances worldwide, with the first official recognition of this category occurring in Europe in 2022. Until then, these categories of compounds were not considered substances of very high concern (SVHC), nor were any hazard statements included for them under the REACH regulation (Reemtsma et al., 2016). On October 19th 2022, an amendment was made to the Classification, Labelling and Packaging (CLP) regulation (European Commission, 2023). This revision ensures that the hazards associated with chemical substances and mixtures are effectively communicated to consumers, workers, and any other individuals who may come into contact with them. The update specifically introduced PMT/vPvM compounds as highly hazardous substances and established the abovementioned criteria for their classification. It should be noted that although this classification is relatively recent, many extensively studied pollutants would fall into this category due to their physicochemical properties, such as venlafaxine or diclofenac, which have been already included in different Watch List over the years (European Commission, 2022, 2015).

5.3 Occurrence in the environment

With the introduction of the new hazard categories into European regulation, a wave of scientific interest was sparked around PMT and vPvM substances. Early contributions created screening lists based on physicochemical properties that helped pinpoint candidate compounds requiring closer examination (Huang et al., 2021; Montes et al., 2022; Neuwald et al., 2021). Montes et al. (2022), for instance, focused on Spain and Portugal, monitoring surface waters and detecting 112 PMT compounds—10 of which appeared in at least 80% of analyzed samples. Among the most prevalent were industrial substances such as di-o-tolylguanidine (DTG) and diphenylguanidine (DPG), commonly used in rubber production, and pharmaceuticals frequently prescribed in Spain such as venlafaxine, its metabolite O-desmethylvenlafaxine, and sulpiride. Previously, a collaborative international study (Schulze et al., 2019) performed a monitoring in Spain, France, and Germany. Across drinking water, surface, groundwater, and wastewater, they found common industrial reagents, including methyl sulfate and trifluoromethanesulfonic acid persisting across various water matrices. Same authors (Neuwald et al., 2021) in Germany identified 14 sulfonate compounds, such as 3-nitrobenzenesulfonate, used in textiles and chemical manufacturing, 1,3-diazabicyclo-(2.2.2)-octane, used in adhesives and pH regulators and 20 ionic substances, which dissociate in water into fluorinated anions and cations such as BF_4^- and PF_6^-, both widely used in industry. Another study conducted in China (Huang et al., 2021) monitored multiple water sources (from surface to drinking water) shown that around 45% of the monitored CECs, particularly those found in drinking and groundwater, would fall under the categories of PMT or vPvM substances.

References

Atasoy Aydin, A., Gören, İ.E., Yavuz Guzel, E., Daglioglu, N., 2023. Method development, validation, and application for simultaneous determination of 56 new psychoactive substances in surface water by LC-MS/MS. Environ. Sci. Pollut. Res. 16147499. 30 (36), 85920–85929. Available from: https://www.springer.com/journal/11356. doi:10.1007/s11356-023-28495-9.

Baabish, A., Sobhanei, S., Fiedler, H., 2021. Priority perfluoroalkyl substances in surface waters—a snapshot survey from 22 developing countries. Chemosphere. 00456535. 273, 129612. doi:10.1016/j.chemosphere.2021.129612.

Bade, R., Rousis, N., Adhikari, S., Baduel, C., Bijlsma, L., Bizani, E., Boogaerts, T., Burgard, D.A., Castiglioni, S., Chappell, A., Covaci, A., Driver, E.M., Sodre, F.F., Fatta-Kassinos, D., Galani, A., Gerber, C., Gracia-Lor, E., Gracia-Marín, E., Halden, R.U., Heath, E., Hernandez, F., Jaunay, E., Lai, F.Y., Lee, H.J., Laimou-Geraniou, M., Oh, J.E., Olafsdottir, K., Phung, K., Castro, M.P., Psichoudaki, M., Shao, X., Salgueiro-Gonzalez, N., Feitosa, R.S., Gomes, C.S., Subedi, B., Löve, A.S.C., Thomaidis, N., Tran, D., van Nuijs, A., Verovšek, T., Wang, D., White, J.M., Yargeau, V., Zuccato, E., Mueller, J.F., 2023. Three years of wastewater surveillance for new psychoactive substances from 16 countries. Water Res. X. 25899147. (19). Available from: https://www.journals.elsevier.com/water-research-x. doi:10.1016/j.wroa.2023.100179.

Basapuram, G., Dutta, A., Duttagupta, S., 2025. Biotransformation of pesticides across biological systems: molecular mechanisms, omics insights, and biotechnological advances for environmental sustainability. ACS Omega. 2470-1343. doi:10.1021/acsomega.5c06484.

Behnami, A., Zoroufchi Benis, K., Pourakbar, M., Yeganeh, M., Esrafili, A., Gholami, M., 2024. Biosolids, an important route for transporting poly- and perfluoroalkyl substances from wastewater treatment plants into the environment: a systematic review. Sci. Total Environ. 00489697. 925, 171559. doi:10.1016/j.scitotenv.2024.171559.

Boles, T.H., Wells, M.J.M., 2010. Analysis of amphetamine and methamphetamine as emerging pollutants in wastewater and wastewater-impacted streams. J. Chromatogr. A. 00219673. 1217 (16), 2561–2568. doi:10.1016/j.chroma.2010.01.014.

Boxall, A.B.A., Rudd, M.A., Brooks, B.W., Caldwell, D.J., Choi, K., Hickmann, S., Innes, E., Ostapyk, K., Staveley, J.P., Verslycke, T., Ankley, G.T., Beazley, K.F., Belanger, S.E., Berninger, J.P., Carriquiriborde, P., Coors, A., DeLeo, P.C., Dyer, S.D., Ericson, J.F., Gagné, F., Giesy, J.P., Gouin, T., Hallstrom, L., Karlsson, M.V., Joakim Larsson, D.G., Lazorchak, J.M., Mastrocco, F., McLaughlin, A., McMaster, M.E., Meyerhoff, R.D., Moore, R., Parrott, J.L., Snape, J.R., Murray-Smith, R., Servos, M.R., Sibley, P.K., Straub, J.O., Szabo, N.D., Topp, E., Tetreault, G.R., Trudeau, V.L., Van Der Kraak, G., 2012. Pharmaceuticals and personal care products in the environment: what are the big questions? Environ. Health Perspect. 15529924. 120 (9), 1221–1229. Available from: http://ehp.niehs.nih.gov/article/fetchObjectAttachment.action?uri=info%3Adoi%2F10.1289%2Fehp.1104477&representation=PDF. doi:10.1289/ehp.1104477.

Brown, S., Kennedy, L., Cullington, M., Mihle, A., Lono-Batura, M., 2019. Relating pharmaceuticals and personal care products in biosolids to home exposure. Urban Agric. Reg. Food Syst. 2575-1220. 4 (1), 1–14. doi:10.2134/urbanag2018.12.0005.

Brüggemann, M., Mayer, S., Brown, D., Terry, A., Rüdiger, J., Hoffmann, T., 2024. Measuring pesticides in the atmosphere: current status, emerging trends and future perspectives. Environ. Sci. Eur. 2190-4715. 36 (1). doi:10.1186/s12302-024-00870-4.

Buck, R.C., Franklin, J., Berger, U., Conder, J.M., Cousins, I.T., Voogt, P.D., Jensen, A.A., Kannan, K., Mabury, S.A., van Leeuwen, S.P.J., 2011. Perfluoroalkyl and polyfluoroalkyl substances in the environment: terminology, classification, and origins. Integr. Environ. Assess. Manag. 15513793. 7 (4), 513–541. doi:10.1002/ieam.258.

Campos-Mañas, M.C., Ferrer, I., Thurman, E.M., Agüera, A., 2018. Opioid occurrence in environmental water samples—a review. Trends Environ. Anal. Chem. 22141588. 20. Available from: http://www.journals.elsevier.com/trends-in-environmental-analytical-chemistry/. doi:10.1016/j.teac.2018.e00059.

Carrasco Cabrera, L., Medina Pastor, P., 2022. The 2020 European Union report on pesticide residues in food. EFSA J. 18314732. 20 (3). doi:10.2903/j.efsa.2022.7215.

Castellani, F., Vitali, M., Antonucci, A., Del Morrone, G., Cofone, L., D'Ancona, G., Pindinello, I., Andolfi, R., Mattiucci, S., Protano, C., 2025. Bisphenols and phthalates contamination in edible mussels (Mytilus galloprovincialis) from central Italy coastline: occurrence, spatial distribution, and human health risk assessment. J. Food Compos. Anal. 08891575. 139, 107174. doi:10.1016/j.jfca.2024.107174.

Castillo-Zacarías, C., Barocio, M.E., Hidalgo-Vázquez, E., Sosa-Hernández, J.E., Parra-Arroyo, L., López-Pacheco, I.Y., Barceló, D., Iqbal, H.N.M., Parra-Saldívar, R., 2021. Antidepressant drugs as emerging contaminants: occurrence in urban and non-urban waters and analytical methods for their detection. Sci. Total Environ. 18791026. 757. Available from: www.elsevier.com/locate/scitotenv. doi:10.1016/j.scitotenv.2020.143722.

Ccanccapa, A., Masiá, A., Navarro-Ortega, A., Picó, Y., Barceló, D., 2016. Pesticides in the Ebro River basin: occurrence and risk assessment. Environ. Pollut. 02697491. 211, 414–424. doi:10.1016/j.envpol.2015.12.059.

Cecinato, A., Balducci, C., Perilli, M., 2016. Illicit psychotropic substances in the air: the state-of-art. Sci. Total Environ. 00489697. 539, 1–6. doi:10.1016/j.scitotenv.2015.08.051.

Chakraborty, A., Adhikary, S., Bhattacharya, S., Dutta, S., Chatterjee, S., Banerjee, D., Ganguly, A., Rajak, P., 2023. Pharmaceuticals and personal care products as emerging environmental contaminants: prevalence, toxicity, and remedial approaches. ACS Chem. Health Saf. 1871-5532. 30 (6), 362–388. doi:10.1021/acs.chas.3c00071.

Chen, J., Feng, D., Xu, J., Ren, G., Guo, C., Cao, X., Yang, A., Wang, L., Wang, M., Wang, R., Chen, L., Wu, X., 2025. Occurrence, distribution, and potential ecological risks of psychoactive substances in the mangrove environment, Hainan Island of South China. Environ. Sci. Eur. 2190-4715. 37 (1). doi:10.1186/s12302-024-01049-7.

Czarny-Krzymińska, K., Krawczyk, B., Szczukocki, D., 2023. Bisphenol A and its substitutes in the aquatic environment: occurrence and toxicity assessment. Chemosphere. 00456535. 315, 137763. doi:10.1016/j.chemosphere.2023.137763.

De Felice, B., Parolini, M., 2020. Effects of single and combined exposure to cocaine and benzoylecgonine on the oxidative status of Mytilus galloprovincialis. Environ. Toxicol. Pharmacol. 13826689. 80, 103475. doi:10.1016/j.etap.2020.103475.

de Souza, R.M., Seibert, D., Quesada, H.B., de Jesus Bassetti, F., Fagundes-Klen, M.R., Bergamasco, R., 2020. Occurrence, impacts and general aspects of pesticides in surface water: a review. Process Saf. Environ. Prot. 09575820. 135, 22–37. Available from: http://www.elsevier.com/wps/find/journaldescription.cws_home/713889/description#description. doi:10.1016/j.psep.2019.12.035.

Disley, J., Gil-Ramirez, G., Gonzalez-Rodriguez, J., 2025. A review of different removal techniques for pesticides found in European waters identified by the European Environmental Agency. Environ. Sci. Policy. 14629011. 172, 104194. doi:10.1016/j.envsci.2025.104194.

Ebele, A.J., Abou-elwafa Abdallah, M., Harrad, S., 2017. Pharmaceuticals and personal care products (PPCPs) in the freshwater aquatic environment. Emerg. Contam. 24056642. 3 (1), 1–16. Available from: www.keaipublishing.com/en/journals/emerging-contaminants/. doi:10.1016/j.emcon.2016.12.004.

Elbert, A., Haas, M., Springer, B., Thielert, W., Nauen, R., 2008. Applied aspects of neonicotinoid uses in crop protection. Pest Manag. Sci. 1526-498X. 64 (11), 1099–1105. doi:10.1002/ps.1616.

Environmental Protection Agency, 2022. Drinking water contaminant candidate list 5—final rule. Fed. Regist. 87 (218), 68060–68077.

Environmental Protection Agency, 2024. Per- and polyfluoroalkyl substances (PFAS) final PFAS national primary drinking water regulation. 89 FR 32532, pp. 32532-32757. Available from: https://www.federalregister.gov/d/2024-07773.

European Commission, 2009. Regulation (EC) No 1107/2009 of the European Parliament and of the Council of 21 October 2009 concerning the placing of plant protection products on the market and repealing Council Directives 79/117/EEC and 91/414/EEC. OJ L309, 1–50.

European Commission, 2015. Commission Implementing Decision (EU) 2015/495 of 20 March 2015 establishing a watch list of substances for Union-wide monitoring in the field of water policy pursuant to Directive 2008/105/EC of the European Parliament and of the Council. OJ L78, 40–42.

European Commission, 2018. Commission Implementing Regulation (EU) 2018/783 of 29 May 2018 amending Implementing Regulation (EU) No 540/2011 as regards the conditions of approval of the active substance imidacloprid. OJ L132, 31–34.

European Commission, 2022. Commission Implementing Decision (EU) 2022/1307 of 22 July 2022 establishing a watch list of substances for Union-wide monitoring in the field of water policy pursuant to Directive 2008/105/EC of the European Parliament and of the Council. OJ L197, 117–121.

European Commission, 2023. Commission Delegated Regulation (EU) 2023/707 of 19 December 2022 amending Regulation (EC) No 1272/2008 as regards hazard classes and criteria for the classification, labelling and of substances and mixtures. OJ L93, 1–64.

European Commission, 2024. Commission Regulation (EU) 2024/2462 of 19 September 2024 amending Annex XVII to Regulation (EC) No 1907/2006 of the European Parliament and of the Council as regards undecafluorohexanoic acid (PFHxA), its salts and PFHxA-related substances. OJ L, 2024/2462, 20.9.2024, ELI: http://data.europa.eu/eli/reg/2024/2462/oj.

European Commission, 2025. Selection of Substances for the 5th Watch List under the Water Framework Directive. European Commission.

European Monitoring Centre for Drugs and Drug Addiction and Europol, 2024. EU drug market: new psychoactive substances—IN-depth analysis. Available from: https://www.euda.europa.eu/publications/eu-drug-markets/new-psychoactive-substances_en. doi:10.2810/338463.

European Union, 2000. Directive 2000/60/EC of the European Parliament and of the Council of 23 October 2000 establishing a framework for Community action in the field of water policy. OJ L327, 1–73.

European Union, 2006. European Union, E.U Regulation (EC) No 1907/2006 concerning the registration, evaluation, authorisation and restriction of chemicals (REACH), annex XVII—restrictions on the manufacture, placing on the market and use of certain dangerous substances, mixtures and articles. OJ L396, pp. 1–849.

European Union, 2013. European Union, E.U Directive 2013/39/EU of the European Parliament and of the Council of 12 August 2013 amending Directives 2000/60/EC and 2008/105/EC as regards priority substances in the field of water policy. OJ 226, 1–17.

European Union, 2019. Regulation (EU) 2019/1021 of the European Parliament and of the Council of 20 June 2019 on persistent organic pollutants. OJ 169, 45–77.

European Union, 2020. European Union, E.U Directive (EU) 2020/2184 of the European Parliament and of the Council of 16 December 2020 on the quality of water intended for human consumption. OJ 435, 1–62.

European Union, 2024. European Parliament & Council of the European Union. Directive (EU) 2024/3019 of the European Parliament and of the Council of 27 November 2024 concerning urban wastewater treatment (recast). OJ L3019, pp. 1–7.

Fan, X., Zhang, J., Fu, X., Zhou, B., Xu, Z., Huang, H., Han, S., Li, X., 2022. Analysis of synthetic cannabinoids in wastewater of major cities in China. Sci. Total Environ. 00489697. 827, 154267. doi:10.1016/j.scitotenv.2022.154267.

Fenoll, J., Vela, N., Navarro, G., Pérez-Lucas, G., Navarro, S., 2014. Assessment of agro-industrial and composted organic wastes for reducing the potential leaching of triazine herbicide residues through the soil. Sci. Total Environ. 00489697. 493, 124–132. doi:10.1016/j.scitotenv.2014.05.098.

Fontanals, N., Marcé, R.M., Borrull, F., 2017. Solid-phase extraction followed by liquid chromatography-high resolution mass spectrometry to determine synthetic cathinones in different types of environmental water samples. J. Chromatogr. A. 00219673. 1524, 66–73. doi:10.1016/j.chroma.2017.10.002.

Fontes, M.K., de Campos, B.G., Cortez, F.S., Pusceddu, F.H., Nobre, C.R., Moreno, B.B., Lebre, D.T., Maranho, L.A., Pereira, C.D.S., 2021. Mussels get higher: a study on the occurrence of cocaine and benzoylecgonine in seawater, sediment and mussels from a subtropical ecosystem (Santos Bay, Brazil). Sci. Total Environ. 18791026. 757. Available from: www.elsevier.com/locate/scitotenv. doi:10.1016/j.scitotenv.2020.143808.

Food and Agricultural Organization (FAO), 2023. Pesticides use and trade (1990-2022). Available from: https://www.fao.org/statistics/highlights-archive/highlights-detail/pesticides-use-and-trade-1990-2022/en.

Food and Drug Administration (FDA), 2025. Fact sheet: FDA at a glance. Available from: https://www.fda.gov/about-fda/fda-basics/fact-sheet-fda-glance.

Food and Drug Administration, 2017. Guidance document classification of products as drugs and devices and additional product classification issues. Available from: https://www.fda.gov/media/80384/download.

Gandhi, K., Khan, S., Patrikar, M., Markad, A., Kumar, N., Choudhari, A., Sagar, P., Indurkar, S., 2021. Exposure risk and environmental impacts of glyphosate: highlights on the toxicity of herbicide co-formulants. Environ. Chall. 26670100. 4, 100149. doi:10.1016/j.envc.2021.100149.

Gao, T., Du, P., Xu, Z., Li, X., 2017. Occurrence of new psychoactive substances in wastewater of major Chinese cities. Sci. Total Environ. 00489697. 575, 963–969. doi:10.1016/j.scitotenv.2016.09.152.

Gebbink, W.A., van Leeuwen, S.P.J., 2020. Environmental contamination and human exposure to PFASs near a fluorochemical production plant: review of historic and current PFOA and GenX contamination in the Netherlands. Environ. Int. 18736750. 137. Available from: www.elsevier.com/locate/envint. doi:10.1016/j.envint.2020.105583.

Glüge, J., Scheringer, M., Cousins, I.T., Dewitt, J.C., Goldenman, G., Herzke, D., Lohmann, R., Ng, C.A., Trier, X., Wang, Z., 2020. An overview of the uses of per- and polyfluoroalkyl substances (PFAS). Environ. Sci. Process. Impacts 22 (12), 2345–2373. doi:10.1039/D0EM00291G.

Guo, C., Hu, S., Cheng, P., Cheng, K., Yang, Y., Chen, G., Wang, Q., Wang, Y., Liu, T., 2024. Speciation and biogeochemical behavior of perfluoroalkyl acids in soils and their environmental implications: a review. Eco-Environ. Health. 27729850. 3 (4), 505–515. doi:10.1016/j.eehl.2024.05.005.

Hale, S.E., Arp, H.P.H., Schliebner, I., Neumann, M., 2020. What's in a name: persistent, mobile, and toxic (PMT) and very persistent and very mobile (vPvM) substances. Environ. Sci. Technol. 15205851. 54 (23), 14790–14792. Available from: http://pubs.acs.org/journal/esthag. doi:10.1021/acs.est.0c05257.

Haman, C., Dauchy, X., Rosin, C., Munoz, J.F., 2015. Occurrence, fate and behavior of parabens in aquatic environments: a review. Water Res. 18792448. 68, 1–11. Available from: www.elsevier.com/locate/watres. doi:10.1016/j.watres.2014.09.030.

Hartz, W.F., Björnsdotter, M.K., Yeung, L.W.Y., Humby, J.D., Eckhardt, S., Evangeliou, N., Jogsten, I.E., Kärrman, A., Kallenborn, R., 2024. Sources and seasonal variations of per- and polyfluoroalkyl substances (PFAS) in surface snow in the Arctic. Environ. Sci. Technol. 15205851. 58 (49), 21817–21828. Available from: http://pubs.acs.org/journal/esthag. doi:10.1021/acs.est.4c08854.

He, K., Yuan, J., Lu, H., Asada, Y., Echigo, S., Wu, Q., Wang, H., 2024. Evaluating the behavior and environmental risks of carbamazepine and its metabolites in soil aquifer treatment: insights from deconjugation dynamics and toxicity assessment. J. Hazard. Mater. 03043894. 480, 135788. doi:10.1016/j.jhazmat.2024.135788.

Health Canada, 2022. Guidelines for Canadian drinking water quality: guideline technical document—2,4-dichlorophenoxyacetic acid. Available from: https://www.canada.ca/en/health-canada/services/publications/healthy-living/guidelines-canadian-drinking-water-quality-guideline-technical-document-2-4-dichlorophenoxyacetic-acid.html.

Hidalgo-Serrano, M., Borrull, F., Marcé, R.M., Pocurull, E., 2022. Phthalate esters in marine ecosystems: analytical methods, occurrence and distribution. TrAC Trends Anal. Chem. 18793142. 151. Available from: www.elsevier.com/locate/trac. doi:10.1016/j.trac.2022.116598.

How, Z.T., Gamal El-din, M., 2021. A critical review on the detection, occurrence, fate, toxicity, and removal of cannabinoids in the water system and the environment. Environ. Pollut. 02697491. 268 (Pt A), 115642. doi:10.1016/j.envpol.2020.115642.

Huang, C., Jin, B., Han, M., Yu, Y., Zhang, G., Arp, H.P.H., 2021. The distribution of persistent, mobile and toxic (PMT) pharmaceuticals and personal care products monitored across Chinese water resources. J. Hazard. Mater. Lett. 26669110. 2, 100026. doi:10.1016/j.hazl.2021.100026.

Huber, S., Remberger, M., Kaj, L., Schlabach, M., Jörundsdóttir, H.Ó., Vester, J., Arnórsson, M., Mortensen, I., Schwartson, R., Dam, M., 2016. A first screening and risk assessment of pharmaceuticals and additives in personal care products in waste water, sludge, recipient water and sediment from Faroe Islands, Iceland and Greenland. Sci. Total Environ. 00489697. 562, 13–25. doi:10.1016/j.scitotenv.2016.03.063.

Huizer, M., ter Laak, T.L., de Voogt, P., van Wezel, A.P., 2021. Wastewater-based epidemiology for illicit drugs: a critical review on global data. Water Res. 18792448. 207. Available from: www.elsevier.com/locate/watres. doi:10.1016/j.watres.2021.117789.

Janousek, R.M., Mayer, J., Knepper, T.P., 2019. Is the phase-out of long-chain PFASs measurable as fingerprint in a defined area? Comparison of global PFAS concentrations and a monitoring study performed in Hesse, Germany from 2014 to 2018. TrAC Trends Anal. Chem. 18793142. 120. Available from: www.elsevier.com/locate/trac. doi:10.1016/j.trac.2019.01.017.

Jeschke, P., Nauen, R., 2008. Neonicotinoids—from zero to hero in insecticide chemistry. Pest Manag. Sci. 1526-498X. 64 (11), 1084–1098. doi:10.1002/ps.1631.

Kaboré, H.A., Vo Duy, S., Munoz, G., Méité, L., Desrosiers, M., Liu, J., Sory, T.K., Sauvé, S., 2018. Worldwide drinking water occurrence and levels of newly-identified perfluoroalkyl and polyfluoroalkyl substances. Sci. Total Environ. 18791026. 616–617, 1089–1100. Available from: www.elsevier.com/locate/scitotenv. doi:10.1016/j.scitotenv.2017.10.210.

Khurshid, C., Silva, V., Gai, L., Osman, R., Mol, H., Alaoui, A., Christ, F., Schlünssen, V., Vested, A., Abrantes, N., Campos, I., Baldi, I., Robelot, E., Bureau, M., Pasković, I., Polić Pasković, M., Glavan, M., Hofman, J., Harkes, P., Lwanga, E.H., Norgaard, T., Ritsema, C.J., Geissen, V., 2024. Pesticide residues in European sediments: a significant concern for the aquatic systems? Environ. Res. 00139351. 261, 119754. doi:10.1016/j.envres.2024.119754.

Kim, D.-H., Ait Bamai, Y., Belova, L., Bessems, J., Poma, G., Covaci, A., 2023. Human exposure to persistent and mobile chemicals: A review of sources, internal levels and health implications. Sci. Total Environ. 893, 164764. doi:10.1016/j.scitotenv.2023.164764.

Kumar, A., Jasrotia, S., Dutta, J., Kyzas, G.Z., 2023. Pyrethroids toxicity in vertebrates and invertebrates and amelioration by bioactive compounds: a review. Pestic. Biochem. Physiol. 00483575. 196, 105615. doi:10.1016/j.pestbp.2023.105615.

Larsson, D.G.J., Flach, C.F., 2022. Antibiotic resistance in the environment. Nat. Rev. Microbiol. 17401534. 20 (5), 257–269. Available from: http://www.nature.com/nrmicro/index.html. doi:10.1038/s41579-021-00649-x.

Lee, J.-S., Lee, J.-S., Kim, H.S., 2024. Toxic effects of triclosan in aquatic organisms: a review focusing on single and combined exposure of environmental conditions and pollutants. Sci. Total Environ. 00489697. 920, 170902. doi:10.1016/j.scitotenv.2024.170902.

Leite, F.G., Sampaio, C.F., Pires, J.A.C., de Oliveira, D.P., Dorta, D.J., 2024. Toxicological impact of strobilurin fungicides on human and environmental health: a literature review. J. Environ. Sci. Health B 15324109. 59 (4), 142–151. Available from: www.tandf.co.uk/journals/titles/03601234.asp. doi:10.1080/03601234.2024.2312786.

Li, H., Cheng, F., Wei, Y., Lydy, M.J., You, J., 2017. Global occurrence of pyrethroid insecticides in sediment and the associated toxicological effects on benthic invertebrates: an overview. J. Hazard. Mater. 18733336. 324, 258–271. Available from: www.elsevier.com/locate/jhazmat. doi:10.1016/j.jhazmat.2016.10.056.

Liang, J., Li, C., Dang, Y., Feng, X., Ji, X., Liu, X., Zhao, X., Zhang, Q., Ren, Z., Wang, Y., Li, Y., Qu, G., Liu, R., 2024. Occurrence of bisphenol A analogues in the aquatic environment and their behaviors and toxicity effects in plants. Environ. Int. 01604120. 193, 109105. doi:10.1016/j.envint.2024.109105.

Liao, C., Kim, U.J., Kannan, K., 2018. A review of environmental occurrence, fate, exposure, and toxicity of benzothiazoles. Environ. Sci. Technol. 15205851. 52 (9), 5007–5026. Available from: http://pubs.acs.org/journal/esthag. doi:10.1021/acs.est.7b05493.

Lin, J.-Y., Zhang, Y., Bian, Y., Zhang, Y.-X., Du, R.-Z., Li, M., Tan, Y., Feng, X.-S., 2023. Non-steroidal anti-inflammatory drugs (NSAIDs) in the environment: recent updates on the occurrence, fate, hazards and removal technologies. Sci. Total Environ. 00489697. 904, 166897. doi:10.1016/j.scitotenv.2023.166897.

Lin, Y.C., Hsiao, T.C., Lin, A.Y.C., 2020. Urban wastewater treatment plants as a potential source of ketamine and methamphetamine emissions to air. Water Res. 18792448. 172. Available from: www.elsevier.com/locate/watres. doi:10.1016/j.watres.2020.115495.

López-García, E., Mastroianni, N., Ponsà-Borau, N., Barceló, D., Postigo, C., López de Alda, M., 2021. Drugs of abuse and their metabolites in river sediments: analysis, occurrence in four Spanish river basins and environmental risk assessment. J. Hazard. Mater. 03043894. 401, 123312. doi:10.1016/j.jhazmat.2020.123312.

Lukić Bilela, L., Matijošytė, I., Krutkevičius, J., Alexandrino, D.A.M., Safarik, I., Burlakovs, J., Gaudêncio, S.P., Carvalho, M.F., 2023. Impact of per- and polyfluorinated alkyl substances (PFAS) on the marine environment: raising awareness, challenges, legislation, and mitigation approaches under the one health concept. Mar. Pollut. Bull. 18793363. 194. Available from: www.elsevier.com/locate/marpolbul. doi:10.1016/j.marpolbul.2023.115309.

Maceira, A., Pecikoza, I., Marcé, R.M., Borrull, F., 2020. Multi-residue analysis of several high-production-volume chemicals present in the particulate matter from outdoor air. A preliminary human exposure estimation. Chemosphere. 00456535. 252, 126514. doi:10.1016/j.chemosphere.2020.126514.

Madani, N.A., Carpenter, D.O., 2022. Effects of glyphosate and glyphosate-based herbicides like Roundup on the mammalian nervous system: a review. Environ. Res. 10960953. 214. Available from: http://www.elsevier.com/inca/publications/store/6/2/2/8/2/1/index.htt. doi:10.1016/j.envres.2022.113933.

Mali, H., Shah, C., Raghunandan, B.H., Prajapati, A.S., Patel, D.H., Trivedi, U., Subramanian, R.B., 2023. Organophosphate pesticides an emerging environmental contaminant: pollution, toxicity, bioremediation progress, and remaining challenges. J. Environ. Sci. (China) 18787320. 127, 234–250. Available from: http://www.journals.elsevier.com/journal-of-environmental-sciences/. doi:10.1016/j.jes.2022.04.023.

Mastroianni, N., Postigo, C., López de Alda, M., Viana, M., Rodríguez, A., Alastuey, A., Querol, X., Barceló, D., 2015. Comprehensive monitoring of the occurrence of 22 drugs of abuse and transformation products in airborne particulate matter in the city of Barcelona. Sci. Total Environ. 00489697. 532, 344–352. doi:10.1016/j.scitotenv.2015.06.027.

Mei, X.Y., Hong, Y.Q., Chen, G.H., 2016. Review on analysis methodology of phenoxy acid herbicide residues. Food Anal. Methods. 1936976X. 9 (6), 1532–1561. Available from: http://www.springer.com/life+sci/food+science/journal/12161. doi:10.1007/s12161-015-0327-z.

Mishra, A., Goel, D., Shankar, S., 2023. Bisphenol A contamination in aquatic environments: a review of sources, environmental concerns, and microbial remediation. Environ. Monit. Assess. 0167-6369. 195 (11). doi:10.1007/s10661-023-11977-1.

Moermond, C.T.A., Smit, C.E., 2016. Derivation of water quality standards for carbamazepine, metoprolol, and metformin and comparison with monitoring data. Environ. Toxicol. Chem. 15528618. 35 (4), 882–888. Available from: http://www.interscience.wiley.com/jpages/0730-7268. doi:10.1002/etc.3178.

Montes, R., Méndez, S., Carro, N., Cobas, J., Alves, N., Neuparth, T., Santos, M.M., Quintana, J.B., Rodil, R., 2022. Screening of contaminants of emerging concern in surface water and wastewater effluents, assisted by the persistency-mobility-toxicity criteria. Molecules. 14203049. 27 (12). Available from: https://www.mdpi.com/1420-3049/27/12/3915/pdf?version=1655545448. doi:10.3390/molecules27123915.

Morrissey, C.A., Mineau, P., Devries, J.H., Sanchez-Bayo, F., Liess, M., Cavallaro, M.C., Liber, K., 2015. Neonicotinoid contamination of global surface waters and associated risk to aquatic invertebrates: a review. Environ. Int. 18736750. 74, 291–303. Available from: www.elsevier.com/locate/envint. doi:10.1016/j.envint.2014.10.024.

Narevski, A.C., Novaković, M.I., Petrović, M.Z., Mihajlović, I.J., Maoduš, N.B., Vujić, G.V., 2021. Occurrence of bisphenol A and microplastics in landfill leachate: lessons from South East Europe. Environ. Sci. Pollut. Res. 16147499. 28 (31), 42196–42203. Available from: https://link.springer.com/journal/11356. doi:10.1007/s11356-021-13705-z.

National Health and Medical Research Council, 2011. Australian drinking water guidelines 6. Available from: https://guidelines.nhmrc.gov.au/australian-drinking-water-guidelines/part-5/physical-chemical-characteristics/2-4-d-2-4-dichlorophenoxy-acetic-acid.

Net, S., Delmont, A., Sempéré, R., Paluselli, A., Ouddane, B., 2015. Reliable quantification of phthalates in environmental matrices (air, water, sludge, sediment and soil): a review. Sci. Total Environ. 00489697. 515-516, 162–180. doi:10.1016/j.scitotenv.2015.02.013.

Neuwald, I., Muschket, M., Zahn, D., Berger, U., Seiwert, B., Meier, T., Kuckelkorn, J., Strobel, C., Knepper, T.P., Reemtsma, T., 2021. Filling the knowledge gap: a suspect screening study for 1310 potentially persistent and mobile chemicals with SFC- and HILIC-HRMS in two German river systems. Water Res. 00431354. 204, 117645. doi:10.1016/j.watres.2021.117645.

Oldenkamp, R., Hamers, T., Wilkinson, J., Slootweg, J., Posthuma, L., 2024. Regulatory risk assessment of pharmaceuticals in the environment: current practice and future priorities. Environ. Toxicol. Chem. 0730-7268. 43 (3), 611–622. doi:10.1002/etc.5535.

Olivero-Verbel, R., Eljarrat, E., Johnson-Restrepo, B., 2025. Organophosphate ester flame retardants in sediments and marine fish species in Colombia: occurrence, distribution, and implications for human risk assessment. Mar. Pollut. Bull. 0025326X. 213, 117654. doi:10.1016/j.marpolbul.2025.117654.

Ore, O.T., Adeola, A.O., Bayode, A.A., Adedipe, D.T., Nomngongo, P.N., 2023. Organophosphate pesticide residues in environmental and biological matrices: occurrence, distribution and potential remedial approaches. Environ. Chem. Ecotoxicol. 25901826. 5, 9–23. Available from: https://www.sciencedirect.com/journal/environmental-chemistry-and-ecotoxicology. doi:10.1016/j.enceco.2022.10.004.

Osuoha, J.O., Anyanwu, B.O., Ejileugha, C., 2023. Pharmaceuticals and personal care products as emerging contaminants: need for combined treatment strategy. J. Hazard. Mater. Adv. 27724166. 9. Available from: https://www.journals.elsevier.com/journal-of-hazardous-materials-advances. doi:10.1016/j.hazadv.2022.100206.

Pai, C.W., Leong, D., Chen, C.Y., Wang, G.S., 2020. Occurrences of pharmaceuticals and personal care products in the drinking water of Taiwan and their removal in conventional water treatment processes. Chemosphere. 18791298. 256. Available from: www.elsevier.com/locate/chemosphere. doi:10.1016/j.chemosphere.2020.127002.

Pal, R., Megharaj, M., Kirkbride, K.P., Naidu, R., 2013. Illicit drugs and the environment—a review. Sci. Total Environ. 18791026. 463-464, 1079–1092. doi:10.1016/j.scitotenv.2012.05.086.

Peng, X., Fan, Y., Jin, J., Xiong, S., Liu, J., Tang, C., 2017. Bioaccumulation and biomagnification of ultraviolet absorbents in marine wildlife of the Pearl River Estuarine, South China Sea. Environ. Pollut. 02697491. 225, 55–65. doi:10.1016/j.envpol.2017.03.035.

Peng, Y., Gautam, L., Hall, S.W., 2019. The detection of drugs of abuse and pharmaceuticals in drinking water using solid-phase extraction and liquid chromatography-mass spectrometry. Chemosphere. 18791298. 223, 438–447. Available from: www.elsevier.com/locate/chemosphere. doi:10.1016/j.chemosphere.2019.02.040.

Postigo, C., López de Alda, M.J., Barceló, D., 2010. Drugs of abuse and their metabolites in the Ebro River basin: occurrence in sewage and surface water, sewage treatment plants removal efficiency, and collective drug usage estimation. Environ. Int. 18736750. 36 (1), 75–84. Available from: www.elsevier.com/locate/envint. doi:10.1016/j.envint.2009.10.004.

Ramos, S., Homem, V., Alves, A., Santos, L., 2015. Advances in analytical methods and occurrence of organic UV-filters in the environment—a review. Sci. Total Environ. 00489697. 526, 278–311. doi:10.1016/j.scitotenv.2015.04.055.

Reemtsma, T., Berger, U., Arp, H.P.H., Gallard, H., Knepper, T.P., Neumann, M., Quintana, J.B., Voogt, P.D., 2016. Mind the gap: persistent and mobile organic compounds—water contaminants that slip through. Environ. Sci. Technol. 15205851. 50 (19), 10308–10315. Available from: http://pubs.acs.org/journal/esthag. doi:10.1021/acs.est.6b03338.

Ren, C.-J., Zhang, Y., Bian, Y., Zhu, J., Feng, X.-S., Liu, K., 2025. Phenoxycarboxylic acid herbicides in environment: recent updates on pretreatment and analysis methods. Ecotoxicol. Environ. Saf. 01476513. 302, 118715. doi:10.1016/j.ecoenv.2025.118715.

Reyes, N.J.D.G., Geronimo, F.K.F., Yano, K.A.V., Guerra, H.B., Kim, L.H., 2021. Pharmaceutical and personal care products in different matrices: occurrence, pathways, and treatment processes. Water (Switzerland) 20734441. 13 (9). Available from: https://www.mdpi.com/2073-4441/13/9/1159/pdf. doi:10.3390/w13091159.

Rizzi, C., Finizio, A., Maggi, V., Villa, S., 2019. Spatial-temporal analysis and risk characterisation of pesticides in Alpine glacial streams. Environ. Pollut. 02697491. 248, 659–666. doi:10.1016/j.envpol.2019.02.067.

Rodayan, A., Afana, S., Segura, P.A., Sultana, T., Metcalfe, C.D., Yargeau, V., 2016. Linking drugs of abuse in wastewater to contamination of surface and drinking water. Environ. Toxicol. Chem. 15528618. 35 (4), 843–849. Available from: http://www.interscience.wiley.com/jpages/0730-7268. doi:10.1002/etc.3085.

Salgueiro-Gonzalez, N., Béen, F., Bijlsma, L., Boogaerts, T., Covaci, A., Baz-Lomba, J.A., Kasprzyk-Hordern, B., Matias, J., Ort, C., Bodík, I., Heath, E., Styszko, K., Emke, E., Hernández, F., van Nuijs, A.L.N., Castiglioni, S., 2024. Influent wastewater analysis to investigate emerging trends of new psychoactive substances use in Europe. Water Res. 18792448. 254. Available from: https://www.sciencedirect.com/science/journal/00431354. doi:10.1016/j.watres.2024.121390.

Santana-Viera, S., Pintado-Herrera, M.G., Sosa-Ferrera, Z., Santana-Rodríguez, J.J., 2023. Analysis of psychoactive substances and metabolites in sludges, soils, sediments and biota: a review. Environ. Chem. Lett. 16103661. 21 (4), 2311–2335. Available from: https://www.springer.com/journal/10311. doi:10.1007/s10311-023-01586-2.

Schulze, S., Zahn, D., Montes, R., Rodil, R., Quintana, J.B., Knepper, T.P., Reemtsma, T., Berger, U., 2019. Occurrence of emerging persistent and mobile organic contaminants in European water samples. Water Res. 18792448. 153, 80–90. Available from: www.elsevier.com/locate/watres. doi:10.1016/j.watres.2019.01.008.

Selim, R.E., Khalil, M.S., 2021. Strobilurins: new group of fungicides. J. Plant Sci. Phytopathol. 25750135. 5 (2), 063–064. doi:10.29328/journal.jpsp.1001062.

Shafer, T.J., Meyer, D.A., Crofton, K.M., 2005. Developmental neurotoxicity of pyrethroid insecticides: critical review and future research needs. Environ. Health Perspect. 113 (2), 123–136. doi:10.1289/ehp.7254.

Shi, Z.Q., Liu, Y.S., Xiong, Q., Cai, W.W., Ying, G.G., 2019. Occurrence, toxicity and transformation of six typical benzotriazoles in the environment: a review. Sci. Total Environ. 18791026. 661, 407–421. Available from: www.elsevier.com/locate/scitotenv. doi:10.1016/j.scitotenv.2019.01.138.

Shuang, Y., Zhang, T., Zhong, H., Li, L., 2021. Simultaneous enantiomeric determination of multiple triazole fungicides in fruits and vegetables by chiral liquid chromatography/tandem mass spectrometry on a bridged bis(β-cyclodextrin)-bonded chiral stationary phase. Food Chem. 03088146. 345, 128842. doi:10.1016/j.foodchem.2020.128842.

Sidhu, G.K., Singh, S., Kumar, V., Dhanjal, D.S., Datta, S., Singh, J., 2019. Toxicity, monitoring and biodegradation of organophosphate pesticides: a review. Crit. Rev. Environ. Sci. Technol.

15476537. 49 (13), 1135–1187. Available from: www.tandf.co.uk/journals/titles/10643389.asp. doi:10.1080/10643389.2019.1565554.

Singh, S., Kashyap, H., Shehata, N., Jammu, G.S., Kapahi, Y., Khan, N.A., Ramamurthy, P.C., Singh, J., Rtimi, S., 2025. Carbamates-based pesticides: detection, toxicity, biodegradation, and sustainable development challenges. Chem. Eng. J. 13858947. 515. Available from: https://www.sciencedirect.com/science/journal/13858947. doi:10.1016/j.cej.2025.163800.

Sokołowski, A., Kończak, M., Oleszczuk, P., Gao, Y., Czech, B., 2024. Environmental and food contamination by phthalic acid esters (PAEs): overview. Water Air Soil Pollut. 0049-6979. 235 (5). doi:10.1007/s11270-024-07121-5.

Subedi, M., Bajaj, S., Kumar, M.S., YC, M., 2019. An overview of tramadol and its usage in pain management and future perspective. Biomed. Pharmacother. 19506007. 111, 443–451. Available from: https://www.journals.elsevier.com/biomedicine-and-pharmacotherapy. doi:10.1016/j.biopha.2018.12.085.

Sumon, K.A., Rashid, H., Peeters, E.T.H.M., Bosma, R.H., Van den Brink, P.J., 2018. Environmental monitoring and risk assessment of organophosphate pesticides in aquatic ecosystems of north-west Bangladesh. Chemosphere. 18791298. 206, 92–100. Available from: www.elsevier.com/locate/chemosphere. doi:10.1016/j.chemosphere.2018.04.167.

Syafrudin, M., Kristanti, R.A., Yuniarto, A., Hadibarata, T., Rhee, J., Al-Onazi, W.A., Algarni, T.S., Almarri, A.H., Al-Mohaimeed, A.M., 2021. Pesticides in drinking water-a review. Int. J. Environ. Res. Public Health. 16604601. 18 (2), 1–15. Available from: https://www.mdpi.com/1660-4601/18/2/468/pdf. doi:10.3390/ijerph18020468.

Tang, W., Wang, D., Wang, J., Wu, Z., Li, L., Huang, M., Xu, S., Yan, D., 2018. Pyrethroid pesticide residues in the global environment: an overview. Chemosphere. 00456535. 191, 990–1007. doi:10.1016/j.chemosphere.2017.10.115.

Tarekegn, M.M., Duga, D.L., Alemayehu, Y.A., Worku, M.A., 2025. Assessment of pesticide residues in fish and associated human health risks in Lake Ziway, Ethiopia. Discover Environ. 27319431. 3 (1). Available from: https://link.springer.com/journal/44274. doi:10.1007/s44274-025-00362-z.

United Nations Environment Programme, 2010. New POPs SC-4/17: listing of perfluorooctane sulfonic acid, its salts and perfluorooctane sulfonyl fluoride. In: Stockholm Convention on Persistent Organic Pollutants (POPs). Secretariat of the Stockholm Convention, pp. 1–4.

United Nations Office on Drugs and Crime (UNODC), 2022. World Drug Report 2022. United Nations Publication.

United Nations Office on Drugs and Crime (UNODC), 2025. Early warning advisory on new psychoactive substances. Available from: https://www.unodc.org/LSS/Home/NPS.

United Nations, 1961. Single convention of narcotic drugs, 1961, as amended by the 1972 protocol amending the single convention on narcotic drugs. Available from: https://www.unodc.org/pdf/convention_1961_en.pdf.

United Nations, 1971. Convention on psychotropic substances. United Nations Treaty Series. 1019, 175.

USEPA, 2024. Per- and Polyfluoroalkyl Substances (PFAS) Final PFAS National Primary Drinking Water Regulation. Washington, DC: U.S. Environmental Protection Agency.

van der Veen, I., de Boer, J., 2012. Phosphorus flame retardants: properties, production, environmental occurrence, toxicity and analysis. Chemosphere. 00456535. 88 (10), 1119–1153. doi:10.1016/j.chemosphere.2012.03.067.

Wan, Y., Xue, J., Kannan, K., 2016. Benzothiazoles in indoor air from Albany, New York, USA, and its implications for inhalation exposure. J. Hazard. Mater. 311, 37–42. doi:10.1016/j.jhazmat.2016.02.057.

Wang, M., Lv, J., Deng, H., Liu, Q., Liang, S., 2022. Occurrence and removal of triazine herbicides during wastewater treatment processes and their environmental impact on aquatic life. Int. J. Environ. Res. Public Health. 1660-4601. 19 (8), 4557. doi:10.3390/ijerph19084557.

Wang, W., Park, S., Choi, B., Oh, J.-E., 2023. Occurrence and removal of benzotriazole and benzothiazole in drinking water treatment plants. Environ. Pollut. 316, 120563. doi:10.1016/j.envpol.2022.120563.

Wang, X., Zhu, Q., Yan, X., Wang, Y., Liao, C., Jiang, G., 2020a. A review of organophosphate flame retardants and plasticizers in the environment: analysis, occurrence and risk assessment. Sci. Total Environ. 00489697. 731, 139071. doi:10.1016/j.scitotenv.2020.139071.

Wang, Y., Li, Y., Hu, A., Rashid, A., Ashfaq, M., Wang, Y., Wang, H., Luo, H., Yu, C.-P., Sun, Q., 2018. Monitoring, mass balance and fate of pharmaceuticals and personal care products in seven wastewater treatment plants in Xiamen City, China. J. Hazard. Mater. 03043894. 354, 81–90. doi:10.1016/j.jhazmat.2018.04.064.

Wang, Z., Han, S., Cai, M., Du, P., Zhang, Z., Li, X., 2020b. Environmental behavior of methamphetamine and ketamine in aquatic ecosystem: degradation, bioaccumulation, distribution, and associated shift in toxicity and bacterial community. Water Res. 00431354. 174, 115585. doi:10.1016/j.watres.2020.115585.

Wong, F., Shoeib, M., Katsoyiannis, A., Eckhardt, S., Stohl, A., Bohlin-Nizzetto, P., Li, H., Fellin, P., Su, Y., Hung, H., 2018. Assessing temporal trends and source regions of per- and polyfluoroalkyl substances (PFASs) in air under the arctic monitoring and assessment programme (AMAP). Atmos. Environ. 13522310. 172, 65–73. doi:10.1016/j.atmosenv.2017.10.028.

World Health Organization (WHO), 2008. Guidelines for Drinking-water Quality, vol. 1, 3rd ed. Incorporating the First and Second Addenda, p. 1.

World Health Organization (WHO), 2025. Anatomical therapeutic chemical (ATC) classification. WHO. Available from: https://www.who.int/tools/atc-ddd-toolkit/atc-classification.

Wu, X., Dodgen, L.K., Conkle, J.L., Gan, J., 2015. Plant uptake of pharmaceutical and personal care products from recycled water and biosolids: a review. Sci. Total Environ. 18791026. 536, 655–666. Available from: www.elsevier.com/locate/scitotenv. doi:10.1016/j.scitotenv.2015.07.129.

Yadav, M.K., Short, M.D., Aryal, R., Gerber, C., van den Akker, B., Saint, C.P., 2017. Occurrence of illicit drugs in water and wastewater and their removal during wastewater treatment. Water Res. 18792448. 124, 713–727. Available from: www.elsevier.com/locate/watres. doi:10.1016/j.watres.2017.07.068.

Yao, T., Sun, P., Zhao, W., 2023. Triazine herbicides risk management strategies on environmental and human health aspects using in silico methods. Int. J. Mol. Sci. 1422-0067. 24 (6), 5691. doi:10.3390/ijms24065691.

Yedhu Krishnan, R., Manikandan, S., Subbaiya, R., Biruntha, M., Balachandar, R., Karmegam, N., 2023. Origin, transport and ecological risk assessment of illicit drugs in the environment—a review. Chemosphere. 00456535. 311 (Pt 2), 137091. doi:10.1016/j.chemosphere.2022.137091.

Young, C.J., Furdui, V.I., Franklin, J., Koerner, R.M., Muir, D.C.G., Mabury, S.A., 2007. Perfluorinated acids in Arctic snow: new evidence for atmospheric formation. Environ. Sci. Technol. 0013936X. 41 (10), 3455–3461. doi:10.1021/es0626234.

Yuan, X., Luo, L., Li, X., Lu, Y., Chen, S., Luan, T., 2024. Recent advances in the removal of psychoactive substances from aquatic environments: a review. Sci. Total Environ. 00489697. 953, 176156. doi:10.1016/j.scitotenv.2024.176156.

Zacs, D., Perkons, I., Sire, J., Bartkevics, V., 2025. Occurrence levels of perfluoroalkyl carboxylic acids (PFCA) and perfluoroalkyl sulfonic acids (PFSA) in European perch (Perca fluviatilis) samples collected from inland waters in Latvia: component profiles, spatial distribution and dietary exposure to consumers. Environ. Res. 00139351. 269, 120882. doi:10.1016/j.envres.2025.120882.

Zergui, A., Nkontcheu Kenko, D.B., 2024. Drugs of abuse and psychoactive pharmaceuticals in wastewater as worldwide contaminants of emerging concern: a systematic review. Environ. Pollut. Manag. 29503051. 1, 71–86. Available from: https://www.keaipublishing.com/en/journals/environmental-pollution-and-management/. doi:10.1016/j.epm.2024.08.001.

Zhang, H., Xu, H., Qin, B., Fu, Y., Yao, Y., Zhao, Y., Qin, C., 2025. Review on the sources, distribution and treatment of per- and polyfluoroalkyl substances in global groundwater. Environ. Res. 00139351. 275, 121387. doi:10.1016/j.envres.2025.121387.

Zhang, Y., Guo, C., Wu, R., Hou, S., Liu, Y., Zhao, J., Jiang, M., Xu, J., Wu, F., 2024. Global occurrence, distribution, and ecological risk assessment of psychopharmaceuticals and illicit drugs in surface water environment: a meta-analysis. Water Res. 00431354. 263, 122165. doi:10.1016/j.watres.2024.122165.

Zhao, J., Lu, J., Zhao, H., Yan, Y., Dong, H., 2023. In five wastewater treatment plants in Xinjiang, China: removal processes for illicit drugs, their occurrence in receiving river waters, and ecological risk assessment. Chemosphere. 00456535. 339, 139668. doi:10.1016/j.chemosphere.2023.139668.

Zheng, Y., Shao, Y., Zhang, Y., Liu, Z., Zhao, Z., Xu, R., Ding, J., Li, W., Wang, B., Zhang, H., 2024. Metformin as an emerging pollutant in the aquatic environment: occurrence, analysis, and toxicity. Toxics. 2305-6304. 12 (7), 483. doi:10.3390/toxics12070483.

Zhou, Y., Guo, J., Wang, Z., Zhang, B., Sun, Z., Yun, X., Zhang, J., 2020. Levels and inhalation health risk of neonicotinoid insecticides in fine particulate matter (PM2.5) in urban and rural areas of China. Environ. Int. 01604120. 142, 105822. doi:10.1016/j.envint.2020.105822.

Zhu, H., Xia, Y., Zhang, Y., Kang, Y., Ding, Y., Chen, R., Feng, H., 2024. Distribution characteristics and transformation mechanism of per- and polyfluoroalkyl substances in drinking water sources: a review. Sci. Total Environ. 00489697. 916, 169566. doi:10.1016/j.scitotenv.2023.169566.

Zuccato, E., Chiabrando, C., Castiglioni, S., Bagnati, R., Fanelli, R., 2008. Estimating community drug abuse by wastewater analysis. Environ. Health Perspect. 0091-6765. 116 (8), 1027–1032. doi:10.1289/ehp.11022.

Part 2

Sampling and sample preparation techniques for environmental samples

Chapter 2

Sampling and extraction of atmospheric samples

Laura Vallecillos and Laura Solé-Domènech
Department of Analytical Chemistry and Organic Chemistry, Rovira i Virgili University, Tarragona, Spain

1 Introduction

Air pollution is one of today's major problems and is a matter of global concern, as it is directly related to the social and economic activities that take place in any environment. Due to the large number of emission sources of atmospheric contaminants, interest in monitoring their presence and the environmental and health effects they can cause has also increased exponentially (García-Garcinuño et al., 2025; Sánchez-Piñero et al., 2022; Wang et al., 2023). Some of the primary effects of atmospheric contaminants, and organic pollutants in particular, are their contribution to the depletion of the ozone layer, the formation of the photochemical smog affecting metropolises, and the increase of the greenhouse effect and global warming, among others (Xuan et al., 2021; Gu et al., 2021; Rovira et al., 2024). Short-term exposure to these contaminants can cause eye and throat irritations and respiratory illnesses (e.g., asthma), while long-term exposure is associated to nervous system damage and cancer (Jia et al., 2021; Sakizadeh, 2020; Gallego et al., 2016). Consequently, a growing number of studies are focused on the need to regulate and control the levels of pollution, which the general population is subjected to.

According to the World Health Organization (WHO), air pollution occurs when there is an alteration in the chemical composition of the air, either by the introduction of a foreign substance or by an abnormal increase in the concentration of a naturally occurring substance (World Health Organization, 2021). Weather, topography, and chemical reactivity are uncontrollable and, combined with natural or anthropogenic emissions of contaminants, can lead to increased pollution levels (Li et al., 2021; Desauziers, 2004). Major air pollutants include nitrogen oxide (NO_x) and sulfur oxide (SO_2) compounds, carbon monoxide (CO), ozone (O_3), some heavy metals, and particulate matter (PM) of different particle size (United States Environmental Protection Agency, 2024). Volatile organic compounds (VOCs) and semi-VOCs (SVOCs) are considered minor

Contaminants of Emerging Concern in the Environment. https://doi.org/10.1016/B978-0-443-33821-2.00003-7

air pollutants, as they are present in much lower concentrations (Karri et al., 2024). The main sources of anthropogenic emissions are combustion processes, as well as fugitive and accidental emissions, which occur in various industrial sectors, such as the petrochemical and steel industries, waste incineration, and wastewater treatment plants (WWTPs). In urban environments, emissions are mainly generated by transport—including land, air, and maritime—as well as domestic sources such as heating systems, cooking, and construction activities (Vallecillos et al., 2024a; Montero-Montoya et al., 2018). These combustion processes release various types of VOCs and SVOCs, including polycyclic aromatic hydrocarbons (PAHs) and organochlorine substances, for example, polychlorinated dibenzodioxins (PCDDs) and polychlorinated dibenzofurans (PCDFs), and polychlorinated biphenyls (PCBs), among others (Abad et al., 2022; García-Garcinuño et al., 2025).

Therefore, due to the wide range of emission sources and the diverse effects that atmospheric pollutants can have on both human health and the environment, several European directives have been developed, including Directive 2004/107/EC (European Parliament and Council of the European Union, 2004), Directive 2008/50/EC (European Parliament and Council of the European Union, 2008), and more recently, Directive 2024/2881/EC (European Parliament and Council of the European Union, 2024), concerning ambient air quality and cleaner air across Europe. These regulations have evolved over time, becoming progressively more stringent, and currently establish maximum concentration levels for major pollutants such as CO, SO_2, NO_2, O_3, PM with diameters <2.5 μm ($PM_{2.5}$) and 10 μm (PM_{10}) and certain heavy metals including lead, arsenic, cadmium, and nickel in urban atmospheres (immission zones). The monitoring of the VOCs known as ozone precursors is also recommended. In addition, an annual average concentration limit of 5.0 $\mu g\ m^{-3}$ is established for benzene and 1 $ng\ m^{-3}$ for benzo(a)pyrene. However, there are other atmospheric pollutants that are commonly present in the air for which no regulated ambient air concentration limits have yet been established, despite being legislated in other types of matrices (Abad et al., 2022; Maceira et al., 2019) and having demonstrated toxic effects on the environment. Examples include PCDDs, PCDFs, pesticides, and PCBs, among others. Other compounds, such as 1,3-butadiene, are regulated in ambient air in specific regions—such as Ontario (Canada) (Ontario Ministry of the Environment, Conservation and Parks, 2025) but are not currently included in Directive 2024/2881/EC (European Parliament and Council of the European Union, 2024).

In addition to the major air pollutants and organic contaminants listed in the Directive 2024/2881/EC (European Parliament and Council of the European Union, 2024), increasing attention is being directed toward the monitoring of VOCs of emerging concern—such as 1,3-butadiene, 1,2-dichloroethane, ethylene oxide, and odorous substances—particularly in urban areas near chemical and petrochemical facilities. Their presence at relatively high concentrations has been related to poor air quality, discomfort, odorous nuisances, and potential risk to human health (Vallecillos et al., 2024b; Yao et al., 2021; Gallego et al., 2023; Dalton et al., 2020). Among the most recently determined emerging SVOCs are high production volume chemicals (HPVCs), which include the following families of compounds: benzothiazoles (BTHs), benzosulfonamides (BSAs), benzotriazoles (BTRs), phthalates esters (PAEs), organophosphate esters (OPEs), ultraviolet stabilizers (Tinuvins), and phenolics antioxidants (PAs). The Organisation for Economic Co-operation and Development (OECD) defines HPVCs as chemicals produced or imported in quantities exceeding 1000 tons per year in at least one European Union member country or European Sate (Organisation for Economic Co-Operation and Development, 2025). These compounds are widely used in industry as corrosion inhibitors, flame retardants, plasticizers, and in the production of plastics, electronics, and fungicides (Li et al., 2020; Zhang et al., 2014; Maceira et al., 2018). The growing concern about indoor air quality (e.g., workplaces, houses, and schools) highlights the need to study fragrances, preservatives, perfluoroalkylated and polyfluoroalkylated substances (PFASs), flame retardants, and plasticizers, which are mainly emitted from personal care products, cleaning agents and building materials (Lucattini et al., 2018; Sakhi et al., 2019; Ninyà et al., 2022; Madureira et al., 2016; Even et al., 2020). The high use and exposure to these compounds make it essential to monitor their presence in the air and assess their potential impacts on both human health and the environment (Zhang et al., 2014; Nuñez et al., 2020; García-Garcinuño et al., 2024b).

In this context, the main objective of this chapter is to provide an overview of the sampling techniques commonly used to collect air samples from both outdoor and indoor environments, as well as the most extensively used pretreatment protocols applied prior to determination by chromatographic techniques. The list of contaminants of emerging concern (CECs) includes HPVCs—BTHs, BTRs, BSAs, PAEs, OPEs, Tinuvins and PAs—as well as fragrances, preservatives, PFASs, 1,3-butadiene, 1,2-dichloroethane, and odorous contaminants. PAHs, PCDDs, PCDFs, and PCBs are also included due to their bioaccumulation in the environment. The atmosphere is a complex mixture of liquids, solids, and gases, and the CECs present in the air can be distributed across all these phases. The sampling techniques described include both the gaseous phase and the PM in the air. Through the various sample protocols, it will be possible to observe their evolution over time, with a clear trend toward simplification and increased automation. The state of the art is completed by an overview of future trends in air sampling and data acquisition.

2 Sampling

The atmosphere is a complex and heterogeneous matrix mainly composed of a mixture of gases and particles. The CECs present can be associated with the gas phase, the particulate phase, or both. Due to their low vapor pressures, VOCs are generally associated with the gas phase of the air. In contrast, SVOCs can be present in both phases. This distribution depends on several factors, such as the different physicochemical properties of the compound (molecular weight, vapor pressure, heat of condensation, among others), meteorological conditions, and the aerosol surface (Su et al., 2018; Barbas et al., 2018).

VOCs are found in the gas phase of air at concentrations on the order of $\mu g\ m^{-3}$, while SVOCs are in the ng m^{-3} range (García-Garcinuño et al., 2025; Maceira et al., 2022). To increase the sensitivity of the method, it is necessary to concentrate, focus, or trap the target compounds before their analysis. To choose the most suitable sampling technique, several parameters must be taken into account (Kumar and Víden, 2007): the type of compound to be analyzed (volatility, molecular weight, polarity, etc.), the expected concentration level, the aim of the measurement (emission or immission), the duration of the sampling period (short-term or long-term), and the information to be obtained (average daily concentration, instantaneous, sequential, etc.). Additional factors include system portability and cost. In this regard, the current trend in air analysis is the development of methods that include compounds with diverse characteristics, with techniques that are faster, more sensitive, easy to use, cost-effective, and environmentally friendly. Therefore, an overview of the most commonly used techniques for active and passive sampling of CECs present in both the gas and particulate phases of air is provided below.

2.1 Gaseous phase

Some of the most common techniques used over the years to sample the gas phase of air consist of collecting the gas in special containers, such as polymeric bags and metal containers or canisters. However, in recent years, most of the analytical methods to determine CECs in the gas phase of air include a sampling step based on enrichment by trapping on solid sorbents. This approach allows sampling and preconcentration of CECs in a single step (Ras et al., 2009).

2.1.1 Sampling containers

The sampling of the gas phase of air in polymeric bags or canisters is the simplest method. Although they can be made from other plastic materials, the bags are typically made of Tedlar (polyvinyl fluoride) and are available in volumes between 0.5 and 100 L (Fig. 2.1A). On the contrary, stainless-steel canisters, as shown in Fig. 2.1B, can be of different shapes and volumes (Ras et al., 2009). Once collected, the sample is usually analyzed by gas chromatography, which can be combined with a preconcentration step depending on the concentration of the target compounds. These methods allow whole-air sampling, enable the collection of several aliquots for repeated analyses and can provide time-integrated samples if controlled-flow pumps are used to introduce the sample into the container (Ras et al., 2009; Wang and Austin, 2006). Therefore, they are widely used in official methods for air sampling from emission sources and for the determination of the most volatile CECs. Tedlar bags are recommended

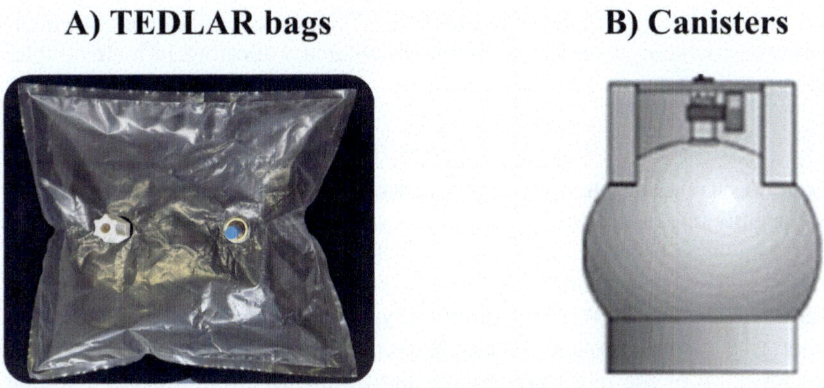

FIG. 2.1 Whole-air containers: (A) TEDLAR bags and (B) stainless-steel canisters. *(Modified from Ras, M.R., Borrull, F., Marcé, R.M., 2009. Sampling and preconcentration techniques for determination of volatile organic compounds in air samples. TrAC Trends Anal. Chem. 0165-9936. 28, 347. doi:10.1016/j.trac.2008.10.009.)*

for the sampling of odorous substances (e.g., sulfur compounds) and hazardous VOCs (e.g., benzene, allyl chloride and 1,2-dichloroethane) in ambient air in the ASTM method D5504-20 (Advancing Standards Transforming Markets (ASTM), 2024) and the USEPA method 0040 (US Environmental Protection Agency (USEPA), 1996), respectively. As specified in USEPA method TO-15A, canisters can also be used to collect air samples for the measurement of ethylene oxide and other VOCs of emerging concern in ambient air (US Environmental Protection Agency (USEPA), 2019b; Hoisington and Herrington, 2021). USEPA method OTM-50 proposes canisters of up to 6 L for monitoring 30 volatile fluorinated compounds (VFCs) in air emission from stationary sources (US Environmental Protection Agency (USEPA), 2025). This method includes multiple short-chain VFCs, which may be monitored as indicators of incomplete decomposition of PFASs.

Tedlar bags and canisters are mainly used for the monitoring of highly volatile CECs in industrial environments. Specifically, these sampling methods are applied in routine analyses to evaluate atmospheric emissions from production plants and fugitive emissions. For instance, Toledo and Muñoz (2023) used Tedlar bags for the monitoring of odorous sulfur compounds and hazardous VOCs at a WWTP in Valladolid, Spain. Time-integrated 24 h air samples were collected at different points in the WWTP to evaluate the effectiveness of new odor prevention strategies (Toledo and Muñoz, 2023). Similarly, 24 h canister sampling was used to assess the presence of ethylene oxide at a sterilization facility in Michigan (Olaguer et al., 2020). Canisters can also be used for air sampling in urban areas (immission). The National Air Pollution Surveillance program of Canada comprised the monitoring of up to 63 VOCs, including ethylene oxide, in various urban areas across the country by collecting 24 h canister air samples (Galarneau et al., 2016). Ramboll US Corporation (2019) studied the background concentrations of ethylene oxide in various neighborhoods surrounding the Chicago metropolitan area, collecting 12 h air samples using canisters.

2.1.2 Enrichment on solid adsorbents

Sampling using solid adsorbent enrichment, either in active or passive mode, is currently one of the most widely employed techniques for VOC monitoring, CECs included (Maceira et al., 2017). It allows sampling and preconcentration in a single step, facilitating the determination of compounds at low concentration levels. Fig. 2.2A shows different types of commercially available samplers used for this purpose. To ensure proper sampling and to guarantee both the retention of the target compounds during the sampling process and their desorption and recovery before analysis, it is essential to select the most suitable adsorbent or combination of adsorbents. This selection depends on the properties of the compounds to be determined (e.g., volatility, polarity, and molecular weight), the sample matrix and the subsequent desorption, analysis and detection technique to be applied. The properties of the adsorbent material, such as surface area, adsorption capacity, hydrophobicity and thermal stability, are also taken into account (Kumar and Víden, 2007; Maceira et al., 2017).

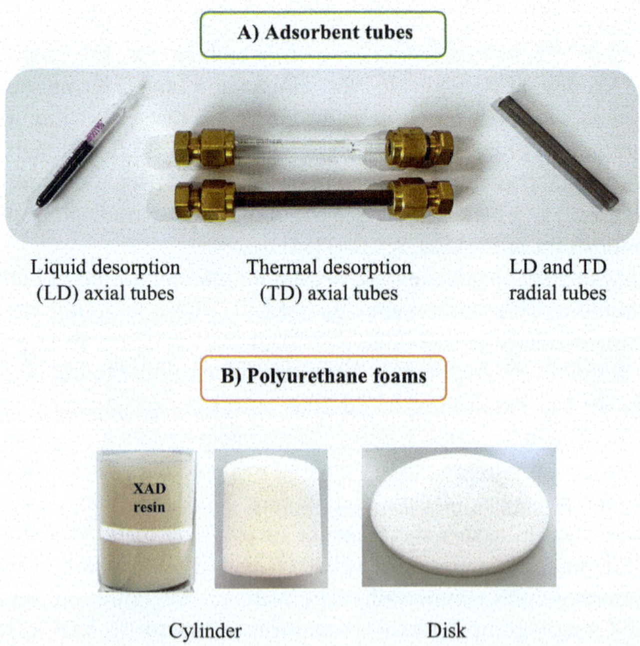

FIG. 2.2 Different samplers for the gas phase of air: (A) adsorbent tubes and (B) polyurethane foams.

Carbon-based adsorbents can be made of different materials, such as graphitized carbon, activated charcoal or carbon forming a molecular sieve. They are the adsorbents with the highest retention capacity, due to surface areas of up to 1500 m^2 g^{-1}, establishing moderate to strong interactions with the target compounds. All of them are stable at temperatures above 400°C and are therefore compatible with thermal desorption (TD). Official methods UNE-EN 14662-1 (Asociación Española de Normalización (AENOR), 2023) and USEPA method TO-17 (US Environmental Protection Agency (USEPA), 1999) applied carbon-based adsorbent tubes, similar than those in Fig. 2.2A, followed by TD, for the measurement of benzene and VOCs of emerging concern, such as 1,3-butadiene, 1,2-dichloroethane, and other ozone precursors, in air. These official methods are the preferred ones for the monitoring of the above-mentioned VOCs in outdoor air samples from urban areas, specifically near chemical and petrochemical facilities (Han et al., 2019; Dörter et al., 2020; Mukerjee et al., 2020; Ghaffari et al., 2021; da Rocha et al., 2023). They have also been increasingly applied to assess indoor air quality in places where people spend numerous hours every day (e.g., workplaces, houses, and schools). The list of target VOCs in indoor studies is broader, including fragrances, solvents, and preservatives, compounds commonly found in personal care and cleaning products (Sakhi et al., 2019; Jeong et al., 2019; Dugheri et al., 2018; Vallecillos et al., 2021). Gallego et al. (2012) and Besis et al. (2021) have also used carbon-based tubes for the determination of odorous VOCs, sulfur compounds and aromatic aldehydes, in air samples from a waste treatment facility (Barcelona, Spain) and the seafront of the city of Thessaloniki (Greece), respectively. Furthermore, molecular sieve can be combined with liquid desorption (LD) with organic solvents, though activated charcoal is by far the most used adsorbent for this application. Although, they are used in official methods UNE-EN 14662-2 (Asociación Española de Normalización y Certificación, 2006) and NIOSH 1501 (Manual of Analytical Methods (NMAM), 2003) for the determination of benzene and other VOCs, their adsorption efficiency is limited in air samples with high humidity because of their hydrophilic nature (Maceira et al., 2017; Vallecillos et al., 2018).

The most representative polymeric adsorbents are those based on styrene-divinylbenzene or polyvinylpyrrolidone, known as XADs, and those based on phenyl phenylene oxide, known as Tenax. Both adsorbents are hydrophobic and, due to their moderate surface areas (300–800 m^2 g^{-1}), they can establish medium–weak interactions with the target compounds. These types of adsorbents are useful for trapping SVOCs. USEPA method 0023A proposed XAD tubes for the sampling of PCDDs and PCDFs from stationary emission sources (US Environmental Protection Agency (USEPA), 2018). This adsorbent is also the most widely used to assess the presence of pesticides in ambient air (Yusà et al., 2009). More recent applications for XAD tubes have focused on evaluating occupational exposure to the aforementioned compounds in specific actions during the workday. For instance, Fent et al. (2020) evaluated the exposure to PCDDs, PCDFs, and brominated flame retardants (BFRs) during controlled residential firefighting. Tenax tubes, on the contrary, are the preferred option for the sampling of the most volatile PFASs (Robbins et al., 2023; Markes International, 2022) and PAHs (Jia et al., 2023; Nargis et al., 2022) directly from emission sources and in urban environments. In addition, Aragón et al. (2013) monitored the presence of PAEs and OPEs in air samples from various urban areas in the Tarragona region (Spain) and the city's port using Tenax tubes.

The use of a single adsorbent is often not suitable for pollutants with different characteristics, such as volatility. Therefore, tubes with multiple adsorbents of different adsorption strength are the best option to capture compounds with a wide range of volatility and to avoid losses during sampling. The most commonly used adsorbent combinations typically include a polymeric adsorbent combined with a graphitized carbon, graphitized carbons with different adsorption strength, or graphitized carbons combined with molecular sieves (Ras et al., 2009). Maceira et al. (2017) tested two multisorbent bed tubes, Tenax/Carbograph and Carbotrap B/Carbopack X/Carboxen 569 for the determination of 100 VOCs in air samples with a high humidity content. Carbotrap B/Carbopack X/Carboxen 569 tubes have also proven to be suitable for the sampling of 1,3-butadiene (Vallecillos et al., 2018) and VOCs responsible of odorous nuisance problems (Gallego et al., 2012). Meanwhile, a Tenax/Carbograph/Carboxen 1003 tube has successfully been applied for the sampling of up to 130 VOCs in tobacco smoke (Savareear et al., 2017). Lee et al. (2019) tested combinations of Carbopack (C + B + X) for the sampling of the 16 priority PAHs. In addition, as with the sampling of aldehydes or ethylene oxide, a derivatization agent must be added to the adsorbent to assure the stability of the sampled compounds. The official method NIOSH 1614 suggested HBr-coated activated charcoal tubes for the sampling of ethylene oxide from stationary sources and the subsequent determination of 2-bromoethanol (Manual of Analytical Methods (NMAM), 1994). Florisil tubes used for aldehyde sampling are coated with 2,4-dinitrophenylhydrazine (2,4-DNPH) (Ninyà et al., 2022).

As shown in Fig. 2.2B, adsorbents based on polyurethane foam (PUF) or XAD resins are also employed in cylindrical or disc form for active and passive sampling, respectively. These adsorbents can establish medium-wake interactions with the target compounds and are followed by LD prior to the analysis. They are suitable for the sampling of SVOCs, such as PAHs, PCDDs, and PCDFs as well as HPVCs. Anh et al. (2020) used PUF disk for the simultaneous monitoring of polybrominated diphenyl ethers (PBDEs), PCBs, BFRs, and methylated-PAHs (Me-PAHs) in urban and industrial environments in Vietnam. The study included indoor, outdoor, and semiopen air samples. Abad et al. (2022) applied the same type of disk for a three-year monitoring of PCDDs and PCDFs in several urban areas in developing countries. García-Garcinuño et al. (2024b, 2024a) used cylindrical and disk-shaped PUFs for the monitoring of PAHs and HPVCs, including PAEs, OPEs, BTHs, BSAs, Tinuvins and PAs, in urban areas from Tarragona (Spain). PUF are the preferred samplers for trapping PFASs and pesticides in ambient air studies around the world. To increase the retention of the most volatile compounds during sampling, some authors have added a layer of XAD resin into the PUF (Yusà et al., 2009; Wallace et al., 2024). These types of adsorbents have also been used for assessing the presence of synthetic musk compounds (SMCs) in air, more specifically in indoors, as the main emission sources of these compounds are personal care and cleaning products (Lucattini et al., 2018; Balci et al., 2020).

Samples collected using solid adsorbents can be obtained through active or passive sampling techniques. Active sampling involves passing a constant flow of air through an adsorbent to retain the target compounds and obtain a defined air volume (Vallecillos et al., 2019). A key parameter to optimize is the sample volume, which should be as large as possible to maximize preconcentration without exceeding the breakthrough volume (air volume that can pass through the adsorbent tube without losses of the target compound). This breakthrough volume depends on the target compound, the type and amount of adsorbent, and the temperature and humidity of the sampled air (Maceira et al., 2017). It is a versatile technique, widely used in official methods, such as EPA TO-17 (US Environmental Protection Agency (USEPA), 1999), NIOSH 1614 (Manual of Analytical Methods (NMAM), 1994), and UNE-EN 14662-1 (Asociación Española de Normalización (AENOR), 2023), among others. It is particularly effective for detecting compounds present at low concentration levels in the air, as it allows the preconcentration of large air volumes. Fig. 2.2A shows examples of adsorbent tubes used in such methods.

A wide range of air volumes can be sampled, and therefore different equipment is available to sample larger or smaller volumes depending on the characteristics and concentrations of the pollutants. Low-volume sampling is used for compounds present in the gas phase of air in concentrations of μg or ng m^{-3}, such as VOCs of emerging concern (odorous compounds, fragrances, aldehydes, 1,3-butadiene, etc.) (Vallecillos et al., 2021; Gallego et al., 2012; Besis et al., 2021). Standard adsorbent tubes, usually made of glass or stainless steel, are between 6 and 9 cm length and have a diameter of approximately 6 mm. These tubes are suitable for sampling flow-rates between 10 and 200 mL min^{-1} using a sample pump and total sampled volumes in the range of 0.1 and 150 L (Kumar and Víden, 2007). Fig. 2.3 shows two of the most common air sampling devices: (a) a sampling pump connected to an adsorbent tube and (b) a sequential autosampler that allows the collection of up to 32 different samples. Sampling pumps are used to collect episodic samples of both outdoor and indoor air during odor events or incidents that may impact air quality (e.g., fires and industrial maintenance activities) (Gallego et al., 2012; Vallecillos et al., 2018). In contrast, sequential autosamplers enable the monitoring of air pollutants over time, providing pollution profiles. They are particularly used to assess the pollution levels of a specific procedure or urban area and to evaluate indoor air quality throughout the day in indoor environments (Vallecillos et al., 2019, 2021).

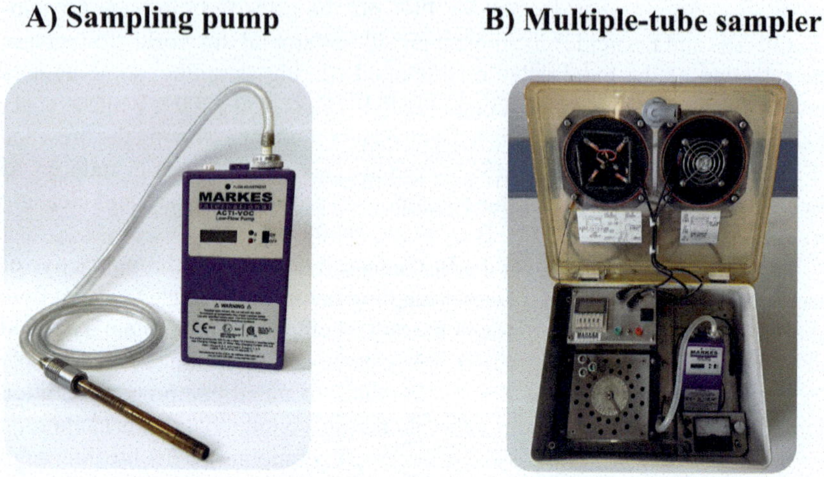

FIG. 2.3 Active air devices with adsorbent tubes: (A) low-flow sampling pump and (B) multiple-tube sampler.

Active sampling of large volumes of air in the gas phase can be performed with different types of samplers and tubes. One option involves the use of glass tubes containing activated charcoal (Fig. 2.2A) and the sampler in Fig. 2.4A. This equipment allows the collection of air samples of up to 1400 L during one day with flow-rates of up to 1000 mL min^{-1}. It also enables the sequential sampling of several tubes during a week. This technique is commonly used for monitoring VOCs, for example, 1,3-butadiene, benzene, 1,2-dichloroetane, odorous compounds, etc., in urban atmospheres (Vallecillos et al., 2018; Ramírez et al., 2012). On the contrary, high volume samplers, such as the one in Fig. 2.4B, operate at flow rates between 100 to 500 L min^{-1} to collect air volumes between 300 and 2000 m^3 over periods from 24 h to two weeks. The sampler in Fig. 2.4B enables the sampling of both gaseous phase and PM of air. PM is deposited on filters, while weak adsorbents, such as PUFs, XAD resins or combinations of both, are used to retain the compounds in the gaseous phase. This type of sampler is mainly used to study the presence of SVOCs, including PFASs, PAEs, OPEs, PAHs, etc., which are at low concentrations levels (<ng m^{-3}) in urban environments and can be detected both in the gaseous phase and PM of air (García-Garcinuño et al., 2024a; Wallace et al., 2024).

In recent years, the use of passive samplers for the determination of organic contaminants in the gaseous phase of air has increased due to their simplicity, low cost, and ease of use (Wallace et al., 2024; Vallecillos et al., 2019; US Environmental Protection Agency (USEPA), 2019a). Enrichment in passive samplers occurs through the diffusion of nearby compounds, which are retained on the adsorbent with which they have a high affinity. These sampling devices are relatively small, do not require sample pumps, measuring equipment or

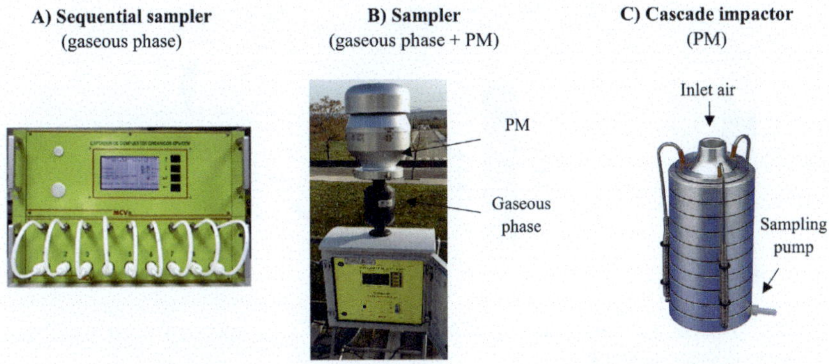

FIG. 2.4 Active air devices of high volume: (A) sequential sampler for gaseous phase, (B) sampler for gaseous phase and PM and (C) cascade impactor for PM of different sizes.

power sources. Passive sampling has been proven to be suitable for monitoring average concentrations of contaminants over a defined exposure period of 7 to 14 days for VOCs and several weeks to months for SVOCs. For instance, they have been widely used for local and global scale monitoring of PCDDs and PCDFs (Bogdal et al., 2013; Abad et al., 2007) and long-term sampling studies focused on assessing the inhalation risk associated with the exposure to HPVCs or VOCs of emerging concern (García-Garcinuño et al., 2025). However, passive sampling does not allow for the detection of short-term variations in the concentration of the pollutants and the episodic peaks in concentration are smoothed out. The lack of diffusive uptake-rates commercially available is the main limitation of this kind of sampling and their calculation often needs to be included as part of the study planning (García-Garcinuño et al., 2024b; Vallecillos et al., 2019). Moreover, as these are long-term samplings, weather conditions have to be controlled, especially fluctuations in temperature and wind patterns (Wania and Shunthirasingham, 2020).

Passive sampling with adsorbent tubes can be of two types, axial or radial, depending on their geometry. As Fig. 2.5A shows, axial tubes use a single adsorbent, have a long axial diffusion path and a small cross-sectional area. In contrast, radial passive samplers have a larger cross-sectional area (Fig. 2.5B). Although they usually contain a single adsorbent, a derivatizing agent can be added to enable the sampling and derivatization of highly reactive compounds in situ. Regardless of the configuration, passive sampling with adsorbent tubes has been applied for the monitoring of VOCs of emerging concern in different kinds of outdoor environments throughout the world, including industrial areas (Gallego et al., 2018; Vallecillos et al., 2019), urban areas (Vallecillos et al., 2024a; Healy et al., 2018), or remotes rural zones (Villanueva et al., 2018; Civan et al., 2015). For instance, Vallecillos et al. (2024a) performed a three-year monitoring of VOCs to assess the impact of petrochemical parks in Tarragona

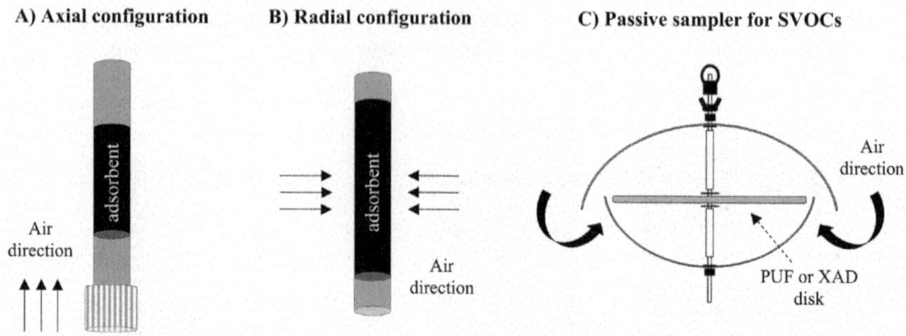

FIG. 2.5 Passive air devices for the gaseous phase: (A) axial tube, (B) radial tubes, and (C) passive sampler for SVOCs.

on the air quality of nearby urban areas. Dörter et al. (2020) monitored a similar group of VOCs in up to 60 sites across the Bolu plateau (Turkey) to study their spatial and seasonal distribution as well as emission sources. Radial tubes with 2,4-DNPH as derivatizing agent have been applied for the monitoring of aldehydes, specifically formaldehyde, in several outdoor and indoor environments (Ninyà et al., 2022; Villanueva et al., 2018; Li et al., 2018). In recent years, passive sampling has also been used to monitor the presence of fragrances, solvents, and preservatives in indoor environments, the limiting factor being the lack of diffusive uptake-rates available for these compounds (Ninyà et al., 2022; Vallecillos et al., 2021; de Gennaro et al., 2013; Hoang et al., 2017).

Passive sampling of SVOCs involves the use of an aluminum double-bowl air sampler containing a PUF or PUF/XAD disk to retain the compounds of interest. Although commercially available samplers may vary in shape, internal volume and assembly configuration (fixed or suspended), they provide similar results (Abad et al., 2022; Wania and Shunthirasingham, 2020). The most commonly used configuration is the one in Fig. 2.5C. Passive sampling is currently a well-established technique for the monitoring of persistent organic pollutants (POPs) such as PCDDs, PCDFs, PCBs and PAHs in outdoor and indoor environments worldwide, allowing interstudy comparisons (Abad et al., 2022; Anh et al., 2020; Degrendele et al., 2020). Due to the simplicity of the sampling device, these studies cover wide geographical areas, sometimes different countries, with numerous sampling sites. This approach allows for representative data on the presence of these compounds in air and their temporal trends. However, as reported by Wallace et al. (2024), passive sampling has not yet been the preferred technique for the sampling of PFASs. Although it has been successfully applied for the monitoring of PFASs in both indoor and outdoor environments, only 21 of the 84 studies included in the review employed passive sampling. The same tendency has been observed with pesticides and HPVCs (PAEs, OPEs, BTHs, BSAs, Tinuvins, and PAs) (García-Garcinuño et al., 2024b; Liaud et al., 2017; Lévy et al., 2020; Chakraborty et al., 2021). However, in recent years, studies such as that of Lévy et al. (2020) have emerged, aiming to evaluate the performance of passive sampling and to calculate the diffusive uptake-rates for pesticides. In the same line García-Garcinuño et al. (2024b) focused their study on the calculation of the diffusive uptake-rates of the above-mentioned HPVCs.

2.2 Particulate matter

Due to its complex composition and low concentrations of semivolatile CECs, atmospheric particulate matter (PM) requires careful sampling, which is the most critical step in ensuring reliable analytical results. Sampling and transportation of the samples need to be performed carefully to avoid losses of the target compounds, contaminations, or artefact formation (Galvão et al., 2018; Maceira et al., 2020b).

The sampling of the PM of air is generally performed in active mode and involves a suction pump, an air inlet selected depending on the size of the PM of interest, and a retaining element, usually a filter, to collect the particles (Galvão et al., 2018). The filter must have several characteristics depending on its chemical, mechanical and thermal stability, inertness, flow resistance, loading capacity, sampling efficiency and the concentrations of the compounds of interest in the filter itself (Galvão et al., 2018; Maceira et al., 2020b). Glass fiber filters (GFFs) and quartz fiber filters (QFFs), available in round and rectangular shapes and in various dimensions, are the preferred materials for the sampling of PM when analyzing semivolatile CECs (Galvão et al., 2018). For instance, García-Garcinuño et al. (2024a) used QFFs to sample PM_{10} for the determination of PAHs and HPVCs, including PAEs, OPEs, BTHs, BSAs, Tinuvins and PAs, in outdoor air samples from Tarragona (Spain). Wallace et al. (2024) and Gent and Paul (2021) reported that QFFs are also the preferred option to sample PM of different sizes when monitoring the presence of PFASs and illicit drugs, including new psychoactive substances, respectively, in the air. Although GFFs are also suitable, QFFs are the most widely used for sampling PM in industrial and urban areas to control the presence of PCDDs and PCDFs (Maceira et al., 2020b). GFFs have also been successfully applied for the determination of PBDEs, PCBs, OPEs, pesticides, and halogenated PAHs in PM samples from various environments worldwide (Yusà et al., 2009; Maceira et al., 2020b). The above-mentioned filters can contain traces of the compounds of interest that may be in the samples or may react with other species during analysis. The most commonly used pretreatment method consists of thermal conditioning of the filters in a muffle at temperatures between 450°C and 600°C for several hours or overnight (Galvão et al., 2018; Maceira et al., 2020b).

Variables such as sampling flow-rate and duration can be set according to the objectives of the study, as they determine the volume of air that passes through the retention filter and consequently the concentration of the analyte. There are various types of air PM samplers, based on the flow-rates at which they operate. The most commonly used are low- and high-volume samplers that operate from a few to 24 or 48 h. Low-volume samplers work at flow-rates of between 1 and 2 $m^3\ h^{-1}$ (Maceira et al., 2020b), while high-volume samplers operate at flow-rates of 25 to 100 $m^3\ h^{-1}$ (García-Garcinuño et al., 2024a). Maceira et al. (2018, 2019, 2020a) and García-Garcinuño et al. (2024a) collected 24 h PM_{10} samples at flow-rates of 30 $m^3\ h^{-1}$ and 50 $m^3\ h^{-1}$ for the determination of HPVCs and PAHs in outdoor air. Both types of samplers can be applied for the determination of PFASs and illicit drugs in PM_{10} and $PM_{2.5}$. Low-volume samplers are typically used in indoor environments, working at 0.12 $m^3\ h^{-1}$–2.3 $m^3\ h^{-1}$ flow-rates for 3.5–96 h. In contrast, due to the low levels of PFASs and drugs expected in outdoor air, larger volumes of up to 1000 m^3 are required and sampling is performed with high-volume samplers for 24–48 h (Wallace et al., 2024; Gent and Paul, 2021).

A compromise has to be made between the volume of air sampled and the PM collected to ensure the representativeness of the sample. The air inlet of the sampler determines the size of the particles to be retained in the filter; however, cascade impactors (see Fig. 2.4C) can also be used to determine the presence of semivolatile EOCs in PM of different size fractions within a single sampling event. Wallace et al. (2024) reported the use of cascade impactors to assess the mass balance distribution of PFASs in indoors and outdoors. For example, Lin et al. (2022) employed a four stages cascade impactor ($d < 1$ μm, $1 < d < 2.5$ μm, $2.5 < d < 10$ μm and $d > 10$ μm) to evaluate the particle size distribution of PFASs in different cities of Japan (Lin et al., 2022). Indoor and outdoor air samples have been taken at a low flow rate (1.2 $m^3\ h^{-1}$) for 48 h. This type of sampler has also been used to evaluate the size distribution of PAHs, PCBs, PCDDs, PCDFs, and PBDEs in various environments. Degrendele et al. (2014) employed a cascade impactor to analyze the above-mentioned compounds across six particle size ranges in different urban and rural areas of Brno (Czech Republic), $10 \geq d \geq 7.2$ μm, $7.2 \geq d \geq 3$ μm, $3 \geq d \geq 1.5$ μm, $1.5. \geq d \geq 0.95$ μm, $0.95. \geq d \geq 0.49$ μm, and $d < 0.49$ μm, and samples were taken at 68 $m^3\ h^{-1}$ for 1 week. On the contrary, Cai et al. (2018) monitor the presence of PBDEs in PM from a pyrolysis process with cascade impactor of five stages ($d > 9$ μm, $9 \geq d \geq 4.7$ μm, $4.7 \geq d \geq 2.1$ μm, $2.1 \geq d \geq 0.43$ μm and $d < 0.43$ μm), and samples were obtained at a flow-rate of 1.69 $m^3\ h^{-1}$ for 10 min. There are also indoor studies assessing the presence of different families of compounds, for example, PCBs, pesticides, PBDEs, BFRs and PAHs, in PM, using cascade impactors ($d \geq 2.5$ μm, $2.5 \geq d \geq 1$ μm, $1 \geq d \geq 0.5$ μm, $0.5 \geq d \geq 0.25$ μm and $d < 0.25$ μm), to sample over long periods of time (minimum one week) at low flow-rates (0.54 $m^3\ h^{-1}$) (Melymuk et al., 2016).

Some authors have employed passive sampling methods to collect PM to overcome some limitations of active sampling, such as the need for specific equipment, dependence on a power supply, and limited sampling area. These passive samplers are similar to active samplers that can collect both gas and particulate phase. The two most commonly used design consists of an aluminum double-bowl or cylindrical air samples (Fig. 2.6), which contains a PUF disk at

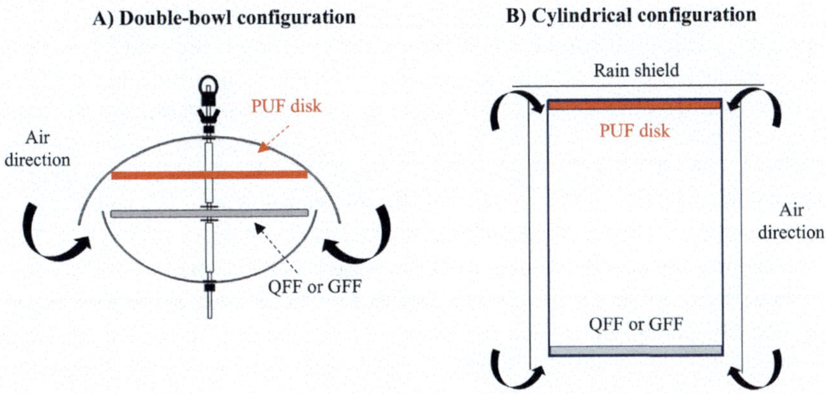

FIG. 2.6 Passive air devices for the gaseous phase and PM: (A) double-bowl and (B) cylindrical configuration for SVOCs.

the top to collect the gaseous phase and a filter at the bottom of the sampling chamber to deposit PM for analysis. Tao et al. (2007) calibrated a cylindrical-based passive sampler for the determination of PAHs in indoors (e.g., residence apartments and business offices). The same passive sampler has been used for Wang et al. (2017) to assess the presence of PAHs in outdoor air from Lanzhou Valley (Northwest China), a petrochemical industrialized area. Abdallah and Harrad (2010) modified and calibrated a double-bowl passive sampler for monitoring BFRs in various indoor environments.

3 Extraction

Due to the aforementioned complexity and heterogeneity of the atmosphere as an analytical matrix, the sample preparation is a key part of the analytical method. If the focus in sampling was placed on increasing the sensitivity of the method, then sample preparation will be essential for cleaning-up the sample and enhancing selectivity.

There is a variety of techniques used to selectively extract the desired compounds from the solid adsorbent onto which they were concentrated, as well as to remove compounds which could interfere with the subsequent analysis. They can be classified according to the nature of this extraction, distinguishing between techniques using thermal or liquid desorption.

3.1 Thermal desorption

Thermal Desorption (TD) is a highly automated technique based on the heating of the solid adsorbent so that the analytes retained on it are volatized and extracted by a flow of an inert gas. It is an established technique, widely used due to its compatibility with both canister and solid adsorbent-based sampling, and its easy coupling to gas chromatography analysis (Ras et al., 2009). It is especially adequate for compounds found in trace concentrations in air, since it allows for preconcentration of the analytes onto a cold trap, lowering the limits of detection. In single-stage TD the whole-air sample is passed through the trap, while in two-stage TD there is a first thermal desorption of the solid adsorbent sampling tube after which the analytes are retained on the cold trap. Similarly to the solid adsorbents used for active and passive sampling of VOCs and SVOCs, there are different types of commercially available cold traps used in TD. The same factors considered in the sampling step, such as the properties of the target compounds, must be taken into account to select the most appropriate solid adsorbent for the trap.

In the official method USEPA TO-15A (US Environmental Protection Agency (USEPA), 2019b), single-step TD is used after canister whole-air sampling for the desorption of ethylene oxide, 1,3-butadiene and other VOCs of emerging concern. The trap used in that case is a multisorbent graphitized carbon trap, containing Carbopack B and Carbosieve S-III, which retains the desired VOCs and allows some water vapor to pass through when working with humid samples. Other traps proposed in this method include both polymeric adsorbents such as Tenax or Ambersorb 340 and charcoal, or cryogenic traps cooled to −150°C for the retention of the analytes. A dryer system or two-stage concentration is proposed in this method to avoid the introduction of water in the gas chromatography system. The traps specified in this method have successfully been applied for the monitoring of VOCs of emerging concern, including ethylene oxide, in outdoor air samples collected with canisters in the USA (Hoisington and Herrington, 2021), Canada (Galarneau et al., 2016), and China (Han et al., 2019). On the contrary, in official method USEPA TO-17 (US Environmental Protection Agency (USEPA), 1999), two-step TD is used for the desorption of benzene, 1,3-butadiene, and other VOCs of interest. In that case, the proposed traps depend on the compounds to determine and the selected sampling adsorbent material. Tenax/Graphitized carbon traps are recommended after sampling with adsorbents of the same nature (e.g., Tenax, Carbotrap C, and Carbopack C). This combination was applied by Ramírez et al. (2012) for the monitoring of VOCs in urban areas surrounding the largest petrochemical park in southern Europe (Tarragona, Spain). However, Carbopack B/Molecular sieve traps are recommended after sampling with adsorbents with larger specific surface area but lower temperature tolerance, such as Chromosorb 102 and Porapak N (US Environmental Protection Agency (USEPA), 1999). These types of sorbents are particularly useful for the sampling and subsequent TD of odorous VOCs characteristic of waste water treatment plants (Sakizadeh, 2020). The official method UNE-EN 14662-1 (Asociación Española de Normalización (AENOR), 2023) proposes a similar procedure for the determination of benzene after active sampling onto adsorbent tubes. This approach has also been widely used for the determination of episodic concentrations of VOCs of emerging concern in different environments (Vallecillos et al., 2024b; Maceira et al., 2017; Vallecillos et al., 2018). Furthermore, USEPA Method 325A (US Environmental Protection Agency (USEPA), 2019a) proposes passive sampling with graphitized carbon adsorbent tubs (e.g., Carbopack X) followed by a two-stage TD using a general-purpose graphitized carbon trap for long-term monitoring of VOCs of emerging concern in both industrial and urban atmospheres (Vallecillos et al., 2024a; García-Garcinuño et al., 2025; Mukerjee et al., 2020).

These four methods have been frequently applied to sampling and subsequent desorption for the determination of VOCs in outdoor air samples (see Table 2.1) (Vallecillos et al., 2024a; Gallego et al., 2016; Even et al., 2020). Similar methods have been applied for the determination of certain families such as solvents, fragrances, and preservatives in indoor environments, using a general-purpose graphitized carbon trap (Ninyà et al., 2022; Vallecillos et al., 2021). Gallego et al. (2012) and Besis et al. (2021) applied different traps, one of Tenax/Carbotrap and another specifically designed for sulfur compounds, for the determination of these compounds in industrial (waste treatment plant) and urban air samples, respectively. The most volatile PFASs, fluorotelomer alcohols (FTOHs), can be thermally desorbed from adsorbent tubes using Gerstel instrumentation with a Cooled Injected System (CIS) for cryotrapping (Robbins et al., 2023) or focused onto a Tenax/graphitized carbon trap specifically designed for these compounds (Markes International, 2022).

In the case of SVOCs, different types of cryogenic traps have been used depending on the families of compounds to be determined. For PAHs, even if Tenax tubes are recommended for sampling in outdoor environments, certain studies have used general-purpose graphitized carbon traps because temperatures up to 350°C are required for their complete desorption, which could degrade Tenax (Jia et al., 2023). On the contrary, a Tenax trap has been used for the determination of PAEs and OPEs, with slightly lower desorption temperatures at 320°C, giving better results than a mixed Tenax/Carbograph trap (Aragón et al., 2013).

3.2 Liquid desorption

Liquid desorption (LD), using organic solvents, is a solid-liquid extraction typically employed for desorption of SVOCs and VOCs from solid adsorbents, in cases where one cannot employ TD. It is a hands-on technique which includes a dilution of the sample, which can introduce contamination and make the sample preparation a more arduous process. To counteract this dilution, one might include a solvent evaporation step, which can lead to a loss of the most volatile target compounds (Ras et al., 2009). The most common solvents used for LD are carbon disulfide (CS_2) for VOCs (Ras et al., 2009) and dichloromethane (DCM) or toluene for SVOCs (Maceira et al., 2020b), low-boiling point solvents that can solubilize many analytes. However, their toxicity poses a high risk to both human health and the environment.

TABLE 2.1 Determination of VOCs in indoor and outdoor air samples using TD as extraction technique.

Compounds	Sample	Sampling technique	Cryogenic trap	Ref.
VOCs	Outdoor air, Urban areas close to petrochemical parks	Passive adsorbent tubes	General-purpose graphitized carbon	(García-Garcinuño et al., 2025)
VOCs	Outdoor air, Urban areas close to waste treatment plants	Active and passive adsorbent tubes	USEPA TO-15/TO-17	(Gallego et al., 2016)
VOCs	Outdoor air, Urban areas close to petrochemical parks	Passive adsorbent tubes	General-purpose graphitized carbon	(Vallecillos et al., 2024a)
1,3-Butadiene	Outdoor air, Urban areas close to petrochemical parks	Active and passive adsorbent tubes	General-purpose graphitized carbon	(Vallecillos et al., 2024b)
VOCs	Indoor air, Conventional school	Passive adsorbent tubes	General-purpose graphitized carbon	(Ninyà et al., 2022)
VOCs	Indoor air, Emission chamber	Active adsorbent tubes	Gerstel CIS	(Even et al., 2020)
VOCs, Ethylene oxide	Outdoor air, Urban areas close to sterilizing facilities	Canisters	USEPA TO-15/TO-17	(Hoisington and Herrington, 2021)
VOCs, Ethylene oxide	Outdoor air, Urban areas	Canisters	Cryotrapping	(Galarneau et al., 2016)
VOCs	Outdoor air, Urban areas close to petrochemical parks	Active adsorbent tubes	Hydrophobic Tenax/ Carbograph	(Maceira et al., 2017)
VOCs	Outdoor air, Urban areas	Canisters	USEPA TO-15	(Han et al., 2019)
VOCs	Outdoor air, Urban areas close to petrochemical parks	Passive adsorbent tubes	USEPA 325A	(Mukerjee et al., 2020)
Solvents, fragrances, preservatives	Indoor air, Conventional and vocational training schools	Passive adsorbent tubes	General-purpose graphitized carbon	(Vallecillos et al., 2021)

VOCs, VSCs	Outdoor air, Industrial area, waste treatment plant	Active adsorbent tubes	Hydrophobic, Tenax/ Carbotrap	(Gallego et al., 2012)
VOCs, VSCs	Outdoor air, Urban areas close to petrochemical parks	Active adsorbent tubes	Sulfur compounds	(Besis et al., 2021)
1,3-Butadiene	Outdoor air, Urban areas close to petrochemical parks	Active adsorbent tubes	General-purpose graphitized carbon trap	(Vallecillos et al., 2018)
Volatile PFASs	Indoor air, Emission chamber	Active adsorbent tubes	Gerstel CIS	(Robbins et al., 2023)
Volatile PFASs	Outdoor air, Urban areas	Active adsorbent tubes	PFASs	(Markes International, 2022)
PAHs	Outdoor air, Urban areas	Active adsorbent tubes	General-purpose graphitized carbon trap	(Jia et al., 2023)
PAEs, OPEs	Outdoor air, Urban and Industrial (Harbour) areas	Active adsorbent tubes	Tenax	(Aragón et al., 2013)
VOCs	Outdoor air, Urban areas close to petrochemical parks	Active adsorbent tubes	Hydrophobic Tenax/ Carbograph	(Ramírez et al., 2012)

The official method UNE-EN 14662-2 (Asociación Española de Normalización y Certificación, 2006) uses CS_2 for the desorption of benzene from activated carbon, with occasional agitation during a period of 30 min. NIOSH method 1501 (Manual of Analytical Methods (NMAM), 2003) uses CS_2 in the same conditions for desorption of volatile aromatic hydrocarbons, including benzene, xylenes, and styrene among others. As Table 2.2 shows, the abovementioned official methods have been used as the basis for determining a broad range of VOCs, including all ozone precursors and specific compounds such as 1,3-butadiene, in outdoor and indoor air samples. Moreover, using CS_2 as the extraction solvent, the presence of VOCs has been monitored in urban, rural and industrial atmospheres with different casuistry such as the presence of petrochemical parks or waste treatment plants (Ghaffari et al., 2021; Vallecillos et al., 2018; Ramírez et al., 2012; Villanueva et al., 2018). Other solvents have also been proposed for the selective desorption of specific VOC families. For instance, dimethylformamide (DMF) is used for the desorption of the derivatized product formed by the retention of ethylene oxide on petroleum charcoal tubes coated with hydrogen bromide, as described in NIOSH method 1614 (Manual of Analytical Methods (NMAM), 1994). Acetonitrile is commonly used for the desorption of formaldehyde and other carbonyl compounds, which are derivatized with 2,4-DNPH to facilitate their retention on the sampling adsorbent during sampling in outdoor and indoor atmospheres (Gallego et al., 2016; Ninyà et al., 2022; da Rocha et al., 2023) (Table 2.2).

For the extraction of SVOCs using different organic solvents, LD has evolved into techniques such as Soxhlet, Ultrasound-Assisted Extraction (UAE) or Pressurized Liquid Extraction (PLE), where high temperature, physical phenomena, high pressure, or a combination of either of the three are used to assist in the desorption of the target compounds from the solid adsorbent, usually QFFs and PUFs, into the organic solvent (Wallace et al., 2024; Gent and Paul, 2021; Maceira et al., 2020b).

3.2.1 Soxhlet

Soxhlet is an extraction technique based on the use of large volumes of organic solvents at their boiling point and at ambient pressure during long periods of time. The choice of solvent will depend on the solubility of the target compounds and its capacity for their extraction from the solid adsorbent on which they were retained.

Nowadays, Soxhlet is commonly used for the extraction of PCDDs and furans PCDFs from solid adsorbents and filters, according to official method USEPA 1613B (US Environmental Protection Agency (USEPA), 1994). In this method, toluene is used as the Soxhlet extracting solvent. The extract is then cleaned up by gel permeation chromatography (GPC), using either a silica column, an alumina column, or a carbon column with different organic solvents. PCBs are also extracted from solid adsorbents using Soxhlet, according to USEPA 1668A (US Environmental Protection Agency (USEPA), 2003), with toluene as the

TABLE 2.2 Determination of SVOCs in indoor and outdoor air samples using various LD extraction technique.

Compounds	Sample	Sampling technique	Extraction		Ref
			Technique	Conditions	
Formaldehyde	Outdoor air, Urban areas close to waste treatment plants	Active and passive adsorbent tubes	LD	Acetonitrile, 30 min	(Gallego et al., 2016)
VOCs	Outdoor and Indoor air, Urban, rural and industrial areas	Active adsorbent tubes	LD	CS_2, 20 min	(Ghaffari et al., 2021)
1,3-Butadiene	Outdoor air, Urban areas close to petrochemical parks	Active adsorbent tubes	LD	CS_2, 30 min	(Vallecillos et al., 2018)
VOCs	Outdoor air, Urban areas close to petrochemical parks	Active adsorbent tubes	LD	CS_2, 30 min	(Ramírez et al., 2012)
VOCs, Aldehydes	Outdoor and Indoor air, Urban, rural and industrial areas	Passive adsorbent tubes	LD	VOCs: CS_2/ Aldehydes: Acetonitrile 30 min	(Villanueva et al., 2018)
PCDDs, PCDFs, PCBs	Outdoor air, Urban areas	Active PUFs	Soxhlet	Toluene, 24 h	(Abad et al., 2022)
PCDDs, PCDFs, PCBs	Outdoor air, Suburban area	Active QFFs, PUFs	Soxhlet	Toluene, 24 h	(Barbas et al., 2018)
PAHs, PCDDs, PCDFs, PCBs, pesticides, illicit drugs	Outdoor air, Urban areas	Active QFFs, PUFs	Soxhlet	PAHs, illicit drugs, pesticides: Cyclohexane/ PCDDs, PCDFs, PCBs: Toluene/ All: Cyclohexane:Toluene 24 h	(Galarneau et al., 2016)
PBDEs	Air during controlled residential firefighting	Passive PUFs	Soxhlet	DCM, 24 h	(Fent et al., 2020)
PAHs, PCBs, PBDEs, BFRs	Outdoor air, Urban and industrial (waste treatment) areas	Passive PUFs	Soxhlet	Acetone, 16 h	(Anh et al., 2020)

Continued

TABLE 2.2 Determination of SVOCs in indoor and outdoor air samples using various LD extraction technique—cont'd

Compounds	Sample	Sampling technique	Extraction		Ref
			Technique	Conditions	
PCDDs, PCDFs, PPCBs, pesticides	Outdoor air, Urban areas	Passive PUFs	Soxhlet	Toluene, 24 h	(Bogdal et al., 2013)
PCDDs, PCDFs	Outdoor air, Urban areas	Active QFFs, PUFs	Soxhlet	Toluene, 24 h	(Abad et al., 2007)
PAHs, PCDDs, PCDFs, PCBs	Outdoor air, Suburban area	Active QFFs, PUFs	Soxhlet	Toluene, 1.5 h	(Degrendele et al., 2020)
PAHs, PAEs, PCDDs, PCDFs, PCBs	Outdoor air, Urban and industrial (electronic waste recycling) areas	Passive PUFs	Soxhlet	Toluene, 24 h	(Chakraborty et al., 2021)
PAHs, PCDDs, PCDFs, PCBs,	Outdoor air, Urban and rural areas	Active QFFs	Soxhlet	Toluene, 24 h	(Degrendele et al., 2014)
PAHs, PCBs, PBDEs, flame retardants, pesticides	Indoor air, Residential bedroom	Passive: PUFs/ Active: QFFs	Soxhlet	DCM, 24 h	(Melymuk et al., 2016)
PAHs	Indoor air, Residence apartments and business offices	Active and passive QFFs, PUFs	Soxhlet	Hexane:Cyclohexane/ PUFs: 4 h/ QFFs: 10 h	(Tao et al., 2007)
PAHs	Outdoor and Indoor air, Urban, rural and industrial areas	Active and passive QFFs, PUFs	Soxhlet	DCM, 24 h	(Wang et al., 2017)
BFRs	Indoor air, Officies	Passive QFFs, PUFs	Soxhlet	DCM, 8 h	(Abdallah and Harrad, 2010)
PAHs, PAEs, flame retardants, SMCs, bisphenols	Outdoor air, Industrial area (automotive manufacturing plant)	Active QFFs	UAE	Hexane:Acetone, 30 min	(Sánchez-Piñero et al., 2022)

PAEs, PCBs, PBDEs, flame retardants, paraffins	Indoor air, Homes and schools	Active QFFs, PUFs	UAE	Hexane, 30 min	(Sakhi et al., 2019)
Aldehydes	Outdoor and Indoor air, Urban area and school	Passive adsorbent tubes	UAE	Acetonitrile, 30 min	(Ninyà et al., 2022)
BTHs, BTRs, BSAs	Oudoor air, Industrial area (harbour)	Active QFFs	UAE	Ethyl acetate, 15 min	(Nuñez et al., 2020)
VOCs, Aldehydes	Outdoor air, Urban areas	Passive adsorbent tubes	UAE	VOCs: CS_2, 10 min/ Aldehydes: Acetonitrile, 30 min	(da Rocha et al., 2023)
SMCs	Indoor air, Home	Active QFFs, PUFs	UAE	Hexane:Acetone, 30 min	(Balci et al., 2020)
PFASs	Outdoor and Indoor air, Urban areas and office, laboratory and occupational setting	Active QFFs, PUFs	UAE	Neutral: Ethyl acetate:DCM/ Ionic: Methanol/ 1 h	(Lin et al., 2022)
BTHs, BSAs, PAEs, OPEs, Tinuvins, PAs, PAHs	Outdoor air, Urban areas close to petrochemical parks	Passive PUFs	PLE	DCM, 100°C, 1500 psi, 5 min	(García-Garcinuño et al., 2025)
PAs, Tinuvins	Outdoor air, Urban areas close to a petrochemical park or a harbour	Active QFFs	PLE	Ethyl acetate, 80°C, 1500 psi, 5 min	(Maceira et al., 2019)
BTHs, BTRs, BSAs	Outdoor air, Urban areas close to a petrochemical park or a harbour	Active QFFs	PLE	Ethyl acetate, 70°C, 1500 psi, 5 min	(Maceira et al., 2018)

Continued

TABLE 2.2 Determination of SVOCs in indoor and outdoor air samples using various LD extraction technique—cont'd

Compounds	Sample	Sampling technique	Extraction		Ref
			Technique	Conditions	
BTHs, BSAs, PAEs, OPEs, Tinuvins, PAs, PAHs	Outdoor air, Urban areas close to a petrochemical park or a harbour	Passive PUFs/ Active QFFs, PUFs	PLE	PUFs: DCM, 80°C/ QFFs: DCM, 100°C, 1500 psi, 5 min	(García-Garcinuño et al., 2024b)
BTHs, BSAs, PAEs, OPEs, Tinuvins, PAs, PAHs	Outdoor air, Urban area close to a petrochemical park	Active QFFs, PUFs	PLE	PUFs: DCM, 80°C/ QFFs: DCM, 100°C, 1500 psi, 5 min	(García-Garcinuño et al., 2024a)
Pesticides	Outdoor air/Rural area	Passive PUFs, XADs/ Active QFFs, PUFs	PLE	Acetonitrile, 150°C, 1500 psi, 5 min	(Lévy et al., 2020)
Pesticides	Outdoor air, Urban and rural areas	Passive XADs	PLE	Acetonitrile, 150°C, 1500 psi, 5 min	(de Gennaro et al., 2013)
BTRs, BTHs, BSAs, OPEs, PAEs	Outdoor air, Urban areas close to a petrochemical park or a harbour	Active QFFs	PLE	Ethyl acetate, 70°C, 1500 psi, 5 min	(Maceira et al., 2020a)

extraction solvent and GPC cleanup. Abad et al. (2022, 2007) performed the extraction of actively sampled QFFs and PUFs samples using 400 mL of toluene during 24 h, after spiking with deuterated surrogates. In this case, the cleanup was performed using silica gel-alumina GPC, with hexane and dichloromethane in different percentages, and solid-liquid extraction from a carbon column with toluene. Barbas et al. (2018) also applied this method with a similar cleanup procedure for the determination of PCDFs and PCDDs in passive PUF samples. As shown in Table 2.2, regardless of the environment evaluated and the sampling performed, Soxhlet extraction is the preferred technique for the determination of these compounds in air. It is also the reference method used for a better comparison between studies around the world (Bogdal et al., 2013; Degrendele et al., 2020; Chakraborty et al., 2021; Degrendele et al., 2014).

Soxhlet has also been used for extraction of PAHs from PUFs, using deuterated surrogates and DCM as the extraction solvent, also with GPC cleanup (Fent et al., 2020; Wang et al., 2017), or hexane: cyclohexane (Tao et al., 2007). Other studies also included Soxhlet extraction of flame retardants, BFRs and chlorinated pesticides for the monitoring of their presence in both the gaseous phase and PM of air (Anh et al., 2020; Bogdal et al., 2013; Degrendele et al., 2020; Melymuk et al., 2016; Abdallah and Harrad, 2010). Gent and Paul (2021) and Wallace et al. (2024) reported Soxhlet extraction with varying solvents, extraction times, and volumes as the most widely used technique for the determination of illicit drugs (QFFs) and PFASs (PUFs and QFFs) in air. Galarneau et al. (2016) adapted the official methods to also include these compounds, by using different extraction solvents for targeting various compounds retained in QFFs and PUFs, including: cyclohexane for PAHs, illicit drugs and pesticides, toluene for PCDDs, PCDFs and PCBs, and cyclohexane: toluene for all target compounds. Anh et al. (2020), on the contrary, used acetone for extraction of both PAHs and POPs (PCBs, PBDEs, and BFRs) in PUF samples from urban and industrial (waste treatment facility) air samples collected passively in Northern-Vietnam.

3.2.2 *Ultrasound-Assisted Extraction (UAE)*

In recent years, attempts have been made to replace Soxhlet extraction by UAE in order to minimize both the consumption of organic solvents and the time required for extraction. Furthermore, working with smaller volumes of solvents makes the clean-up and evaporation steps less laborious (Wallace et al., 2024; Gent and Paul, 2021; Maceira et al., 2020b). Ninyà et al. (2022) and da Rocha et al. (2023) employed UAE with acetonitrile and CS_2 to enhance the LD of aldehydes and VOCs from radial passive sampling tubes applied for long-term monitoring of their presence in outdoor air and different indoor school environments. As Table 2.2 shows, some authors developed multiclass methods based on UAE to extract SVOCs such as PAHs, PAEs, flame retardants, SMCs, bisphenols, PCBs, PBDEs, and paraffins retained in QFFs and PUFs samples collected actively in indoor environments (e.g., schools and homes) and/or industrial areas (Sánchez-Piñero et al., 2022; Sakhi et al., 2019; Balci et al., 2020). In the case of UAE, some authors applied a single solvent, such as hexane (Sakhi et al., 2019), or solvent mixtures, such as hexane: acetone (Sánchez-Piñero et al., 2022; Balci et al., 2020), to improve the recovery of the families of compounds under study. A 30-min extraction time is the most commonly applied duration. BTHs, BTRs and BSAs have also been extracted from PM after 1-month active sampling with small QFFs, allowing for the use of low extraction volumes (5 mL) of ethyl acetate (Nuñez et al., 2020). Lin et al. (2022) used ethyl acetate: DCM (1:1) for extraction of neutral PFASs and methanol for extraction of ionic PFASs in PM QFF samples, and the same solvents with basified methanol for extraction from PUFs.

3.2.3 Pressurized Liquid Extraction (PLE)

PLE, also known as Accelerated Solvent Extraction (ASE), uses organic solvents at high temperatures and pressures to ensure they remain liquid through the whole extraction process, increasing the solubility of the target compounds in the used solvent. It gives recoveries comparable to those obtained with Soxhlet and other available techniques, while shortening the extraction time and increasing the automatization (Yusà et al., 2009). Although it is not the most commonly used technique due to the specific equipment required, PLE has been successfully employed to extract certain families of SVOCs, including BTHs, BTRs, BSAs, Tinuvins, PAs, OPEs, and PAEs, after active sampling of PM from urban atmospheres near petrochemical parks or a harbor in QFFs, using ethyl acetate as the extracting solvent at 70°C or 80°C and 1500 psi (Maceira et al., 2018, 2019, 2020a). García-Garcinuño et al. (2024a) found that, when PAHs were included along with the previously mentioned families, recoveries for PAHs were significantly higher when DCM was used as the extracting solvent at 100°C. They applied the same extraction methodology to QFFs and PUFs collected through active and passive urban air sampling (García-Garcinuño et al., 2024a, 2024b, 2025). Lévy et al. (2020) and Liaud et al. (2017) also reported the application of PLE for the extraction of pesticides, using acetonitrile at 150°C as the extraction solvent after sampling on PUF and XAD resin passive samplers. Moreover, Gent and Paul (2021) suggested PLE as one of the suitable extraction techniques for the determination of illicit drugs, such as cocaine, heroin, amphetamine, and methamphetamine in PM deposited in QFFs.

4 Future trends

The evolution of sample preparation is aligned with the current trends in analytical chemistry, moving toward high automatization and sustainability. In many instances, sampling and part of the preparation procedure are becoming integrated, as it occurred with established air sampling techniques where the preconcentration occurred on site, but every day more techniques including the detection itself on site are being developed and used (Maceira et al., 2020b; Brüggemann et al., 2024).

Online GC coupled to photo-ionization or flame-ionization detection (PID, FID) is an established on-site sampling, preparation, separation, and detection technique for the determination of regulated VOCs, such as benzene (Asociación Española de Normalización (AENOR), 2016). In the last years, it has been applied for the determination of other VOCs with varying analysis times depending on the amount of target compounds, from the 15 min required for sampling of benzene, toluene, ethylbenzene, and xylenes (BTEX) (Healy et al., 2018) to the 1 h required for analysis of up to 56 VOCS (Xuan et al., 2021). In certain instances, longer cycle times are required not for separation but for the proper preconcentration of the target compounds in immission sites, such as 1,3-butadiene which was studied in urban areas with a sampling/analysis period of 30 min (Vallecillos et al., 2024b). Recently, online GC-PID and Tunable Infrared Laser Direct Absorption Spectroscopy (TILDAS) has been used for the in situ determination of ethylene oxide (Galarneau et al., 2023). GC-PID offers concentration values every 30 min for the proper preconcentration of ethylene oxide from ambient air. Moreover, its detection limits restrict its application mainly to industrial environments. In contrast, TILDAS does not require any preconcentration step, provides real-time concentration data at 1 s intervals, and achieves detection limits suitable for monitoring of both industrial and urban atmospheres (Galarneau et al., 2023; Yacovitch et al., 2023). Low-cost sensors are also being used, especially for approximation of total VOCs concentrations and the calculation of the air quality index from PM concentrations. Existing technologies include electrochemical sensors (EC), metal oxide semiconductor sensors (MOS), photo-ionization detectors (PID), and light scattering (Ródenas García et al., 2022). While certain EC sensors can be used to target specific compounds, MOS and PID are mainly used for total VOCs, while light scattering is used for PM.

To date, in situ determination of SVOCs remains unfeasible. Current efforts are toward the automation of the sampling procedure and developing standardized protocols for the analysis of SVOCs that can be applied globally. The ultimate goal is to enable comparability across studies conducted in different regions (Wallace et al., 2024; Maceira et al., 2020b; Gent and Paul, 2021; Brüggemann et al., 2024). In this context, passive sampling is prioritized due to its simplicity, low cost, and ease of implementation. These advantages enable its simultaneous deployment across multiple locations and over extended periods, providing SVOC concentration data suitable for intercomparison studies and risk assessments (Wallace et al., 2024; Gent and Paul, 2021; Brüggemann et al., 2024).

References

Abad, E., Abalos, M., Fiedler, H., 2022. Air monitoring with passive samplers for dioxin-like persistent organic pollutants in developing countries (2017–2019). Chemosphere. 287. doi:10.1016/J.CHEMOSPHERE.2021.131931

Abad, E., Martínez, K., Gustems, L., Gómez, R., Guinart, X., Hernández, I., Rivera, J., 2007. Ten years measuring PCDDs/PCDFs in ambient air in Catalonia (Spain). Chemosphere. 00456535.

67, 1709. Available from: http://www.elsevier.com/locate/chemosphere. doi:10.1016/j.chemosphere.2006.05.083.

Abdallah, M.A.E., Harrad, S., 2010. Modification and calibration of a passive air sampler for monitoring vapor and particulate phase brominated flame retardants in indoor air: application to car interiors. Environ. Sci. Technol. 15205851. 44, 3059. doi:10.1021/es100146r.

Advancing Standards Transforming Markets (ASTM), 2024. Method D5504-20, standard test method for determination of sulfur compounds in natural gas and gaseous fuels by gas chromatography and chemiluminescence. ASTM Int. doi:10.1520/D5504-20.

Anh, H.Q., Watanabe, I., Tue, N.M., Tuyen, L.H., Viet, P.H., Chi, N.K., Minh, T.B., Takahashi, S., 2020. Polyurethane foam-based passive air sampling for simultaneous determination of POP- and PAH-related compounds: a case study in informal waste processing and urban areas, northern Vietnam. Chemosphere. 18791298. 247. Available from: http://www.elsevier.com/locate/chemosphere. doi:10.1016/j.chemosphere.2020.125991.

Aragón, M., Borrull, F., Marcé, R.M., 2013. Thermal desorption-gas chromatography-mass spectrometry method to determine phthalate and organophosphate esters from air samples. J. Chromatogr. A. 18733778. 1303, 76. doi:10.1016/j.chroma.2013.06.025.

Asociación Española de Normalización (AENOR), 2016. UNE-EN 14662-3:2016 Calidad del aire ambiente. Método normalizado para la medición de las concentraciones de benceno. Parte 3: muestreo automático por aspiración con cromatografía de gases in situ. AENOR.

Asociación Española de Normalización (AENOR), 2023. UNE-EN 14662-1:2023 Calidad del aire ambiente. Método normalizado de medida de las concentraciones de benceno. Parte 1: muestreo por aspiración seguido de desorción térmica y cromatografía de gases. AENOR.

Asociación Española de Normalización y Certificación, 2006. UNE-EN 14662-2:2006, Calidad del aire ambiente. Método normalizado de medida de las concentraciones de benceno. Parte 2: muestreo por aspiración seguido de desorción por disolvente y cromatografía de gases. AENOR.

Balci, E., Genisoglu, M., Sofuoglu, S.C., Sofuoglu, A., 2020. Indoor air partitioning of synthetic musk compounds: gas, particulate matter, house dust, and window film. Sci. Total Environ. 00489697. 729, 138798. doi:10.1016/j.scitotenv.2020.138798.

Barbas, B., de la Torre, A., Sanz, P., Navarro, I., Artíñano, B., Martínez, M.A., 2018. Gas/particle partitioning and particle size distribution of PCDD/Fs and PCBs in urban ambient air. Sci. Total Environ. 00489697. 624, 170. doi:10.1016/j.scitotenv.2017.12.114.

Besis, A., Georgiadou, E., Samara, C., 2021. Odor-active volatile organic compounds along the seafront of Thessaloniki, Greece. Implications for sources of nuisance odor. Sci. Total Environ. 00489697. 799, 149388. doi:10.1016/j.scitotenv.2021.149388.

Bogdal, C., Scheringer, M., Abad, E., Abalos, M., van Bavel, B., Hagberg, J., Fiedler, H., 2013. Worldwide distribution of persistent organic pollutants in air, including results of air monitoring by passive air sampling in five continents. TrAC Trends Anal. Chem. 01659936. 46, 150. doi:10.1016/j.trac.2012.05.011.

Brüggemann, M., Mayer, S., Brown, D., Terry, A., Rüdiger, J., Hoffmann, T., 2024. Measuring pesticides in the atmosphere: current status, emerging trends and future perspectives. Environ. Sci. Eur. 21904715. 36. Available from: https://enveurope.springeropen.com. doi:10.1186/s12302-024-00870-4.

Cai, C.Y., Yu, S.Y., Liu, Y., Tao, S., Liu, W.X., 2018. PBDE emission from E-wastes during the pyrolytic process: emission factor, compositional profile, size distribution, and gas-particle partitioning. Environ. Pollut. 18736424. 235, 419. Available from: http://www.elsevier.com/inca/publications/store/4/0/5/8/5/6. doi:10.1016/j.envpol.2017.12.068.

Chakraborty, P., Gadhavi, H., Prithiviraj, B., Mukhopadhyay, M., Khuman, S.N., Nakamura, M., Spak, S.N., 2021. Passive air sampling of PCDD/Fs, PCBs, PAEs, DEHA, and PAHs from informal electronic waste recycling and allied sectors in Indian megacities. Environ. Sci.

Technol. 15205851. 55, 9469. Available from: http://pubs.acs.org/journal/esthag. doi:10.1021/acs.est.1c01460.

Civan, M.Y., Elbir, T., Seyfioglu, R., Kuntasal, Ö.O., Bayram, A., Doğan, G., Yurdakul, S., Andiç, Ö., Müezzinoğlu, A., Sofuoglu, S.C., Pekey, H., Pekey, B., Bozlaker, A., Odabasi, M., Tuncel, G., 2015. Spatial and temporal variations in atmospheric VOCs, NO_2, SO_2, and O_3 concentrations at a heavily industrialized region in Western Turkey, and assessment of the carcinogenic risk levels of benzene. Atmos. Environ. 18732844. 103, 102. Available from: http://www.elsevier.com/locate/atmosenv. doi:10.1016/j.atmosenv.2014.12.031.

da Rocha, F.O.C., Campos, V.P., da Rocha, G.O., Brito, A.V.S., Dos Santos Sampaio, I., 2023. Characterization of airborne gaseous BTEX and aldehydes from populated Brazilian cities as representative atmospheres of typical urban areas from the Southern Hemisphere. Air Qual. Atmos. Health. 18739326. 16, 1271. Available from: https://www.springer.com/journal/11869. doi:10.1007/s11869-023-01341-0.

Dalton, P., Claeson, A.S., Horenziak, S., 2020. The impact of indoor malodor: historical perspective, modern challenges, negative effects, and approaches for mitigation. Atmosphere. 20734433. 11. Available from: http://res.mdpi.com/d_attachment/atmosphere/atmosphere-11-00126/article_deploy/atmosphere-11-00126-v2.pdf. doi:10.3390/atmos11020126.

de Gennaro, G., Farella, G., Marzocca, A., Mazzone, A., Tutino, M., 2013. Indoor and outdoor monitoring of volatile organic compounds in school buildings: indicators based on health risk assessment to single out critical issues. Int. J. Environ. Res. Public Health. 16604601. 10, 6273. Available from: http://www.mdpi.com/1660-4601/10/12/6273/pdf. doi:10.3390/ijerph10126273.

Degrendele, C., Fiedler, H., Kočan, A., Kukučka, P., Přibylová, P., Prokeš, R., Klánová, J., Lammel, G., 2020. Multiyear levels of PCDD/Fs, dl-PCBs and PAHs in background air in central Europe and implications for deposition. Chemosphere. 00456535. 240, 124852. doi:10.1016/j.chemosphere.2019.124852.

Degrendele, C., Okonski, K., Melymuk, L., Landlová, L., Kukučka, P., Čupr, P., Klánová, J., 2014. Size specific distribution of the atmospheric particulate PCDD/Fs, dl-PCBs and PAHs on a seasonal scale: implications for cancer risks from inhalation. Atmos. Environ. 18732844. 98, 410. Available from: http://www.elsevier.com/locate/atmosenv. doi:10.1016/j.atmosenv.2014.09.001.

Desauziers, V., 2004. Traceability of pollutant measurements for ambient air monitoring. TrAC Trends Anal. Chem. 01659936. 23, 252. doi:10.1016/s0165-9936(04)00310-3.

Dörter, M., Odabasi, M., Yenisoy-Karakaş, S., 2020. Source apportionment of biogenic and anthropogenic VOCs in Bolu plateau. Sci. Total Environ. 00489697. 731, 139201. doi:10.1016/j.scitotenv.2020.139201.

Dugheri, S., Bonari, A., Pompilio, I., Colpo, M., Mucci, N., Arcangeli, G., 2018. An integrated air monitoring approach for assessment of formaldehyde in the workplace. Saf. Health Work. 20937997. 9, 479. Available from: http://www.e-shaw.org/. doi:10.1016/j.shaw.2018.05.002.

European Parliament and Council of the European Union, 2004. Directive 2004/107/EC of the European Parliament and of the council of 15 December 2004 relating to arsenic, cadmium, mercury, nickel and polycyclic aromatic hydrocarbons in ambient air. Off. J. Eur. Comm. L23, 3–16. Available from: https://eur-lex.europa.eu/legal-content/ES/TXT/?uri=celex:32004L0107.

European Parliament and Council of the European Union, 2008. Directive 2008/50/EC of the European Parliament and of the Council of 21 May 2008 on ambient air quality and cleaner air for Europe. Off. J. Eur. Comm. L152, 1–44. Available from: http://news.cleartheair.org.hk/wp-content/uploads/2013/02/LexUriServ.pdf.

European Parliament and Council of the European Union, 2024. Directive 2024/2881/EC of the European Parliamen and the Council of 23 October 2024 on ambient air quality and cleaner air for Europe. Off. J. Eur. Comm. L1, 1–70. Available from: https://eur-lex.europa.eu/eli/dir/2024/2881/oj.

Even, M., Hutzler, C., Wilke, O., Luch, A., 2020. Emissions of volatile organic compounds from polymer-based consumer products: comparison of three emission chamber sizes. Indoor Air. 16000668. 30, 40. Available from: http://onlinelibrary.wiley.com/journal/10.1111/(ISSN)1600-0668. doi:10.1111/ina.12605.

Fent, K.W., LaGuardia, M., Luellen, D., McCormick, S., Mayer, A., Chen, I.C., Kerber, S., Smith, D., Horn, G.P., 2020. Flame retardants, dioxins, and furans in air and on firefighters' protective ensembles during controlled residential firefighting. Environ. Int. 18736750. 140. Available from: http://www.elsevier.com/locate/envint. doi:10.1016/j.envint.2020.105756.

Galarneau, E., Wang, D., Dabek-Zlotorzynska, E., Siu, M., Celo, V., Tardif, M., Harnish, D., Jiang, Y., 2016. Air toxics in Canada measured by the National Air Pollution Surveillance (NAPS) program and their relation to ambient air quality guidelines. J. Air Waste Manag. Assoc. 21622906. 66, 184. Available from: http://www.tandfonline.com/loi/uawm20?. doi:10.1080/10962247.2015.1096863.

Galarneau, E., Yacovitch, T.I., Lerner, B., Sheppard, A., Quach, B.T., Kuang, W., Rai, H., Staebler, R., Mihele, C., Vogel, F., 2023. From hotspots to background: high-resolution mapping of ethylene oxide in urban air. Atmos. Environ. 18732844. 307. Available from: http://www.elsevier.com/locate/atmosenv. doi:10.1016/j.atmosenv.2023.119828.

Gallego, E., Perales, J.F., Calaf, J.M., 2023. Continuous monitoring of volatile organic compounds through sensorization. Automatic sampling during pollution/odour/nuisance episodic events. Atmos. Environ. 18732844. 299. Available from: http://www.elsevier.com/locate/atmosenv. doi:10.1016/j.atmosenv.2023.119657.

Gallego, E., Roca, F.J., Perales, J.F., Guardino, X., Gadea, E., Garrote, P., 2016. Impact of formaldehyde and VOCs from waste treatment plants upon the ambient air nearby an urban area (Spain). Sci. Total Environ. 18791026. 568, 369. Available from: http://www.elsevier.com/locate/scitotenv. doi:10.1016/j.scitotenv.2016.06.007.

Gallego, E., Roca, F.J., Perales, J.F., Sánchez, G., Esplugas, P., 2012. Characterization and determination of the odorous charge in the indoor air of a waste treatment facility through the evaluation of volatile organic compounds (VOCs) using TD–GC/MS. Waste Manag. 0956053X. 32, 2469. doi:10.1016/j.wasman.2012.07.010.

Gallego, E., Teixidor, P., Roca, F.J., Perales, J.F., Gadea, E., 2018. Outdoor air 1,3-butadiene monitoring: comparison of performance of Radiello passive samplers and active multi-sorbent bed tubes. Atmos. Environ. 18732844. 182, 9. Available from: http://www.elsevier.com/locate/atmosenv. doi:10.1016/j.atmosenv.2018.03.022.

Galvão, E.S., Santos, J.M., Lima, A.T., Reis, N.C., Orlando, M.T.D.A., Stuetz, R.M., 2018. Trends in analytical techniques applied to particulate matter characterization: a critical review of fundaments and applications. Chemosphere. 18791298. 199, 546. Available from: http://www.elsevier.com/locate/chemosphere. doi:10.1016/j.chemosphere.2018.02.034.

García-Garcinuño, R., Marcé, R.M., Vallecillos, L., Borrull, F., 2024b. Passive sampling of high production volume chemicals and polycyclic aromatic hydrocarbons in urban atmospheres near petrochemical sites: uptake rate determination and application. Environ. Pollut. 02697491. 361, 124697. doi:10.1016/j.envpol.2024.124697.

García-Garcinuño, R., Solé-Domènech, L., Marcé, R.M., Vallecillos, L., Borrull, F., 2025. Assessment of the presence and risk associated with volatile and semi-volatile organic compounds in urban areas close to petrochemical parks. Chemosphere. 00456535. 379, 144432. doi:10.1016/j.chemosphere.2025.144432.

García-Garcinuño, R., Vallecillos, L., Marcé, R.M., Borrull, F., 2024a. Occurrence of high production volume chemicals and polycyclic aromatic hydrocarbons in urban sites close to industrial areas. Human exposure and risk assessment. Chemosphere. 00456535. 351, 141167. doi:10.1016/j.chemosphere.2024.141167.

Gent, L., Paul, R., 2021. Air monitoring for illegal drugs including new psychoactive substances: a review of trends, techniques and thermal degradation products. Drug Test. Anal. 19427611. 13, 1078. Available from: http://onlinelibrary.wiley.com/journal/10.1002/(ISSN)1942-7611. doi:10.1002/dta.3051.

Ghaffari, H.R., Kamari, Z., Hassanvand, M.S., Fazlzadeh, M., Heidari, M., 2021. Level of air BTEX in urban, rural and industrial regions of Bandar Abbas, Iran; indoor-outdoor relationships and probabilistic health risk assessment. Environ. Res. 10960953. 200. Available from: http://www.elsevier.com/inca/publications/store/6/2/2/8/2/1/index.htt. doi:10.1016/j.envres.2021.111745.

Gu, S., Guenther, A., Faiola, C., 2021. Effects of anthropogenic and biogenic volatile organic compounds on Los Angeles air quality. Environ. Sci. Technol. 15205851. 55, 12191. Available from: http://pubs.acs.org/journal/esthag. doi:10.1021/acs.est.1c01481.

Han, C., Liu, R., Luo, H., Li, G., Ma, S., Chen, J., An, T., 2019. Pollution profiles of volatile organic compounds from different urban functional areas in Guangzhou China based on GC/MS and PTR-TOF-MS: atmospheric environmental implications. Atmos. Environ. 13522310. 214, 116843. doi:10.1016/j.atmosenv.2019.116843.

Healy, R.M., Bennett, J., Wang, J.M., Karellas, N.S., Wong, C., Todd, A., Sofowote, U., Su, Y., Di Federico, L., Munoz, A., Charland, J.P., Herod, D., Siu, M., White, L., 2018. Evaluation of a passive sampling method for long-term continuous monitoring of volatile organic compounds in urban environments. Environ. Sci. Technol. 15205851. 52, 10580. Available from: http://pubs.acs.org/journal/esthag. doi:10.1021/acs.est.8b02792.

Hoang, T., Castorina, R., Gaspar, F., Maddalena, R., Jenkins, P.L., Zhang, Q., McKone, T.E., Benfenati, E., Shi, A.Y., Bradman, A., 2017. VOC exposures in California early childhood education environments. Indoor Air. 09056947. 27, 609. doi:10.1111/ina.12340.

Hoisington, J., Herrington, J.S., 2021. Rapid determination of ethylene oxide and 75 vocs in ambient air with canister sampling and associated growth issues. Separations. 22978739. 8. Available from: https://www.mdpi.com/2297-8739/8/3/35. doi:10.3390/SEPARATIONS8030035.

Jeong, C.H., Salehi, S., Wu, J., North, M.L., Kim, J.S., Chow, C.W., Evans, G.J., 2019. Indoor measurements of air pollutants in residential houses in urban and suburban areas: Indoor versus ambient concentrations. Sci. Total Environ. 18791026. 693. Available from: http://www.elsevier.com/locate/scitotenv. doi:10.1016/j.scitotenv.2019.07.252.

Jia, C., Fu, X., Nored, A., Batbaatar, N., Smith, L., 2023. Comparison of generic Tenax and specialized PAH tubes for monitoring polycyclic aromatic hydrocarbons in the ambient air. Atmos. Pollut. Res. 13091042. 14, 101780. doi:10.1016/j.apr.2023.101780.

Jia, H., Gao, S., Duan, Y., Fu, Q., Che, X., Xu, H., Wang, Z., Cheng, J., 2021. Investigation of health risk assessment and odor pollution of volatile organic compounds from industrial activities in the Yangtze River Delta region, China. Ecotoxicol. Environ. Saf. 01476513. 208, 111474. doi:10.1016/j.ecoenv.2020.111474.

Karri, R.R., Vera, T., Mohamed Hassan, S.K., Khoder, M.I., Dehghani, M.H., Mubarak, N.M., Ravindran, G., 2024. Classification, sources, and occurrence of outdoor air pollutants: a comprehensive overview. In: Dehghani, M.H., Karri, R.R., Vera, T. and Mohamed Hassan, S.K. (Eds.), Health and Environmental Effects of Ambient Air Pollution: Volume 1: Air Pollution, Human Health, and the Environment. Elsevier, Brunei Darussalam, pp. 1–34. Available from: https://www.sciencedirect.com/book/9780443160882. doi:10.1016/C2022-0-02289-3.

Kumar, A., Víden, I., 2007. Volatile organic compounds: sampling methods and their worldwide profile in ambient air. Environ. Monit. Assess. 15732959. 131, 301. doi:10.1007/s10661-006-9477-1.

Lee, Y.S., Maitlo, H.A., Kim, K.H., Kwon, E.E., Lee, M.H., Kim, J.C., Song, H.N., Szulejko, J.E., 2019. Utilization of activated carbon as an effective replacement for a commercialized three-bed sorbent (Carbopack) to quantitate aromatic hydrocarbons in ambient air. Environ. Res. 10960953. 179. Available from: http://www.elsevier.com/inca/publications/store/6/2/2/8/2/1/index.htt. doi:10.1016/j.envres.2019.108802.

Lévy, M., Ba, H., Pallares, C., Pham-Huu, C., Millet, M., 2020. Comparison and calibration of diverse passive samplers used for the air sampling of pesticides during a regional sampling monitoring campaign. Atmos. Pollut. Res. 13091042. 11, 1217. Available from: http://www.atmospolres.com/accesstext.html. doi:10.1016/j.apr.2020.03.014.

Li, M., Li, Q., Nantz, M.H., Fu, X.A., 2018. Analysis of carbonyl compounds in ambient air by a microreactor approach. ACS Omega. 24701343. 3, 6764. Available from: http://pubs.acs.org/journal/acsodf. doi:10.1021/acsomega.8b00503.

Li, W., Li, J., Deng, M., Pan, Y., Zeng, L., 2020. Benzotriazoles and benzothiazoles prevail in indoor dust from an E-waste dismantling area in South China: elevated concentrations and implication for human exposure. Sci. Total Environ. 00489697. 723, 137979. doi:10.1016/j.scitotenv.2020.137979.

Li, Z., Ming, T., Liu, S., Peng, C., de Richter, R., Li, W., Zhang, H., Wen, C.-Y., 2021. Review on pollutant dispersion in urban areas-part A: effects of mechanical factors and urban morphology. Build. Environ. 03601323. 190, 107534. doi:10.1016/j.buildenv.2020.107534.

Liaud, C., Schwartz, J.J., Millet, M., 2017. Comparison of atmospheric concentrations of currently used pesticides between urban and rural areas during intensive application period in Alsace (France) by using XAD-2 based passive samplers. J. Environ. Sci. Health B. 15324109. 52, 458. Available from: http://www.tandf.co.uk/journals/titles/03601234.asp. doi:10.1080/03601234.2017.1301756.

Lin, H., Taniyasu, S., Yamazaki, E., Wu, R., Lam, P.K.S., Eun, H., Yamashita, N., 2022. Fluorine mass balance analysis and per- and polyfluoroalkyl substances in the atmosphere. J. Hazard. Mater. 18733336. 435. Available from: http://www.elsevier.com/locate/jhazmat. doi:10.1016/j.jhazmat.2022.129025.

Lucattini, L., Poma, G., Covaci, A., de Boer, J., Lamoree, M.H., Leonards, P.E.G., 2018. A review of semi-volatile organic compounds (SVOCs) in the indoor environment: occurrence in consumer products, indoor air and dust. Chemosphere. 18791298. 201, 466. Available from: http://www.elsevier.com/locate/chemosphere. doi:10.1016/j.chemosphere.2018.02.161.

Maceira, A., Borrull, F., Marcé, R.M., 2019. Occurrence of plastic additives in outdoor air particulate matters from two industrial parks of Tarragona, Spain: human inhalation intake risk assessment. J. Hazard. Mater. 18733336. 373, 649. Available from: http://www.elsevier.com/locate/jhazmat. doi:10.1016/j.jhazmat.2019.04.014.

Maceira, A., Borrull, F., Marcé, R.M., 2022. Occurrence of organic contaminants bonded to the particulate matter from outdoor air influenced by industrial activities. Environ. Sci. Pollut. Res. 16147499. 29, 76644. Available from: https://www.springer.com/journal/11356. doi:10.1007/s11356-022-23103-8.

Maceira, A., Marcé, R.M., Borrull, F., 2018. Occurrence of benzothiazole, benzotriazole and benzenesulfonamide derivates in outdoor air particulate matter samples and human exposure assessment. Chemosphere. 18791298. 193, 557. Available from: http://www.elsevier.com/locate/chemosphere. doi:10.1016/j.chemosphere.2017.11.073.

Maceira, A., Marcé, R.M., Borrull, F., 2020b. Analytical methods for determining organic compounds present in the particulate matter from outdoor air. TrAC Trends Anal. Chem. 01659936. 122, 115707. doi:10.1016/j.trac.2019.115707.

Maceira, A., Pecikoza, I., Marcé, R.M., Borrull, F., 2020a. Multi-residue analysis of several high-production-volume chemicals present in the particulate matter from outdoor air. A preliminary human exposure estimation. Chemosphere. 00456535. 252, 126514. doi:10.1016/j.chemosphere.2020.126514.

Maceira, A., Vallecillos, L., Borrull, F., Marcé, R.M., 2017. New approach to resolve the humidity problem in VOC determination in outdoor air samples using solid adsorbent tubes followed by TD-GC–MS. Sci. Total Environ. 18791026. 599-600, 1718. Available from: http://www.elsevier.com/locate/scitotenv. doi:10.1016/j.scitotenv.2017.05.141.

Madureira, J., Paciência, I., Pereira, C., Teixeira, J.P., Fernandes, E.D.O., 2016. Indoor air quality in Portuguese schools: levels and sources of pollutants. Indoor Air. 09056947. 26, 526. doi:10.1111/ina.12237.

Manual of Analytical Methods (NMAM), 1994. NIOSH method 1614 ethylene oxide.

Manual of Analytical Methods (NMAM), 2003. NIOSH method 1501 aromatic hydrocarbons.

Markes International, 2022. Measuring PFAS pollution in ambient air using TD-GC-MS/MS.

Melymuk, L., Bohlin-Nizzetto, P., Vojta, Š., Krátká, M., Kukučka, P., Audy, O., Přibylová, P., Klánová, J., 2016. Distribution of legacy and emerging semivolatile organic compounds in five indoor matrices in a residential environment. Chemosphere. 18791298. 153, 179. Available from: http://www.elsevier.com/locate/chemosphere. doi:10.1016/j.chemosphere.2016.03.012.

Montero-Montoya, R., López-Vargas, R., Arellano-Aguilar, O., 2018. Volatile organic compounds in air: sources, distribution, exposure and associated illnesses in children. Ann. Glob. Health 22149996. 84, 225. Available from: https://agh.ubiquitypress.com/articles/10.29024/aogh.910/galley/2309/download/. doi:10.29024/aogh.910.

Mukerjee, S., Smith, L.A., Thoma, E.D., Whitaker, D.A., Oliver, K.D., Duvall, R., Cousett, T.A., 2020. Spatial analysis of volatile organic compounds using passive samplers in the Rubbertown industrial area of Louisville, Kentucky, USA. Atmos. Pollut. Res. 13091042. 11, 81. Available from: http://www.atmospolres.com/accesstext.html. doi:10.1016/j.apr.2020.02.021.

Nargis, A., Habib, A., Zhao, S., Nigar, R., Liu, X., Zhang, G., Cai, M., 2022. Monitoring of atmospheric polycyclic aromatic hydrocarbons by polyurethane foam-passive air samplers in Bangladesh: source apportionment and health risk assessment. Atmos. Environ. 13522310. 289, 119346. doi:10.1016/j.atmosenv.2022.119346.

Ninyà, N., Vallecillos, L., Marcé, R.M., Borrull, F., 2022. Evaluation of air quality in indoor and outdoor environments: impact of anti-COVID-19 measures. Sci. Total Environ. 00489697. 836, 155611. doi:10.1016/j.scitotenv.2022.155611.

Nuñez, A., Vallecillos, L., Marcé, R.M., Borrull, F., 2020. Occurrence and risk assessment of benzothiazole, benzotriazole and benzenesulfonamide derivatives in airborne particulate matter from an industrial area in Spain. Sci. Total Environ. 00489697. 708, 135065. doi:10.1016/j.scitotenv.2019.135065.

Olaguer, E.P., Robinson, A., Kilmer, S., Haywood, J., Lehner, D., 2020. Ethylene oxide exposure attribution and emissions quantification based on ambient air measurements near a sterilization facility. Int. J. Environ. Res. Public Health 16604601. 17. Available from: https://www.mdpi.com/1660-4601/17/1/42/pdf. doi:10.3390/ijerph17010042.

Ontario Ministry of the Environment, Conservation and Parks, 2025. Ontario's Ambient Air Quality Criteria. Ministry of the Environment, Conservation and Parks. Available from: https://www.ontario.ca/page/ontarios-ambient-air-quality-criteria#section-4.

Organisation for Economic Co-Operation and Development, 2025. The 2004 OECD list of high production volume chemicals. Available from: https://hpvchemicals.oecd.org/ui/Default.aspx.

Ramboll US Corporation, 2019. Ambient air monitoring report ethylene oxide background concentrations in Chicago metro area, Sterigenics US, LLC. Available from: https://static1.squarespace.com/static/5b8457322714e5cbd33c889a/t/5c4c5cad1ae6cfaa21e6cd7e/1548508356697/EO+background+study+report+1-24-19.pdf.

Ramírez, N., Cuadras, A., Rovira, E., Borrull, F., Marcé, R.M., 2012. Chronic risk assessment of exposure to volatile organic compounds in the atmosphere near the largest Mediterranean industrial site. Environ. Int. 18736750. 39, 200. Available from: http://www.elsevier.com/locate/envint. doi:10.1016/j.envint.2011.11.002.

Ras, M.R., Borrull, F., Marcé, R.M., 2009. Sampling and preconcentration techniques for determination of volatile organic compounds in air samples. TrAC Trends Anal. Chem. 01659936. 28, 347. doi:10.1016/j.trac.2008.10.009.

Robbins, Z.G., Liu, X., Schumacher, B.A., Smeltz, M.G., Liberatore, H.K., 2023. Method development for thermal desorption-gas chromatography-tandem mass spectrometry (TD-GC–MS/MS) analysis of trace level fluorotelomer alcohols emitted from consumer products. J. Chromatogr. A. 00219673. 1705, 464204. doi:10.1016/j.chroma.2023.464204.

Ródenas García, M., Spinazzé, A., Branco, P.T.B.S., Borghi, F., Villena, G., Cattaneo, A., Di Gilio, A., Mihucz, V.G., Gómez Álvarez, E., Lopes, S.I., Bergmans, B., Orłowski, C., Karatzas, K., Marques, G., Saffell, J., Sousa, S.I.V., 2022. Review of low-cost sensors for indoor air quality: features and applications. Appl. Spectrosc. Rev. 1520569X. 57, 747. Available from: http://www.tandf.co.uk/journals/titles/05704928.asp. doi:10.1080/05704928.2022.2085734.

Rovira, J., González, N., Nadal, M., Domingo, J.L., Schuhmacher, M., 2024. Air concentrations of trace elements in a municipality under the influence of Tarragona petrochemical complex: human health risks. Environ. Res. 00139351. 243, 117859. doi:10.1016/j.envres.2023.117859.

Sakhi, A.K., Cequier, E., Becher, R., Bølling, A.K., Borgen, A.R., Schlabach, M., Schmidbauer, N., Becher, G., Schwarze, P., Thomsen, C., 2019. Concentrations of selected chemicals in indoor air from Norwegian homes and schools. Sci. Total Environ. 18791026. 674, 1. Available from: http://www.elsevier.com/locate/scitotenv. doi:10.1016/j.scitotenv.2019.04.086.

Sakizadeh, M., 2020. Short- and long-term variations, spatial analysis along with cancer health risk assessment associated with 1, 3-butadiene. Atmos. Pollut. Res. 13091042. 11, 755. Available from: http://www.atmospolres.com/accesstext.html. doi:10.1016/j.apr.2020.01.004.

Sánchez-Piñero, J., Novo-Quiza, N., Moreda-Piñeiro, J., Turnes-Carou, I., Muniategui-Lorenzo, S., López-Mahía, P., 2022. Multi-class organic pollutants in atmospheric particulate matter (PM2.5) from a SouthWestern Europe industrial area: levels, sources and human health risk. Environ. Res. 00139351. 214, 114195. doi:10.1016/j.envres.2022.114195.

Savareear, B., Brokl, M., Wright, C., Focant, J.F., 2017. Thermal desorption comprehensive two-dimensional gas chromatography coupled to time of flight mass spectrometry for vapour phase mainstream tobacco smoke analysis. J. Chromatogr. A 18733778. 1525, 126. Available from: http://www.elsevier.com/locate/chroma. doi:10.1016/j.chroma.2017.10.013.

Su, P.H., Tomy, G.T., Hou, C.Y., Yin, F., Feng, D.L., Ding, Y.S., Li, Y.F., 2018. Gas/particle partitioning, particle-size distribution of atmospheric polybrominated diphenyl ethers in southeast Shanghai rural area and size-resolved predicting model. Chemosphere. 18791298. 197, 251. Available from: http://www.elsevier.com/locate/chemosphere. doi:10.1016/j.chemosphere.2018.01.005.

Tao, S., Liu, Y., Xu, W., Lang, C., Liu, S., Dou, H., Liu, W., 2007. Calibration of a passive sampler for both gaseous and particulate phase polycyclic aromatic hydrocarbons. Environ. Sci. Technol. 0013936X. 41, 568. doi:10.1021/es0617486.

Toledo, M., Muñoz, R., 2023. Odour prevention strategies in wastewater treatment plants: a pilot scale study of activated sludge recycling and oxidized nitrogen recycling. J. Environ. Chem. Eng. 22133437. 11, 110366. doi:10.1016/j.jece.2023.110366.

United States Environmental Protection Agency, 2024. Protecting air quality. Criteria Air Pollutants. https://www.epa.gov/criteria-air-pollutants/information-pollutant.

US Environmental Protection Agency (USEPA), 1994. Method 1613B: tetra- through octa-chlorinated dioxins and furans by isotope dilution HRGC/HRMS.

US Environmental Protection Agency (USEPA), 1996. Method 0040: sampling of principal organic hazardous constituents from combustion sources using tedlar bags.

US Environmental Protection Agency (USEPA), 1999. Compendium method TO-17: determination of volatile organic compounds in ambient air using active sampling onto sorbent tubes.

US Environmental Protection Agency (USEPA), 2003. Method 1668A: revision a chlorinated biphenyl congeners in water, soil, sediment, biosolids, and tissue by HRGC/HRMS.

US Environmental Protection Agency (USEPA), 2018. Method 0023A: sampling method for polychlorinated dibenzo-p-Dioxins and polychlorintated dibenzofuran emissions from stationary sources.

US Environmental Protection Agency (USEPA), 2019a. Method 325A: volatile organic compounds from fugitive and area sources: sampler deployment and VOC sample collection.

US Environmental Protection Agency (USEPA), 2019b. Method TO-15A. Determination of volatile organic compounds (VOCs) in air collected in specially prepared canisters and analyzed by gas chromatography-mass spectrometry.

US Environmental Protection Agency (USEPA), 2025. Method OTM-50: sampling and analysis of volatile fluorinated compounds from stationary sources using passivated stainless-steel canister.

Vallecillos, L., Borrull, A., Marcé, R.M., Borrull, F., 2021. Presence of emerging organic contaminants and solvents in schools using passive sampling. Sci. Total Environ. 00489697. 764, 142903. doi:10.1016/j.scitotenv.2020.142903.

Vallecillos, L., Espallargas, E., Allo, R., Marcé, R.M., Borrull, F., 2019. Passive sampling of volatile organic compounds in industrial atmospheres: uptake rate determinations and application. Sci. Total Environ. 18791026. 666, 235. Available from: http://www.elsevier.com/locate/scitotenv. doi:10.1016/j.scitotenv.2019.02.213.

Vallecillos, L., Maceira, A., Marcé, R.M., Borrull, F., 2018. Evaluation of active sampling strategies for the determination of 1,3-butadiene in air. Atmos. Environ. 18732844. 176, 21. Available from: http://www.elsevier.com/locate/atmosenv. doi:10.1016/j.atmosenv.2017.12.012.

Vallecillos, L., Marcé, R.M., Borrull, F., 2024b. Applicability of a gas chromatograph-photoionization detection analyzer for the continuous monitoring of 1,3-butadiene in urban and industrial atmospheres. Atmos. Pollut. Res. 13091042. 15, 101949. doi:10.1016/j.apr.2023.101949.

Vallecillos, L., Riu, J., Marcé, R.M., Borrull, F., 2024a. Air monitoring with passive samplers for volatile organic compounds in atmospheres close to petrochemical industrial areas. The case study of Tarragona. Atmos. Pollut. Res. 15. doi:10.1016/J.APR.2023.101986

Villanueva, F., Tapia, A., Lara, S., Amo-Salas, M., 2018. Indoor and outdoor air concentrations of volatile organic compounds and NO2 in schools of urban, industrial and rural areas in Central-Southern Spain. Sci. Total Environ. 18791026. 622-623, 222. Available from: http://www.elsevier.com/locate/scitotenv. doi:10.1016/j.scitotenv.2017.11.274.

Wallace, M.A.G., Smeltz, M.G., Mattila, J.M., Liberatore, H.K., Jackson, S.R., Shields, E.P., Xhani, X., Li, E.Y., Johansson, J.H., 2024. A review of sample collection and analytical methods for detecting per- and polyfluoroalkyl substances in indoor and outdoor air. Chemosphere. 18791298. 358. Available from: https://www.sciencedirect.com/science/journal/00456535. doi:10.1016/j.chemosphere.2024.142129.

Wang, D.K.W., Austin, C.C., 2006. Determination of complex mixtures of volatile organic compounds in ambient air: an overview. Anal. Bioanal. Chem. 1618-2642. 386, 1089. doi:10.1007/s00216-006-0475-5.

Wang, L., Zhao, Y., Yi, X., Wang, Z., Yi, Y., Huang, T., Gao, H., Ma, J., 2017. Spatial distribution of atmospheric PAHs and their genotoxicity in petrochemical industrialized Lanzhou valley, northwest China. Environ. Sci. Pollut. Res. 16147499. 24, 12820. Available from: http://www.springerlink.com/content/0944-1344. doi:10.1007/s11356-017-8808-9.

Wang, S., Cai, W., Tao, Y., Sun, Q.C., Wong, P.P.Y., Huang, X., Liu, Y., 2023. Unpacking the inter- and intra-urban differences of the association between health and exposure to heat and air quality in Australia using global and local machine learning models. Sci. Total Environ.

18791026. 871. Available from: http://www.elsevier.com/locate/scitotenv. doi:10.1016/j.scitotenv.2023.162005.

Wania, F., Shunthirasingham, C., 2020. Passive air sampling for semi-volatile organic chemicals. Environ. Sci. Process. Impacts 20507895. 22, 1925. Available from: http://pubs.rsc.org/en/journals/journal/em. doi:10.1039/d0em00194e.

World Health Organization, 2021. World Health Organization Global Air Quality Guidelines. Particulate Matter (PM2.5 and PM10), Ozone, Nitrogen Dioxide, Sulfur Dioxide and Carbon Monoxide. World Health Organization. Available from: https://www.who.int/publications/i/item/9789240034228

Xuan, L., Ma, Y., Xing, Y., Meng, Q., Song, J., Chen, T., Wang, H., Wang, P., Zhang, Y., Gao, P., 2021. Source, temporal variation and health risk of volatile organic compounds (VOCs) from urban traffic in harbin, China. Environ. Pollut. 02697491. 270, 116074. doi:10.1016/j.envpol.2020.116074.

Yacovitch, T.I., Dyroff, C., Roscioli, J.R., Daube, C., McManus, J.B., Herndon, S.C., 2023. Ethylene oxide monitor with part-per-trillion precision for in situ measurements. Atmos. Meas. Tech. 18678548. 16, 1915. Available from: http://www.atmos-meas-tech.net/volumes_and_issues.html. doi:10.5194/amt-16-1915-2023.

Yao, S., Wang, Q., Zhang, J., Zhang, R., Gao, Y., Zhang, H., Li, J., Zhou, Z., 2021. Ambient volatile organic compounds in a heavy industrial city: concentration, ozone formation potential, sources, and health risk assessment. Atmos. Pollut. Res. 13091042. 12, 101053. doi:10.1016/j.apr.2021.101053.

Yusà, V., Coscollà, C., Mellouki, W., Pastor, A., de la Guardia, M., 2009. Sampling and analysis of pesticides in ambient air. J. Chromatogr. A. 00219673. 1216, 2972. doi:10.1016/j.chroma.2009.02.019.

Zhang, L., Wang, F., Ji, Y., Jiao, J., Zou, D., Liu, L., Shan, C., Bai, Z., Sun, Z., 2014. Phthalate esters, and human exposure to PAEs via inhalation of indoor air in Tianjin. Atmos. Environ. 85. doi:10.1016/J.ATMOSENV.2013.11.068

Chapter 3

Sampling of environmental water samples

Monika Paszkiewicz, Hanna Lis, and Magda Caban
Faculty of Chemistry, Department of Environmental Analysis, University of Gdansk, Gdansk, Poland

Abbreviations

CECs	contaminants of emerging concern
CNT	carbon nanotubes
CNT-PSD	carbon nanotube passive sampling device
DDD	dichlorodiphenyldichloroethane
DDE	dichlorodiphenyldichloroethylene
DDT	dichlorodiphenyltrichloroethane
FTSA	fluorotelomer sulfonic acid
GF	glass fiber filter
HDPE	high-density polyethylene
HLB	hydrophilic–lipophilic balanced sorbent
HRMS	high-resolution mass spectrometry
LC–MS/MS	liquid chromatography coupled with tandem mass spectrometry
LDPE	low-density polyethylene
LOQ	limit of quantification
MCX	mixed-mode cation exchange sorbent
MDL	method detection limit
NHFRs	novel halogenated flame retardants
NSAIDs	nonsteroidal antiinflammatory drugs
PAHs	polycyclic aromatic hydrocarbons
PBDEs	polybrominated diphenyl ethers
PC	polycarbonate
PCBs	polychlorinated biphenyls
PFAS	per- and polyfluoroalkyl substances
PFCAs	perfluoroalkyl carboxylic acids
PFSAs	perfluoroalkyl sulfonic acids
POCIS	polar organic chemical integrative sampler
POCIS-HLB	POCIS containing hydrophilic–lipophilic balanced sorbent
POCIS-MMA	POCIS with mixed-mode adsorbents

Contaminants of Emerging Concern in the Environment. https://doi.org/10.1016/B978-0-443-33821-2.00002-5

POCIS-WAX	POCIS containing weak anion-exchange sorbent
PRC	performance reference compound
PSD	passive sampling device
SPE	solid phase extraction
TPs	transformation products
UHPLC	ultra-high-performance liquid chromatography
UV filters	ultraviolet light-absorbing organic compounds
WCX	weak cation-exchange sorbent
WWTP	wastewater treatment plant

1 Introduction

Water sampling is one of the most important stages of environmental monitoring, as the accuracy of analytical results depends not only on laboratory equipment but also on how well the samples reflect actual environmental conditions. According to the review (Ort et al., 2010), most articles contain detailed sections on sample preservation (75%) and analytical methods (99%), while detailed sampling protocols are described in only 11% of articles (Ort et al., 2010). Aquatic systems contain a wide range of pollutants with very different chemical and physical properties, including nutrients, inorganic ions, trace metals, hydrocarbons, pesticides, pharmaceuticals, and other emerging pollutants (Richardson and Kimura, 2017; Menger et al., 2020). Given their different properties, a universal sampling method for all pollutants is impossible. Sampling strategies must therefore be carefully planned to preserve the original composition of the water and prevent chemical, physical, or biological changes prior to analysis (Ort et al., 2010; Telgmann and Horn, 2024). The main aim of water sampling is to collect a portion of water that accurately reflects the conditions at a specific location and time. In rivers and streams, samples are typically taken at middepth and in the middle of the channel, where the water is best mixed. In lakes and reservoirs, it is often necessary to take samples vertically to account for stratification and changes in redox conditions with depth. Grab sampling is the primary surface sampling technique, with an analysis of approximately 0.5–1.0 (Carmona et al., 2014; Ng et al., 2020; Oosterhuis et al., 2013). Irregular introduction and removal of contaminants of emerging concern (CECs), along with changing concentrations, make random surface water sampling unrepresentative and highly variable. As a result, composite sampling (often conducted over 24 h) is increasingly used as the norm for CECs monitoring, enabling a slightly larger volume (5–10 L) to be analyzed over an extended period (roughly a day) (Aristizabal-Ciro et al., 2017; Castiglioni et al., 2013; Kosma et al., 2010). Composite samples are usually collected using an automatic sampler operating in one of three modes: (a) volume-proportional mode, that is, collecting samples of equal volume at different intervals, proportional to the wastewater flow; (b) time-proportional mode, that is, sampling equal volumes at fixed intervals; or (c) flow-proportional mode, that is, sampling variable volumes proportional to the wastewater flow at fixed intervals (Castiglioni et al., 2013). Despite composite sampling considering same-day changes, daily fluctuations still affect samples, especially when sampling wastewater. These problems can be solved using passive devices that can continuously monitor the concentration of dissolved analytes in water for weeks, which circumvents the drawbacks of grab and composite sampling (Hawker et al., 2022). Most passive samplers used in aquatic environments are constructed similarly (Alvarez et al., 2004; Godlewska et al., 2021; MacKeown et al., 2024; Tapie et al., 2011). Analytes are retained in a receiving phase (such as a solvent or solid sorbent), separated from the environment (matrix) by a semipermeable membrane (limiting phase). Analyte molecules are transported from the water to the receiving medium in the passive sampling device because of a concentration difference. The device design, analyte properties, and environment affect sampling rates (Rs) for target compounds (Godlewska et al., 2021). Various types of passive samplers are used to monitor pollution levels in surface water and wastewater, including those containing semipermeable membranes (SPMD), passive in situ semipermeable membrane devices (PISCES), ceramic probes (DGT, diffusive gradients in thin films), membrane-enclosed sorptive coating (MESCO), chemcatcher, and polar organic chemical integrative sampler (POCIS) (Godlewska et al., 2021). Passive samplers allow us to determine the time-weighted average concentrations (C_{TWA}) of pollutants. This is a significant improvement compared to grab sampling, in which the concentrations of analytes are often below the limit of detection (LOD) and the limit of quantification (LOQ) of the selected instrumental method, which makes it impossible to determine, in particular, chemical compounds present in water in trace amounts (Godlewska et al., 2022a). Advanced sorption materials are rapidly developing and being applied in passive sampling devices, which is the primary methodological trend from 2015 to 2026. There is growing interest in utilizing carbon nanotubes (SWCNTs [single walled carbon nanotubes] and MWCNTs [multi walled carbon nanotubes]) as receiving phases in passive samplers. The extensive specific surface area, coupled with potent π–π interactions and a hydrophobic quality, allows for the efficient adsorption of diverse compounds (Godlewska et al., 2022a, 2022b; Jakubus et al., 2019). Moreover, metal-organic frameworks (MOFs) are another rapidly developing class of sorbents used in passive sampling and sample preparation. MOFs are crystalline porous materials composed of metal ions or clusters coordinated with organic ligands, offering an exceptionally large surface area, tunable pore size, and a functional chemical environment (Godlewska et al., 2024; Hashemi et al., 2017). To ensure accurate results, sample preservation is also crucial because of analyte degradation. Common residue analysis in water involves cooling, freezing, acidifying, or adding sodium thiosulfate, ascorbic acid, formaldehyde, or others. In summary, effective water sampling requires careful consideration of hydrological conditions, the chemical properties of the target compounds, and practical field constraints. The wide range of contaminants found in aquatic environments makes it necessary to use sampling strategies tailored to specific compounds. Sampling design, container selection, preservation techniques, and quality control measures are therefore critical to obtaining accurate, repeatable, and environmentally meaningful water quality data.

2 Sampling fundamentals and instrumentation

In studies of pharmaceuticals and personal care products (PPCPs), samples collected from surface waters and sewage matrices samples were consistently collected manually using previously cleaned amber glass bottles, typically with a capacity of 500 mL to 1 L (Tran et al., 2013; Hawker et al., 2022; Carmona et al., 2014). The bottles were routinely rinsed with methanol and ultrapure water prior to use, and in several studies, prefired glassware was used to minimize organic contamination. Immediately after collection, samples were placed in cooled containers (4 °C) and transported to the laboratory on the same day. To remove suspended solids and prevent storage-related losses, filtration was done on-site or within a day, using glass fiber filters (GF/F, 0.7 μm) or membrane filters (0.45 μm, cellulose acetate or polytetrafluoroethylene [PTFE]) (Croft et al., 2020; Diamanti et al., 2019). Studies on composite sampling (Aristizabal-Ciro et al., 2017; Kosma et al., 2010) mainly used automatic refrigerated samplers installed on the inflows and outflows of wastewater treatment plants. Sampling strategies included time-proportional samples (e.g., every 10–30 min) and flow-proportional samples integrated over 24 h periods, with some campaigns involving daily samples collected over multiple days. However, sometimes, composite samples of 72 h (Thomas et al., 2017) and 12 h (Ort et al., 2010) were reported. Compound samples were collected in high-density polyethylene (HDPE) or glass containers, kept at 4 °C throughout the sampling period to limit biological activity. After completion of the composite cycle, samples were homogenized, subsampled, and typically filtered through 0.45 μm membrane filters prior to extraction. Chemical preservation varied depending on the target compounds, but typically involved acidification to pH 2–3 with hydrochloric or formic acid, particularly for antibiotics and beta-blockers, to inhibit microbial degradation (Diamanti et al., 2019; Lei et al., 2021; Thomas et al., 2017). However, this led to the formation of acid-catalyzed degradation products and also resulted in lower recoveries of the primary analytes. In studies of labile compounds (e.g., paracetamol, caffeine), samples were extracted by solid phase extraction (SPE) within 24–48 h, while longer storage periods were managed by freezing at −20 °C (Causanilles et al., 2017; Ofrydopoulou et al., 2022). Degradation is also limited by adding sodium thiosulfate (up to 80 mg) or ascorbic acid to the aqueous matrix (Fedorova et al., 2014). Moreover, formaldehyde was used to inhibit bacterial growth without altering the pH, which led to higher recoveries across a wide range of target analytes. Fedorova et al. (2014) conducted stability studies and found that many PPCPs remain stable when stored at 4 °C for a short period (2 days), but recommended storing compounds at −18 °C for longer storage periods (28 days or more). Vanderford and Snyder (2006) observed in their stability study that the recovery of 21 pharmaceuticals and personal care products (PPCPs) was within the acceptable range even after 28 days of storage at 4 °C when sodium azide, ascorbic acid, and amber bottles were used as quenching agents, preservatives, and storage containers, respectively. By adding the right quenching agents and preservatives, and utilizing appropriate storage containers, PPCPs were recovered acceptably. An alternative approach to grab or composite sampling is to use passive probes, for example, POCIS (Godlewska et al., 2020). Generally, the $POCIS_{pharm}$ devices are used for targeting hydrophilic molecules, including PPCPss, and endocrine disrupting compounds (Alvarez et al., 2004). In this device, a hydrophilic-lipophilic balance (HLB) sorbent located between two polyethersulfone (PES) membranes is used as the receiving phase, and the whole is tightly sealed with metal rings. It should be noted that each sampler is configured and designed differently, and the exposure time will vary depending on the sampler used and the analyte being tested. Passive sampling procedures involve placing devices such as POCIS on site, typically for pharmaceuticals, for a period of two to four weeks (Domínguez-García et al., 2025). After collection, the accumulated analytes are extracted from the sorption material under controlled laboratory conditions. Because passive sampling devices do not directly measure concentrations, compound-specific calibration models are needed to convert the accumulated mass into ambient concentrations. Numerous studies show a good correlation between passive and composite sampling of various pharmaceuticals (Bishop et al., 2020). The United States Environmental Protection Agency (US EPA) provides comprehensive guidelines and procedures for the collection and preservation of per- and polyfluoroalkyl substances (PFAS), including protocols for methods 537.1 and 533 for drinking water, method 1633 for 40 PFAS compounds across various sample types, and additional methods (Munoz et al., 2023; US EPA, Environmental Protection Agency, 2019). The United States Geological Survey (USGS) has published a practical guide to PFAS sampling for environmental assessments, and individual state agencies (e.g., Michigan, Virginia) are preparing operating instructions to supplement federal recommendations (US EPA, Environmental Protection Agency, 2019; Pulster et al., 2024; Tokranov et al., 2021). In practice, laboratories frequently do not fully comply with official standards for sampling PFAS, and other CECs. Guidelines are often general and do not encompass all relevant matrices, such as wastewater, water with high suspended solids content, or samples containing analytes at ultra-trace concentrations. Compliance is further restricted by equipment limitations, especially the availability and cost of certified contaminant-free materials. In addition, standards often address only specific matrices or individual steps within the analytical chain, limiting compatibility with other stages of sampling and handling. Standards are also updated more slowly than procedures described in scientific publications. Field conditions, including limited access to immediate cooling, the presence of fluoropolymer materials in water supply systems, adverse weather, or long transport distances between the sampling site and the laboratory, may further constrain implementation. For PFAS, plastic containers are preferred because of the risk of contamination from fluoropolymer materials commonly used in caps and seals associated with glass bottles. Containers made of HDPE or polypropylene (PP), and less commonly polyethylene (PE), are used, often with PFAS-free certification. Nevertheless, desorption of fluoropolymer degradation products into samples remains a recognized limitation. Metal buckets have also been used for the collection of large volumes of rainwater (Chen et al., 2019a). For PFAS grab sampling, PFAS-free containers and field equipment are used throughout the sampling process. Containers are typically rinsed with methanol and deionized water prior to environmental exposure. Sample volumes usually range from 0.5 to 2 L and are collected directly into HDPE or PP containers. Most samples, particularly drinking water and rainwater, are not filtered. When filtration is applied, PP (Elmoznino et al., 2018), glass (Joerss et al., 2019), or nylon filters (Campo et al., 2016) with a porosity of 0.45 μm are used. Samples are preserved at approximately 4 °C and protected from light, with storage durations ranging from several days to several weeks. Sampling for PFAS is characterized by extensive quality assurance/quality control (QA/QC) measures, including field, bottle, equipment, and procedural blanks. POCIS is also a common technique for PFAS, employing sorbents such as weak anion exchange (WAX or X-AW), HLB, and ENV+ (200–600 mg) (Gardiner et al., 2022; Medon et al., 2023; Gobelius et al., 2019). The sorbents, stabilized by PES membranes (0.1–0.45 μm pore size), are assembled into aluminum or steel housings. Field deployment is typically conducted using metal frames or cages, and samplers are transported in PFAS-free HDPE or PP containers, frequently wrapped in aluminum foil. Diffusion-based passive samplers have also been applied for PFAS monitoring. These include microporous polyethylene diffusion tubes filled with approximately 0.6 g of HLB sorbent (Gardiner et al., 2022) and HDPE diffusion chambers equipped with polycarbonate membranes (0.4 μm pore size) (Medon et al., 2023). Transport to the laboratory is performed in PFAS-free containers under cooled conditions. Exposure times typically range from 14 to 30 days (Medon et al., 2023; Gardiner et al., 2022; Cerveny et al., 2016; Gobelius et al., 2019). After exposure, samplers are rinsed with ultrapure water, transferred to PFAS-free containers, and transported under cooled or frozen conditions. Diffusion samplers for PFAS are similarly deployed for periods ranging from several weeks to over one month. Microporous polyethylene tubes filled with sorbent are sealed prior to deployment and transported after exposure at approximately 4 °C or −20 °C. Diffusion cell samplers are assembled in the laboratory using HDPE bodies and polycarbonate membranes (Medon et al., 2023), filled with approximately 65 mL of ultrapure water, and deployed in flowing environmental waters for 2–7 weeks. After retrieval, samplers are transported to the laboratory under cooled conditions. For polybrominated diphenyl ethers (PBDEs) and other polybrominated flame retardants (PBFRs), international guidance related to the Stockholm Convention (inventory and guidance for POP-PBDE) and US EPA methods specific to PBDEs (e.g., Method 1614A (US EPA, Environmental Protection Agency, 2019) describe recommended sampling vessels (glass), equipment cleaning procedures, filtration decisions (explicit declaration of filtered versus whole-water samples), and sampling approaches applicable to aquatic matrices. For hydrophobic CECs such as PBDEs/PBFRs, equipment used for direct water sampling is generally limited to precleaned glass bottles. These compounds readily adsorb to plastic materials (PP, PTFE, PE), and glass containers minimize analyte losses because of sorption on container walls. For PBDEs/PBFRs grab sampling, glass bottles may be washed with detergents and rinsed with deionized water or organic solvents prior to use (Elmoznino et al., 2018), although cleaning procedures are not consistently reported. Sampling is performed directly into glass bottles, sometimes with the aid of metal buckets. Some studies report filtration using 0.45 μm filters (Chen et al., 2012; Elmoznino et al., 2018; Yang et al., 2014) whereas others analyze whole-water samples without filtration. Preservation is primarily achieved by cooling samples to approximately 4°C in the dark, with analysis typically conducted within 24–48 h. Chemical preservation is generally not applied to PBDEs/PBFRs. For PBDE/PBFRs, the most commonly used passive samplers are low-density polyethylene (LDPE) strips (typically 10 × 40 cm, 50 μm thick), which are cleaned with organic solvents such as acetone or hexane prior to deployment (McDonough et al., 2016; Ruge et al., 2015). Passive sampling for hydrophobic CECs often includes the use of performance reference compounds (PRCs), such as deuterated polycyclic aromatic hydrocarbons (PAH-d), bromophenyls, and FBDEs, which are introduced into the samplers before exposure (Ruge et al., 2015; Zhao et al., 2018; Perron et al., 2013; McDonough et al., 2016). Polyoxymethylene (POM) strips have also been applied for PBDEs/PBFRs sampling, often deployed in parallel with LDPE samplers (Perron et al., 2013). POM strips (approximately 75 μm thick, 6 × 40 cm) are mounted in steel cages and suspended using nylon or PP lines. Reported exposure times range from 21 to 38 days, with assembly, deployment, and transport procedures comparable to those used for LDPE samplers. Ceramic passive samplers have been described for selected flame retardants, comprising a sorbent phase enclosed within a porous ceramic housing that provides mechanical protection and resistance to biofouling (Cristale et al., 2013). For passive sampling of hydrophobic CECs using LDPE strips, samplers are prepared, optionally cleaned, preloaded with PRCs, and suspended in the water column. Exposure durations typically range from 14 to 60 days, with 14–30 days being the most common. Polychlorinated biphenyls (PCBs) occur in aquatic environments at trace concentrations (fg–ng L^{-1}), which makes sampling and sample handling critical for reliable determination. In surface waters, especially in uncontaminated areas, atmospheric dry and wet deposition is a major PCBs source because of their volatility (Wolkersdorfer et al., 2025). PCBs concentrations in water are reduced by strong sorption to suspended particulate matter, dissolved organic matter (DOM), biofilm, microplastics, and biota. Therefore, sampling strategies must distinguish between freely dissolved, particle-bound, and whole-water concentrations. For particle-bound species, standard sampling procedures involve the separation of a particles from the liquid phase by centrifugation or filtration. Filter material selection is critical, as cellulose nitrate (or ester) filters can sorb up to 100% of PCBs depending on the congener (Wolkersdorfer, 2021), potentially biasing results; glass fiber filters are therefore preferred. Plastic sampling materials should be avoided because adsorption to polymer surfaces has been reported (Andersson et al., 2012). All sampling materials should be prechecked for contamination. Reference sampling protocols specify the use of glass bottles. EPA Method 608 recommends collecting 1 L water samples in amber glass bottles with Teflon-lined screw caps. Bottles must be thoroughly cleaned using detergents and organic solvents and dried prior to use. Samples should be kept at 4 °C during transport and storage, and a pH change to 5–9 is recommended if extraction is delayed beyond 72 h. This method applies whole-water analysis without filtration. EPA Method 1628 extends sampling guidance to all 209 PCB congeners. Samples are collected in precleaned amber glass bottles with fluoropolymer-lined caps and stored below 6 °C. For samples containing more than ~1% solids, centrifugation applies to separate the liquid phase from particulate matter. HELCOM sampling guidelines for persistent organic pollutants (POPs) in seawater emphasize strict contamination control, including the avoidance of plastic materials and the use of thoroughly cleaned glassware. Cleaning procedures include detergent washing, water rinsing, solvent rinsing, and optional heating above 250 °C. Field blanks should be collected during each campaign. Whole-water samples typically require no pretreatment, while filtration or centrifugation is applied when dissolved and particulate phases are analyzed separately. Recent studies on PCBs report the collection of unfiltered whole water samples in glass bottles, preserved by refrigeration above freezing temperatures (Iniaghe and Kpomah, 2022; Gorshkov et al., 2022). In one study, sodium azide was applied as a chemical (Gorshkov et al., 2022). Sampling of chlorinated paraffins generally follows procedures similar to those for PCBs. Glassware cleaning includes detergent, deionized water, and organic solvent washing, with thermal treatment at 250 °C recommended. According to ISO 12010:2019, short-chain chlorinated paraffins (SCCPs) from water samples are usually extracted without filtration unless suspended solids exceed approximately 200 mg L^{-1}, in which case filtration through glass fiber filters is recommended. For SCCPs, reported preservation is limited to storage at low temperatures, with no chemical preservation specified in ISO 12010:2019. Environmental monitoring studies of synthetic dyes typically apply grab water sampling using glass bottles. For synthetic dyes, whole-water sampling has been applied, with preservation by cooling or, in one case, acidification to pH 3.5 and transport on ice (Zocolo et al., 2015). Sometimes suspended particles are filtered and extracted separately. Novel psychoactive substances are primarily monitored in wastewater, where the sampling strategy influences reported concentrations. Whole-water sampling without filtration has been recommended to reduce underestimation caused by partitioning to suspended particles. Filtration or centrifugation is routinely applied for wastewater preservation. Studies of novel psychoactive substances show that composite (24 h) sampling can reduce measured concentrations because of compound instability, emphasizing the influence of sampling duration and preservation. Novel pesticides are mainly sampled in surface and groundwater using grab sampling. Samples may be filtered or analyzed as whole water, and preservation is typically achieved by cooling or freezing. Because of their lipophilicity, novel flame retardants are infrequently analyzed in water. When sampled in wastewater, whole-water analysis without filtration has been reported to prevent analyte losses associated with sorption to particulate matter. Glass bottles are generally used. Sampling data for alternative nonphthalate plasticizers in water is limited. In the available study, wastewater samples were collected in brown glass bottles. Studies reporting UV filters in water typically use grab sampling with glass bottles, with preservation by acidification, filtration, freezing, or combinations thereof.

3 Templates for sampling procedures

Study by Medon et al. (2023), which presents a novel application of an equilibrium-based passive sampler, structured as a diffusion cell equipped with a polycarbonate membrane. The authors developed a simple equilibrium sampler comprising a neutral aqueous receiving phase enclosed by a flow-limiting membrane, and evaluated its performance through a coordinated series of laboratory calibrations and field experiments. The sampler was deployed both in sediment porewaters and in surface water at a site heavily contaminated with PFAS. Before deployment, isotopically labelled PFAS were introduced as performance reference compounds, enabling calibration of diffusive mass transfer during exposure. During environmental deployment, the samplers were inserted directly into the matrix and left in place for periods ranging from several weeks to more than a month, allowing PFAS to diffuse across the membrane into the internal aqueous phase. Analysis of the PRC concentration profiles showed porewater exposures of approximately six to seven weeks were required to approach 90% of equilibrium, whereas shorter deployments of two to four weeks still permitted reliable estimation of equilibrium concentrations through PRC-based corrections. Parallel grab samples of water and porewater were analyzed using the same instrumental method. The concentrations obtained from the passive samplers closely matched those from direct sampling, generally within a factor of two, providing strong evidence of the quantitative robustness of the approach across the entire workflow. A second exemplary study is the work of Gardiner et al. (2022), which investigates the performance of a microporous polyethylene tube (MPT) filled with HLB sorbent for passive PFAS monitoring. The sampler comprises a relatively thick, approximately 2 mm microporous PE wall that functions as an intrinsic diffusion barrier, closely conforming to the sorbent contained within. This design is particularly well-suited to highly mobile and polar substances such as PFAS, for which controlling hydrodynamic effects is a persistent challenge. MPTs were deployed for 24–36 days in both treated wastewater and estuarine surface waters as part of an extensive field validation. Sampling rates were (Chen et al., 2018) calculated for nine PFAS, demonstrating a wide and well-defined kinetic uptake region. Importantly, the thickness and porosity of the PE wall attenuated the influence of external flow conditions by limiting mass transfer through the membrane rather than through the boundary layer, thereby minimizing artifacts arising from turbulence—an inherent limitation in membrane-based samplers such as POCIS. Concentrations estimated from MPTs were generally of the same order of magnitude as those obtained from grab samples, supporting the method's applicability for time-weighted average monitoring. Vrana et al. (2021) published in situ calibration results for 76 pharmaceuticals and their metabolites in the Svratka and Blanice rivers, considering varying temperatures and water flows. Water temperature and flow rate did not affect the RS values for any of the tested substances. Based on the results obtained, the presence of a significant number of compounds in the waters of both rivers was confirmed, including antibiotics and their metabolites (clarithromycin, clindamycin, clindamycin sulfoxide, sulfamethoxazole, sulfapyridine, trimethoprim), antiepileptics (carbamazepine and its metabolites), antihistamines (fexofenadine), blockers and their metabolites (atenolol, bisoprolol, metoprolol, metoprolol acid), cardiovascular drugs (irbesartan, telmisartan, valsartan), psychoactive drugs (oxazepam, tramadol, venlafaxine, and metabolite) and nonsteroidal antiinflammatory drugs (diclofenac). Chen et al. (2018) developed an innovative DGT method for detecting endocrine-disrupting chemicals. Three different resins were used as the binding phase in DGT devices, including hydrophilic-lipophilic balance (HLB), XAD18, and StrataXL-A (SXL-A). The results showed that DGT with HLB and XAD18 was more stable compared to SXL-A. However, commonly used binding gels can adsorb various organic contaminants, making them susceptible to interference from nontarget compounds such as dissolved organic matter. Cui et al. (2020) used a molecularly imprinted polymer (MIP)-based DGT to achieve selective adsorption of quinolones and sulfonamide antibiotics. Criquet et al. (2017), who studied pesticides and pharmaceuticals, including carbamazepine, in wastewater, suggested that POCIS have certain advantages over grab sampling because of the possibility of obtaining time-integrated results. Regarding the determination of psychoactive drugs and illicit drugs, a comparison between POCIS and 24 h automatic sampling has been published, which showed similar concentrations in the wastewater samples collected. Monitoring 181 priority pesticides across 32 Australian water reservoirs was done using a combined passive sampling and mass spectrometry methodology, as presented by Serasinghe et al. (2024). Two types of passive samplers were used for this purpose: POCIS and Chemcatcher, and liquid chromatography coupled with high-resolution mass spectrometry (LC-HRMS) in data-independent acquisition mode as a final determination technique. From 181 pesticides, 21 were found at 22 sites initially, with pentiopyrad, bistrifluron, fluazifop-*p*-butyl, and fluopicolide showing up most often. Alternatively, Tapie et al. (2011) detail the application of a pharm-POCIS sampler for tracking pharmaceutical, phenol, polycyclic aromatic hydrocarbon, hormone, and pesticide levels in water systems. Results show POCIS effectively extracts various compounds from surface water, enabling C_{TWA} determination.

4 Updated sampling methodologies: Fundamentals, instrumentation, procedure templates, and future trends

Across compound groups, updated sampling methodologies emphasize the importance of defining the sampled fraction (whole water, dissolved, or particle-bound) and minimizing contamination and analyte loss during sampling and handling. Filtration and centrifugation are applied selectively, with glass fiber filters and glassware preferred to reduce sorption artifacts. Cooling at 4 °C–6 °C is the most frequently applied preservation method, while freezing is used for longer storage. US EPA and ISO protocols provide structured procedure templates specifying sample volume, container material, cleaning procedures, storage temperature, and conditions for phase separation. HELCOM guidelines further emphasize contamination control, use of blanks, and verification of separation efficiency. Prior to 2015, sampling of hydrophobic CECs such as PBFRs in surface waters relied predominantly on grab sampling into glass bottles with limited QA/QC, minimal use of blanks, and preservation by cooling to 4 °C (Chen et al., 2012; Chokwe et al., 2015). This approach provided short-term concentration snapshots. Updated methodologies increasingly emphasize passive sampling techniques, which provide time-weighted average concentrations, such as LDPE strips for PBFRs (Ruge et al., 2015). Early PFAS monitoring studies similarly relied on grab sampling using PP or HDPE containers (Loos et al., 2013), with limited contamination control and sporadic use of blanks. Between 2015 and 2025, PFAS sampling methodologies underwent substantial modernization. Current approaches emphasize comprehensive PFAS-free protocols, including exclusive use of HDPE/PP materials, complete elimination of fluoropolymers, multistage container rinsing, extensive use of blanks (field, travel, equipment, fabrication), and detailed documentation of sampling cleanliness. An example is the “clean hands/dirty hands” field sampling approach described by Woodward et al. (2024), which enforces strict separation of clean and potentially contaminated handling tasks during sampling. Moreover, for PPCs, more studies now deploy multiple passive samplers side-by-side (e.g., POCIS and o-DGT-type formats) to show where agreement holds and where compound-family-dependent deviations persist under real river conditions. Stroski et al. (2018) developed an innovative configuration combining a polyacrylamide diffusion gel and SepraTM ZT sorbent for 31 pharmaceuticals and pesticides over 25 days and calibrated it for different pH values. Linear uptake ($r^2 > 0.9$) was observed at pH 5. The sampler performed consistently with the agarose/HLB o-DGT configuration used simultaneously. Compared to simple agarose o-DGT, polyacrylamide diffusion gel exhibits a more robust and stable character. Furthermore, Sepra ZT diffusion gels served as an effective binding resin, offering a commercially available sorbent. Furthermore, CNTs are a promising sorbent in passive techniques because of their unique sorption properties, regeneration and reuse potential, which is helpful from analytical and economic viewpoints. Using passive sampling devices based on carbon nanotubes (CNTs-PSDs) was proposed by Godlewska et al. as a novel approach for the passive extraction of pharmaceuticals and other micropollutants, redirecting attention from standard polymer sorbents to advanced nanocarbon materials (Godlewska and Paszkiewicz, 2023; Godlewska et al., 2022a, 2022b). In a calibration study on CNTs-PSDs, the effects of salinity, pH, and dissolved humic acids on sampling rates were systematically evaluated, demonstrating that environmental control over sampling depends on the analyte. The results obtained indicate that water pH, the presence of dissolved humic acids, or salinity do not affect the uptake rates of hormones, cytostatic drugs, phenol derivatives, nonsteroidal antiinflammatory drugs, and tricyclic antidepressants by CNT-PSDs. However, in the case of sulfonamide uptake by COOH-CNT-PSD, all the abovementioned environmental factors impacted the Rs values. In the case of β-blockers, only water salinity above 7 PSU affected the Rs values of these analytes (Godlewska et al., 2022b). Godlewska et al. also showed that CNTs sorbents can be chemically and thermally regenerated and reused in passive samplers (at least three cycles) while maintaining sorption efficiency, clearly linking CNT-PSD to the goals of green chemistry and sustainable development (Godlewska and Paszkiewicz, 2023). Meanwhile, MOFs are being assessed as sorbents within passive sampling devices, using their adjustable porosity and functional adsorption sites to enhance selectivity and capacity for polar and ionizable contaminants related to surface waters. Studies on MOFs assessed NH_2-MIL-125 (Ti) and UiO-66 (Ce) as sorbents for surface water pollutants, involving laboratory calibration and sampling frequency (20). In addition, field tests were performed in the Elbląg River and Vistula Lagoon, comparing MOF-PSD and CNT-PSD, using chemometric analysis to connect CEC properties to sorption and showing regeneration and reuse possibilities. These new sorbents enhance both sensitivity and chemical coverage. Despite this, calibration and comparability are still an issue. To counter this, we need harmonized protocols, field blanks, and interstudy benchmarks to enable reliable surface-water trend analyses. Current methods increasingly integrate passive sampling with LC-HRMS for broad screening of targeted, suspected, and unknown substances, moving beyond predefined lists to better identify and prioritize emerging river and lake contaminants. At the same time, critical reviews emphasize that credible trend interpretation in surface water depends on tighter reporting of deployment conditions (hydrodynamics, temperature, fouling control, etc.) and on transparent uncertainty handling rather than single-number C_{TWA} reporting.

5 Case studies and examples of sampling applications to contaminants

Comprehensive details regarding the application of spot, composite, and passive sampling methods for surface waters and wastewater are presented in Tables 3.1 and 3.2.

TABLE 3.1 Grab and composite sampling strategies for the assessment of contaminants in surface waters and wastewaters.

Compounds and matrix	Method of sampling/volume	Preparation of lab equipment	Filtration	Blank sample	Preservation and storage	Ref.
Polychlorinated biphenyls (PCBs)						
28 PCBs in river water	Grab sampling (200 mL), 1 m below surface	Precleaned brown glass bottles, rinsed by sampled water	Not specified	Procedural blanks	Stored at 4 °C	(Iniaghe and Kpomah, 2022)
12 PCBs in lake water	Grab sampling (1 L)	Glass bottles lid with an aluminum foil gasket	Not filtered (LLE extraction)	Not stated	0.5 mL of a 1 M aqueous solution of sodium azide, stored at 5 °C	(Gorshkov et al., 2022)
Chlorinated paraffins (CPs) and short-chain chlorinated paraffins (SCCPs)						
Three standard SCCPs solutions in surface water	Grab sampling	Amber glass	Microfiber filter (GF/C 0.7 μm)	Procedural blank	Stored at 4 °C, dark	(Wang et al., 2018)
24 homologue groups of SCCPs (C10–13, Cl5–10) in groundwater	Grab sampling	Amber glass bottles solvent-rinsed, heated at 450 °C, 6 h	Glass fiber filter GF/F, 0.7 μm	Procedural blank	Stored at 4 °C	(Wang et al., 2025)
Three technical mixtures of SCCPs (C10–C13, with a 51.5%, 55.5%, and 63% chlorine content)	Grab sampling	Amber glass bottles, precleaned	Glass fiber filter 1.2 μm	Procedural blank	Stored at 4 °C, analyzed within 48 h	(Rubirola et al., 2018)

Continued

Table 3.1 Grab and composite sampling strategies for the assessment of contaminants in surface waters and wastewaters—cont'd

Compounds and matrix	Method of sampling/volume	Preparation of lab equipment	Filtration	Blank sample	Preservation and storage	Ref.
Synthetic dyes and pigments						
C.I. Disperse Red 1, C.I. Disperse Orange 1, C.I. Disperse Orange 3, C.I. Disperse Orange 25, C.I Disperse Yellow 3, C.I. Disperse Yellow 7 and C.I. Disperse Red 13 in effluents and receiving water	Grab sampling	Glass bottles, washed, rinsed with acetone	Glass fiber filters, filters washed with 2 mL of MeOH	Not specified	Transport in ice; pH adjusted to 3.5	(Zocolo et al., 2015)
Disperse blue 291, disperse blue 373, disperse red 1, disperse orange 30, disperse violet 93, disperse yellow 3 in river water	Grab sampling	Amber glass	No need, LLE extraction	Solvent blank	No conservation	(Vacchi et al., 2017)
Novel psychoactive substances						
54 substances in wastewaters	24 h composite sample	No data	Filtration before SPE (no data on filter type)	Procedural blank, instrumental blank	Acidification to pH 2 (2 M HCl); storage at −20 °C	(Nadarajan et al., 2026)
35 novel psychoactive substances, cocaine and cannabis in wastewater	24 h composite samples (influents) Grab sampling (effluents)	1 L polypropylene bottles	No filtration	No data	Frozen until analysis, thaw in a refrigerator per 1 day	(Davies et al., 2024)
Suspect screening of 3600 substances in the SWGDRUG library	Grab sampling	Polyethylene terephthalate bottles	Glass filter membrane, 0.5 μm	No data	Sodium azide (100 mg/L), stored in 4 °C	(Chen et al., 2023)

Neonicotinoids						
Clothianidin, imidacloprid, thiamethoxam in drinking water	Grab sampling	1 L amber glass jars (prebaked at 550 °C)	No data	Laboratory blank	Stored at 11 °C, 30 days	(Klarich et al., 2017)
Acetamiprid, clothianidin, dinotefuran, thiacloprid thiamethoxam in tap water	Grab sampling	Polypropylene bottles precleaned	No data	Procedural blanks, instrumental blanks	Stored at −20 °C	(He et al., 2021)
Acetamiprid, clothianidin, imidacloprid, thiamethoxam in river and groundwater	Grab sampling	1 L polycarbonate bottles	Glass fiber filter GF/F, ≤ 0.7 μm;	Method blanks, field blanks, equipment blanks	Stored at 4 °C	(Berens et al., 2021)
Acetamiprid, imidacloprid, thiamethoxam in river and groundwater	Grab sampling	2 L plastic bottles	No data	Blank matrix	Stored at −20 °C	(Toumi et al., 2025)
Acetamiprid, clothianidin, dinotefuran, imidacloprid, thiacloprid thiamethoxam, nitenpyram, sulfoxaflor in river water	Grab sampling	1 L high-density polyethylene bottles	Filter, 0.2 μm	Field blank	Stored at 4 °C, dark, 1 to 2 weeks	(Watt et al., 2026)
Diamides pesticides						
Chlorantraniliprole among other pesticides in river and groundwater	Grab sampling	2 L plastic bottles	No data	Blank matrix	Stored at −20 °C.	(Toumi et al., 2025)
Chlorantraniliprole, cyantraniliprole in river water	Grab sampling	1 L high-density polyethylene bottles	Filter, 0.2 μm	Field blank	Stored at 4 °C, dark, 1 to 2 weeks	(Watt et al., 2026)

Continued

Table 3.1 Grab and composite sampling strategies for the assessment of contaminants in surface waters and wastewaters—cont'd

Compounds and matrix	Method of sampling/volume	Preparation of lab equipment	Filtration	Blank sample	Preservation and storage	Ref.
Novel brominated frame retardant						
Decabromodiphenylethane (DBDPE), 1,2-bis(2,4,6-tribromophenoxy) ethane (BTBPE), pentabromoethylbenzene (PBEB), hexabromobenzene (HBB) in wastewater	Composite sample (refrigerated autosamplers)	Amber glass, precleaned	No need, LLE extraction	Equipment blanks, laboratory blanks,	No data	(Kim et al., 2014)
1,2,5,6,9,10-hexabromocy-clododecane (HBCD), PBEB, BTBPE, and DBDPE in wastewater	Composite sample	Brown bottles	Glass fiber filters, 0.45 μm	Method blank	No data	(Li et al., 2016)
Alternative plasticizers (nonphthalate plasticizers)						
27 nonphthalate plasticizers; including family edipates, citrates, phosphates, sebacates in river, well, rainwater, spring and swamp water), drinking water, wastewater treatment plant	Grab sampling	Amber glass bottles	No data	Solvent blank	Stored at 4 °C	(Capilla-Flores et al., 2025)
UV-filters						
Octocrylene (OCR), avobenzone (BMDBM, for butyl methoxydibenzoylmethane), octyl methoxycinnamate (OMC)in marine cost water	Grab sampling (three sample per day)	250 mL amber glass bottle	No filtration	No data	Stored at 3 °C, dark	(Thallinger et al., 2023)

Homosalate (HMS), 4-MBC, BP3, drometrizole trisiloxane (DTS), octocrylene (OC), butyl methoxydibenzoylmethane (BMDBM), isoamyl *p*-methoxicinnamate (IMC), and methylene bis-benzotriazolyltetramethyl butylphenol (MBP)in marine cost water; wastewater	Grab sampling	2 L amber glass bottles	Cellulose filters, 0.45 μm (wastewater) Membrane filters, 0.22 μm (seawater)	No data	Acidified to pH 2–3	(Cadena-Aizaga et al., 2022)
Pharmaceuticals						
Carbamazepine, diclofenac, ibuprofen, naproxen, sulfamethoxazole, trimethoprim in river water	Grab sampling, manual collection, 500 mL–1 L	Amber glass bottles, solvent-rinsed with methanol and ultrapure water	Membrane filters, 0.45 μm, Glass fiber filters GF/F, 0.7 μm	Field blanks, procedural blanks	Stored at −20 °C	(Verlicchi et al., 2012)
83 pharmaceuticals (analgesics, antibiotics, β-blockers, antidepressants, antiepileptics) in WWTP influent and effluent	24 h time-composite sampling using automatic refrigerated autosampler, 1–2 L	Sampling bottles washed with laboratory detergent, rinsed with tap water and ultrapure water	Glass fiber filter (0.7 μm)	Field blank, procedural blank	Refrigerated at 4 °C during collection, transported cold	(Paíga et al., 2019)
Emerging contaminants including pharmaceuticals (antibiotics, analgesics, psychiatric drugs, industrial CECs) in WWTP effluent and receiving rivers	24 h composite sampling, grab sampling, 1 L	Sampling bottles precleaned and rinsed with ultrapure water	Filtration (0.45 μm)	Field blank, procedural blank	Stored at 4 °C	(Beltrán de Heredia et al., 2024)

Continued

Table 3.1 Grab and composite sampling strategies for the assessment of contaminants in surface waters and wastewaters—cont'd

Compounds and matrix	Method of sampling/volume	Preparation of lab equipment	Filtration	Blank sample	Preservation and storage	Ref.
Pharmaceuticals and other CECs in reclaimed wastewater and reuse system waters	Composite sampling, 1–2 L	Glass or HDPE bottles precleaned with detergent and ultrapure water)	Membrane filter (0.45 μm)	Method blank, field blank	Stored at 4 °C	(Lin et al., 2020)
Antibiotics in hospital wastewater	Grab sampling; midstream collection, 1 L	Precleaned amber glass bottles, rinsed with methanol and ultrapure water	Membrane filter (0.45 μm)	Method blanks, procedural blanks, matrix spikes	Storage at −20 °C	(Wilk et al., 2025)
Antiepileptic drugs, analgesics in surface water	Grab sampling, 500 mL–1 L	Precleaned glass bottles, rinsed with methanol and ultrapure water	Filtration	Field blanks, procedural blanks	Refrigerated transport, storage at 4 °C or −20 °C	(Ferencik et al., 2022)
Carbamazepine, propranolol, sertraline and other pharmaceuticals in river water and WWTP influent	Grab sampling, 1 L	Glass bottles, solvent-rinsed; laboratory equipment precleaned	Filtration	Field blanks, analytical blanks	Storage at 4 °C	(Kasprzyk-Hordern et al., 2021)
CECs including pharmaceuticals and transformation products in WWTP effluent and river water	Grab sampling combined with passive sampling	Precleaned sampling bottles	Filtration	Field blanks (grab) and sampler blanks (passive)	Storage at 4 °C	(Tadić et al., 2022)
Pharmaceuticals in wastewater	24 h composite sampling (time-/ flow-proportional), autosampler	Refrigerated autosamplers, cleaned glass or HDPE bottles	Filtration (0.45 μm)	Autosampler blanks, procedural blanks	Refrigerated during collection	(Gago-Ferrero et al., 2020)

Antivirals, antibiotics, analgesics in lakes and WWTP–river–estuary systems	Grab sampling, 1 L	Precleaned amber glass bottles	Filtration	Field blanks, method blanks	Storage at 4 °C or −20 °C	(Chen et al., 2021)
68 pharmaceuticals in river, bay water	Grab sampling	Solvent-cleaned glass bottles	Filtration	Field blanks, laboratory blanks	Refrigerated transport, storage at 4 °C	(Castaño-Ortiz et al., 2024)
Per- and polyfluoroalkyl substances (PFASs)						
PFAS (EPA-monitored + emerging PFAS) in rainwater	Grab samplings; 500 mL, 1 L	HDPE tubs or HDPE funnel, carboy, HDPE bottles	None (samples loaded directly onto SPE)	Site blanks Ride-along blank	Stored 4 °C in HDPE	(Kim et al., 2022)
14 PFAAs in WWTF final effluent (aqueous + SPM fractions)	24 h composite samples Grab samplings	PP or HDPE carboys, washed with hot tap water, triple ultrapure water rinse rinsed with methanol	PP filters 0.45 μm (MeOH-rinsed)	Procedural blanks: reagent blanks, PP filter blanks, ENVI-carb blanks, nylon syringe filter blanks. Field blanks	Stored at 4 °C	(Elmoznino et al., 2018)
11 PFAA (PFCAs + PFSAs) + 6:2 FTS precursor; WWTP influent	24 h composite samples, flow-proportional autosampler	HDPE bottles prerinsed with methanol and Milli-Q water	Centrifugation only (20 min)	Laboratory blank, interbatch sample,	Acidified to pH 2 (HCl 2 M), stored at −20 °C	(Nguyen et al., 2019)

Continued

Table 3.1 Grab and composite sampling strategies for the assessment of contaminants in surface waters and wastewaters—cont'd

Compounds and matrix	Method of sampling/volume	Preparation of lab equipment	Filtration	Blank sample	Preservation and storage	Ref.
22 PFASs incl. C2–C12 PFCAs, C4/C6/C8 PFSAs, 6:2 FTSA, 6:2 Cl-PFAES, diPAPs; precipitation (rain)	Grab sampling	Stainless steel Buckets and PP bottles precleansed with Milli-Q water and methanol	Unfiltered	No described	Stored at −20 °C	(Chen et al., 2019b)
29 PFASs in surface water	Grab sampling, 1 L, 0.5 m depth	PP bottles rinsed with site water and methanol	GF/F glass microfiber filters (0.7 μm) prior heated to 450 °C	Field blanks, procedural blanks, instrumental methanol blanks	Stored at 4 °C, max. 1 month	(Joerss et al., 2019)
56 PFAS in raw water and produced drinking water	Grab sampling; 2 L	HDPE bottles precleaned with water and methanol	None (unfiltered); 1 L aliquot used for SPE	Procedural blank, instrumental blanks	Stored at 4 °C, max. 1 month	(Sadia et al., 2023)
23 PFAS in drainage water (run-off) from WMF subsections	Grab sampling, 500 mL Composite samples: pooled from 16 filtered subsamples, each 400 mL	PP containers methanol-washed before sampling; duplicate samples ultrasonicated; container walls rinsed with MeOH and transferred to beaker	GF/B glass fiber filter (1.0 μm, (composite subsamples before pooling)	Procedural blanks for target PFAS	Stored at 8 °C, dark	(Pettersson et al., 2024)

10 PFAS (PFBA, PFPeA, PFHxA, PFHpA, PFOA, PFNA, PFDA, PFBS, PFHxS, PFOS) in drinking water	Tap water: Grab samplings, Bottled water: 26 brands. 500 mL PP bottles used for both.	PP bottles precleaned with pure water and methanol; bottles rinsed repeatedly with sample before tap-water collection	Not filtered	Procedural blanks, travel blanks, instrumental blanks	Stored at 4 °C	(Ünlü Endirlik et al., 2019)
25 PFAS (EPA 533 list + PFDS, PFOSA, PFBA–PFDoDA, FTSA, HFPO-DA, etc.) in tap water	Grab sampling	PP bottles prerinsed with methanol and high-purity water, then rinsed with sample, PFAS-free materials used	Not filtered	Field blanks laboratory blanks, method blanks, spiked recovery samples	Stored at 4 °C	(Cimpean et al., 2025)
30 PFAS in surface water and tap water	Grab sampling, swing-arm sampler, 500 mL	All bottles, vials, tubing cleaned with methylene chloride, hexane, acetone, MeOH and water	Not filtered	LC–MS blanks, procedural blanks, field/trip blanks	Stored at 4 °C	(Li et al., 2022)
40 PFAS in surface water	Grab sampling (equal-width, depth-integrated, USGS protocol)	Clean-hands/dirty-hands procedure; equipment selection to avoid PFAS all equipment cleaned with additional PFAS-avoidance steps	Not filtered	None	Samples frozen immediately, shipped on ice (EPA 1633 & TOPA)	(Woodward et al., 2024)

TABLE 3.2 Passive sampling strategies for the assessment of contaminants in surface waters and wastewaters.

Compounds and matrix	Type/amount of receiving phase	Sampler exposure time	Blank sample	Preservation and storage	Ref.
12 PBDEs, 9 NHFRs in surface water	LDPE sheets (50 μm), precleaned and PRC-loaded (PAHs, PBBs)	4–30 weeks	Method blanks, field blanks	Returned on ice; stored at −20 °C	(McDonough et al., 2016)
16 PFAS in surface water	POCIS with Isolute ENV + + Ambersorb 1500 dispersed on S-X3 Bio Beads	21 days	Method blanks during extraction; no PFAS detected in blanks	Rinsed, transported on ice; stored at −18 °C	(Cerveny et al., 2016)
PFAS (C4–C9 acids, C4–C8 sulfonates) in surface water	Diffusion cell (65 mL Milli-Q); 0.40 μm PC membrane; PRC-spiked	46–47 days	Sampler materials tested PFAS-free; no field blanks reported	Transported on ice; stored at 4 °C	(Medon et al., 2023)
PFAS (C4–C14 PFCAs, C4–C10 PFSAs, precursors) in estuary and surface water	Microporous PE tube (7 cm) with 0.6 g HLB	2–29 days	Field blank, process blank, instrument blank	Stored at −15 °C	(Gardiner et al., 2022)
26 PFAS in surface water	POCIS-HLB and POCIS-WAX (200 mg)	14 days	Laboratory blank, field blank water blank	Vacuum-sealed, stored at −20 °C predeployment; 4 °C −8 °C postretrieval	(Gobelius et al., 2019)
34 PFAS in surface water	POCIS with Oasis HLB; PES membranes (41 cm^2)	26–38 days	Field blank	Sealed in metal cans; transported on ice	(Loken et al., 2025)
22 PAHs, 12 PBDEs in surface water	LDPE sheets (51 μm), PRC-loaded	4–5 weeks	Field blank, method blank	Wiped clean; storage temperature not specified	(Zhao et al., 2018)
PCBs in surface water	Silicone rubber (AlteSil, 0.5 mm; 10 × 6 cm)	4–6 weeks	Procedural blank and field blank	Wrapped in aluminum foil; stored at −20 °C	(Silva-Barni et al., 2019)
PCBs in surface water	LDPE sheets (25.4 μm; 0.5–3 g)	≥28 days	Field blank	Wrapped in aluminum foil, stored at −20 °C	(Burgess et al., 2023)

Neonicotinoids and TPs in surface water	POCIS-MMA (HLB:WAX:WCX = 2:1:1; 200 mg)	14 days	Procedural blank, matrix blank	Not reported	(Xiong et al., 2021)
Psychoactive substances in surface water	Homemade POCIS (HLB + MCX)	48 h	Matrix blank	Not reported	(Anies et al., 2025)
UV filters in seawater and wastewater	POCIS (HLB, 220 mg; 45.8 cm^2)	14 days	Not reported	Wrapped in aluminum foil, stored at −20 °C	(MacKeown et al., 2024)
UV filters in marine coast	POCIS (Attract-HLB)	18 days	Procedural blank	Stored at −20 °C	(Gandar et al., 2025)
49 pharmaceuticals in WWTP effluent	POCIS (Oasis HLB, 200 mg)	14–28 days	Field and laboratory blanks	Transported on ice; stored at −20 °C	(Domínguez-García et al., 2025)
Licit and illicit drugs in river and drinking water	Ceramic passive sampler	13 days	Field blank	Stored at 4 °C, dark	(Fontanals et al., 2023)
Pharmaceuticals in river and lake water	CNT-based passive sampler (~150 mg CNT)	7–21 days	Procedural, transport and field blanks	Stored at −20 °C	(Godlewska et al., 2022a)
Pharmaceuticals in wastewater and surface water	Reusable CNT passive sampler (~100 mg CNT)	10–20 days	Field blanks	Stored at −20 °C	(Godlewska and Paszkiewicz, 2023)
47 pharmaceuticals; surface water	POCIS (Oasis HLB, standard design)	~14–21 days	Field and lab blanks	Samplers retrieved, rinsed; sorbent extracted and stored cold until analysis	(Vrana et al., 2021)
63 pharmaceuticals + other residues; surface and wastewater	POCIS integrated with nontarget analysis	~days to weeks (deployment)	Field blanks and replicates	Samplers retrieved and processed; extracts analyzed	(Brunelle et al., 2022)

References

Alvarez, D.A., Petty, J.D., Huckins, J.N., Jones-Lepp, T.L., Getting, D.T., Goddard, J.P., Manahan, S.E., 2004. Development of a passive, in situ, integrative sampler for hydrophilic organic contaminants in aquatic environments. Environ. Toxicol. Chem. 23 (7), 1640–1648. doi:10.1897/03-603.

Andersson, M., Ottesen, R.T., Jartun, M., Eggen, O., Enqvist, A.C., 2012. PCB contamination from sampling equipment and packaging. Appl. Geochem. 08832927. 27 (1), 146–150. doi:10.1016/j.apgeochem.2011.09.026.

Anies, A.J., He, J., Lefever, S., Boogaerts, T., Covaci, A., Blust, R., Town, R.M., 2025. Sorption of psychoactive compounds on passive sampling sorbents in the presence of wastewater suspended particulate matter. Environ. Pollut. 02697491. 385, 127104. doi:10.1016/j.envpol.2025.127104.

Aristizabal-Ciro, C., Botero-Coy, A.M., López, F.J., Peñuela, G.A., 2017. Monitoring pharmaceuticals and personal care products in reservoir water used for drinking water supply. Environ. Sci. Pollut. Res. 16147499. 24 (8), 7335–7347. Available from: http://www.springerlink.com/content/0944-1344. doi:10.1007/s11356-016-8253-1.

Beltrán de Heredia, I., González-Gaya, B., Zuloaga, O., Garrido, I., Acosta, T., Etxebarria, N., Ruiz-Romera, E., 2024. Occurrence of emerging contaminants in three river basins impacted by wastewater treatment plant effluents: spatio-seasonal patterns and environmental risk assessment. Sci. Total Environ. 00489697. 946, 174062. doi:10.1016/j.scitotenv.2024.174062.

Berens, M.J., Capel, P.D., Arnold, W.A., 2021. Neonicotinoid insecticides in surface water, groundwater, and wastewater across land-use gradients and potential effects. Environ. Toxicol. Chem. 0730-7268. 40 (4), 1017–1033. doi:10.1002/etc.4959.

Bishop, N., Jones-Lepp, T., Margetts, M., Sykes, J., Alvarez, D., Keil, D.E., 2020. Wastewater-based epidemiology pilot study to examine drug use in the Western United States. Sci. Total Environ. 00489697. 745, 140697. doi:10.1016/j.scitotenv.2020.140697.

Brunelle, L.D., Huang, I.J., Angeles, L.F., Running, L.S., Sirotkin, H.I., McElroy, A.E., Aga, D.S., 2022. Comprehensive assessment of chemical residues in surface and wastewater using passive sampling, chemical, biological, and fish behavioral assays. Sci. Total Environ. 18791026. 828. Available from: www.elsevier.com/locate/scitotenv. doi:10.1016/j.scitotenv.2022.154176.

Burgess, R.M., Cantwell, M.G., Dong, Z., Grundy, J.S., Joyce, A.S., 2023. Comparing equilibrium concentrations of polychlorinated biphenyls based on passive sampling and bioaccumulation in water column deployments. Environ. Toxicol. Chem. 15528618. 42 (2), 317–332. Available from: http://onlinelibrary.wiley.com/journal/10.1002/(ISSN)1552-8618. doi:10.1002/etc.5536.

Cadena-Aizaga, M.I., Montesdeoca-Esponda, S., Sosa-Ferrera, Z., Santana-Rodríguez, J.J., 2022. Occurrence and environmental hazard of organic UV filters in seawater and wastewater from Gran Canaria Island (Canary Islands, Spain). Environ. Pollut. 18736424. 300. Available from: https://www.journals.elsevier.com/environmental-pollution. doi:10.1016/j.envpol.2022.118843.

Campo, J., Lorenzo, M., Pérez, F., Picó, Y., Farré, M., Barceló, D., 2016. Analysis of the presence of perfluoroalkyl substances in water, sediment and biota of the Jucar River (E Spain). Sources, partitioning and relationships with water physical characteristics. Environ. Res. 00139351. 147, 503–512. doi:10.1016/j.envres.2016.03.010.

Capilla-Flores, R., López-Ruíz, R., Romero-González, R., Garrido Frenich, A., 2025. Innovative extraction methods for non-phthalate plastic additives determination in water using GC and LC coupled to Q-Orbitrap. J. Hazard. Mater. 03043894. 489, 137623. doi:10.1016/j.jhazmat.2025.137623.

Carmona, E., Andreu, V., Picó, Y., 2014. Occurrence of acidic pharmaceuticals and personal care products in Turia River Basin: from waste to drinking water. Sci. Total Environ.

18791026. 484 (1), 53–63. Available from: www.elsevier.com/locate/scitotenv. doi:10.1016/j.scitotenv.2014.02.085.

Castaño-Ortiz, J.M., Gil-Solsona, R., Ospina-Álvarez, N., Alcaraz-Hernández, J.D., Farré, M., León, V.M., Barceló, D., Santos, L.H.M.L.M., Rodríguez-Mozaz, S., 2024. Fate of pharmaceuticals in the Ebro River Delta region: the combined evaluation of water, sediment, plastic litter, and biomonitoring. Sci. Total Environ. 00489697. 906, 167467. doi:10.1016/j.scitotenv.2023.167467.

Castiglioni, S., Bijlsma, L., Covaci, A., Emke, E., Hernández, F., Reid, M., Ort, C., Thomas, K.V., Van Nuijs, A.L.N., De Voogt, P., Zuccato, E., 2013. Evaluation of uncertainties associated with the determination of community drug use through the measurement of sewage drug biomarkers. Environ. Sci. Technol. 15205851. 47 (3), 1452–1460. doi:10.1021/es302722f.

Causanilles, A., Kinyua, J., Ruttkies, C., van Nuijs, A.L.N., Emke, E., Covaci, A., de Voogt, P., 2017. Qualitative screening for new psychoactive substances in wastewater collected during a city festival using liquid chromatography coupled to high-resolution mass spectrometry. Chemosphere. 18791298. 184, 1186–1193. Available from: www.elsevier.com/locate/chemosphere. doi:10.1016/j.chemosphere.2017.06.101.

Cerveny, D., Grabic, R., Fedorova, G., Grabicova, K., Turek, J., Kodes, V., Golovko, O., Zlabek, V., Randak, T., 2016. Perfluoroalkyl substances in aquatic environment-comparison of fish and passive sampling approaches. Environ. Res. 10960953. 144, 92–98. Available from: http://www.elsevier.com/inca/publications/store/6/2/2/8/2/1/index.htt. doi:10.1016/j.envres.2015.11.010.

Chen, H., Zhang, L., Li, M., Yao, Y., Zhao, Z., Munoz, G., Sun, H., 2019a. Per- and polyfluoroalkyl substances (PFASs) in precipitation from mainland China: contributions of unknown precursors and short-chain (C2–C3) perfluoroalkyl carboxylic acids. Water Res. 18792448. 153, 169–177. Available from: www.elsevier.com/locate/watres. doi:10.1016/j.watres.2019.01.019.

Chen, H., Zhang, L., Li, M., Yao, Y., Zhao, Z., Munoz, G., Sun, H., 2019b. Per- and polyfluoroalkyl substances (PFASs) in precipitation from mainland China: contributions of unknown precursors and short-chain (C2–C3) perfluoroalkyl carboxylic acids. Water Res. 18792448. 153, 169–177. Available from: www.elsevier.com/locate/watres. doi:10.1016/j.watres.2019.01.019.

Chen, W., Pan, S., Cheng, H., Sweetman, A.J., Zhang, H., Jones, K.C., 2018. Diffusive gradients in thin-films (DGT) for in situ sampling of selected endocrine disrupting chemicals (EDCs) in waters. Water Res. 18792448. 137, 211–219. Available from: www.elsevier.com/locate/watres. doi:10.1016/j.watres.2018.03.029.

Chen, X.F., Cheng, C.G., Wang, X., Zhao, R.S., 2012. Sensitive determination of polybrominated diphenyl ethers in environmental water samples with etched stainless steel wire based on solid-phase microextraction prior to gas chromatography-mass spectrometry. Anal. Methods. 17599679. 4 (9), 2908–2913. Available from: http://www.rsc.org/Publishing/Journals/AY/About.asp. doi:10.1039/c2ay25483b.

Chen, X., Lei, L., Liu, S., Han, J., Li, R., Men, J., Li, L., Wei, L., Sheng, Y., Yang, L., Zhou, B., Zhu, L., 2021. Occurrence and risk assessment of pharmaceuticals and personal care products (PPCPs) against COVID-19 in lakes and WWTP-river-estuary system in Wuhan, China. Sci. Total Environ. 18791026. 792. Available from: www.elsevier.com/locate/scitotenv. doi:10.1016/j.scitotenv.2021.148352.

Chen, Y.C., Hsu, J.Y., Chang, C.W., Chen, P.Y., Lin, Y.C., Hsu, I.L., Chu, C.J., Lin, Y.P., Liao, P.C., 2023. Investigation of new psychoactive substances (NPS), other illicit drugs, and drug-related compounds in a taiwanese wastewater sample using high-resolution mass-spectrometry-based targeted and suspect screening. Molecules. 14203049. 28 (13). Available from: http://www.mdpi.com/journal/molecules. doi:10.3390/molecules28135040.

Chokwe, T.B., Okonkwo, J.O., Sibali, L.L., Ncube, E.J., 2015. Alkylphenol ethoxylates and brominated flame retardants in water, fish (carp) and sediment samples from the Vaal River, South

Africa. Environ. Sci. Pollut. Res. 16147499. 22 (15), 11922–11929. Available from: http://www.springerlink.com/content/0944-1344. doi:10.1007/s11356-015-4430-x.

Cimpean, I.A., Paun, I., Pirvu, F., Iancu, V.I., Chiriac, F.L., 2025. Unregulated and regulated PFASs in bottled and tap water: occurrence, co-occurrence patterns, and implications for human health and regulatory frameworks. J. Xenobiot. 20394713. 15 (3). Available from: https://www.mdpi.com/2039-4713/11. doi:10.3390/jox15030081.

Criquet, J., Dumoulin, D., Howsam, M., Mondamert, L., Goossens, J.F., Prygiel, J., Billon, G., 2017. Comparison of POCIS passive samplers vs. composite water sampling: a case study. Sci. Total Environ. 18791026. 609, 982–991. Available from: www.elsevier.com/locate/scitotenv. doi:10.1016/j.scitotenv.2017.07.227.

Cristale, J., Katsoyiannis, A., Chen, C., Jones, K.C., Lacorte, S., 2013. Assessment of flame retardants in river water using a ceramic dosimeter passive sampler. Environ. Pollut. 18736424. 172, 163–169. doi:10.1016/j.envpol.2012.08.014.

Croft, T.L., Huffines, R.A., Pathak, M., Subedi, B., 2020. Prevalence of illicit and prescribed neuropsychiatric drugs in three communities in Kentucky using wastewater-based epidemiology and Monte Carlo simulation for the estimation of associated uncertainties. J. Hazard. Mater. 03043894. 384, 121306. doi:10.1016/j.jhazmat.2019.121306.

Cui, Y., Tan, F., Wang, Y., Ren, S., Chen, J., 2020. Diffusive gradients in thin films using molecularly imprinted polymer binding gels for in situ measurements of antibiotics in urban wastewaters. Front. Environ. Sci. Eng. 2095221X. 14 (6). Available from: http://www.springerlink.com/content/2095-2201/. doi:10.1007/s11783-020-1290-5.

Davies, B., Paul, R., Osselton, D., Woolley, T., 2024. Analysis of crude wastewater from two treatment plants in South Wales for 35 new psychoactive substances and cocaine, and cannabis. Sci. Rep. 2045-2322. 14 (1). doi:10.1038/s41598-024-70378-7.

Diamanti, K., Aalizadeh, R., Alygizakis, N., Galani, A., Mardal, M., Thomaidis, N.S., 2019. Wide-scope target and suspect screening methodologies to investigate the occurrence of new psychoactive substances in influent wastewater from Athens. Sci. Total Environ. 18791026. 685, 1058–1065. Available from: www.elsevier.com/locate/scitotenv. doi:10.1016/j.scitotenv.2019.06.173.

Domínguez-García, P., Almirall, X.O., Gómez-Canela, C., 2025. A POCIS-based approach for the monitoring of pharmaceuticals in wastewater treatment plants: calibration and deployment challenges. Environ. Pollut. 02697491. 367, 125641. doi:10.1016/j.envpol.2025.125641.

Elmoznino, J., Vlahos, P., Whitney, M., 2018. Occurrence and partitioning behavior of perfluoroalkyl acids in wastewater effluent discharging into the Long Island Sound. Environ. Pollut. 18736424. 243, 453–461. Available from: www.elsevier.com/inca/publications/store/4/0/5/8/5/6. doi:10.1016/j.envpol.2018.07.076.

Fedorova, G., Golovko, O., Randak, T., Grabic, R., 2014. Storage effect on the analysis of pharmaceuticals and personal care products in wastewater. Chemosphere. 00456535. 111, 55–60. doi:10.1016/j.chemosphere.2014.02.067.

Ferencik, M., Blahova, J., Schovankova, J., Siroka, Z., Svobodova, Z., Kodes, V., Stepankova, K., Lakdawala, P., 2022. Residues of selected anticonvulsive drugs in surface waters of the Elbe River Basin (Czech Republic). Water. 2073-4441. 14 (24), 4122. doi:10.3390/w14244122.

Fontanals, N., Boleda, M.R., Borrull, F., Marcé, R.M., Lacorte, S., 2023. Ceramic passive samplers for determining pharmaceuticals and drugs of abuse in river and drinking water. Sci. Total Environ. 18791026. 889. Available from: www.elsevier.com/locate/scitotenv. doi:10.1016/j.scitotenv.2023.164267.

Gago-Ferrero, P., Bletsou, A.A., Damalas, D.E., Aalizadeh, R., Alygizakis, N.A., Singer, H.P., Hollender, J., Thomaidis, N.S., 2020. Wide-scope target screening of >2000 emerging contaminants in wastewater samples with UPLC-Q-ToF-HRMS/MS and smart evaluation of its performance through the validation of 195 selected representative analytes. J. Hazard. Mater. 18733336. 387. Available from: www.elsevier.com/locate/jhazmat. doi:10.1016/j.jhazmat.2019.121712.

Gandar, A., Giraudo, M., Perion, T., Houël, E., Noguer, T., Rodrigues, A.M.S., Calas-Blanchard, C., Stien, D., 2025. Targeted and untargeted discovery of UV filters and emerging contaminants with environmental risk assessment on the Northwestern Mediterranean coast. Mar. Pollut. Bull. 18793363. 212. Available from: https://www.sciencedirect.com/science/journal/0025326X. doi:10.1016/j.marpolbul.2025.117567.

Gardiner, C., Robuck, A., Becanova, J., Cantwell, M., Kaserzon, S., Katz, D., Mueller, J., Lohmann, R., 2022. Field validation of a novel passive sampler for dissolved PFAS in surface waters. Environ. Toxicol. Chem. 15528618. 41 (10), 2375–2385. Available from: http://onlinelibrary.wiley.com/journal/10.1002/(ISSN)1552-8618. doi:10.1002/etc.5431.

Gobelius, L., Persson, C., Wiberg, K., Ahrens, L., 2019. Calibration and application of passive sampling for per- and polyfluoroalkyl substances in a drinking water treatment plant. J. Hazard. Mater. 18733336. 362, 230–237. Available from: www.elsevier.com/locate/jhazmat. doi:10.1016/j.jhazmat.2018.09.005.

Godlewska, K., Białk-Bielińska, A., Mazierski, P., Zdybel, S., Sosnowska, A., Górzyński, D., Puzyn, T., Zaleska-Medynska, A., Klimczuk, T., Paszkiewicz, M., 2024. Assessment of the application of selected metal-organic frameworks as advanced sorbents in passive extraction used in the monitoring of contaminants of emerging concern in surface waters. Sci. Total Environ. 00489697. 927, 172215. doi:10.1016/j.scitotenv.2024.172215.

Godlewska, K., Paszkiewicz, M., 2023. Reusable passive sampler with carbon nanotubes for monitoring contaminants in wastewater: application, regeneration and reuse. Chemosphere. 00456535. 332, 138855. doi:10.1016/j.chemosphere.2023.138855.

Godlewska, K., Stepnowski, P., Paszkiewicz, M., 2020. Application of the polar organic chemical integrative sampler for isolation of environmental micropollutants—a review. Crit. Rev. Anal. Chem. 15476510. 50 (1), 1–28. Available from: www.tandf.co.uk/journals/titles/10408347.asp. doi:10.1080/10408347.2019.1565983.

Godlewska, K., Stepnowski, P., Paszkiewicz, M., 2021. Pollutant analysis using passive samplers: principles, sorbents, calibration and applications. A review. Environ. Chem. Lett. 1610-3653. 19 (1), 465–520. doi:10.1007/s10311-020-01079-6.

Godlewska, K., Stepnowski, P., Paszkiewicz, M., 2022a. Carbon nanotube-passive samplers as novel tools for sampling and determining micropollutants in the aquatic environment. Sci. Total Environ. 00489697. 836, 155551. doi:10.1016/j.scitotenv.2022.155551.

Godlewska, K., Stepnowski, P., Paszkiewicz, M., 2022b. Carbon nanotubes, activated carbon and Oasis HLB as sorbents of passive samplers for extraction of selected micropollutants—comparison of sampling rates and extraction efficiency. Microchem. J. 0026265X. 172, 106975. doi:10.1016/j.microc.2021.106975.

Gorshkov, A.G., Kustova, O.V., Bukin, Y.S., 2022. Assessment of PCBs in surface waters at ultratrace levels: traditional approaches and biomonitoring (Lake Baikal, Russia). Appl. Sci. 2076-3417. 12 (4), 2145. doi:10.3390/app12042145.

Hashemi, B., Zohrabi, P., Raza, N., Kim, K.H., 2017. Metal-organic frameworks as advanced sorbents for the extraction and determination of pollutants from environmental, biological, and food media. TrAC Trends Anal. Chem. 18793142. 97, 65–82. Available from: www.elsevier.com/locate/trac. doi:10.1016/j.trac.2017.08.015.

Hawker, D.W., Clokey, J., Gorji, S.G., Verhagen, R., Kaserzon, S.L., 2022. Monitoring techniques-Grab and passive sampling. In: Dalu, T., Tavengwa, N.T. (Eds.), Emerging Freshwater Pollutants: Analysis, Fate and Regulations. Elsevier, Australia, pp. 25–48. Available from: https://www.sciencedirect.com/book/9780128228500. doi:10.1016/B978-0-12-822850-0.00014-4.

He, Y., Zhang, B., Wu, Y., Ouyang, J., Huang, M., Lu, S., Sun, H., Zhang, T., 2021. A pilot nationwide baseline survey on the concentrations of Neonicotinoid insecticides in tap water from China: implication for human exposure. Environ. Pollut. 02697491. 291, 118117. doi:10.1016/j.envpol.2021.118117.

Iniaghe, P.O., Kpomah, E.D., 2022. Polychlorinated biphenyls (PCBs) in water and sediments from the Udu River, Niger Delta, Nigeria: concentration, distribution and risk assessment. J. Environ. Expo. Assess. 27715949. 1 (3), 20. doi:10.20517/jeea.2022.19.

Jakubus, A., Godlewska, K., Gromelski, M., Jagiello, K., Puzyn, T., Stepnowski, P., Paszkiewicz, M., 2019. The possibility to use multi-walled carbon nanotubes as a sorbent for dispersive solid phase extraction of selected pharmaceuticals and their metabolites: effect of extraction condition. Microchem. J. 0026265X. 146, 1113–1125. doi:10.1016/j.microc.2019.02.051.

Joerss, H., Apel, C., Ebinghaus, R., 2019. Emerging per- and polyfluoroalkyl substances (PFASs) in surface water and sediment of the North and Baltic Seas. Sci. Total Environ. 18791026. 686, 360–369. Available from: www.elsevier.com/locate/scitotenv. doi:10.1016/j.scitotenv.2019.05.363.

Kasprzyk-Hordern, B., Proctor, K., Jagadeesan, K., Watkins, S., Standerwick, R., Barden, R., Barnett, J., 2021. Diagnosing down-the-drain disposal of unused pharmaceuticals at a river catchment level: unrecognized sources of environmental contamination that require nontechnological solutions. Environ. Sci. Technol. 15205851. 55 (17), 11657–11666. Available from: http://pubs.acs.org/journal/esthag. doi:10.1021/acs.est.1c01274.

Kim, M., Guerra, P., Alaee, M., Smyth, S.A., 2014. Occurrence and fate of four novel brominated flame retardants in wastewater treatment plants. Environ. Sci. Pollut. Res. 0944-1344. 21 (23), 13394–13404. doi:10.1007/s11356-014-3262-4.

Kim, Y., Pike, K.A., Gray, R., Sprankle, J.W., Faust, J.A., Edmiston, P.L., 2022. Non-targeted identification and semi-quantitation of emerging per- and polyfluoroalkyl substances (PFAS) in US rainwater. Environ. Sci. Process. Impacts. 20507895. 25 (11), 1771–1787. Available from: http://pubs.rsc.org/en/journals/journal/em. doi:10.1039/d2em00349j.

Klarich, K.L., Pflug, N.C., DeWald, E.M., Hladik, M.L., Kolpin, D.W., Cwiertny, D.M., LeFevre, G.H., 2017. Occurrence of neonicotinoid insecticides in finished drinking water and fate during drinking water treatment. Environ. Sci. Technol. Lett. 23288930. 4 (5), 168–173. Available from: http://pubs.acs.org/page/estlcu/about.html. doi:10.1021/acs.estlett.7b00081.

Kosma, C.I., Lambropoulou, D.A., Albanis, T.A., 2010. Occurrence and removal of PPCPs in municipal and hospital wastewaters in Greece. J. Hazard. Mater. 03043894. 179 (1–3), 804–817. doi:10.1016/j.jhazmat.2010.03.075.

Lei, H.J., Yang, B., Ye, P., Yang, Y.Y., Zhao, J.L., Liu, Y.S., Xie, L., Ying, G.G., 2021. Occurrence, fate and mass loading of benzodiazepines and their transformation products in eleven wastewater treatment plants in Guangdong province, China. Sci. Total Environ. 18791026. 755. Available from: www.elsevier.com/locate/scitotenv. doi:10.1016/j.scitotenv.2020.142648.

Li, B., Sun, S.J., Huo, C.Y., Li, W.L., Zhu, N.Z., Qi, H., Kong, L.J., Li, Y.F., Ma, W.L., 2016. Occurrence and fate of PBDEs and novel brominated flame retardants in a wastewater treatment plant in Harbin, China. Environ. Sci. Pollut. Res. 16147499. 23 (19), 19246–19256. Available from: http://www.springerlink.com/content/0944-1344. doi:10.1007/s11356-016-7142-y.

Li, X., Fatowe, M., Lemos, L., Quinete, N., 2022. Spatial distribution of per- and polyfluoroalkyl substances (PFAS) in waters from Central and South Florida. Environ. Sci. Pollut. Res. 16147499. 29 (56), 84383–84395. Available from: https://www.springer.com/journal/11356. doi:10.1007/s11356-022-21589-w.

Lin, X., Xu, J., Keller, A.A., He, L., Gu, Y., Zheng, W., Sun, D., Lu, Z., Huang, J., Huang, X., Li, G., 2020. Occurrence and risk assessment of emerging contaminants in a water reclamation and ecological reuse project. Sci. Total Environ. 00489697. 744, 140977. doi:10.1016/j.scitotenv.2020.140977.

Loken, L.C., Corsi, S.R., Alvarez, D.A., Pronschinske, M.A., Lenaker, P.L., Nott, M., Zhang, C., Mani, E., Ankley, G.T., 2025. Per- and polyfluoroalkyl substances in surface water and sediment in Great Lakes tributaries and relations with watershed attributes. Environ. Toxicol. Chem. 15528618. 44 (6), 1503–1524. Available from: https://global.oup.com/academic/product/environmental-toxicology-and-chemistry-9780195117134?lang=en&cc=us#. doi:10.1093/etojnl/vgaf077.

Loos, R., Carvalho, R., António, D.C., Comero, S., Locoro, G., Tavazzi, S., Paracchini, B., Ghiani, M., Lettieri, T., Blaha, L., Jarosova, B., Voorspoels, S., Servaes, K., Haglund, P., Fick, J., Lindberg, R.H., Schwesig, D., Gawlik, B.M., 2013. EU-wide monitoring survey on emerging polar organic contaminants in wastewater treatment plant effluents. Water Res. 18792448. 47 (17), 6475–6487. Available from: www.elsevier.com/locate/watres. doi:10.1016/j.watres.2013.08.024.

MacKeown, H., Scapuzzi, C., Baglietto, M., Benedetti, B., Di Carro, M., Magi, E., 2024. Wastewater and seawater monitoring in Antarctica: passive sampling as a powerful strategy to evaluate emerging pollution. Sci. Total Environ. 00489697. 925, 171755. doi:10.1016/j.scitotenv.2024.171755.

McDonough, C.A., Puggioni, G., Helm, P.A., Muir, D., Lohmann, R., 2016. Spatial distribution and air-water exchange of organic flame retardants in the lower great lakes. Environ. Sci. Technol. 15205851. 50 (17), 9133–9141. Available from: http://pubs.acs.org/journal/esthag. doi:10.1021/acs.est.6b02496.

Medon, B., Pautler, B.G., Sweett, A., Roberts, J., Risacher, F.F., D'Agostino, L.A., Conder, J., Gauthier, J.R., Mabury, S.A., Patterson, A., McIsaac, P., Mitzel, R., Hakimabadi, S.G., Pham, A.L.T., 2023. A field-validated equilibrium passive sampler for the monitoring of per- and polyfluoroalkyl substances (PFAS) in sediment pore water and surface water. Environ. Sci. Process. Impacts. 20507895. 25 (5), 980–995. Available from: http://pubs.rsc.org/en/journals/journal/em. doi:10.1039/d2em00483f.

Menger, F., Gago-Ferrero, P., Wiberg, K., Ahrens, L., 2020. Wide-scope screening of polar contaminants of concern in water: a critical review of liquid chromatography-high resolution mass spectrometry-based strategies. Trends Environ. Anal. Chem. 22141588. 28, e00102. doi:10.1016/j.teac.2020.e00102.

Munoz, G., Liu, M., Vo Duy, S., Liu, J., Sauvé, S., 2023. Target and nontarget screening of PFAS in drinking water for a large-scale survey of urban and rural communities in Québec, Canada. Water Res. 00431354. 233, 119750. doi:10.1016/j.watres.2023.119750.

Nadarajan, D., Laimou-Geraniou, M., Chappell, A., Baduel, C., Bijlsma, L., Boogaerts, T., Burgard, D.A., Castiglioni, S., Ceolotto, N., Driver, E.M., Sodre, F.F., Kassinos, D.F., Quintana, H.F., Gerber, C., Gracia-Lor, E., Gracia-Marín, E., Halden, R.U., Heath, E., Huchthausen, J., Kasprzyk-Hordern, B., Keller, E.L., Lai, F.Y., Löve, A.S.C., Matias, J., Ocenaskova, V., Oh, J.E., Oluseyi, T., Phung, K., Pineda-Castro, M., Psichoudaki, M., Salgueiro-Gonzalez, N., Silvino-Gomes, C., Nascimento, F.M.S., Subedi, B., Thomaidis, N., Van Wichelen, N., Wang, D., Yargeau, V., Cresswell, S., Mueller, J.F., Bade, R., 2026. Wastewater analysis as a global toxicovigilance tool for the monitoring of new psychoactive substances. Water Res. 18792448. 289. Available from: https://www.sciencedirect.com/science/journal/00431354. doi:10.1016/j.watres.2025.124847.

Ng, K.T., Rapp-Wright, H., Egli, M., Hartmann, A., Steele, J.C., Sosa-Hernández, J.E., Melchor-Martínez, E.M., Jacobs, M., White, B., Regan, F., Parra-Saldivar, R., Couchman, L., Halden, R.U., Barron, L.P., 2020. High-throughput multi-residue quantification of contaminants of emerging concern in wastewaters enabled using direct injection liquid chromatography-tandem mass spectrometry. J. Hazard. Mater. 18733336. 398. Available from: www.elsevier.com/locate/jhazmat. doi:10.1016/j.jhazmat.2020.122933.

Nguyen, H.T., Kaserzon, S.L., Thai, P.K., Vijayasarathy, S., Bräunig, J., Crosbie, N.D., Bignert, A., Mueller, J.F., 2019. Temporal trends of per- and polyfluoroalkyl substances (PFAS) in the influent of two of the largest wastewater treatment plants in Australia. Emerg. Contam. 24056642. 5, 211–218. Available from: www.keaipublishing.com/en/journals/emerging-contaminants/. doi:10.1016/j.emcon.2019.05.006.

Ofrydopoulou, A., Nannou, C., Evgenidou, E., Christodoulou, A., Lambropoulou, D., 2022. Assessment of a wide array of organic micropollutants of emerging concern in wastewater treatment plants in Greece: occurrence, removals, mass loading and potential risks. Sci. Total Environ. 00489697. 802, 149860. doi:10.1016/j.scitotenv.2021.149860.

Oosterhuis, M., Sacher, F., ter Laak, T.L., 2013. Prediction of concentration levels of metformin and other high consumption pharmaceuticals in wastewater and regional surface water based on sales data. Sci. Total Environ. 18791026. 442, 380–388. doi:10.1016/j.scitotenv.2012.10.046.

Ort, C., Lawrence, M.G., Rieckermann, J., Joss, A., 2010. Sampling for pharmaceuticals and personal care products (PPCPs) and illicit drugs in wastewater systems: are your conclusions valid? A critical review. Environ. Sci. Technol. 15205851. 44 (16), 6024–6035. doi:10.1021/es100779n.

Paíga, P., Correia, M., Fernandes, M.J., Silva, A., Carvalho, M., Vieira, J., Jorge, S., Silva, J.G., Freire, C., Delerue-Matos, C., 2019. Assessment of 83 pharmaceuticals in WWTP influent and effluent samples by UHPLC-MS/MS: hourly variation. Sci. Total Environ. 18791026. 648, 582–600. Available from: www.elsevier.com/locate/scitotenv. doi:10.1016/j.scitotenv.2018.08.129.

Perron, M.M., Burgess, R.M., Suuberg, E.M., Cantwell, M.G., Pennell, K.G., 2013. Performance of passive samplers for monitoring estuarine water column concentrations: 2. Emerging contaminants. Environ. Toxicol. Chem. 15528618. 32 (10), 2190–2196. doi:10.1002/etc.2248.

Pettersson, M., Jogsten, I.E., van Hees, P., Karlsson, P., Axelsson, M., Yeung, L.W.Y., 2024. Sampling of per- and polyfluoroalkyl substances in drainage water from a waste management facility. Chemosphere. 18791298. 364. Available from: https://www.sciencedirect.com/science/journal/00456535. doi:10.1016/j.chemosphere.2024.143031.

Pulster, E.L., Bowman, S.R., Keele, L., Steevens, J., 2024. Guide to per-and polyfluoroalkyl substances (PFAS) sampling within natural resource damage assessment and restoration. In: US Geological Survey Open-File Report 2024–1001, pp. 1–57. US Geological Survey, Reston, VA. 23311258. Available from: https://pubs.usgs.gov/of/2024/1001/ofr20241001.pdf. doi:10.3133/ofr20241001.

Richardson, S.D., Kimura, S.Y., 2017. Emerging environmental contaminants: challenges facing our next generation and potential engineering solutions. Environ. Technol. Innov. 23521864. 8, 40–56. Available from: http://www.journals.elsevier.com/environmental-technology-and-innovation/. doi:10.1016/j.eti.2017.04.002.

Rubirola, A., Santos, F.J., Boleda, M.R., Galceran, M.T., 2018. Routine method for the analysis of short-chain chlorinated paraffins in surface water and wastewater. Clean—Soil Air Water. 18630669. 46 (2). Available from: http://onlinelibrary.wiley.com/journal/10.1002/%28ISSN%291863-0669. doi:10.1002/clen.201600151.

Ruge, Z., Muir, D., Helm, P., Lohmann, R., 2015. Concentrations, trends, and air–water exchange of PAHs and PBDEs derived from passive samplers in lake superior in 2011. Environ. Sci. Technol. 0013-936X. 49 (23), 13777–13786. doi:10.1021/acs.est.5b02611.

Sadia, M., Nollen, I., Helmus, R., Ter Laak, T.L., Béen, F., Praetorius, A., Van Wezel, A.P., 2023. Occurrence, fate, and related health risks of PFAS in raw and produced drinking water. Environ. Sci. Technol. 15205851. 57 (8), 3062–3074. Available from: http://pubs.acs.org/journal/esthag. doi:10.1021/acs.est.2c06015.

Serasinghe, P., Nguyen, H.T.K., Hepburn, C., Nugegoda, D., Pettigrove, V., 2024. Use of passive sampling and high-resolution mass spectrometry for screening emerging pesticides of concern within surface waters. J. Hazard. Mater. Adv. 27724166. 13, 100408. doi:10.1016/j.hazadv.2024.100408.

Silva-Barni, M.F., Smedes, F., Fillmann, G., Miglioranza, K.S.B., 2019. Passive sampling of pesticides and polychlorinated biphenyls along the Quequén Grande River watershed, Argentina. Environ. Toxicol. Chem. 15528618. 38 (2), 340–349. Available from: http://onlinelibrary.wiley.com/journal/10.1002/(ISSN)1552-8618. doi:10.1002/etc.4325.

Stroski, K.M., Challis, J.K., Wong, C.S., 2018. The influence of pH on sampler uptake for an improved configuration of the organic-diffusive gradients in thin films passive sampler. Anal. Chim. Acta. 18734324. 1018, 45–53. Available from: http://www.journals.elsevier.com/analytica-chimica-acta/. doi:10.1016/j.aca.2018.02.074.

Tadić, Đ., Manasfi, R., Bertrand, M., Sauvêtre, A., Chiron, S., 2022. Use of passive and grab sampling and high-resolution mass spectrometry for non-targeted analysis of emerging contaminants and their semi-quantification in water. Molecules. 1420-3049. 27 (10), 3167. doi:10.3390/molecules27103167.

Tapie, N., Devier, M.H., Soulier, C., Creusot, N., Le Menach, K., Aït-Aïssa, S., Vrana, B., Budzinski, H., 2011. Passive samplers for chemical substance monitoring and associated toxicity assessment in water. Water Sci. Technol. 0273-1223. 63 (10), 2418–2426. doi:10.2166/wst.2011.129.

Telgmann, L., Horn, H., 2024. The behavior of pharmaceutically active compounds and contrast agents during wastewater treatment—combining sampling strategies and analytical techniques: a critical review. Sci. Total Environ. 00489697. 946, 174344. doi:10.1016/j.scitotenv.2024.174344.

Thallinger, D., Labille, J., Milinkovitch, T., Boudenne, J., Loosli, F., Slomberg, D., Angeletti, B., Lefrançois, C., 2023. UV filter occurrence in beach water of the Mediterranean coast—a field survey over 2 years in Palavas-les-Flots, France. Int. J. Cosmet. Sci. 0142-5463. 45 (S1), 67–83. doi:10.1111/ics.12904.

Thomas, K.V., Amador, A., Baz-Lomba, J.A., Reid, M., 2017. Use of mobile device data to better estimate dynamic population size for wastewater-based epidemiology. Environ. Sci. Technol. 15205851. 51 (19), 11363–11370. Available from: http://pubs.acs.org/journal/esthag. doi:10.1021/acs.est.7b02538.

Tokranov, A.K., Bradley, P.M., Focazio, M.J., Kent, D.B., LeBlanc, D.R., McCoy, J.W., Smalling, K.L., Steevens, J.A., Toccalino, P.L., 2021. Integrated science for the study of perfluoroalkyl and polyfluoroalkyl substances (PFAS) in the environment a strategic science vision for the U.S. geological survey. In: Circular. US Geological Survey, pp. 1–60. 23305703. Available from: https://pubs.er.usgs.gov/publication/cir1490. doi:10.3133/cir1490.

Toumi, K., Arbi, A., Soudani, N., Lomadze, A., Haouas, D., Bertuzzi, T., Cardinali, A., Lamastra, L., Capri, E., Suciu, N.A., 2025. Ecotoxicological risk assessment and monitoring of pesticide residues in soil, surface water, and groundwater in Northwestern Tunisia. Water (Switzerland) 20734441. 17 (16). Available from: http://www.mdpi.com/journal/water. doi:10.3390/w17162387.

Tran, N.H., Hu, J., Ong, S.L., 2013. Simultaneous determination of PPCPs, EDCs, and artificial sweeteners in environmental water samples using a single-step SPE coupled with HPLC-MS/MS and isotope dilution. Talanta. 00399140. 113, 82–92. Available from: https://www.journals.elsevier.com/talanta. doi:10.1016/j.talanta.2013.03.072.

Ünlü Endirlik, B., Bakır, E., Boşgelmez, İ.İ., Eken, A., Narin, İ., Gürbay, A., 2019. Assessment of perfluoroalkyl substances levels in tap and bottled water samples from Turkey. Chemosphere. 18791298. 235, 1162–1171. Available from: www.elsevier.com/locate/chemosphere. doi:10.1016/j.chemosphere.2019.06.228.

US EPA, Environmental Protection Agency, 2019. Water research on per- and polyfluoroalkyl substances (PFAS). Available from: https://www.epa.gov/water-research/water-research-and-polyfluoroalkyl-substances-pfas.

Vacchi, F.I., Vendemiatti, J.A.D.S., da Silva, B.F., Zanoni, M.V.B., Umbuzeiro, G.D.A., 2017. Quantifying the contribution of dyes to the mutagenicity of waters under the influence of textile activities. Sci. Total Environ. 18791026. 601-602, 230–236. Available from: www.elsevier.com/locate/scitotenv. doi:10.1016/j.scitotenv.2017.05.103.

Vanderford, B.J., Snyder, S.A., 2006. Analysis of pharmaceuticals in water by isotope dilution liquid chromatography/tandem mass spectrometry. Environ. Sci. Technol. 0013936X. 40 (23), 7312–7320. doi:10.1021/es0613198.

Verlicchi, P., Al Aukidy, M., Zambello, E., 2012. Occurrence of pharmaceutical compounds in urban wastewater: removal, mass load and environmental risk after a secondary treatment-A review. Sci. Total Environ. 18791026. 429, 123–155. doi:10.1016/j.scitotenv.2012.04.028.

Vrana, B., Urík, J., Fedorova, G., Švecová, H., Grabicová, K., Golovko, O., Randák, T., Grabic, R., 2021. In situ calibration of polar organic chemical integrative sampler (POCIS) for monitoring of pharmaceuticals in surface waters. Environ. Pollut. 02697491. 269, 116121. doi:10.1016/j.envpol.2020.116121.

Wang, J., Wang, X., Zhu, Z., Ying, S., Hayat, K., Bai, X., Liu, S., Xiao, X., Jin, C., Liu, W., 2025. Distribution, homologue pattern, sources, and environmental behavior of short-chain chlorinated paraffins in multi-media environments within the petrochemical industry of the Yangtze River Delta. Environ. Sci. Adv. 2754-7000. 4 (9), 1477–1487. doi:10.1039/d5va00052a.

Wang, Y.J., Wang, Y.W., Jiang, G.B., 2018. Solid-phase extraction for analysis of short-chain chlorinated paraffins in water samples. Chin. J. Anal. Chem. 18722040. 46 (7), 1102–1108. Available from: https://www.journals.elsevier.com/chinese-journal-of-analytical-chemistry. doi:10.1016/S1872-2040(18)61097-7.

Watt, C., Davies, J.M., Sheedy, C., Challis, J.K., 2026. Not-so-simple patterns of neonicotinoids and diamides in small Prairie streams : implications for assessing risk and understanding pesticide dynamics. Environ. Toxicol. Chem. 45 (1), 24–36. doi:10.1093/etojnl/vgaf253.

Wilk, J., Sowik, P., Felis, E., Harnisz, M., Korzeniewska, E., Bajkacz, S., 2025. Development SPE-LC-MS/MS method for determination of WHO AWaRe reserve antibiotics in hospital wastewater. Sci. Rep. 2045-2322. 15 (1). doi:10.1038/s41598-025-04951-z.

Wolkersdorfer, C., 2021. Reinigungsverfahren Für Grubenwasser. Springer, Heidelberg.

Wolkersdorfer, C., Mahlangu, O.T., Wiltschka, K., Böhm, L., 2025. Analysis, occurrence and removal of polychlorinated biphenyls (PCBs) in mine water—a review. Environ. Sci. Eur. 21904715. 37 (1). Available from: https://enveurope.springeropen.com. doi:10.1186/s12302-025-01197-4.

Woodward, E.E., Senior, L.A., Fleck, J.A., Barber, L.B., Hansen, A.M., Duris, J.W., 2024. Using a time-of-travel sampling approach to quantify per- and polyfluoroalkyl substances (PFAS) stream loading and source inputs in a mixed-source, urban catchment. ACS ES T Water. 26900637. 4 (10), 4356–4367. Available from: https://pubs.acs.org/page/aewcaa/about.html. doi:10.1021/acsestwater.4c00288.

Xiong, J., Tan, B., Ma, X., Li, H., You, J., 2021. Tracing neonicotinoid insecticides and their transformation products from paddy field to receiving waters using polar organic chemical integrative samplers. J. Hazard. Mater. 03043894. 413, 125421. doi:10.1016/j.jhazmat.2021.125421.

Yang, J., Li, J.Y., Qiao, J.Q., Cui, S.H., Lian, H.Z., Chen, H.Y., 2014. Magnetic solid phase extraction of brominated flame retardants and pentachlorophenol from environmental waters with carbon doped Fe3O4 nanoparticles. Appl. Surf. Sci. 01694332. 321, 126–135. Available from: http://www.journals.elsevier.com/applied-surface-science/. doi:10.1016/j.apsusc.2014.09.175.

Zhao, W., Cai, M., Adelman, D., Khairy, M., August, P., Lohmann, R., 2018. Land-use-based sources and trends of dissolved PBDEs and PAHs in an urbanized watershed using passive polyethylene samplers. Environ. Pollut. 18736424. 238, 573–580. Available from: www.elsevier.com/inca/publications/store/4/0/5/8/5/6. doi:10.1016/j.envpol.2018.02.057.

Zocolo, G.J., Pilon Dos Santos, G., Vendemiatti, J., Vacchi, F.I., Umbuzeiro, G.D.A., Zanoni, M.V.B., 2015. Using SPE-LC-ESI-MS/MS analysis to assess disperse dyes in environmental water samples. J. Chromatogr. Sci. 1945239X. 53 (8), 1257–1264. doi:10.1093/chromsci/bmu221.

Chapter 4

Extraction of environmental water samples

Somandla Ncube[a], Lawrence Mzukisi Madikizela[b], and Luke Chimuka[c]
[a]*Department of Chemistry, Durban University of Technology, Durban, South Africa*
[b]*College of Science, Engineering and Technology, Institute for Nanotechnology and Water Sustainability, University of South Africa, Florida Science Campus, Roodepoort, South Africa*
[c]*Department of Chemistry, University of the Witwatersrand, Johannesburg, South Africa*

1 Introduction

Contaminants of emerging concern (CECs) in the aqueous environment exist mostly in trace amounts that fall below the quantitation capabilities of most instruments including those based on chromatography hyphenated with mass spectrometry (Madikizela et al., 2022). In addition, the targeted compounds exist as a mixture with other complex compounds that can interfere with the instrument detectors if not eliminated. Compounds such as salts and humic acids can contaminate the analytical instruments if not removed from the sample. These matrix interferents can either enhance or suppress the instrument response signal leading to distorted results (Williams et al., 2023). Inaccurate quantitative results may have serious consequences especially in drafting policies and risk assessment based on predictive ecotoxicological models. Accurate measurements are also important for studies that use live organisms to observe behaviors after exposure to certain quantities over a certain period of time. Usually the outcome of predictive models and exposure studies are utilized in drafting environmental and ecotoxicological policies that define permissible levels of CECs in aqueous environments.

While there are international directives on pollution policies, usually each country has its own policies that are mostly derived from the pollution trends recorded in that country. Localized policies made based on inaccurate reporting of environmental concentration of contaminants can have a domino effect on public health and impact on the environment and aquatic organisms. Against this background, CECs require extraction from environmental water samples before determination. Various methods have been used depending on the physicochemical properties of the target contaminant including its polarity. This chapter looks at the conventional methods that are commonly used for extracting CECs

Contaminants of Emerging Concern in the Environment. https://doi.org/10.1016/B978-0-443-33821-2.00007-4

from environmental water samples focusing on pharmaceuticals and personal care products (PPCPs), pesticides, illicit drugs and psychoactive substances, and industrial chemicals, such as plasticizers, brominated and emerging flame retardants, or per- and polyfluoroalkyl substances (PFAS), among others. Solid-phase extraction (SPE) and liquid–liquid extraction (LLE) are identified as the two conventional extraction methods in aqueous environmental pollution studies. The chapter discusses their principles, applicability and how they remain the reference point in the development of advanced technologies.

2 Conventional extraction methods of CECs in surface water

The choice of extraction technique for CECs in environmental water samples depends largely on the polarity of the target residue as it is throughly discussed in Chapter 1. For example, PPCPs have a wide polarity spectrum ranging from highly polar to moderately nonpolar. The relatively nonpolar PPCPs are characterized by large multiring structures with very few polar groups such as ethinylestradiol, a synthetic hormone used as the birth control pill, and the majority of antipsychotics that require sufficient lipophilicity to cross the blood-brain barrier (Ben Fredj et al., 2015; Escudero et al., 2021). Those with smaller molecular masses and multiple polar functional groups such as penicillin and sulfonamide antibiotics as well as metformin (antidiabetic) and gabapentin (anticonvulsant) are highly polar and require more attention during extraction. Generally, most PPCPs residues have a large cyclic nonpolar backbone with one or more polar functional groups attached making them moderately polar. These amphipathic residues are the most studied pharmaceuticals in environmental samples especially antibiotics and the nonsteroidal antiinflammatory drugs (Lin et al., 2023; Yang et al., 2021).

The polarity of industrial chemicals, such as plasticizers, brominated and emerging flame retardants, or per- and PFAS, among others varies less compared to PPCPs which makes multiresidue extraction less challenging. Common plasticizers can be classified under phthalates, phosphates, and esters of carboxylic acids, with a diester group typically attached to a benzene ring and an aliphatic chain, making them relatively nonpolar (Godwin, 2024). Among flame retardants, the brominated flame retardants (BFRs), especially the polybrominated diphenyl ethers (PBDEs), are probably the most reported in the aqueous environment (Ohoro et al., 2021). The PBDEs have two phenol groups in their structure with 10 bromine atoms attached at various positions. The bromine effect on the overall polarity is minimum which makes BFRs relatively nonpolar. All PFAS residues have a hydrophilic head and a lipophilic tail which makes them amphiphilic (Lyu et al., 2022). Pesticides can be classified into herbicides, insecticides, fungicides, and other pest-specific ones such as rodenticides and acaricides. Organophosphates and organochlorines are widely used as insecticides but environmental concerns are on organochlorines because they are highly persistent in the environment and are more likely to bioaccumulate in living organisms (Jayaraj et al., 2016; Pirsaheb et al., 2017).

The diverse polarity of CECs in addition to their trace existence and the complex nature of environmental samples makes multiresidue extraction a bit challenging. In this regard, various extraction techniques have been reported but the conventional methods especially the SPE technique is a preferred choice for a wide range of organic contaminants (Daniels et al., 2020). With the field of analytical chemistry constantly evolving, researchers have developed numerous other innovative and specialized techniques to address specific challenges like specificity, miniaturization, automation, and in-field sampling (Badawy et al., 2022; Peris-Pastor et al., 2024). Although these advanced miniaturized techniques (see information in Chapter 5) offer benefits such as improved specificity and reduced sample and reagent consumption, conventional analytical methods remain widely used and continue to serve as the benchmark against which the newly developed miniaturized approaches are evaluated and validated. This chapter focuses on these conventional techniques providing a detailed review of their principles and applications in extraction of different classes of organic contaminants in the aqueous environment.

3 Solid-phase extraction

3.1 Principle

The SPE is an extraction technique where components of a liquid sample are retained based on their affinity to a solid sorbent (Andrade-Eiroa et al., 2016; Badawy et al., 2022). Its three main purposes are to remove matrix interferents that could distort instrument's response to the target analyte; analyte preconcentration from a larger volume into a smaller volume with concentration above the instrument's detection limit; and solvent exchange where the analyte needs to be reconstituted into a solvent compatible with an analytical instrument. Before extraction, the sample may require pretreatment to enhance optimum retention of the analytes on the SPE sorbent.

Common pretreatment methods include filtering, dilution, and pH adjustment. Dilution is usually done for complex samples such as wastewater to minimize the concentration of interferents that tend to compete with the target analyte on the binding sites (Badawy et al., 2022). It can also be done to prevent overloading the cartridge since high concentrations of the analyte in the sample can lead to breakthrough. The extent of sample dilution depends on the complexity of the wastewater sample and the concentration of the target analyte, and the specific dilution factor may need to be optimized. The pH of the sample may need to be adjusted to either ionize or neutralize the target analytes depending on the type of sorbent used. The pH adjustments are necessary in the ion-exchange extraction mode where the separation mode is based on the charge of the analyte (Fontanals et al., 2020). The choice of SPE sorbent is determined by the physicochemical properties of the target analyte especially its polarity and ionic properties (pK_a value).

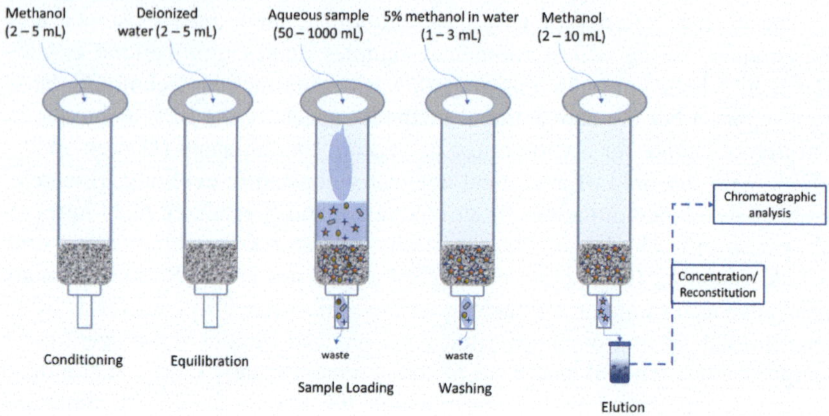

FIG. 4.1 Illustration of the steps involved in solid-phase extraction technique.

There are five major steps involved in SPE as illustrated in Fig. 4.1 and Table 4.1. First is the conditioning step using about 3–5 mL organic solvent to activate the functional groups of the sorbent. This is followed by equilibration of the sorbent phase to prepare it before loading the water samples. Ideally, the solvent used for the equilibration step must be the aqueous media of the target analyte or at least have similar properties to the sample matrix to ensure maximum retention of analytes. For aqueous samples, deionized water is used. The equilibration step serves to prepare the sorbent for the sample thus ensuring optimum adsorption performance. The filtered environmental water sample is then loaded through the sorbent, often between 100 mL and 1 L, allowing the targets to bind on the sorbent while salts and interferents pass into waste. Interferents that weakly bind to the sorbent can be washed away using 1–5 mL of weak solvent usually 5% methanol in water. Retention of target analytes on the SPE sorbent during the washing step can be enhanced by increasing or decreasing the pH or changing the ionic strength. After the washing step, the sorbent can then be dried by purging for a few minutes using an inert gas (usually nitrogen) to remove all water molecules.

Finally, 3–5 mL of a strong solvent (either methanol or acetonitrile) is used to elute the target analytes from the sorbent. The organic eluting phase disrupts the sorbent–analyte interactions.

The eluent may need to be reconstituted with an appropriate solvent for gas chromatography (GC) analysis. For liquid chromatography (LC) analysis, the eluent may be reconstituted with a solvent composition that matches the initial mobile phase composition. For instance, 10% methanol in water is preferred but analytes with lower polarity may require a solvent with up to 30%–40% methanol in water for complete dissolution. In these cases, the injection volume should be kept low (about 1–2 μL) to preserve chromatographic resolution. Furthermore, about 0.1% formic acid or ammonium hydroxide should be added to improve the chromatographic separation and peak shape, and to enhance the

TABLE 4.1 Main steps during solid-phase extraction.

Step	Description	Typical solvent	Typical volume (mL)
1. Conditioning of the sorbent	The SPE sorbent is first conditioned with an organic solvent (mainly methanol) to activate the sorbent surface	Methanol	3–5
2. Equilibration of the sorbent	The SPE cartridge is then equilibrated with the sample solvent typically deionized water to prepare the sorbent for the sample	Deionized water	3–5
3. Loading the sample	The liquid sample is passed through the cartridge. Target analytes bind to the sorbent	Aqueous samples	50–1000
4. Washing interferents	A washing solvent is added to remove interfering impurities from the SPE sorbent while keeping the analytes retained	5% methanol in water	3–5
5. Elution of analytes	A stronger solvent is used to release (elute) the analytes from the sorbent into a clean collection vial.	Methanol	3–5

ionization efficiency in the mass spectrometer (Fabregat-Safont et al., 2023; Nasiri et al., 2021). Formic acid and ammonium hydroxide are volatile and will easily evaporate with the mobile phase.

3.2 Solid-phase extraction sorbents

The main types of SPE sorbents used in environmental pollution studies are summarized in Fig. 4.2 and Table 4.2. The modes can be classified into six distinct groups: (1) the reversed phase SPE (commonly C18 and phenyl sorbents) for extraction of nonpolar to moderately polar compounds; (2) the normal phase (containing silica, diol, or amino sorbents) for extraction of polar analytes in nonpolar matrices; (3) the ion-exchange SPE (such as Strong Anion Exchange-SAX, and Strong Cation Exchange-SCX) for extraction of ionizable acidic and basic compounds; (4) the polymeric (such as Oasis HLB and Strata-X) for extracting both polar and nonpolar compounds; (5) the mixed mode that combines both reversed-phase and ion-exchange properties (such as the mixed-mode strong cation exchange—SCX, the mixed-mode strong anion exchange—SAX, or weak cation exchange—WCX); and (6) the molecularly imprinted polymer SPE (MISPE) which are custom-made for specific analytes from complex samples such as wastewater (Kaya et al., 2023). Only MISPE cartridges are less commercially available because they have to be customized for a specific target which becomes a challenge considering that there is no universal target

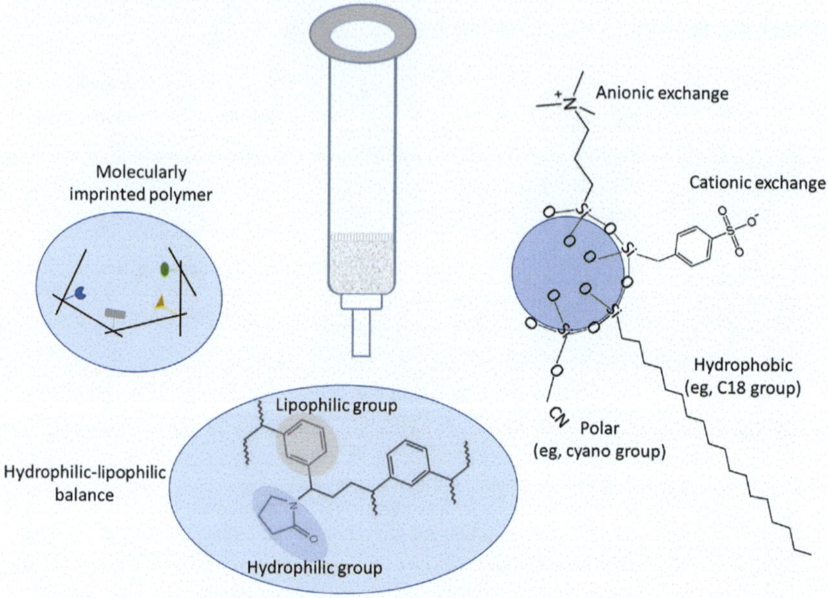

FIG. 4.2 Different types of sorbents for solid-phase extraction sorbents.

analyte (Ncube et al., 2019). However, these smart polymers can still be easily synthesized in the laboratory and have the advantage of reusability.

The mixed mode sorbents present the advantage of combining different modes of extraction, but the lipophilic–hydrophilic polymeric sorbents remain very popular (Thavarayan and Moodley, 2025). The Waters' Oasis HLB and Phenomenex's Strata-X are commonly used in environmental water analysis because a copolymer structure consisting of N-vinylpyrrolidone (the hydrophilic part) and divinylbenzene (the lipophilic part) offers a hydrophilic–lipophilic balance needed for targeting both polar and nonpolar contaminants. They are also stable in the entire pH range and resistant to drying out effects which makes them robust, reproducible, and automation-friendly. They are more flexible and suitable for complex aqueous samples such wastewater.

3.3 Case studies

The polymeric sorbents (Oasis HLB and Strata-X) are the most commonly used SPE modes among organic environmental contaminants because their features allow them to capture a wide range of contaminant residues of varying polarity in a single extraction. These are also the SPE modes of choice in the simultaneous extraction of different classes of organic contaminants. Some of these examples of applications can be found in Table 4.3. For example, one study used the Oasis HLB to determine 21 pharmaceuticals, 12 pesticides, and 6 organic UV filters in river water samples (Gorito et al., 2025). Pesticides are

TABLE 4.2 Summary of SPE modes and their applications.

SPE mode	Retention mechanism	Typical sorbent	Application	Key notes
Reversed phase	Hydrophobic interactions	C18, C8, phenyl	Nonpolar to moderately polar analytes in polar matrices	This is the conventional SPE mode for nonpolar analytes
Normal phase	Polar interactions (dipole–dipole, H-bonding)	Silica, diol, amino, cyano	Polar analytes in nonpolar matrices	Useful for separating polar compounds from hydrophobic media (e.g., oils, organic solvents).
Polymeric	Combination of hydrophilic–lipophilic interactions	N-vinylpyrrolidone (hydrophilic part) and divinylbenzene (lipophilic part)	Both polar and nonpolar analytes	The most commonly used SPE mode for environmental samples
Ion-exchange	Electrostatic attraction between charged analyte and oppositely charged sorbent	Anion exchange typically quaternary amine sorbentCation exchange typically sulfonic acid sorbent	Ionized (charged) analytes. Anion-exchange—for acidic analytes Cation-exchange—for basic compounds such as amines	pH control is essential. Adjust the sample pH so that the analyte is charged.
Mixed-mode	Combination of hydrophobic and ion-exchange interaction	Mixed functional groups	Molecules with both hydrophobic and ionizable groups such as pharmaceuticals	Increased selectivity. pH control is essential
Molecularly imprinted polymer	Shape-specific binding sites customized for the target analyte	MIP cavities with shape, size and functional groups complementary to the specific analyte(s)	Specific analytes requiring high selectivity	Highly selective extraction technique. Excellent for complex samples like wastewater

TABLE 4.3 Applications of SPE in analysis of contaminants of emerging concern.

Case study	Class	Matrix	SPE sorbent	SPE method	Reference
Pharmaceuticals	64 antibiotics	Wastewater effluent	Oasis MCX (mixed mode Cation-Exchange)	Condition with 3 mL MeOH. Equilibrate with 3 mL H_2O; Load 100 mL acidified to pH 3. Elute with 5 mL of 5% NH_4OH in MeOH.	(Fernández-Fernández et al., 2022)
	4 antiretrovirals		MISPE—molecularly imprinted polymer specific for ARVDs	Condition with 1 mL MeOH. Load 50 mL of sample. Wash with 3 mL of 2% MeOH in water. Elute with 2 mL ACN.	(Xolo and Mahlambi, 2024)
Illicit drugs	10 illicit drugs including cocaine	Wastewater	Oasis HLB (polymeric)	Condition with 2 mL MeOH. Equilibrate with 2 mL water. Load 100 mL sample. Elute with 4 mL MeOH. Evaporate to dryness and reconstitute to 0.5 mL of MeOH.	(Lawrence et al., 2023)
Pesticides	65 different pesticides	Surface water	Oasis HLB (polymeric)	Condition with 5 mL ethyl acetate, 5 mL n-hexane, 5 mL ACN and 5 mL MeOH sequentially. Equilibrate with 5 mL water. Load 100 mL; Wash with 5 mL water. Elute with ethyl acetate/n-hexane/acetone (31:38:31).	(Nakhjavan et al., 2021)
	24 different pesticides	Portable water	Oasis HLB (polymeric)	Condition with 5 mL MeOH. Equilibrate with 5 mL water. Load 300 mL sample. Wash with 3 mL water. Elute with 5 mL MeOH/ethyl acetate (70:30 *v/v*)). Reconstitute to 1 mL.	(Drimaropoulou et al., 2024)
	8 different pesticides	Mountain reservoir and river water	SAX/SCX/HLB (sequential-mode SPE)	Condition with 5 mL MeOH and 5 mL water. Load 500 mL sample acidified to pH 3. Elute with 6 mL 0.5% formic acid in methanol/ethyl acetate (1:1). Evaporate to dryness and reconstitute to 2 mL of MeOH.	(Wang et al., 2019)

Multiresidue microcontaminants	21 pharmaceuticals, 12 pesticides, 6 UV filters	River water	Oasis HLB (polymeric)	No conditioning as per supplier. Load 375 mL sample acidified to pH 3–4. Elute with 3.5 mL of ethanol. Evaporate to dryness and reconstitute to 0.25 mL of MeOH/water (80:20, *v*/*v*).	(Gorito et al., 2025)
	15 psychoactives, 3 drugs of abuse, 2 antidepressants	Drinking water	Strata-X	Condition with 2 mL MeOH and 2 mL 0.1 M HCl. Load 200 mL sample acidified to pH 2. Elute with 6 mL isopropanol/ethyl acetate (15:85). Evaporate to dryness and reconstitute to 0.1 mL of 0.5% formic acid/5% ACN/94.5% water (LCMS injection solvent).	(Peng et al., 2019)
PFAS compounds	80 PFAS	Drinking water	Oasis WAX	Condition with 4 mL MeOH. Equilibrate with 4 mL water. Load 50 mL sample. Elute with 4 mL 0.1% NH_4OH in MeOH/ACN followed by 4 mL ACN.	(Alghamdi et al., 2025)
	26 PFAS	Drinking water	Bond Elut PFAS WAX	Condition with 5 mL 3% NH_4OH in 60/40 ACN/MeOH. Equilibrate with 5 mL of 0.3% acetic acid in water. Load 250 mL sample. Wash with 5 mL water. Elute with 5 mL 3% NH_4OH in 60/40 ACN/MeOH. Evaporate to dryness and reconstitute to 0.5 mL with ACN/water (1:1 *v*/*v*).	(Martínez et al., 2024)
	18 PFAS	Wastewater and river water	ENVI-18	Condition with 3 mL MeOH. Equilibrate with 3 mL water. Load 250 mL sample. Elute with 5 mL MeOH. Evaporate to dryness and reconstitute to 0.95 mL with MeOH.	(Okwuosa et al., 2025)

Continued

TABLE 4.3 Applications of SPE in analysis of contaminants of emerging concern—cont'd

Case study	Class	Matrix	SPE sorbent	SPE method	Reference
	17 PFAS	Drinking water	Oasis WAX	Washed with water and MeOH. Load 1000 mL sample. Elute with eluted with 1% NH_4OH in 9:1 methyl tert-butyl ether:MeOH. Concentrated to 250 μL.	(Boone et al., 2019)
	30 PFAS	Tap water	Strata-XL-AW	Preconditioned. Load 250 mL sample. Elute with 10 mL of 0.3% NH_4OH in MeOH. Reconstitute with 1 mL 5 mM ammonium formate/MeOH (90:10 *v/v*).	(Li et al., 2022)
Flame retardants	6 BFRs	River and lake water	PS/DVB polymer	Condition with 6 mL of ethyl acetate. Equilibrate with 5 mL of water. Load 1 L sample. Elute with 2 × 5 mL of ethyl acetate.	(Kowalski and Mazur, 2014)
	17 BFRs			Condition with 6 mL ACN. Equilibrate with 2 × 5 mL water. Load 1 L sample. Wash with 10 mL deionized water. Elute with 18 mL of ACN.	(Han et al., 2019)
	10 OPFRs	Bottled drinking water	Oasis HLB (polymeric)	Condition with 2 × 6 mL ACN. Equilibrate with 2 × 6 mL water. Load 500 mL sample. Wash with 10 mL deionized water. Elute with 6 mL of ACN and 6 mL of MeOH.	(Nguyen et al., 2025)

usually determined in farmed products with studies on their presence in aquatic environments limited, but those that report them preferentially used the polymeric sorbents (Table 4.3) Sequential-mode SPE that combines various modes such as the SAX/SCX/HLB in which the different sorbents are layered on the same cartridge have also been reported for pesticide determination in aqueous samples (Wang et al., 2019). This mode has both the polymeric and the ion-exchange components packed in different layers.

The determination of PFAS in environmental samples almost always utilizes the weak anion exchange (WAX) SPE cartridges (Table 4.3). This is a mixed-mode sorbent that has both reversed phase and WAX modes. The determination of PFAS is prone to contamination and the commercially available Oasis WAX cartridges designed specifically for PFAS minimize the risk of cross contamination which is essential in PFAS analysis (Alghamdi et al., 2025). The Strata-XL-AW, also a polymeric weak anion mixed mode cartridge, has been reported in the determination of 30 PFAS compounds in Li et al. (2022). These cartridges require addition of 1%–5% NH_4OH in the elution solvent to ensure complete recovery of the negatively charged PFAS from the sorbent. The presence of NH_4OH disrupts the PFAS-WAX ion exchange interaction by deprotonating/neutralizing the WAX functional groups while keeping the PFAS anionic.

Flame retardants and plasticizers are lipophilic in nature and as such studies in the aqueous environment are few compared to soils/sediments and the biota (Enyoh et al., 2024; Xiong et al., 2019). In this regard, researchers preferentially use advanced techniques but the SPE approach has also been utilized. For example, Kowalski and Mazur investigated 6 SPE cartridges of different modes and observed that the Bond Elut ENV+ consisting of the polystyrene-divinylbenzene polymer performed better as an extraction sorbent for 6 BFRs in a study that involved 5 other SPE sorbents including the C_{18} and the HLB (Kowalski and Mazur, 2014). On the other hand, a recent study observed that the Oasis HLB was better compared to Sep-Pak C_{18}, MAX, and Bond Elut C_{18} for simultaneous extraction of 10 organophosphate flame retardants (OPFRs) (Nguyen et al., 2025).

4 Liquid–liquid extraction

This is a legacy technique that has largely been replaced in modern laboratories for routine analysis. Its principle involves vigorously shaking a water sample with a water-immiscible organic solvent (like dichloromethane or hexane) in a separatory funnel (Malhotra, 2023). The nonpolar analytes partition into the organic layer, which is then collected, dried, and concentrated. LLE efficiency depends on several key factors, including the partition coefficient, which determines how well the analyte distributes between the two immiscible phases. The choice of solvent should ensure good selectivity, immiscibility with the aqueous phase, and be safe and easy to remove. The pH of the aqueous phase is especially important for ionizable compounds, as neutral forms extract better into the organic phase. Proper mixing promotes mass transfer, while excessive agitation should be avoided to prevent emulsions.

Performing multiple small-volume extractions is generally more effective than one large extraction. The distribution of a solute between two immiscible phases is governed by the partition coefficient and therefore reaches equilibrium during each extraction step. A single extraction transfers only a fraction of the analyte into the extractant phase, leaving a residual amount in the original matrix. By applying additional extraction steps with fresh solvent, the remaining analyte is repeatedly redistributed, progressively reducing its concentration in the initial phase. Consequently, the cumulative recovery from several smaller-volume extractions exceeds that obtained from a single extraction of equivalent total solvent volume. This approach is supported by the principles of the Nernst distribution law, which demonstrates that repeated partitioning increases total analyte transfer and thereby improves extraction efficiency (Akhter and Alan, 2023). Finally, matrix composition, phase separation and salting out also influence the overall recovery and purity of the extracted analyte. High salt concentration can salt out organic compounds and improve extraction efficiency while the presence of colloids may trap analytes, reducing extraction unless pretreated (Gezahegn et al., 2019). The addition of salts such as sodium chloride and magnesium sulfate, ammonium sulfate, ammonium acetate, and sodium acetate can facilitate the transfer of especially hydrophilic analytes from the aqueous media into the organic phase (Khatibi et al., 2022).

The main disadvantage of the LLE approach is that it is nonselective and quite a large amount of matrix effects are coextracted (Zhang et al., 2013). Moreover, it requires huge volumes of hazardous solvents, it is labor-intensive, and may form emulsions that complicate the separation. While the conventional LLE was very common some years ago (Hii and Lee, 2010), recent studies preferentially utilize the miniaturized and pressurized versions (Alvarez-Rivera et al., 2020; Seidi et al., 2018), with some of these utilizing ionic liquids and deep eutectic solvents (Fortunato et al., 2023; Ishtaweera and Baker, 2024; Lee et al., 2020; Sereshti et al., 2021; Sibiya et al., 2024), a selection of which are detailed in Chapter 5. In recent years, only a few studies have utilized the conventional LLE technique for determination of organic contaminants. For example, the LLE method has been used to determine organic contaminants in aqueous environment samples such as estrogen hormones (Ben Fredj et al., 2015), cytostatics (Gouveia et al., 2020), PFAS (Yang et al., 2024), and pesticides (Rösch et al., 2019).

5 Future perspectives

While the conventional LLE technique has largely been discontinued by research laboratories, the SPE continues to play a crucial role in environmental monitoring. The polymeric Oasis HLB and Strata-X cartridges remain the preferred choice for monitoring pharmaceuticals, pesticides, and flame retardants while the mixed mode WAX sorbents are critical for ultra-trace analysis of PFAS, enabling efficient extraction while minimizing matrix interferences. However, recent developments indicate a shift toward sorbents with enhanced selectivity, miniaturized formats, and greener workflows. The conventional SPE sorbents are increasingly being complemented by advanced materials, molecularly imprinted polymers (MIPs), and nano-sized sorbents which offer improved recognition and retention of the important classes of polar, ionic, and structurally complex contaminants commonly encountered in wastewaters and surface waters (Jagirani and Soylak, 2024; Ma et al., 2023).

In addition to sorbent material innovation, SPE is becoming increasingly miniaturized and solvent-efficient, with emerging formats such as micro-SPE, magnetic SPE, and solid-phase microextraction (SPME) supporting the principles of green analytical chemistry by reducing reagent volumes and waste generation (Badawy et al., 2022). The reader is directed to Chapter 5 for more information about these microextraction techniques. Another notable emerging trend is the integration of SPE into on-site and field-deployable systems, including passive samplers and portable SPE manifolds, which helps reduce analyte transformation during transport and allows more representative temporal sampling (Sajid, 2022). Overall, these advancements suggest that SPE will continue to evolve toward automation, selectivity, scalability, and environmental sustainability, supporting the increasing need for ultimate ultra-trace detection of emerging contaminants in complex matrices. However, its core operational principles are expected to remain central to analytical workflows until commercially viable and widely accessible alternative sample-preparation technologies become established.

6 Conclusions

While LLE represents one of the earliest and simplest techniques for isolating organic compounds, its reliance on large volumes of hazardous solvents, limited selectivity, and susceptibility to matrix coextraction have significantly reduced its use in modern routine analysis. In contrast, SPE remains the most widely adopted approach due to its versatility, selectivity, and compatibility with modern chromatographic systems. Polymeric sorbents such as Oasis HLB and Strata-X remain central to multiresidue analysis across diverse contaminant classes, while mixed mode WAX sorbents are indispensable for ultra-trace PFAS determination. Although emerging miniaturized and advanced formats including μSPE, SPME, and MIPs offer enhanced specificity and greener workflows, conventional SPE continues to serve as the benchmark for method development and validation. Looking forward, improvements in sorbent design, automation, and sustainable extraction practices are expected to enhance its analytical efficiency and greenness, but its fundamental principles will remain relevant until newer alternatives become widely standardized and commercially accessible.

References

Akhter, M., Alan, M.M., 2023. The distribution law. In: Akhter, M., Alan, M.M. (Eds.), Physical Pharmacy and Instrumental Methods of Analysis the Distribution Law. Switzerland: Springer Nature , pp. 45–54. doi:10.1007/978-3-031-36777-9_4.

Alghamdi, W., Partington, J.M., Leung, I.K.H., Clarke, B.O., 2025. Premium ultra-trace analytical method for part per quadrillion (ppq) PFAS quantification in drinking water. Anal. Chim. Acta 18734324. 1368. Available from: https://www.sciencedirect.com/science/journal/00032670. doi:10.1016/j.aca.2025.344333.

Alvarez-Rivera, G., Bueno, M., Ballesteros-Vivas, D., Mendiola, J.A., Ibañez, E., 2020. Pressurized liquid extraction. In: Handbooks in Separation Science. Elsevier, pp.375–398. doi:10.1016/B978-0-12-816911-7.00013-X.

Andrade-Eiroa, A., Canle, M., Leroy-Cancellieri, V., Cerdà, V., 2016. Solid-phase extraction of organic compounds: a critical review (Part I). TrAC Trends Anal. Chem. 01659936. 80, 641–654. doi:10.1016/j.trac.2015.08.015.

Badawy, M.E.I., El-Nouby, M.A.M., Kimani, P.K., Lim, L.W., Rabea, E.I., 2022. A review of the modern principles and applications of solid-phase extraction techniques in chromatographic analysis. Anal. Sci. 13482246. 38 (12), 1457–1487. Available from: https://www.springer.com/journal/44211. doi:10.1007/s44211-022-00190-8.

Ben Fredj, S., Nobbs, J., Tizaoui, C., Monser, L., 2015. Removal of estrone (E1), 17β-estradiol (E2), and 17α-ethinylestradiol (EE2) from wastewater by liquid-liquid extraction. Chem. Eng. J. 262, 417–426. doi:10.1016/j.cej.2014.10.007.

Boone, J.S., Vigo, C., Boone, T., Byrne, C., Ferrario, J., Benson, R., Donohue, J., Simmons, J.E., Kolpin, D.W., Furlong, E.T., Glassmeyer, S.T., 2019. Per- and polyfluoroalkyl substances in source and treated drinking waters of the United States. Sci. Total Environ. 18791026. 653, 359–369. Available from: www.elsevier.com/locate/scitotenv. doi:10.1016/j.scitotenv.2018.10.245.

Daniels, K.D., Park, M., Huang, Z., Jia, A., Flores, G.S., Lee, H.K., Snyder, S.A., 2020. A review of extraction methods for the analysis of pharmaceuticals in environmental waters. Crit. Rev. Environ. Sci. Technol. 15476537. 50 (21), 2271–2299. Available from: www.tandf.co.uk/journals/titles/10643389.asp. doi:10.1080/10643389.2019.1705723.

Drimaropoulou, G., Christophoridis, C., Zacharis, C.K., Fytianos, K., 2024. Chemometric optimization of SPE for the determination of multiclass pesticides in portable water samples by UHPLC-MS/MS. Bull. Environ. Contam. Toxicol. 14320800. 112 (3). Available from: https://www.springer.com/journal/128. doi:10.1007/s00128-024-03873-2.

Enyoh, C.E., Maduka, T.O., Rana, M.S., Osigwe, S.C., Ihenetu, S.C., Wang, Q., 2024. Chemicals from brominated flame retardants: analytical methods, occurrence, transport and risks. Appl. Sci. 20763417. 14 (17). Available from: https://www.mdpi.com/journal/applsci/. doi:10.3390/app14177892.

Escudero, J., Muñoz, J.L., Morera-Herreras, T., Hernandez, R., Medrano, J., Domingo-Echaburu, S., Barceló, D., Orive, G., Lertxundi, U., 2021. Antipsychotics as environmental pollutants: an underrated threat? Sci. Total Environ. 00489697. 769, 144634. doi:10.1016/j.scitotenv.2020.144634.

Fabregat-Safont, D., Gracia-Marín, E., Ibáñez, M., Pitarch, E., Hernández, F., 2023. Analytical key issues and challenges in the LC-MS/MS determination of antibiotics in wastewater. Anal. Chim. Acta. 00032670. 1239, 340739. doi:10.1016/j.aca.2022.340739.

Fernández-Fernández, V., Ramil, M., Cela, R., Rodríguez, I., 2022. Solid-phase extraction and fractionation of multiclass pollutants from wastewater followed by liquid chromatography tandem-mass spectrometry analysis. Anal. Bioanal. Chem. 1618-2642. 414 (14), 4149–4165. doi:10.1007/s00216-022-04066-8.

Fontanals, N., Borrull, F., Marcé, R.M., 2020. Overview of mixed-mode ion-exchange materials in the extraction of organic compounds. Anal. Chim. Acta 18734324. 1117, 89–107. Available from: http://www.journals.elsevier.com/analytica-chimica-acta/. doi:10.1016/j.aca.2020.03.053.

Fortunato, L., Fuhaid, L.A., Murgolo, S., De Ceglie, C., Mascolo, G., Falivene, L., Vrouwenvelder, J.S., Witkamp, G.J., Farinha, A., 2023. Removal of polyfluoroalkyl substances (PFAS) from water using hydrophobic natural deep eutectic solvents (NADES): a proof of concept study. J. Water Process Eng. 22147144. 56. Available from: http://www.journals.elsevier.com/journal-of-water-process-engineering/. doi:10.1016/j.jwpe.2023.104401.

Gezahegn, T., Tegegne, B., Zewge, F., Chandravanshi, B.S., 2019. Salting-out assisted liquid-liquid extraction for the determination of ciprofloxacin residues in water samples by high performance liquid chromatography-diode array detector. BMC Chem. 2661801X. 13 (3). Available from: https://bmcchem.biomedcentral.com/. doi:10.1186/s13065-019-0543-5.

Godwin, A.D., 2024. Plasticizers. In: Applied Plastics Engineering Handbook. Elsevier, pp. 595–618. doi:10.1016/B978-0-323-88667-3.00031-X.

Gorito, A.M., Mota, B.R., Silva, A.M.T., Ribeiro, A.R.L., 2025. SPE-LC-MS/MS optimisation using response surface methodology for quantification of pharmaceuticals, pesticides and organic UV filters included in water protection guidelines. J. Chromatogr. A 1756. doi:10.1016/j.chroma.2025.466052

Gouveia, T.I.A., Silva, A.M.T., Ribeiro, A.R., Alves, A., Santos, M.S.F., 2020. Liquid-liquid extraction as a simple tool to quickly quantify fourteen cytostatics in urban wastewaters and access their impact in aquatic biota. Sci. Total Environ. 18791026. 740. Available from: www.elsevier.com/locate/scitotenv. doi:10.1016/j.scitotenv.2020.139995.

Han, F., Gao, Y., Hu, F., Yu, X., Xie, H., Li, H., Zhao, Y., Kimura, S.Y., Zhang, Y., Zubizarreta, M.E., Xiao, S., Zhan, M., Zheng, W., 2019. Solid-phase extraction of seventeen alternative flame retardants in water as determined by ultra-high-performance liquid chromatography-tandem mass spectrometry. J. Chromatogr. A. 00219673. 1602, 64–73. doi:10.1016/j.chroma.2019.05.048.

Hii, T.M., Lee, H.K., 2010. Liquid—liquid extraction in environmental analysis. In: Pawliszyn, J., Lord, H.L. (Eds.), Handbook of Sample Preparation. New jersey, USA: Wiley, pp. 39–51. doi:10.1002/9780813823621.ch3.

Ishtaweera, P., Baker, G.A., 2024. Progress in the application of ionic liquids and deep eutectic solvents for the separation and quantification of per- and polyfluoroalkyl substances. J. Hazard. Mater. 03043894. 465, 132959. doi:10.1016/j.jhazmat.2023.132959.

Jagirani, M.S., Soylak, M., 2024. Green sorbents for the solid phase extraction of trace species. Curr. Opin. Green Sustain. Chem. 24522236. 47, 100899. doi:10.1016/j.cogsc.2024.100899.

Jayaraj, R., Megha, P., Sreedev, P., 2016. Review Article. Organochlorine pesticides, their toxic effects on living organisms and their fate in the environment. Interdiscip. Toxicol. 13379569. 9 (3–4), 90–100. Available from: http://versita.com/science/medicine/it/. doi:10.1515/intox-2016-0012.

Kaya, S.I., Cetinkaya, A., Ozkan, S.A., 2023. Molecularly imprinted polymers as highly selective sorbents in sample preparation techniques and their applications in environmental water analysis. Trends Environ. Anal. Chem. 22141588. 37. Available from: http://www.journals.elsevier.com/trends-in-environmental-analytical-chemistry/. doi:10.1016/j.teac.2022.e00193.

Khatibi, S.A., Hamidi, S., Siahi-Shadbad, M.R., 2022. Application of liquid-liquid extraction for the determination of antibiotics in the foodstuff: recent trends and developments. Crit. Rev. Anal. Chem. 15476510. 52 (2), 327–342. Available from: www.tandf.co.uk/journals/titles/10408347.asp. doi:10.1080/10408347.2020.1798211.

Kowalski, B., Mazur, M., 2014. The simultaneous determination of six flame retardants in water samples using SPE pre-concentration and UHPLC-UV method. Water, Air, & Soil Pollution. 0049-6979. 225 (3). doi:10.1007/s11270-014-1866-4.

Lawrence, T.I., Sims, N., Kasprzyk-Hordern, B., Jonnalagadda, S.B., Martincigh, B.S., 2023. Wastewater profiling of illicit drugs, an estimation of community consumption: a case study of eThekwini Metropolitan Municipality, South Africa. Environ. Pollut. 18736424. 335. Available from: https://www.journals.elsevier.com/environmental-pollution. doi:10.1016/j.envpol.2023.122270.

Lee, J., Kim, H., Kang, S., Baik, N., Hwang, I., Chung, D.S., 2020. Applications of deep eutectic solvents to quantitative analyses of pharmaceuticals and pesticides in various matrices: a brief review. Arch. Pharm. Res. 0253-6269. 43 (9), 900–919. doi:10.1007/s12272-020-01266-7.

Li, X., Fatowe, M., Lemos, L., Quinete, N., 2022. Spatial distribution of per- and polyfluoroalkyl substances (PFAS) in waters from Central and South Florida. Environ. Sci. Pollut. Res. 0944-1344. 29 (56), 84383–84395. doi:10.1007/s11356-022-21589-w.

Lin, J.-Y., Zhang, Y., Bian, Y., Zhang, Y.-X., Du, R.-Z., Li, M., Tan, Y., Feng, X.-S., 2023. Non-steroidal anti-inflammatory drugs (NSAIDs) in the environment: recent updates on the occurrence, fate, hazards and removal technologies. Sci. Total Environ. 00489697. 904, 166897. doi:10.1016/j.scitotenv.2023.166897.

Lyu, X., Xiao, F., Shen, C., Chen, J., Park, C.M., Sun, Y., Flury, M., Wang, D., 2022. Per- and polyfluoroalkyl substances (PFAS) in subsurface environments: occurrence, fate, transport, and research prospect. Rev. Geophys. 19449208. 60 (3). Available from: http://agupubs.onlinelibrary.wiley.com/agu/journal/10.1002/(ISSN)1944-9208/. doi:10.1029/2021RG000765.

Ma, X., Wang, L., He, Q., Sun, Q., Yin, D., Zhang, Y., 2023. A review on recent developments and applications of green sorbents-based solid phase extraction techniques. Adv. Sample Prep. 27725820. 6, 100065. doi:10.1016/j.sampre.2023.100065.

Madikizela, L.M., Rimayi, C., Khulu, S., Chimuka, L., Ncube, S., 2022. Pharmaceuticals and personal care products. In: Dalu, T., Tavengwa, N.T. (Eds.), Emerging Freshwater Pollutants: Analysis, Fate and Regulations. Elsevier, Amsterdam, pp. 171–190. Available from: https://www.sciencedirect.com/book/9780128228500. doi:10.1016/B978-0-12-822850-0.00009-0.

Malhotra, P., 2023. Solvent extraction. In: Malhotra, P. (Ed.), Analytical Chemistry: Basic Techniques and Methods. Berlin: Springer International Publishing. doi:10.1007/978-3-031-26757-4_12.

Martínez, J., Picardo, M., Peñalver, A., Fabregas, J., Aguilar, C., Borrull, F., 2024. Occurrence and health risk assessment of PFAS and possible precursors: a case study in a drinking water treatment plant and bottled water (south Catalonia, Spain). Environ. Sci. Pollut. Res. 1614-7499. 31 (45), 56536–56549. doi:10.1007/s11356-024-34805-6.

Nakhjavan, B., Bland, J., Khosravifard, M., 2021. Optimization of a multiresidue analysis of 65 pesticides in surface water using solid-phase extraction by LC-MS/MS. Molecules. 1420-3049. 26 (21), 6627. doi:10.3390/molecules26216627.

Nasiri, A., Jahani, R., Mokhtari, S., Yazdanpanah, H., Daraei, B., Faizi, M., Kobarfard, F., 2021. Overview, consequences, and strategies for overcoming matrix effects in LC-MS analysis: a critical review. Analyst. 0003-2654. 146 (20), 6049–6063. doi:10.1039/d1an01047f.

Ncube, S., Madikizela, L.M., Nindi, M.M., Chimuka, L., 2019. Solid phase extraction technique as a general field of application of molecularly imprinted polymer materials. In: Marć, M. (Ed.), MIP Synthesis, Characteristics and Analytical Application. Comprehensive Analytical Chemistry. Amterdam: Elsevier, Vol. 86, pp. 41–76. doi:10.1016/bs.coac.2019.05.001.

Nguyen, M.K., Ahn, J., Han, T.G., Tung, N.K.T., Lee, J.Y., Kang, J.S., Kim, H.M., 2025. Determination of organophosphate flame retardants in bottled drinking water via solid-phase extraction and gas chromatography-mass spectrometry. J. Food Compos. Anal. 08891575. 141. Available from: https://www.sciencedirect.com/science/journal/08891575. doi:10.1016/j.jfca.2025.107322.

Ohoro, C.R., Adeniji, A.O., Okoh, A.I., Okoh, O.O., 2021. Polybrominated diphenyl ethers in the environmental systems: a review. J. Environ. Health Sci. Eng. 2052336X. 19 (1), 1229–1247. Available from: https://rd.springer.com/journal/40201. doi:10.1007/s40201-021-00656-3.

Okwuosa, R., Nomngongo, P.N., Olatunji, O.S., Okonkwo, O.J., 2025. Legacy and emerging poly and perfluoroalkyl substances (PFAS) in surface water, sediments, and treated effluent: a case study in Pretoria (South Africa). Water Air Soil Pollut. 0049-6979. 236 (5). doi:10.1007/s11270-025-07984-2.

Peng, Y., Gautam, L., Hall, S.W., 2019. The detection of drugs of abuse and pharmaceuticals in drinking water using solid-phase extraction and liquid chromatography-mass spectrometry. Chemosphere. 18791298. 223, 438–447. Available from: www.elsevier.com/locate/chemosphere. doi:10.1016/j.chemosphere.2019.02.040.

Peris-Pastor, G., Azorín, C., Grau, J., Benedé, J.L., Chisvert, A., 2024. Miniaturization as a smart strategy to achieve greener sample preparation approaches: a view through greenness assessment. TrAC Trends Anal. Chem. 01659936. 170, 117434. doi:10.1016/j.trac.2023.117434.

Pirsaheb, M., Hossini, H., Asadi, F., Janjani, H., 2017. A systematic review on organochlorine and organophosphorus pesticides content in water resources. Toxin Rev. 15569551. 36 (3), 210–221. doi:10.1080/15569543.2016.1269810.

Rösch, A., Beck, B., Hollender, J., Singer, H., 2019. Picogram per liter quantification of pyrethroid and organophosphate insecticides in surface waters: a result of large enrichment with liquid–liquid extraction and gas chromatography coupled to mass spectrometry using atmospheric pressure chemical ionization. Anal. Bioanal. Chem. 1618-2642. 411 (14), 3151–3164. doi:10.1007/s00216-019-01787-1.

Sajid, M., 2022. Advances in on-site analytical sample preparation for analysis of environmental waters: a review. Trends Environ. Anal. Chem. 22141588. 36, e00175. doi:10.1016/j.teac.2022.e00175.

Seidi, S., Rezazadeh, M., Yamini, Y., 2018. Pharmaceutical applications of liquid-phase microextraction. TrAC Trends Anal. Chem. 01659936. 108, 296–305. doi:10.1016/j.trac.2018.09.014.

Sereshti, H., Zarei-Hosseinabadi, M., Soltani, S., Jamshidi, F., Shojaee Aliabadi, M.H., 2021. Hydrophobic liquid-polymer-based deep eutectic solvent for extraction and multi-residue analysis of pesticides in water samples. Microchem. J. 0026265X. 167, 106314. doi:10.1016/j.microc.2021.106314.

Sibiya, I.V., Daso, A.P., Okonkwo, O.J., 2024. Synthesis and application of choline chloride based deep eutectic solvents in liquid-liquid and solid-liquid assisted extraction of organophosphorus flame retardants from landfill leachate and sediment. Int. J. Environ. Anal. Chem. 10290397. 104 (17), 5668–5686. Available from: www.tandf.co.uk/journals/titles/03067319.asp. doi:10.1080/03067319.2022.2128793.

Thavarayan, L., Moodley, B., 2025. Development of solid phase extraction method for the LC-PDA detection of selected pharmaceuticals and their metabolites in surface water and sediment from the Isipingo River, KZN, South Africa. Sci. Afr. 24682276. 28, e02655. doi:10.1016/j.sciaf.2025.e02655.

Wang, X., Jia, R., Song, Y., Wang, M., Zhao, Q., Sun, S., 2019. Determination of pesticides and their degradation products in water samples by solid-phase extraction coupled with liquid chromatography-mass spectrometry. Microchem. J. 0026265X. 149, 104013. doi:10.1016/j.microc.2019.104013.

Williams, M.L., Olomukoro, A.A., Emmons, R.V., Godage, N.H., Gionfriddo, E., 2023. Matrix effects demystified: strategies for resolving challenges in analytical separations of complex samples. J. Sep. Sci. 16159314. 46 (23). Available from: http://onlinelibrary.wiley.com/journal/10.1002/(ISSN)1615-9314. doi:10.1002/jssc.202300571.

Xiong, P., Yan, X., Zhu, Q., Qu, G., Shi, J., Liao, C., Jiang, G., 2019. A review of environmental occurrence, fate, and toxicity of novel brominated flame retardants. Environ. Sci. Technol. 0013-936X. 53 (23), 13551–13569. doi:10.1021/acs.est.9b03159.

Xolo, T., Mahlambi, P., 2024. Molecularly imprinted polymers as solid-phase and dispersive solid-phase extraction sorbents in the extraction of antiretroviral drugs in water: adsorption, selectivity and reusability studies. J. Anal. Sci. Technol. 20933371. 15 (1). Available from: https://jast-journal.springeropen.com. doi:10.1186/s40543-024-00418-4.

Yang, Q., Gao, Y., Ke, J., Show, P.L., Ge, Y., Liu, Y., Guo, R., Chen, J., 2021. Antibiotics: an overview on the environmental occurrence, toxicity, degradation, and removal methods. Bioengineered. 21655987. 12 (1), 7376–7416. Available from: http://www.tandfonline.com/loi/kbie20. doi:10.1080/21655979.2021.1974657.

Yang, Y., He, H., Chen, Y., Chen, B., Esfahani, E.B., Mohseni, M., Xu, N., Nesterenko, P., 2024. Towards elevated perfluorooctanoic acid (PFOA) enrichment in water: sequential liquid-liquid extraction pretreatment for ion chromatography detection. Chemosphere. 00456535. 358, 142227. doi:10.1016/j.chemosphere.2024.142227.

Zhang, Y., Cao, H., Jiang, H., 2013. Supported liquid extraction versus liquid–liquid extraction for sample preparation in LC–MS/MS-based bioanalysis. Bioanalysis. 1757-6180. 5 (3), 285–288. doi:10.4155/bio.12.335.

Chapter 5

Miniaturized and sustainable extraction techniques for environmental water samples

Irina Guerra-Martín[a,b], Norberto Medina-Rodríguez[a,b], María José Trujillo-Rodríguez[a,b], and Verónica Pino[a,b,c]
[a]*Laboratorio de Materiales para Análisis Químico (MAT4LL), Departamento de Química, Facultad de Ciencias, Universidad de La Laguna, San Cristóbal de La Laguna, Spain*
[b]*Unidad de Investigación de Bioanalítica y Medio Ambiente, Instituto Universitario de Enfermedades Tropicales y Salud Pública de Canarias (IUETSPC), Universidad de La Laguna, San Cristóbal de La Laguna, Spain*
[c]*Centro de Investigación Biomédica en Red de Enfermedades Infecciosas (Ciberinfec), Instituto de Salud Carlos III, Madrid, Spain*

Abbreviations

µ-dSPE	dispersive micro-solid phase extraction
µ-SPE	miniaturized solid-phase extraction
AGREEprep	analytical greenness metric for sample preparation
BAGI	blue applicability grade index
BTEX	benzene, toluene, ethylbenzene, and xylene
Car	carboxen
CEC	contaminant of emerging concern
CNT	carbon nanotube
COF	covalent-organic framework
CVD	chemical vapor deposition
DES	deep eutectic solvent
DI	direct immersion
DLLME	dispersive liquid–liquid microextraction
DSDME	directly suspended droplet microextraction
DVB	divinylbenzene
EME	electromembrane extraction
FID	flame ionization detection
FLM	free liquid membrane
G	graphene
GAC	green analytical chemistry
GC	gas chromatography

Contaminants of Emerging Concern in the Environment. https://doi.org/10.1016/B978-0-443-33821-2.00011-6

GSP	green sample preparation
HBA	hydrogen bond acceptor
HBD	hydrogen bond donor
HF	hollow fiber
HPLC	high-performance liquid chromatography
HS	headspace
IL	ionic liquid
IT	in tube
LLE	liquid–liquid extraction
LPME	liquid-phase microextraction
m-μdSPE	magnetic dispersive micro-solid phase extraction
MHS	multiple headspace
MIP	molecularly imprinted polymer
MNM	metallic nanomaterial
MNP	magnetic nanoparticle
MNW	magnetic nanowire
MO	metal oxide
MOF	metal-organic framework
MS	mass spectrometry
mSBSE	multi-stir bar sorptive extraction
NADES	natural deep eutectic solvent
PA	polyacrylate
PAH	polycyclic aromatic hydrocarbon
PAN	polyacrylonitrile
PDMS	polydimethylsiloxane
PEG	polyethylene glicol
POP	porous organic polymer
RSD	relative standard deviation
SBSDME	stir bar sorptive dispersive microextraction
SBSE	stir bar sorptive extraction
SDME	single drop microextraction
SLM	supported liquid membrane
SPE	solid-phase extraction
SPME	solid-phase microextraction
SPMS	sample preparation metric of sustainability
SUPRAS	supramolecular solvent
TF	thin film
WAC	white analytical chemistry

1 Introduction

Analytical microextraction comprises a group of extraction techniques characterized by a significant reduction of the requirements of both sample and extraction material, particularly when compared to conventional extraction methods like liquid–liquid extraction (LLE), Soxhlet extraction, and solid-phase extraction (SPE) (Zhang et al., 2024). For example, microextraction techniques applied to aqueous samples often require sample volumes ranging from microliters to a few milliliters, compared to volumes from several milliliters to liters in conventional techniques. Similarly, the amounts of extraction phases (either liquid solvents or solid materials as sorbents) are reduced to nanoliters/nanograms up to few milliliters/milligrams, whereas several milliliters/milligrams or liters/grams are demanded in conventional techniques (Trujillo-Rodríguez et al., 2025).

As it happens in conventional extraction, analyte enrichment during microextraction is the result of the analyte partitioning from the sample to the extraction phase. However, due to the low amount of material employed, microextraction techniques normally perform in a nonexhaustive fashion and, as a result, the overall extraction efficiency is lower than in conventional extraction. In fact, it is normal to report extraction efficiency values of ~20% or even lower when microextraction is used. In any case, as there is a high sample-to-extraction phase volume ratio, impressive preconcentration can be achieved. As a result, analytical microextraction procedures can provide higher sensitivity than traditional extraction workflows despite this nonexhaustive character, in turn making possible the detection of certain compounds that are nondetectable with conventional techniques (unless exhaustive additional evaporation and reconstitution steps are added, negatively affecting the overall sustainability) (Psillakis and Kalogerakis, 2003).

At the same time, it is evident that the significant reduction of both sample and extraction phase has several advantages from the point of view of sustainability, simplicity, and reduction of costs. In this sense, analytical microextraction techniques are clearly in alignment with current trends in green analytical chemistry (GAC) (Gałuszka et al., 2013) and white analytical chemistry (WAC) (Nowak et al., 2021). However, it is important to point out that most of the microextraction techniques that are widely used today were developed and consolidated as extraction and preconcentration strategies before the appearance of the first articles discussing GAC by Armenta et al. (2008)). In fact, the first studies of certain microextraction modes were published in 1995–1999 (Liu and Dasgupta, 1995; Arthur and Pawliszyn, 1990). Since then, the binomial microextraction-GAC is clear and well-recognized.

Depending on the type of the extraction material (liquid or solid), two main types of microextraction procedures can be distinguished: liquid-phase microextraction (LPME) (Rutkowska et al., 2019) and sorbent-based microextraction (Płotka-Wasylka et al., 2015), respectively. In addition, each main type of microextraction has several sub-modes according to the configuration and/or the way the extraction material is exposed to the sample, clearly dependent on the geometry of the device.

In terms of analytical applications, microextraction has been extensively used in the analysis of environmental samples, including soil (Asensio-Ramos et al., 2012), sediments (Kong et al., 2022), dust (Dueñas-Mas et al., 2019), and waters from different origin (e.g., seas, oceans, rivers, lakes, ponds, reservoirs, groundwaters, recycled waters, and wastewaters) (Lorenzo-Parodi et al.,

2019) and for the determination of wide variety of analytes including metals (Habibollahi et al., 2019) and organic compounds (Kenessov and Muratuly, 2024).

This chapter covers the analytical microextraction techniques most widely used in environmental water analysis, with particular focus on the monitoring of contaminants of emerging concern (CECs) over the period from 2010 to 2025. Attention is paid to the sustainable features of such analytical methods involving microextraction techniques, not limiting to the fact that there is a clear reduction in the amount of sample volume and/or extraction phase.

2 Miniaturized extraction techniques

2.1 Liquid-phase microextraction

2.1.1 Microextraction modes

The most important LPME modes applied up to date to environmental water analysis include single drop microextraction (SDME) (Tang et al., 2018), hollow fiber liquid-phase microextraction (HF-LPME) (Pawliszyn, 2012), electromembrane extraction (EME) (Huang et al., 2017), and dispersive liquid-liquid microextraction (DLLME) (Zgoła-Grześkowiak and Grześkowiak, 2011). All these techniques are based on the use of few microliters of an extraction solvent, which is typically immiscible with water to ensure analyte extraction from water samples. This section describes these LPME modes in detail.

Single drop microextraction

SDME is a type of LPME mode introduced almost simultaneously by Liu and Dasgupta, and Jeannot and Cantwell, in 1995 (Liu and Dasgupta, 1995; Jeannot and Cantwell, 1996). This technique consists in the exposure of a microdroplet (in general <5 μL) of extraction solvent to the sample. This droplet is normally hanging on the needle tip of a microsyringe (Guo et al., 2012), although there are alternative possibilities: the droplet can be directly exposed to the sample when using an extraction solvent with lower density than water (a procedure known as directly suspended droplet microextraction or DSDME) (Yıldırım et al., 2022) or by using a rod magnet when the extraction solvent exhibits paramagnetic properties (Mafra et al., 2019). SDME can be performed in two different configurations, depending on how the droplet is exposed to the sample: direct immersion (DI-), where the microdroplet is in direct contact with the sample (the microsyringe needle, if used, is immersed in the sample), and headspace (HS), where the microdroplet is suspended in the headspace of the extraction vial. The selection of the extraction mode mainly depends on the nature of the target analytes and the sample. Thus DI-SDME is used for the determination of nonvolatile analytes whereas HS-SDME is used for volatiles or semi-volatiles, and especially when highly complex water samples are analyzed since there is not direct contact between the sample and the extraction solvent (Yıldırım et al., 2022; Abolghasemi et al., 2020). Another important aspect to point out is that the HS mode extends the range of suitable of extraction phases. In fact, certain studies used water soluble extraction solvents for HS-SDME, as there is no direct contact between the aqueous sample and the extraction solvent (Li and Row, 2022).

Another classification for SDME techniques is based on the number of phases involved in the microextraction procedure. When using this criterion, SDME can be divided in two phases and three phases modes. Two phases SDME is the equivalent to DI-SDME, as the analyte mass transfer is accomplished directly from the sample to the extraction phase (Yıldırım et al., 2022). In three phases SDME, there is a third phase involved in the microextraction. In these cases, analytes undergo two partition processes: (i) from the sample (normally termed donor phase) to an intermediate extraction phase, and (ii) from this intermediate phase to a final acceptor phase. Thus HS-SDME can be considered as a three-phases SDME procedure where the intermediate phase is the HS (Abolghasemi et al., 2020), although there are applications where both the intermediate and acceptor phases are solvents (Li and Row, 2022). In addition to these main SDME modes and configurations, there are several other variants that have been described in different reviews (Jain and Verma, 2011; Trujillo-Rodríguez et al., 2025; Li and Row, 2022).

SDME offers advantages in terms of miniaturization and sustainability, including low consumption of solvents, simplicity, low cost, and high preconcentration capabilities. However, this technique has drawbacks, especially in terms of the operation due to the low stability of the microdroplet, requirement of a prolonged equilibrium time to ensure adequate partitioning, as well as matrix issues problems due to the SDME operational mode (Tang et al., 2018).

An example of the application of this SDME technique is the determination of polycyclic musks in surface waters of the Pearl River Estuary and South China Sea (Wang et al., 2014). In this study, only 2 μL of extraction solvent were required for extraction, which was carried out via HS-SDME but performed in a dynamic fashion. This dynamic procedure consisted of exposing and retracting the microdroplet during 50 dynamic cycles, thus achieving extraction in 25 min. To ensure droplet stability and to increase sample throughput, all these steps of the analytical method, including the microextraction and the subsequent injection of the extraction in gas chromatography-mass spectrometry (GC-MS), were performed automatically using an autosampler equipped with a robotic arm with syringe holder. Limits of detection were as low as $11\ ng \cdot L^{-1}$.

Hollow fiber liquid phase microextraction

HF-LPME was proposed by Pedersen-Bjergaard and Rasmussen (Pedersen-Bjergaard and Rasmussen, 1999), and involved a thin phase of extraction solvent (1-octanol) inside the pores of a hollow fiber (HF), which serves to enhance both the volume and stability of the extraction solvent when exposed to the sample. The gaining on stability and volume of extraction solvent (in turn involving higher extraction efficiency) is an advantage of HF-LPME over SDME. HF is generally made of polypropylene, with pores impregnated with the extraction solvent, and filled with an acceptor solution (Olasupo and Suah, 2022). To achieve extraction, analyte mass transfer occurs from the sample to the extraction phase and finally to the acceptor solution. Thus the HF impregnated with the extraction phase in this system acts as a supported liquid membrane (SLM), and it serves as a barrier between the other two phases (Gjelstad, 2019). The HF can be exposed to the sample with the help of microsyringes connected to one or both ends of the system and/or by sealing the ends of the HF. These different options generate alternative configurations for extraction like the U-shaped or the rod setups. The selection of the best configuration depends on the dimensions of the HF, especially its length, as well as the extraction kinetics (Trine et al., 2001).

Depending on the nature of the acceptor solution, the HF-LPME technique can operate in two phases and three phases modes. In two phases, both the extraction solvent and the acceptor solution have the same nature (Esrafili et al., 2018). In contrast, in three phases HF-LPME, the acceptor solution has a different nature than the extraction solvent. Aqueous solutions at controlled pH values are frequently employed as acceptor solutions, thus fostering selective extractions of anions or cations with transfer driven by a gradient of pH (Gjelstad, 2019).

HF-LPME has the inherit properties of other microextraction procedures (e.g., low organic solvent consumption, and simplicity). However, in terms of operation, it has difficulties in automation due to the fragility of the membrane (Madikizela et al., 2020).

Gonzalez-Salamo et al. applied two-phase HF-LPME for the extraction of 9 phthalic acid esters (plasticizers) from wastewater, tap water, pond water, and mineral water (González-Sálamo et al., 2018). Authors used 1-octanol as extraction solvent (40 μL). This analytical method provided high preconcentration factors and low cost due to the minimum solvent consumption and the price of the HFs. Moreover, when the microextraction procedure was combined with GC-MS/MS, it provided acceptable recoveries (74%–120%) and relative standard deviation (RSD) lower than 20%.

Electromembrane extraction

EME is a membrane-based LPME procedure that was described by Gjelstad et al. (2006)). The fundamentals of the technique are the same as those of HF-LPME, unlike during EME an electric potential is applied to enhance the migration of the analytes from the sample through the SLM to the acceptor phase. The electric potential occurs due to the use of two inert electrodes. One of the electrodes is immersed in the sample, and the second one in the lumen of the HF (Drouin et al., 2019). Thus extraction is the result of both analyte partition (diffusive migration) and electrokinetic migration of ions through the SLM. Apart from the traditional system based on HFs (Ghambarian et al., 2012), alternative EME procedures use porous flat membranes, normally with polymeric nature, impregnated with an organic solvent as SLMs (Huang et al., 2014). Another possibility is to perform the extraction via the use of free liquid membranes (FLMs), i.e., through liquid systems without any physical support for the organic solvent (Kubáň and Boček, 2014). Several reviews describe the wide variety of EME configurations that are the results of combining different numbers of SLMs, electrodes, and the further miniaturization of the technique (Drouin et al., 2019; Huang et al., 2017). It is important to mention that, as separation is the result of electromigration, the application of EME is limited to charged compounds. Apart from that, EME requires complex adjustment of parameters such as voltage, pH, and membrane composition. However, high selectivity and higher speed than HF-LPME (i.e., 5–15 min of extraction time) is achieved. At the same time, EME is characterized by a low organic solvent consumption, and it can be easily miniaturized (Gjelstad and Pedersen-Bjergaard, 2013). In fact, several studies have proposed EME systems based on small chips (Ramos Payán et al., 2018).

In terms of the analysis of environmental waters, Villar-Navarro et al. (2016) used a EME system based on a HF impregnated with 1-octanol as SLM and an aqueous acceptor solution (45 μL) at pH 12 for the extraction of different parabens in surface waters (river and lake).

Dispersive liquid–liquid microextraction

DLLME is the most widely used LPME approach due to its simplicity and high extraction efficiency. It involves three simple steps: (i) the dispersion of few microliters of the water-immiscible extraction solvent in few milliliters of the aqueous sample, (ii) the separation of both phases, and (iii) the collection of the microdroplet formed by the extraction solvent—containing the extracted and preconcentrated analytes (and subsequent treatment if needed). The first step involves a direct contact between both the extraction solvent and the sample and, at the same time, a strong dispersion of the extraction solvent into the aqueous sample. This creates a cloudy solution composed by a multitude of tiny microdroplets that maximizes analyte mass transfer. Due to this feature, DLLME procedures are characterized by higher enrichment factors compared with other LPME modes (Pacheco-Fernández et al., 2021).

When it was introduced by Rezaee et al. (2006), the technique used an auxiliary solvent termed dispersive solvent that was soluble with both the aqueous sample and the water-immiscible extraction solvent. Such solvent was responsible of the quick dispersion of the extraction solvent in water (cloudy solution). Today, different strategies can be implemented for dispersion while avoiding the need of a dispersive solvent; e.g., the application of vortex, sonication, microwaves, increasing the temperature of the system, and even carrying out other chemical reactions like in the case of the effervescence DLLME mode (Leong et al., 2014). On the other hand, the most common separation strategy after dispersion involves centrifugation and subsequent recollection of the droplet with a microsyringe (Sereshti et al., 2020). Nevertheless, other approaches have been described to simplify these steps, including (i) magnetic separation, when the extraction solvent has paramagnetic properties (Johnson et al., 2022), (ii) microdroplet isolation by filtration (Hu et al., 2016), (iii) the use of the solidified floating organic droplet approach (SFOD-DLLME) (Sorouraddin et al., 2020), and (iv) the application of special extraction devices designed to facilitate this collection (Lemos et al., 2022), among others.

Depending on the nature of the extraction solvent, different strategies are needed prior to instrumental analysis to ensure its compatibility with the subsequent separation and/or detection technique. This is applicable not only to DLLME but also to SDME and the two-phase HF-LPME mode. For example, it is common to use evaporation/reconstitution or dilution of the final microdroplet (Kokosa, 2019). This aspect probably is the main drawback of this technique, as the usage of a water-immiscible extraction solvent can involve compatibility issues with the further analytical technique. Furthermore, DLLME may present problems related to the formation of emulsions given the nature of the technique, and losses of efficiency can be also present if the microdroplet collection is not exhaustive and it is normally tedious and difficult (Quigley et al., 2016; Jalili et al., 2020). Despite these problems, the technique has been widely accepted due to its high simplicity and high preconcentration capability.

As an example of applicability of DLLME to the analysis of waters to monitor CECs, Cunha et al. (2022)) used DLLME for the analysis of 50 water samples from the estuaries of the Tagus and Douro rivers. This work focused on the evaluation of 37 endocrine disruptive compounds (i.e., pesticides, musk, UV filters, and bisphenols) during 1 year sampling. Authors performed vortex-assisted DLLME and analyte derivatization simultaneously in ~ 7 min, and utilized only 85 μL of extraction solvent, and acetonitrile as dispersive solvent.

2.1.2 Developments in extraction solvents

In the early 2010s, the use of greener solvents in LPME was already a significant trend in the analysis of CECs in environmental waters, intending to increase the sustainability of the entire method. Among such greener solvents, ionic liquids (ILs), deep eutectic solvents (DESs), including those with natural origin (NADES), supramolecular solvents (SUPRASs), and switchable solvents, are those that have been mostly used in all LPME modes (Zhang et al., 2024). Table 5.1 shows the most important characteristics of these solvents, together with representative examples of analytical applications for the monitoring of contaminants in waters. Only ~4 examples per material are shown in Table 5.1,

TABLE 5.1 Types of solvents used in liquid-phase microextraction, together with their main properties of interest in microextraction, green-related aspects, and selected examples of their application in the analysis of environmental waters.

Solvent	Composition	Characteristics of interest in microextraction	Green-related aspects[a]	Examples in the analysis of water
Ionic liquids (ILs)	Nonmolecular compounds based on bulky and asymmetric organic cations and both inorganic or organic anions	• Tunability • Wide solvation capabilities • Modulable viscosity and solubility • High thermal and chemical stability	+ Low volatility + Non inflammability + Possibility of reuse + Safe solvents – Most are nonbiodegradable – Hard to purify, causing low synthetic yields – Certain classes present toxicity	• Sulfonamides (Guo et al., 2012) • Pharmaceutical drugs (Rajendran et al., 2022) • Bisphenol A, antibiotics, and pharmaceutical drugs (Yao et al., 2011) • Polycyclic aromatic hydrocarbons (Maszkowska et al., 2015) • Pesticides (Negussie et al., 2025)
Deep eutectic solvents (DESs) orNatural deep eutectic solvents (NADESs)	Mixture of a hydrogen bond donor (HBD) and a hydrogen bond acceptor (HBA)	• Tunability • Wide solvation capabilities • Easy and low-cost preparation • In general, no purification required	+ Low toxicity + High biodegradability + Easy and green synthesis in most cases + Safe solvents + NADES are prepared from natural compounds – Certain classes may present toxicity – DES prepared from hydrophilic components may be leached to the aqueous sample during microextraction	• Antiinflammatory drugs (Qiao et al., 2021) • Bisphenols, alkylphenols, phthalates, and hormones (Ortega-Zamora et al., 2022) • UV filters, polycyclic aromatic hydrocarbons, and a pesticide (Farooq et al., 2022) • Antibiotics (Li et al., 2020)

Continued

TABLE 5.1 Types of solvents used in liquid-phase microextraction, together with their main properties of interest in microextraction, green-related aspects, and selected examples of their application in the analysis of environmental waters—cont'd

Solvent	Composition	Characteristics of interest in microextraction	Green-related aspects[a]	Examples in the analysis of water
Supramolecular solvents (SUPRASs)	Nanostructured liquids formed by a three-dimensional network of amphiphilic aggregates. SUPRASs are created from colloidal solutions after inducing an external stimulus (coacervation)	• Tunability • Wide solvation capabilities • Smart properties (coacervation is promoted by an external stimulus)	+ Low toxicity + Certain SUPRAS are prepared from natural compounds – Difficult to recover and recycle	• Herbicides (Seebunrueng et al., 2020) • Pyrethroid insecticides (Hai et al., 2024) • Perfluorinated compounds (Liang et al., 2020) • Chlorophenols (Karimiyan and Hadjmohammadi, 2016)

Switchable solvents	Guanidines, amidines, and, in general, species containing functional groups such as amines, that reversibly change their polarity, charge or solubility in response to an external stimuli (e.g., the addition of CO_2)	• Tunability • Smart properties that can be exploited during extraction to change the solubility of the extractant	+ Recyclable and regenerative – Need of CO_2 for changing the solubility of the extractant – Possible toxicity – Sustainability uncertain (still under evaluation)	• Azole antifungal medicines (Arabkhani et al., 2025) • Triazine herbicides (Lasarte-Aragonés et al., 2015) • Insecticides (Wang et al., 2023) • Steroid hormones (Shishov et al., 2017)

[a]*The "+" and "–" symbols denote advantages and disadvantages related to green-aspects, respectively.*

although it is important to mention that these solvents have been widely used for water analysis (Guo et al., 2012; Rajendran et al., 2022; Yao et al., 2011; Maszkowska et al., 2015; Negussie et al., 2025; Qiao et al., 2021; Ortega-Zamora et al., 2022; Farooq et al., 2022; Li et al., 2020; Seebunrueng et al., 2020; Hai et al., 2024; Liang et al., 2020; Karimiyan and Hadjmohammadi, 2016; Arabkhani et al., 2025; Lasarte-Aragonés et al., 2015; Wang et al., 2023; Shishov et al., 2017).

All these materials present interesting features for their use in LPME, including high tunability (especially for ILs and DESs), and different functionalities for the extraction of a wide variety of compounds (from inorganic and small organic compounds to biomolecules). Besides, they are safer solvents, especially in comparison with traditional organic compounds (i.e., hexane and chloroform), that easily generate undesirable volatile organic compounds to the atmosphere.

However, not all these solvents are equally safer and equally sustainable. In terms of synthesis, DESs are probably the greenest solvents. DESs are mixtures of a hydrogen bond donor (HBD) and a hydrogen bond acceptor (HBA) that can be prepared by the simple mixture of the corresponding reagent with heating (Smith et al., 2014). The H-bonds established between HBA and HBD significantly decrease the melting point of the system and, as a result, a liquid stable at room temperature is formed. However, many DESs require higher temperatures of heating for their preparation (in general, 70°C–80°C), and during longer times (in general, around 1 h). Therefore not all DESs fulfill the green synthesis property. Another important feature is the fact that the synthesis of DESs, in general, does not require purification. Besides, DESs can be obtained from precursors of low cost, high availability, and low toxicity. DESs prepared from natural sources, known as natural DESs (NADESs), are also commonly used (Mansour et al., 2024). DESs have been used as extraction solvents in water analysis for antiinflammatory drugs, bisphenols, alkylphenols, phthalates, hormones, UV filters, polycyclic aromatic hydrocarbons (PAHs), and pesticides (Table 5.1).

ILs are nonmolecular compounds formed of bulky and asymmetric organic cations and both inorganic and organic anions (Lei et al., 2017). The synthesis of ILs in general involves quite simple organic reactions. However, the purification of ILs is difficult, thus hindering the overall synthetic yields. Besides, certain ILs based on imidazolium cations (which are the most common ones in ILs) have shown toxicity during in vitro studies with cell lines and also in in vivo studies with bacteria and animal models (Cho et al., 2021). The green character of ILs is mostly linked to the fact that their vapor pressure is negligible at room temperature. ILs have been extensively used in water analysis as extraction phases, for example, for the determination of sulfonamides (Guo et al., 2012) and pharmaceutical drugs (Rajendran et al., 2022; Qiao et al., 2021), among other families of analytes (see Table 5.1).

Another group of emergent solvents with a great future in microextraction techniques are SUPRASs. They are nanostructured liquids formed by a three-dimensional network of amphiphilic aggregates created from colloidal solutions that are altered by an external stimulus (i.e., the addition of a solvent, a salt, or a change in temperature) (Rubio, 2020). SUPRASs are efficient, selective, and sustainable; however, their synthesis may require costly compounds or specific conditions that limit their scalability. In terms of toxicity, although many SUPRASs are less toxic than traditional organic solvents, some may present risks if their effects on health and the environment are not properly assessed. These solvents have been used as extraction phases for the determination of herbicides (Seebunrueng et al., 2020) and insecticides (Hai et al., 2024), among others (see Table 5.1).

Regarding switchable solvents, they are formed by amines or other molecules that contain functional groups which reversibly change their polarity, charge, or solubility, in response to external stimuli (i.e., pH changes or the addition/removal of CO_2) (Ullah and Tuzen, 2023). However, their use may involve additional costs and sensitivity to specific conditions (such as CO_2 or pH) that limit their industrial application. For example, they have been used as extraction phases for the determination of antifungal medicines (Arabkhani et al., 2025) and insecticides (Wang et al., 2023).

2.2 Sorbent-based microextraction

Sorbent-based phase microextraction techniques were developed in the 1990 s as alternatives to SPE. The basic principle is the use of a small amount of sorbent material that will interact with the sample to extract and preconcentrate target analytes. The sorbent material can be used as powder, as in the case of the miniaturized SPE (μ-SPE), or can be coated onto a solid support, as in solid-phase microextraction (SPME) (Płotka-Wasylka et al., 2015). Since sorbent-based microextraction requires low sample volumes (in the order of a few milliliters or even lower) and it eliminates the need for extensive solvent requirements, together with high preconcentration ability, it has gained popularity in environmental analysis.

All sorbent-based microextraction procedures are generally performed in four main steps (Płotka-Wasylka et al., 2015): sorbent conditioning, extraction, washing, and desorption, although certain steps can be adapted or omitted depending on the technique. Sorbent conditioning is performed (if needed) by washing with a solvent and/or through thermal activation. Extraction is promoted by exposing the sorbent to the sample, allowing the analytes to partition between the two phases. During this step, factors like pH and ionic strength of the sample, temperature, extraction time, and stirring may be optimized to improve extraction efficiency. There are two main modes for extraction as other microextraction modes: DI-SPME and HS-SPME. For HS-mode, the sorbent is not put into direct contact with the sample, and thus the final partitioning is from the gas phase to the sorbent. After extraction, the next step is to wash the sorbent (if needed), using a mild solvent to remove undesired compounds and prevent matrix effects during analysis. For desorption, three main options can be used: (i) thermal desorption directly in the GC injection port or in a thermal desorption unit connected to GC, (ii) liquid desorption with a low volume of organic solvent, and (iii) laser desorption in certain direct MS applications. In some cases, the sorbent can be reused if it is properly reconditioned through liquid washing and/or thermal activation.

Different microextraction techniques can be distinguished within sorbent-based microextraction, mostly attending to the geometry of the microextraction device. This section focuses on those techniques that are widely used in environmental analysis. Thus among them it is possible to cite: in fiber SPME, which is the most regular mode of SPME (SPME with fibers), thin film SPME (TF-SPME, which is also found in the scientific literature abbreviated as TFME), in-tube SPME (IT-SPME), stir bar sorptive extraction (SBSE), and miniaturized solid-phase extraction (μ-SPE).

2.2.1 Microextraction modes

In fiber solid-phase microextraction

SPME with fibers, introduced in 1990 by Janusz Pawliszyn's group, undoubtedly marked the beginning of all sorbent-based microextraction techniques, providing an efficient, cost-effective, and versatile alternative to traditional sample preparation methods (Arthur and Pawliszyn, 1990). This technique originally used a thin capillary made of fused silica (further improved to a metallic fiber), with ~1 cm of length, coated with an extracting phase (with maximum 100 μm of thickness) (Feng et al., 2013), and can be put directly in contact with the aqueous samples (DI-SPME) or in its headspace (HS-SPME). Commercially available coatings are based on polyacrylate (PA), polydimethylsiloxane (PDMS), and polyethylene glycol (PEG), or combinations of these polymers with other materials (e.g., carboxen (Car)/PDMS or divinylbenzene (DVB)–Car/PDMS).

SPME is widely used combined with GC, as the conventional and commercially available SPME coatings (mounted on 22–23 ga needles, with volumes lower than ~1 μL) can be directly desorbed into the GC inlet port without any modification of the system. This thermal desorption can ensure a quite efficient desorption of volatile and semi-volatile analytes (without losses in preconcentration) and also reutilization of the SPME fibers. This aspect simplifies the analytical procedure and ensures impressive preconcentration capabilities (e.g., a maximum enrichment factor of 10,000 could be achieved when extracting 10 mL of sample). Other fibers with higher dimensions, such as arrow fibers (with up to 250 μm of sorbent thickness) can be easily combined with GC with minor inlet modifications (Herrington et al., 2020), thereby ensuring a relatively higher extraction efficiency compared to conventional fibers. In addition, when combined with GC, all the microextraction procedure can be performed in an automated way using a robotic autosampler (Crucello et al., 2023).

As aforementioned, SPME can be performed via the DI- (Li et al., 2016) or HS-modes (González-Hernández et al., 2021). Apart from these main options, certain studies have reported modifications during the extraction step. It is interesting to point out multiple HS (MHS)-SPME, where successive extractions of the same extraction vial are promoted to ensure an exhaustive extraction (Alonso et al., 2022), being useful for complex solid samples, and vacuum HS-SPME, in which reduced-pressure conditions are used to speed up the extraction kinetics of volatiles and semi-volatiles with low Henry's law constants (Solomou et al., 2020).

From the point of view of sustainability, SPME is an interesting technique as it can be considered solvent-free, especially when combined with GC. An example of application of SPME in the determination of CECs in environmental waters is the determination of synthetic musks in tap and river waters with an amino-modified graphene coated stainless steel fiber (Li et al., 2016). The developed fiber was evaluated through autosampler-assisted DI-SPME followed by GC-MS analysis and showed low LODs (0.46–5.96 $ng \cdot L^{-1}$), wide linear ranges (5–500 $ng \cdot L^{-1}$), and acceptable one-fiber ($n=6$, RSD < 5.4%) and fiber-to-fiber ($n=3$, RSD < 12%) reproducibility and good recoveries (82.3%–112%).

Thin film solid-phase microextraction

TF-SPME was introduced in 2003 by Bruheim et al. (2003) as an improvement over SPME with fibers. This microextraction procedure employs a thin film (TF) of extraction phase as a microdevice for extraction. This configuration increases the volume of the extraction phase with respect to SPME fibers. In addition, the disposition of the extraction phase as a TF significantly increases the surface area of material that is exposed to the sample but without significantly increasing the extraction times if compared with SPME. As a result, TF-SPME procedures normally provide much higher extraction efficiency but similar extraction times with respect to SPME (Olcer et al., 2019).

TF microdevices normally consist of solid (mostly planar) supports coated with the extraction phase. Flexible substrates can be used as solid supports, including polymeric, fabric, and paper-based membranes (Gutiérrez-Serpa et al., 2022). Less flexible supports like metallic blades and carbon meshes are also widely used due to their compatibility with thermal desorption and GC (Emmons et al., 2019). Regarding the microextraction procedure in TF-SPME, the extraction is generally performed via the DI mode. The TF microdevice can be hanging in the sample with the aid of a cotter pin, or it can be freely dispersed in the aqueous sample with the aid of stirring.

In terms of sustainability, TF-SPME is characterized by the requirement of low amounts of solvent and extracting phase. Its main advantage relates to the ability of attaining higher extraction efficiencies than in SPME with similar extraction times, as previously mentioned (Olcer et al., 2019). However, there are some drawbacks that need to be considered, such as the robustness and stability of TF microdevices, together with the fact that their manufacture is not as mature as in SPME, which can affect the method reproducibility (Emmons et al., 2019).

An example of the application of TF-SPME for the analysis of river waters to monitor pesticides included the use of DVB/PDMS carbon-mesh-supported membranes coupled to an autosampler, prior to GC-MS analysis (Piri-Moghadam et al., 2018). This method outperformed an official US EPA LLE protocol, achieving higher accuracy and lower LODs, while reducing sample volume, simplifying the overall handling, and enabling high-throughput automated sampling.

In tube solid-phase microextraction

IT-SPME was developed by Eisert and Pawliszyn in 1997 as a solution to automate the SPME procedure but, mostly, to increase the compatibility with HPLC applications (Yang et al., 2014). The microdevice originally described in this technique was formed by a thin capillary (as solid support) internally coated with the extraction phase. The internal diameter for the capillary is normally equal or lower than 1 mm, and the length is variable (but in the order of centimeters) (Trujillo-Rodríguez et al., 2020). With respect to the capillary materials, fused silica (the same used for GC columns) and polymeric tubes have been employed (Trujillo-Rodríguez et al., 2020). The coating in these in tube devices can form a TF in the inner capillary walls, thus creating an open tubular system, or it can be a sorbent or polymeric phase completely filling the capillary.

With respect to the extraction procedure in IT-SPME, it is important to note that this technique is considered a dynamic microextraction procedure, and thus it requires the continuous flow of the sample and the desorption solvent through the capillary to ensure adequate extraction and desorption, respectively. Due to the low capillary dimensions, this movement of sample/solvents is achieved when connecting the capillary to a flow injection system. The simplest flow-injection system is based on a single injection valve connected to a syringe pump, although alternative systems based on multiple valves have been described (Kataoka, 2021).

Two main configurations are distinguished during the extraction with these automated systems: draw/eject and flow-through (Grecco et al., 2022). In the draw/eject configuration, the sample is aspirated through the capillary several times (each time is considered a cycle) whereas in the flow-through systems, there is always a fresh portion of sample in contact with the system (Grecco et al., 2022). For desorption, a liquid organic solvent is passed through the capillary. This solvent can flow in a dynamic mode (as it happened during the extraction) or it can remain static. In the last case, the capillary is filled with the solvent and both phases remain in contact for a certain (fixed) period of time (Grecco et al., 2022).

From the point of view of green practices, IT-SPME presents the advantage of the automation of the procedure. This feature favors waste management and increases the safety of the operator (Gałuszka et al., 2013). However, the technique also has drawbacks. In particular, these small microdevices when connected in the flow through systems can generate black pressure, which could make automation and direct hyphenation with chromatography difficult. Besides, memory effects are likely more frequent in this technique than in others, given the difficulties in the cleaning step as the coating is internal.

An example of the use of IT-SPME for environmental water analysis is the determination of tetracycline antibiotics in sea, lake, and treated sewage water samples (Pang et al., 2024). Authors installed a capillary microextraction column coated with an amorphous precursor polymer of COF-300 in the loop of a high performance liquid chromatography (HPLC) injection system working in a flow-through mode (Pang et al., 2024). The validated method showed acceptable linear ranges (200–20,000 $ng \cdot L^{-1}$), low detection limits (200 $ng \cdot L^{-1}$), and low intra-day (<10.2%) and inter-day precision (<9.4%).

Stir bar sorptive extraction

SBSE was introduced by Baltussen et al. (1999)) to achieve an exhaustive extraction despite using a miniaturized procedure. SBSE employs a stir bar coated with a thicker layer (0.5–1 mm) of extraction phase. Commercially available stir bars (called Twister) are made with a metallic wire precoated with a layer of glass that acts as supporting material for the final polymeric phase that is responsible for extraction (i.e., PDMS or PDMS/ethylene glycol, as the commercialized ones).

The extraction procedure in SBSE is normally performed through the DI mode (Sánchez-Rojas et al., 2009). In these cases, the coated stir bar is used as both a stirring system and an extraction microdevice. After extraction, liquid and thermal desorption can be performed (Sánchez-Rojas et al., 2009), using a thermal desorption unit directly connected to GC in the last case (Xiao et al., 2016).

With regard to the variants of SBSE, it is important to mention multi-SBSE (mSBSE) (Marín-san Román et al., 2022). This strategy consists of using two or more coated stir bars for extraction to further increase the extraction performance. Another interesting approach is stir bar sorptive dispersive microextraction (SBSDME), which was developed by Chisvert and co-workers (Benedé et al., 2014). This technique uses a neodymium-core stir bar coated with a hydrophobic magnetic nanosorbent. As the magnetic material interacts with the stir bar due to its magnetism, it can disperse into the sample (under strong stirring) or it can act as a coating when the stirring is stopped, allowing for magnetic separation.

SBSE (and SBSDME) microdevices normally use higher amounts of sorbent than other microdevices like fibers or TFs. Therefore SBSE can achieve higher extraction efficiency (even exhaustive extraction, especially in mSBSE) (Sánchez-Rojas et al., 2009). However, this aspect also carries out limitations. In fact, SBSE in general requires longer equilibration times and often suffer from carry over effects. At the same time, the direct impact of the stir bars onto the vial walls during extraction can decrease the lifetime of the coated stir bars. Moreover, as aforementioned, the restricted availability of commercial coatings limits the range of extractable analytes (Hasan et al., 2020). From the point of view of sustainability, SBSE can require higher volumes of organic solvents for desorption compared with other solid-based microextraction strategies when combined with liquid desorption.

An example of SBSE for environmental analysis is the determination of pesticides in lake and pond waters developed by Xiao et al. (2016)). Authors used PDMS/metal-organic framework (MOF) coated stir bars coupled to GC with flame ionization detection (FID). The resulting method ensured low limits of detection (43–85 $ng \cdot L^{-1}$), an adequate linear range (0.2–100 $ng \cdot L^{-1}$), and acceptable intra- and inter-day RSD (<10.7%, $n = 7$).

Miniaturized solid-phase extraction

μ-SPE resembles the miniaturization of conventional SPE when performed in cartridges. This technique uses low amounts of sorbent (in the order of a few milligrams or lower) for extraction (Płotka-Wasylka et al., 2015). μ-SPE can be operated with the sorbent material packed in a small cartridge or disk, or in any other type of small device (e.g., a micropipette tip, a needle) (Płotka-Wasylka et al., 2015). The aqueous sample and all the solvents required for conditioning, washing, and desorption, are passed by the sorbent in μ-SPE. The microextraction procedure in these applications can be performed in batch or can be carried out using automated systems via flow-through systems (e.g., using sequential injection analysis, flow injection analysis) (Płotka-Wasylka et al., 2015).

In dispersive microsolid-phase extraction (μ-dSPE), the sorbent is directly added to the sample solution in the form of powder. The stirring promoted during the extraction process fosters sorbent dispersion, which enhances analyte partitioning from the aqueous sample to the sorbent, and accelerates extraction kinetics, overall increasing the method sensitivity. After dispersion, sorbent isolation depends on its properties. Nonmagnetic sorbents require centrifugation or filtration steps, while magnetic sorbents enable simpler magnetic separation, a variation known as magnetic dispersive microsolid-phase extraction (m-μdSPE) (Chisvert et al., 2019).

A key advantage of μ-dSPE is that the sorbent does not need to be incorporated into (or onto) a fixed substrate. This avoids time and resource-intensive preparation steps, simplifying the development and testing of new microextraction materials. Researchers can rapidly evaluate the extraction performance of novel sorbents before integrating them into more advanced microextraction devices using this technique (Khezeli and Daneshfar, 2017). This makes μ-dSPE a valuable technique in modern Analytical Chemistry, complementing the development of more sophisticated methods that offer greater ease of operation and enhanced analytical performance. In addition, efforts are being made to automate μ-dSPE for higher-throughput applications, increasing its potential for routine analysis (Nascimento et al., 2021; Yamaguchi et al., 2016).

An example of the use of μ-dSPE for environmental water analysis is the determination of PAHs from ground and river waters using porphyrin-based magnetic nanocomposites, reported by Yu et al. (2018)). The resulting method presented low detection limits (2–10 ng · L^{-1}), a wide linear range (50–10,000 ng · L^{-1}), and low intra- and inter-day precision in terms of RSD (<15%, $n=5$). Another interesting feature that impacted the green character of the methodology was the possibility of reusing the sorbent for at least 12 extraction cycles without observing a loss of extraction efficiency.

2.2.2 Developments in extraction sorbents

In parallel with the development of these sorbent-based microextraction strategies, several materials have been used as novel extraction sorbents. In environmental analysis, sorbents of relevance include polymers, carbon-based materials, metallic nanomaterials (MNMs), MOFs, and covalent-organic frameworks (COFs) (Demirkurt et al., 2018; Liu et al., 2016, Liu et al. (2023); Turiel et al., 2016; Pashayi Sarnaghi and Ayazi, 2025; Allgaier-Díaz et al., 2023; Zhang et al., 2018; Hejabri Kandeh et al., 2022; Naing et al., 2016; Yu et al., 2019; Ortega-Zamora et al., 2024; Ansari and Masoum, 2020; Dil et al., 2021; Li et al., 2016; Vosough et al., 2018; Cheng et al., 2019; Xu et al., 2022; Wang et al., 2015, 2018, Wang et al. (2019); Akkaya et al., 2019; Guo et al., 2015; Xie et al., 2015; Jakavula et al., 2023; Lirio et al., 2016; Liu et al., 2019; Yang et al., 2023; Song et al., 2023; Gao et al., 2022). (see Table 5.2).

As occurred with novel solvents (see section 2.1.2) these materials can be tailored to attain specific functionalities for enhancing the extraction efficiency of certain groups of analytes. This aspect is an advantage, especially in comparison with commercial sorbents like PDMS or PA that often have limited selectivity. From the point of view of sustainability, certain sorbents can be obtained from natural sources whereas other novel sorbents are designed to be highly stable in both aqueous media even at relatively high temperatures, and they can generally be reused, which partially compensates their nonnatural origin. For their use in sorbent-based microextraction, these materials must be processed to attain certain geometry (e.g., a SPME fiber, a TF membrane, a stir bar, an in tube device, a μ-SPE cartridge). The protocols for obtaining these devices with novel materials are often complex, as they may require modifications of the solid support surface to ensure the proper sorbent material-support attachment and their use in microextraction.

With respect to the different types of materials, polymers have been widely used. In fact, several commercially available sorbents (i.e., PDMS, PA) are polymers. Further advancements introduced specialized materials such as molecularly imprinted polymers (MIPs) for high selectivity (Kaya et al., 2023), porous organic polymers (POPs) with large surface areas (Wang et al., 2022), smart polymers able to respond to environmental stimuli (Balcerak-Woźniak et al., 2024), and biopolymers as sustainable alternatives (Allgaier-Díaz et al., 2023). Hybrid materials, including carbon-based, MNMs, MOFs, and COFs composites, have

TABLE 5.2 Types of sorbents used in sorbent-based microextraction, together with the main properties of interest in microextraction, green-related aspects, and selected examples of their application in the analysis of environmental waters.

Sorbent	Composition	Characteristics of interest in microextraction	Green-related aspects[a]	Examples in the analysis of water[b,c]
Molecularly imprinted polymers (MIPs)	Synthetic polymers prepared from monomeric solutions in the presence of a template molecule that will give the selectivity/specificity to the polymer	• High selectivity, as MIPs possess specific recognition sites for the selective extraction of a target analyte or analyte family	+ Certain options of reusability (memory effects) + They can be prepared from natural monomers – Require of intensive cleaning protocols to remove the template from the MIP	• Parabens (Demirkurt et al., 2018) • Bisphenol A (Liu et al., 2023; Liu et al., 2016) • Triazines (Turiel et al., 2016) • Narcotic pharmaceutical (Pashayi Sarnaghi and Ayazi, 2025)
Biopolymers	Natural polymers (i.e., obtained from the purification of natural products)	• Contain functional groups that can be modified or exploited for extraction	+ Possibility of reusability + Obtained from natural sources – Purification protocols can be laborious, and must be adapted depending on each batch of natural product	• PAHs, UV filters, parabens, and triclosan (Allgaier-Díaz et al., 2023) • Bisphenols (Zhang et al., 2018) • Pesticides (Hejabri Kandeh et al., 2022) • Benzene derivatives (Naing et al., 2016)

Stimuli-responsive polymers	Natural and/or synthetic polymers that can respond to external stimuli (e.g., changes in temperature, pH, UV light …)	• Controllable binding and release of analytes • Changes in physicochemical properties during extraction (i.e., viscosity)	+ Possibility of reusability + Energy efficiency + Options for biodegradation – Toxicity of additives and monomers	• Fluoroquinolones (Yu et al., 2019) • PAHs, pesticides, polychlorobiphenyls, and antibacterial agents (Ortega-Zamora et al., 2024) • Drugs (Ansari and Masoum, 2020)Melatonin (Dil et al., 2021)
Carbon-based materials	Materials based on carbon molecules in a hexagonal lattice (graphene, graphene oxide, carbon nanotubes, activated carbon …)	• High surface area • High π–π stacking interactions • Easily functionalized	+ Obtainable from abundant and renewable sources + Relatively cheap – Potentially difficult to recycle (graphene oxide, carbon nanotubes…) – Limited tunability	• Synthetic musk (Li et al., 2016) • UV filters (Vosough et al., 2018) • Chlorobenzenes and polychlorinated biphenyls (Cheng et al., 2019) • Chlorobenzenes (Xu et al., 2022)
Metallic nanomaterials	Nanosized metals (metal nanoparticles, nanowires, oxides, …)	• High surface areas with active sites • Catalytic properties • Magnetic responsiveness	+ Possibility of reusability – High toxicity (especially due to their small size) – Some rely on scarce resources (Au, Pt…) – Low stability when used as pristine sorbents	• UV filters (Wang et al., 2019) • Chlorophenols, triclosan and bisphenol A (Wang et al., 2018) • Endocrine disrupting chemicals (Akkaya et al., 2019) • PAHs (Guo et al., 2015)

Continued

TABLE 5.2 Types of sorbents used in sorbent-based microextraction, together with the main properties of interest in microextraction, green-related aspects, and selected examples of their application in the analysis of environmental waters—cont'd

Sorbent	Composition	Characteristics of interest in microextraction	Green-related aspects[a]	Examples in the analysis of water[b,c]
Metal-organic frameworks (MOFs)	Extended coordination networks formed by metal atoms or metallic clusters and organic linkers	• Ultra-high surface areas • Tunable porosity • Chemical and structural versatility	+ Possibility of reusability – Energy-intensive synthesis – Limited large-scale production	• PAHs and bisphenols (Wang et al., 2015) • BTEX and PAHs (Xie et al., 2015) • Parabens (Jakavula et al., 2023) • Drugs (Lirio et al., 2016)
Covalent-organic frameworks (COFs)	Extended covalent networks formed by light atoms (C, H, O, N, …)	• High surface area • Ordered and tunable porosity • High π–π stacking interactions	+ Possibility of reusability + Obtainable from abundant and renewable sources – Energy-intensive synthesis – Highly limited large-scale production (more limited than in the case of MOFs)	• Polybrominated diphenyl ethers (Liu et al., 2019) • Pesticides (Yang et al., 2023) • PFAS (Song et al., 2023) • Endocrine disruptors (Gao et al., 2022)

[a]*The "+" and "–" symbols denote advantages and disadvantages related to green-aspects, respectively.*

[b]*Certain examples imply composites instead of the pristine sorbents. In these cases, the sorbent evaluated in each entry was the main responsible of the extraction.*

[c]*BTEX, benzene, toluene, ethylbenzene, and xylene; PAH, polycyclic aromatic hydrocarbon.*

further expanded polymers use as supporting and synergistic-extracting material. Table 5.2 summarizes some examples of the use of these materials, as well as hybrid composites, in the analysis of CECs from environmental water samples.

Carbon-based materials include carbon nanotubes (CNTs), graphene (G) and its derivatives, and activated carbon (Sajid et al., 2021). These materials are synthesized via mechanical or chemical exfoliation, pyrolysis, chemical vapor deposition (CVD), and electric arc discharge, among others (Mbayachi et al., 2021; Prasek et al., 2011). Many of these methods can incorporate biomass-derived precursors or green chemistry approaches, enhancing sustainability. Their high surface area and pore volume provide excellent sorption capacity, which can be further improved through surface modifications. Moreover, their stability, reusability, and potential for regeneration, make them attractive for sustainable analytical applications. However, their hydrophobicity often leads to aggregation, reducing efficiency in μ-dSPE unless they are supported on other materials or functionalized with organic moieties (Sajid et al., 2021). In general, main features of C-based materials, along with their eco-friendly potential, have led to their widespread use in analyzing CECs from environmental water samples, as shown in Table 5.2.

MNMs are submicron-scale metal-based structures such as 0D nanoparticles (MNPs), 1D nanowires (MNWs), 2D nanostructures (nanoprisms, nanoplates, nanodisks, nanosheets), and 3D metal oxides (MOs) with controlled porosity (Sun and Zhu, 2024). Due to their high surface energy, tendency to agglomerate, susceptibility to oxidation, and rapid biodegradation, pure MNMs are rarely used as sorbents. These challenges, along with selectivity limitations, are addressed through surface modification and functionalization (Sajid et al., 2021). In this sense, composite materials have been developed to combine MNMs properties, such as magnetism, high surface area, and thermal stability, with the adsorption capacity and selectivity of other extractants. In addition, concerns over their toxicity have driven research into green synthesis strategies (Wei et al., 2016), with environmentally friendly MNMs increasingly applied in pollutant separation and preconcentration in environmental matrices (El-Deen and Hussain, 2023), as illustrated in Table 5.2.

Porous crystalline polymers built from coordination bonds (MOFs) or covalent bonds (COFs) are outstanding materials for analytical applications due to their highly regular, tunable porosity and exceptionally high specific surface area (Zuliani et al., 2023). Their well-defined structures enable selective extraction and preconcentration of target analytes, making them valuable in analytical sample preparation and separation techniques (Zuliani et al., 2023). However, their stability in water and organic solvents can be a limiting factor for long-term use, particularly for certain MOFs that undergo hydrolysis or ligand displacement under humid conditions (Ding et al., 2019). In any case, considering the high number of MOFs already described (Freund et al., 2021), it is clear that many of them are certainly chemical stable and indeed those MOFs have been the most studied for environmental applications.

Sustainability is a critical concern, as their synthesis often relies on solvothermal methods involving hazardous reagents and toxic organic solvents. In addition, the potential degradation of these materials can release harmful organic and inorganic byproducts (Rocío-Bautista et al., 2019). To address these challenges, research has focused on developing green MOFs (El-Sayed and Yuan, 2020) and green COFs (Wang et al., 2021), using environmentally friendly precursors, milder synthesis conditions, and solvent-free or aqueous-phase methods. COFs, in particular, offer advantages in terms of chemical stability, recyclability, and metal-free frameworks, making them attractive alternatives in GAC. Nevertheless, the preparation of these materials at high scale (or a scale enough for their use in analytical methods) often is a challenge, and for that their use in microextraction is more limited compared to MOFs.

3 Analysis of the sustainability and applicability of microextraction techniques

Undoubtedly, GAC practices have broadened the vision with respect to the characteristics that an analytical method must fulfill. In fact, together with achieving a proper analytical performance, additional green and sustainable aspects must be taken into account nowadays. GAC principles include certain measurements that directly impact the way analytical sample preparation is performed, including the miniaturization of the sample size (principle 2) and methods (principle 5), their automation (principle 5), the minimization of the amount of wastes (principle 7), the need to decrease the energy consumption (principle 9), and the use of more biocompatible and less toxic reagents (principle 10) (Gałuszka et al., 2013). Green sample preparation (GSP) philosophy further debates about the features required to make a sample preparation strategy as sustainable as possible (López-Lorente et al., 2022). At the same time, WAC practices imply a more realistic scenario where three main aspects are considered: sustainability, applicability, and analytical performance (commonly referred to as green, blue, and red aspects, respectively) (Nowak et al., 2021).

To evaluate the compliance of these green and blue aspects (as the analytical performance for environmental analysis should be indeed compiled) and to compare different analytical methods based on these criteria, several metrics have been developed in the last years, especially with respect to sustainability (Hammad et al., 2025). The most recent and used metrics normally imply the construction of a visual pictogram with a final score that quantifies the overall green (or blue) character of the analytical method. In addition to this, the pictograms are divided in different subsections, each related to different green (or blue) criteria, and are depicted with a color coding that indicates a higher or lower compliance of each specific criterion. Among the green metrics, it is interesting to point out the Analytical greenness metric for sample preparation (AGREEprep) (Wojnowski et al., 2022) and the Sample preparation metric of sustainability (SPMS) (González-Martín et al., 2023) due to their wide use and because they are specifically focused on sample preparation. AGREEprep gives a final score between 0 and 1 (with 1 being the most ideal green metric based on the studied criteria) and with color coding from red to orange, yellow, and finally green (green indicating a better compliance of the studied criterion). Similarly, SPMS provides a score from 0 to 10, being 10 the greenest sample preparation method, and color coding varying in the same manner as AGREEprep.

With respect to blue metrics, the Blue Applicability Grade Index (BAGI) is worth mentioning (Manousi et al., 2023). This metric considers attributes of the analytical method related to applicability such as the number of analytes that can be determined in a single analysis, the type of sample preparation and instrumentation, the degree of automation, and the sample throughput. The BAGI tool provides a final score between 0 and 100 (being 100 the highest possible applicability for analytical method), and with subsections with color coding from white to different tones of blue (with dark blue indicating the total compliance of each studied attribute).

In this section, the aforementioned three metrics (AGREE-prep, SPMS, and BAGI) are applied to representative analytical methods based on microextraction and devoted to the determination of CECs in environmental waters. Therefore the objective was to compare between the different microextraction strategies previously discussed and to obtain insights about these additional green and applicability aspects of the techniques. Figs. 5.1 and 5.2 include the pictograms obtained for each of the selected analytical methods.

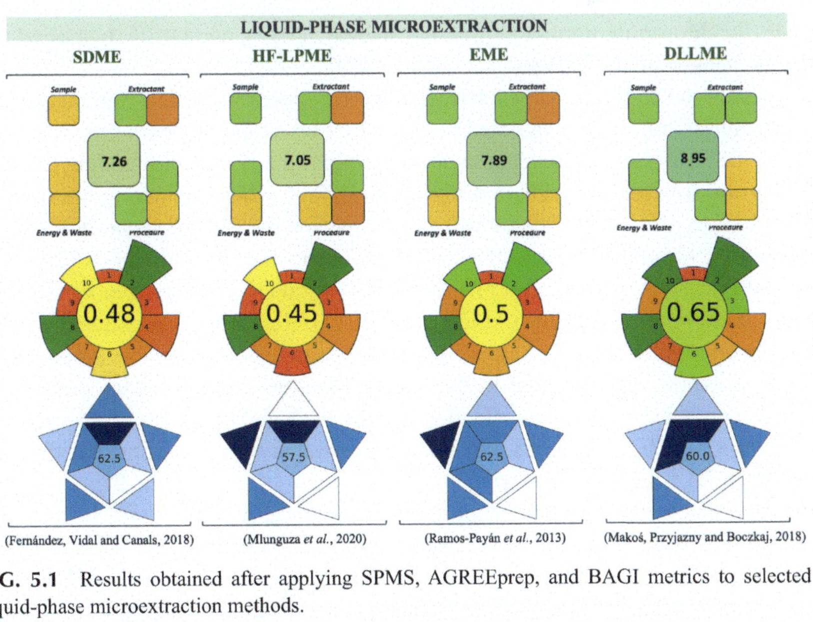

FIG. 5.1 Results obtained after applying SPMS, AGREEprep, and BAGI metrics to selected liquid-phase microextraction methods.

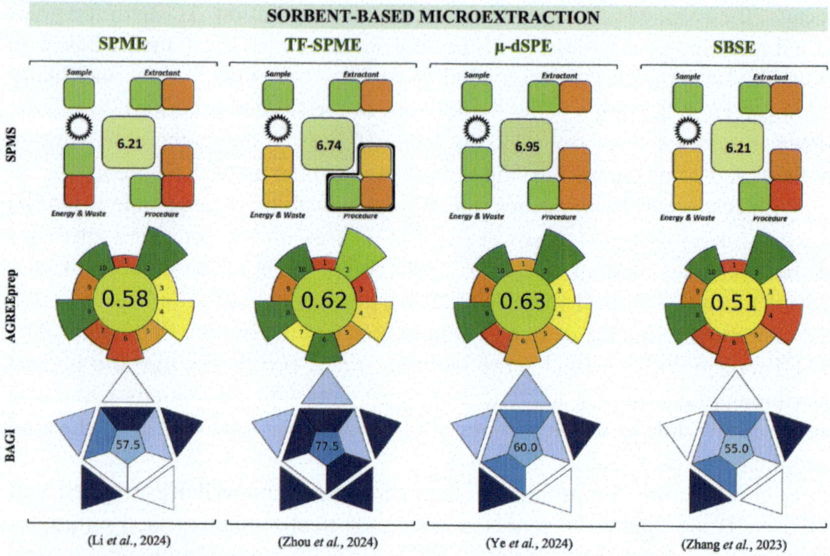

FIG. 5.2 Results obtained after applying SPMS, AGREEprep, and BAGI metrics to selected sorbent-based microextraction methods.

3.1 Liquid-phase microextraction

Fig. 5.1 includes the obtained pictograms for four representative methods, specifically: HS-SDME for determining chlorobenzenes (Fernández et al., 2018), three phases HF-LPME applied for antiretroviral drugs (Mlunguza et al., 2020), EME for fluoroquinolones (Ramos-Payán et al., 2013), and DLLME applied for PAHs (Makoś et al., 2018).

In relation to the SPMS metric, the overall scores indicate a similar degree of greenness for all the evaluated microextraction strategies, with scores higher than 7.05 in all cases. However, a deeper evaluation of the color coding of each pictogram subsection revealed differences. In fact, when evaluating the sample information, the lowest scores were achieved for the SDME technique due to the higher sample volume (20 mL) in relation to the other methods (10 mL). In any case, it is important to mention that several studies have reported lower sample volumes for SDME (Amde et al., 2015). Regarding the extraction step, the observed differences relate to the nature of the solvent. In this sense, most of the solvents implemented in the selected studies were considered as alternative persistent (e.g., magnetic ILs (Mafra et al., 2019), switchable solvents (Cheng et al., 2020) and thus they were penalized in the pictogram. The method based on DLLME provided a greener score due to the nature of the extraction material that was a DES based on thymol, camphor, decanoic, and 10-undecylenic acids (Makoś et al., 2018). Finally, if considering the procedure information, DLLME required the highest number of steps (>3), whereas the HF-LPME required the highest extraction time (60 min) compared to the others (<30 min).

The results of AGREEprep were in general worse in terms of greenness with respect to those related to SPMS, with scores higher than 0.45, likely because this metric takes into consideration aspects related to the overall analytical method (e.g., overall waste and type of analytical instrumentation). The most important differences between the scores were related to criterion 6 (i.e., maximize sample throughput) and criterion 10 (i.e., ensure safe procedures for the operator). The criterion 6 is assessed as the number of samples that can be analyzed per hour and in this sense, HF-LPME had the lower score as it requires a higher extraction time than remaining strategies (i.e., 60 min versus 10–30 min). Criterion 10 is related to the type of reagents and the nature of the extractant, and it is measured by the number of warning and danger pictograms on the reagents data sheet. For this reason, the DES-DLLME procedure provided the higher score, whereas the di-(2-ethylhexyl) phosphoric acid/dihexyl ether employed for the HF-LPME approach scored the lowest.

Finally, all microextraction procedures provided similar scores in the BAGI metric, with values higher than 57.5, indicating similar applicability. Methods with scores above 60 are considered applicable when using this metric (Manousi et al., 2023). In any case, it is important to mention that those aspects that decreased the applicability of the selected analytical methods were mainly related to the absence of quantitative and confirmatory analysis, lack of simultaneous sample preparation, the relatively low sample throughput (<10 samples per hour), and the low degree of automation.

3.2 Sorbent-based microextraction

The obtained scores and pictograms for the selected analytical methods are summarized in Fig. 5.2. The selected methods for the comparison were: HS-SPME for the determination of PAHs (Li et al., 2024), TF-SPME blades for monitoring drugs (Zhou et al., 2024), μ-dSPE applied for extracting glucocorticoids (Ye et al., 2024), and SBSE for the monitoring of polychlorinated biphenyls (Zhang et al., 2023).

The scores obtained in SPMS were similar in all the examples selected, with values higher than 6.21. These scores were slightly lower than those assessed for LPME methods (Fig. 5.1), likely due to the higher energy requirements for performing the procedures. When comparing the SPMS pictograms of the sorbent-based methods, the main differences among the techniques lie in the procedure, energy consumption, and waste generation. Most methods involved the same number of steps and did not require additional procedures after extraction. However, extraction time and sample throughput favor TF-SPME and μ-dSPE. Regarding energy consumption, the SPME procedure was penalized due to the use of heating during extraction (as the specific application chosen for the comparison was HS-SPME), while the use of a stir plate in TF-SPME and SBSE was considered less sustainable compared to the use of an ultrasound in μ-dSPE.

With regard to AGREEprep, the obtained scores were higher than 0.51 in all cases and, in general, higher than those obtained for LPME methods (Fig. 5.1).

Criteria 2 and 10, which assess the safety of the procedure, were met in all cases, likely due to the almost solvent-free character of these techniques. Indeed, the use of toxic solvents is reduced to 100 μL of acetonitrile as desorption solvent in the TF-SPME example, and in many instances, safer alternatives such as methanol can be employed, such as in the example of μ-dSPE. Although the use of sustainable materials is desirable, most of the examples chosen for this comparison did not employ green extractants. In fact, they involved nonbiodegradable polymers with a polyacrylonitrile (PAN) binder (Zhou et al., 2024), polydopamine coated iron tetroxide (Ye et al., 2024), and porous aromatic framework/PDMS materials (Zhang et al., 2023). However, this negative aspect was compensated by their reusability, achieving a medium score in attribute 3. Sample throughput was evaluated in criterion 6, where TF-SPME and μ-dSPE demonstrated superior performance compared to techniques with slower sorption rates.

The applicability of the techniques assessed by BAGI indicated different performance. In this case, most of the studies provided an average score of 60, except for the TF-SPME method, which scored a 77.5. This score is the result of the automation of the procedure and the possibility of performing parallel extraction, which improved the scoring of the criterion related to simultaneous sample preparation, standing out in comparison with remaining techniques. Attribute 6 (samples per hour) also indicated that the selected SPME and SBSE examples were less efficient than TF-SPME and μ-dSPE due to their longer extraction and desorption times. Despite these differences, all techniques achieve the highest score in attribute 10, related to sample volume, due to their miniaturization principles.

4 Conclusions

Analytical microextraction offers several advantages for CEC monitoring in environmental waters. In the last few years, several LPME and solid-based microextraction strategies have been successfully used for monitoring CECs in these samples. All these strategies are considered greener than conventional methods because of their microscale character, although they often implement additional features to further increase sustainability. A common strategy is to introduce novel greener and/or advanced solvents and sorbents. In this sense, research is focused on natural materials, simple synthetic protocols, and biodegradable extraction phases, all in agreement in GAC and WAC.

A simple way to evaluate the sustainable character of a microextraction method and its applicability is through green and blue metrics such as SPMS, AGREEprep, and BAGI. This chapter demonstrated that most microextraction-based analytical methods are generally green, although there are certain aspects that differ from technique to technique and certain attributes like sample throughput and energy consumption should be improved in the future.

Acknowledgments

This work was developed within the scope of the projects ref. PID2023-147246OB-I00 and ref. PID2022-137822NA-I00, both funded by the Spanish Ministry of Science, Innovation, and Universities and the European Regional Development Fund MICIU/AEI/10.13039/501100011033/ERDF/EU. N.M.-R. thanks his FPI predoctoral contract, co-financed by the Research Canary Agency "ACIISI" and by European Social Fund Plus (ESF+), integrated operation program Canary Islands 2021–2027, Axis 3 topic priority 74 (85%). I.G–M. thanks her FPI predoctoral contract linked to the project. ref. PID2023-147246OB-I00, contract funded by MICIU/AEI/10.13039/501100011033/ and the European Social Fund Plus (FSE+), and University of La Laguna. M.J.T.-R. thanks her current Ramón y Cajal contract (ref. RYC2021-032502-I) at Universidad de La Laguna, contract with funding of MICIU/AEI/10.13039/501100011033 (previously MCIN), and the European Union «NextGenerationEU»/PRTR; and both MICIU/AEI and Universidad de La Laguna for the additional funding related to the contract.

References

Abolghasemi, M.M., Piryaei, M., Imani, R.M., 2020. Deep eutectic solvents as extraction phase in head-space single-drop microextraction for determination of pesticides in fruit juice and vegetable samples. Microchem. J. 0026-265X. 158, 105041. doi:10.1016/j.microc.2020.105041.

Akkaya, E., Bozyiğit, G.D., Bakirdere, S., 2019. Simultaneous determination of 4-tert-octylphenol, chlorpyrifos-ethyl and penconazole by GC–MS after sensitive and selective preconcentration with stearic acid coated magnetic nanoparticles. Microchem. J. 0026-265X. 146, 1190–1194. doi:10.1016/j.microc.2019.01.077.

Allgaier-Díaz, D.W., Trujillo-Rodríguez, M.J., Ayala, J.H., Díaz Díaz, D., Pino, V., 2023. Unmodified biopolymers as sustainable microextraction materials for the environmental monitoring of polycyclic aromatic hydrocarbons and personal care products. Microchem. J. 0026-265X. 191, 108873. doi:10.1016/j.microc.2023.108873.

Alonso, M.L., San Román, I., Bartolomé, L., Monfort, N., Alonso, R.M., Ventura, R., 2022. Multiple headspace solid-phase microextraction (MHS-SPME) methodology applied to the determination of volatile metabolites of plasticizers in human urine. Microchem. J. 0026-265X. 180, 107567. doi:10.1016/j.microc.2022.107567.

Amde, M., Tan, Z.Q., Liu, R., Liu, J.F., 2015. Nanofluid of zinc oxide nanoparticles in ionic liquid for single drop liquid microextraction of fungicides in environmental waters prior to high performance liquid chromatographic analysis. J. Chromatogr. A. 1873-3778. 1395, 7–15. doi:10.1016/j.chroma.2015.03.049.

Ansari, S., Masoum, S., 2020. Ultrasound-assisted dispersive solid-phase microextraction of capecitabine by multi-stimuli responsive molecularly imprinted polymer modified with chitosan nanoparticles followed by HPLC analysis. Microchim. Acta. 0026-3672. 187 (6), 366. doi:10.1007/s00604-020-04345-0.

Arabkhani, B., Goudarzi, N., Arab Chamjangali, M., 2025. Development of a sensitive and accurate method for the simultaneous determination of climbazole and clotrimazole using β-cyclodextrin assisted liquid–liquid microextraction based on switchable solvent. Chem. Pap. 0366-6352. 79 (3), 1945–1957. doi:10.1007/s11696-025-03900-4.

Armenta, S., Garrigues, S., de la Guardia, M., 2008. Green analytical chemistry. TrAC. Trends. Anal. Chem. 0165-9936. 27 (6), 497–511. doi:10.1016/j.trac.2008.05.003.

Arthur, C.L., Pawliszyn, J., 1990. Solid phase microextraction with thermal desorption using fused silica optical fibers . Anal. Chem. 0003-2700. 62, 2145–2148. doi:10.1021/ac00218a019.

Asensio-Ramos, M., Hernández-Borges, J., González-Hernández, G., Rodríguez-Delgado, M.Á., 2012. Hollow-fiber liquid-phase microextraction for the determination of pesticides and metabolites in soils and water samples using HPLC and fluorescence detection. Electrophoresis. 1522-2683. 33 (14), 2184–2191. doi:10.1002/elps.201200138.

Balcerak-Woźniak, A., Dzwonkowska-Zarzycka, M., Kabatc-Borcz, J., 2024. A comprehensive review of stimuli-responsive smart polymer materials-recent advances and future perspectives. Materials. 1996-1944. 17 (17), 4255. doi:10.3390/ma17174255.

Baltussen, E., Sandra, P., David, F., Cramers, C., 1999. Stir bar sorptive extraction (SBSE), a novel extraction technique for aqueous samples: theory and principles. J. Microcolumn. Sep. 1040-7685. 11 (10), 737–747. doi:10.1002/(sici)1520-667x(1999)11:10<737::aid-mcs7>3.0.co;2-4.

Benedé, J.L., Chisvert, A., Giokas, D.L., Salvador, A., 2014. Development of stir bar sorptive-dispersive microextraction mediated by magnetic nanoparticles and its analytical application to the determination of hydrophobic organic compounds in aqueous media. J. Chromatogr. A. 1873-3778. 1362, 25–33. doi:10.1016/j.chroma.2014.08.024.

Bruheim, I., Liu, X., Pawliszyn, J., 2003. Thin-film microextraction. Anal. Chem. 0003-2700. 75 (4), 1002–1010. doi:10.1021/ac026162q.

Cheng, H., Wang, F., Bian, Y., Ji, R., Song, Y., Jiang, X., 2019. Co- and self-activated synthesis of tailored multimodal porous carbons for solid-phase microextraction of chlorobenzenes and polychlorinated biphenyls. J. Chromatogr. A. 0021-9673. 1585, 1–9. doi:10.1016/j.chroma.2018.11.047.

Cheng, J., Guo, H., Qiu, Y., Zhang, Z., Mao, Y., Qian, L., Yang, W., Park, J.-Y., 2020. Switchable solvent N, N, N′, N′-tetraethyl-1, 3-propanediamine was dissociated into cationic surfactant to promote cell disruption and lipid extraction from wet microalgae for biodiesel production. Bioresour. Technol. 0960-8524. 312, 123607. doi:10.1016/j.biortech.2020.123607.

Chisvert, A., Cárdenas, S., Lucena, R., 2019. Dispersive micro-solid phase extraction. TrAC. Trends. Anal. Chem. 0165-9936. 112, 226–233. doi:10.1016/j.trac.2018.12.005.

Cho, C.W., Pham, T.P.T., Zhao, Y., Stolte, S., Yun, Y.S., 2021. Review of the toxic effects of ionic liquids. Sci. Total. Environ. 1879-1026. 786, 147309. doi:10.1016/j.scitotenv.2021.147309.

Crucello, J., Sampaio, N.M., Junior, I.M., Carvalho, R.M., Gionfriddo, E., Marriott, P.J., Hantao, L.W., 2023. Automated method using direct-immersion solid-phase microextraction and on-fiber derivatization coupled with comprehensive two-dimensional gas chromatography high-resolution mass spectrometry for profiling naphthenic acids in produced water. J. Chromatogr. A. 1873-3778. 1692, 463844. doi:10.1016/j.chroma.2023.463844.

Cunha, S.C., Ferreira, R., Marmelo, I., Vieira, L.R., Anacleto, P., Maulvault, A., Marques, A., Guilhermino, L., Fernandes, J.O., 2022. Occurrence and seasonal variation of several endocrine disruptor compounds (pesticides, bisphenols, musks and UV-filters) in water and sediments from the estuaries of Tagus and Douro Rivers (NE Atlantic Ocean coast). Sci. Total. Environ. 1879-1026. 838, 155814. doi:10.1016/j.scitotenv.2022.155814.

Demirkurt, M., Olcer, Y.A., Demir, M.M., Eroglu, A.E., 2018. Electrospun polystyrene fibers knitted around imprinted acrylate microspheres as sorbent for paraben derivatives. Anal. Chim. Acta. 0003-2670. 1014, 1–9. doi:10.1016/j.aca.2018.02.016.

Dil, E.A., Doustimotlagh, A.H., Javadian, H., Asfaram, A., Ghaedi, M., 2021. Nano-sized Fe3O4@ SiO2-molecular imprinted polymer as a sorbent for dispersive solid-phase microextraction of melatonin in the methanolic extract of Portulaca oleracea, biological, and water samples. Talanta. 0039-9140. 221, 121620. doi:10.1016/j.talanta.2020.121620.

Ding, M., Cai, X., Jiang, H.L., 2019. Improving MOF stability: approaches and applications. Chem. Sci. 2041-6539. 10 (44), 10209–10230. doi:10.1039/c9sc03916c.

Drouin, N., Kubáň, P., Rudaz, S., Pedersen-Bjergaard, S., Schappler, J., 2019. Electromembrane extraction: overview of the last decade. TrAC. Trends. Anal. Chem. 0165-9936. 113, 357–363. doi:10.1016/j.trac.2018.10.024.

Dueñas-Mas, M.J., Ballesteros-Gómez, A., Rubio, S., 2019. Supramolecular solvent-based microextraction of emerging bisphenol A replacements (colour developers) in indoor dust from public environments. Chemosphere. 1879-1298. 222, 22–28. doi:10.1016/j.chemosphere.2019.01.095.

El-Deen, A.K., Hussain, C.M., 2023. Recent developments in green magnetic nanoparticles for extraction and preconcentration of pollutants from environmental samples. Trends. Environ. Anal. Chem. 2214-1588. 39, e00211. doi:10.1016/j.teac.2023.e00211.

El-Sayed, E.S.M., Yuan, D., 2020. Waste to MOFs: sustainable linker, metal, and solvent sources for value-added MOF synthesis and applications. Green. Chem. 1463-9270. 22 (13), 4082–4104. doi:10.1039/d0gc00353k.

Emmons, R.V., Tajali, R., Gionfriddo, E., 2019. Development, optimization and applications of thin film solid phase microextraction (TF-SPME) devices for thermal desorption: a comprehensive review. Separations. 2297-8739. 6 (3), 39. doi:10.3390/separations6030039.

Esrafili, A., Baharfar, M., Tajik, M., Yamini, Y., Ghambarian, M., 2018. Two-phase hollow fiber liquid-phase microextraction. TrAC. Trends. Anal. Chem. 0165-9936. 108, 314–322. doi:10.1016/j.trac.2018.09.015.

Farooq, M.Q., Ocaña-Rios, I., Anderson, J.L., 2022. Analysis of persistent contaminants and personal care products by dispersive liquid-liquid microextraction using hydrophobic magnetic deep eutectic solvents. J. Chromatogr. A. 1873-3778. 1681, 463429. doi:10.1016/j.chroma.2022.463429.

Feng, J., Qiu, H., Liu, X., Jiang, S., Feng, J., 2013. The development of solid-phase microextraction fibers with metal wires as supporting substrates. TrAC. Trends. Anal. Chem. 0165-9936. 46, 44–58. doi:10.1016/j.trac.2013.01.015.

Fernández, E., Vidal, L., Canals, A., 2018. Hydrophilic magnetic ionic liquid for magnetic headspace single-drop microextraction of chlorobenzenes prior to thermal desorption-gas chromatography-mass spectrometry. Anal. Bioanal. Chem. 1618-2642. 410 (19), 4679–4687. doi:10.1007/s00216-017-0755-2.

Freund, R., Zaremba, O., Arnauts, G., Ameloot, R., Skorupskii, G., Dincă, M., Bavykina, A., Gascon, J., Ejsmont, A., Goscianska, J., Kalmutzki, M., Lächelt, U., Ploetz, E., Diercks, C.S., Wuttke, S., 2021. The current status of MOF and COF applications. Angew. Chem. Int. Ed. 1433-7851. 60 (45), 23975–24001. doi:10.1002/anie.202106259.

Gałuszka, A., Migaszewski, Z., Namieśnik, J., 2013. The 12 principles of green analytical chemistry and the SIGNIFICANCE mnemonic of green analytical practices. TrAC. Trends Anal. Chem. 0165-9936. 50, 78–84. doi:10.1016/j.trac.2013.04.010.

Gao, W., Li, M., Fa, Y., Zhao, Z., Cai, Y., Liang, X., Yu, Y., Jiang, G., 2022. Porous covalent organic frameworks-improved solid phase microextraction ambient mass spectrometry for ultrasensitive analysis of tetrabromobisphenol-A analogs. Chin. Chem. Lett. 1001-8417. 33 (8), 3849–3852. doi:10.1016/j.cclet.2021.10.030.

Ghambarian, M., Yamini, Y., Esrafili, A., 2012. Developments in hollow fiber based liquid-phase microextraction: principles and applications. Microchim. Acta. 0026-3672. 177 (3–4), 271–294. doi:10.1007/s00604-012-0773-x.

Gjelstad, A., 2019. Three-phase hollow fiber liquid-phase microextraction and parallel artificial liquid membrane extraction. TrAC. Trends Anal. Chem. 0165-9936. 113, 25–31. doi:10.1016/j.trac.2019.01.007.

Gjelstad, A., Pedersen-Bjergaard, S., 2013. Recent developments in electromembrane extraction. Anal. Methods. 1759-9679. 5 (18), 4549–4557. doi:10.1039/c3ay40547h.

Gjelstad, A., Rasmussen, K.E., Pedersen-Bjergaard, S., 2006. Electrokinetic migration across artificial liquid membranes. Tuning the membrane chemistry to different types of drug substances. J. Chromatogr. A. 0021-9673. 1124 (1–2), 29–34. doi:10.1016/j.chroma.2006.04.039.

González-Hernández, P., Pacheco-Fernández, I., Bernardo, F., Homem, V., Pasán, J., Ayala, J.H., Ratola, N., Pino, V., 2021. Headspace solid-phase microextraction based on the metal-organic framework CIM-80(Al) coating to determine volatile methylsiloxanes and musk fragrances in water samples using gas chromatography and mass spectrometry. Talanta. 0039-9140. 232, 122440. doi:10.1016/j.talanta.2021.122440.

González-Martín, R., Gutiérrez-Serpa, A., Pino, V., Sajid, M., 2023. A tool to assess analytical sample preparation procedures: sample preparation metric of sustainability. J. Chromatogr. A. 0021-9673. 1707, 464291. doi:10.1016/j.chroma.2023.464291.

González-Sálamo, J., González-Curbelo, M.Á., Socas-Rodríguez, B., Hernández-Borges, J., Rodríguez-Delgado, M.Á., 2018. Determination of phthalic acid esters in water samples by hollow fiber liquid-phase microextraction prior to gas chromatography tandem mass spectrometry. Chemosphere. 1879-1298. 201, 254–261. doi:10.1016/j.chemosphere.2018.02.180.

Grecco, C.F., de Souza, I.D., Oliveira, I.G.C., Queiroz, M.E.C., 2022. In-tube solid-phase microextraction directly coupled to mass spectrometric systems: a review. Separations. 2297-8739. 9 (12), 394. doi:10.3390/separations9120394.

Guo, M., Song, W., Wang, T., Li, Y., Wang, X., Du, X., 2015. Phenyl-functionalization of titanium dioxide-nanosheets coating fabricated on a titanium wire for selective solid-phase microextraction of polycyclic aromatic hydrocarbons from environment water samples. Talanta. 0039-9140. 144, 998–1006. doi:10.1016/j.talanta.2015.07.064.

Guo, X., Yin, D., Peng, J., Hu, X., 2012. Ionic liquid-based single-drop liquid-phase microextraction combined with high-performance liquid chromatography for the determination of sulfonamides in environmental water. J. Sep. Sci. 1615-9306. 35 (3), 452–458. doi:10.1002/jssc.201100777.

Gutiérrez-Serpa, A., Pasán, J., Jiménez-Abizanda, A.I., Kaskel, S., Senkovska, I., Pino, V., 2022. Thin-film microextraction using the metal-organic framework DUT-52 for determining endocrine disrupting chemicals in cosmetics. Microchem. J. 0026-265X. 181, 107685. doi:10.1016/j.microc.2022.107685.

Habibollahi, M.H., Karimyan, K., Arfaeinia, H., Mirzaei, N., Safari, Y., Akramipour, R., Sharafi, H., Fattahi, N., 2019. Extraction and determination of heavy metals in soil and vegetables irrigated with treated municipal wastewater using new mode of dispersive liquid–liquid microextraction based on the solidified deep eutectic solvent followed by GFAAS. J. Sci. Food. Agric. 1097-0010. 99 (2), 656–665. doi:10.1002/jsfa.9230.

Hai, X., Niu, J., Ren, T., Fu, R., Li, H., Zhang, Y., Guo, X., Di, X., 2024. Rhamnolipids-based bio-supramolecular solvents as green and sustainable media for extraction of pyrethroid insecticides in water and food matrices. J. Chromatogr. A. 0021-9673. 1731, 465215. doi:10.1016/j.chroma.2024.465215.

Hammad, S.F., Hamid, M.A.A., Adly, L., Elagamy, S.H., 2025. Comprehensive review of greenness, whiteness, and blueness assessments of analytical methods. Green. Anal. Chem. 2772-5774. 12, 100209. doi:10.1016/j.greeac.2025.100209.

Hasan, C.K., Ghiasvand, A., Lewis, T.W., Nesterenko, P.N., Paull, B., 2020. Recent advances in stir-bar sorptive extraction: coatings, technical improvements, and applications. Anal. Chim. Acta. 1873-4324. 1139, 222–240. doi:10.1016/j.aca.2020.08.021.

Hejabri Kandeh, S., Amini, S., Ebrahimzadeh, H., 2022. Development of poly(vinyl alcohol)/chitosan/aloe vera gel electrospun composite nanofibers as a novel sorbent for thin-film micro-extraction of pesticides in water and food samples followed by HPLC-UV analysis. New. J. Chem. 1144-0546. 46 (5), 2431–2440. doi:10.1039/d1nj05634d.

Herrington, J.S., Gómez-Ríos, G.A., Myers, C., Stidsen, G., Bell, D.S., 2020. Hunting molecules in complex matrices with SPME arrows: a review. Separations. 2297-8739. 7 (1), 12. doi:10.3390/separations7010012.

Hu, L., Wang, H., Qian, H., Liu, C., Lu, R., Zhang, S., Zhou, W., Gao, H., Xu, D., 2016. Centrifuge-less dispersive liquid-liquid microextraction base on the solidification of switchable solvent for rapid on-site extraction of four pyrethroid insecticides in water samples. J. Chromatogr. A. 0021-9673. 1472, 1–9. doi:10.1016/j.chroma.2016.10.013.

Huang, C., Chen, Z., Gjelstad, A., Pedersen-Bjergaard, S., Shen, X., 2017. Electromembrane extraction. TrAC. Trends Anal. Chem. 0165-9936. 95, 47–56. doi:10.1016/j.trac.2017.07.027.

Huang, C., Eibak, L.E.E., Gjelstad, A., Shen, X., Trones, R., Jensen, H., Pedersen-Bjergaard, S., 2014. Development of a flat membrane based device for electromembrane extraction: a new approach for exhaustive extraction of basic drugs from human plasma. J. Chromatogr. A. 1873-3778. 1326, 7–12. doi:10.1016/j.chroma.2013.12.028.

Jain, A., Verma, K.K., 2011. Recent advances in applications of single-drop microextraction: a review. Anal. Chim. Acta. 1873-4324. 706 (1), 37–65. doi:10.1016/j.aca.2011.08.022.

Jakavula, S., Nqombolo, A., Mpupa, A., Ren, J., Nomngongo, P.N., 2023. Hybrid porous material supported in a cellulose acetate polymeric membrane for the direct immersion thin-film microextraction of parabens in water. J. Chromatogr. A. 1873-3778. 1705, 464187. doi:10.1016/j.chroma.2023.464187.

Jalili, V., Barkhordari, A., Ghiasvand, A., 2020. Liquid-phase microextraction of polycyclic aromatic hydrocarbons: a review. Rev. Anal. Chem. 2191-0189. 39 (1), 1–19. doi:10.1515/revac-2020-0101.

Jeannot, M.A., Cantwell, F.F., 1996. Solvent microextraction into a single drop. Anal. Chem. 0003-2700. 68 (13), 2236–2240. doi:10.1021/ac960042z.

Johnson, T.A., Armstrong, M.D.S., de la Mata, A.P., Harynuk, J.J., 2022. Exploration of extraction and separation techniques for routine trace analysis of organic compounds in water: dispersive liquid-liquid microextraction vs liquid-liquid extraction. J. Chromatogr. Open. 2772-3917. 2, 100070. doi:10.1016/j.jcoa.2022.100070.

Karimiyan, H., Hadjmohammadi, M., 2016. Ultrasound-assisted supramolecular-solvent-based microextraction combined with high-performance liquid chromatography for the analysis of chlorophenols in environmental water samples. J. Sep. Sci. 1615-9306. 39 (24), 4740–4747. doi:10.1002/jssc.201600941.

Kataoka, H., 2021. In-tube solid-phase microextraction: current trends and future perspectives. J. Chromatogr. A. 0021-9673. 1636, 461787. doi:10.1016/j.chroma.2020.461787.

Kaya, S.I., Cetinkaya, A., Ozkan, S.A., 2023. Molecularly imprinted polymers as highly selective sorbents in sample preparation techniques and their applications in environmental water analysis. Trends Environ. Anal. Chem. 2214-1588. 37, e00193. doi:10.1016/j.teac.2022.e00193.

Kenessov, B., Muratuly, A., 2024. Evaluation of the vacuum effect on headspace solid-phase microextraction of volatile organic compounds from aqueous samples using finite element analysis modeling. J. Chromatogr. Open. 2772-3917. 6, 100148. doi:10.1016/j.jcoa.2024.100148.

Khezeli, T., Daneshfar, A., 2017. Development of dispersive micro-solid phase extraction based on micro and nano sorbents. TrAC. Trends Anal. Chem. 0165-9936. 89, 99–118. doi:10.1016/j.trac.2017.01.004.

Kokosa, J.M., 2019. Dispersive liquid-liquid microextraction. In: Poole, C.F. (Ed.), Handbooks in Separation Science: Liquid-Phase Extraction. Elsevier, United States, pp. 473–497. doi:10.1016/B978-0-12-816911-7.00016-5. 978-0-12-816911-7.

Kong, J., Gao, Z., Hu, G., Huang, W., Zhou, S., He, H., Xian, Q., Sun, C., 2022. Solid-phase microextraction combined with gas chromatography/triple quadrupole tandem mass spectrometry for determination of nitrated polycyclic aromatic hydrocarbons in sediments. J. Sep. Sci. 1615-9306. 45 (5), 1094–1105. doi:10.1002/jssc.202100644.

Kubáň, P., Boček, P., 2014. Micro-electromembrane extraction across free liquid membranes. Extractions of basic drugs from undiluted biological samples. J. Chromatogr. A. 0021-9673. 1337, 32–39. doi:10.1016/j.chroma.2014.02.046.

Lasarte-Aragonés, G., Lucena, R., Cárdenas, S., Valcárcel, M., 2015. Use of switchable hydrophilicity solvents for the homogeneous liquid–liquid microextraction of triazine herbicides from environmental water samples. J. Sep. Sci. 1615-9306. 38 (6), 990–995. doi:10.1002/jssc.201401224.

Lei, Z., Chen, B., Koo, Y.M., Macfarlane, D.R., 2017. Introduction: ionic liquids. Chem. Rev. 1520-6890. 117 (10), 6633–6635. doi:10.1021/acs.chemrev.7b00246.

Lemos, V.A., Barreto, J.A., Santos, L.B., de Assis, R.D.S., Novaes, C.G., Cassella, R.J., 2022. In-syringe dispersive liquid-liquid microextraction. Talanta. 0039-9140. 238, 123002. doi:10.1016/j.talanta.2021.123002.

Leong, M.I., Fuh, M.R., Huang, S.D., 2014. Beyond dispersive liquid-liquid microextraction. J. Chromatogr. A. 1873-3778. 1335, 2–14. doi:10.1016/j.chroma.2014.02.021.

Li, G., Row, K.H., 2022. Single-drop microextraction technique for the determination of antibiotics in environmental water. J. Sep. Sci. 1615-9314. 45 (4), 883–895. doi:10.1002/jssc.202100682.

Li, K., Jin, Y., Jung, D., Park, K., Kim, H., Lee, J., 2020. In situ formation of thymol-based hydrophobic deep eutectic solvents: application to antibiotics analysis in surface water based on liquid-liquid microextraction followed by liquid chromatography. J. Chromatogr. A. 0021-9673. 1614, 460730. doi:10.1016/j.chroma.2019.460730.

Li, P., Han, Y., Han, D., Yan, H., 2024. Development of a high-efficiency and environmentally friendly melamine–formaldehyde resin-based solid phase microextraction fiber for enhanced extraction of polycyclic aromatic hydrocarbons from environmental water. Green. Chem. 1463-9262. 26 (6), 3211–3220. doi:10.1039/d3gc04764d.

Li, S., Zhu, F., Jiang, R., Ouyang, G., 2016. Preparation and evaluation of amino modified graphene solid-phase microextraction fiber and its application to the determination of synthetic musks in water samples. J. Chromatogr. A. 0021-9673. 1429, 1–7. doi:10.1016/j.chroma.2015.11.025.

Liang, M., Xian, Y., Wang, B., Hou, X., Wang, L., Guo, X., Wu, Y., Dong, H., 2020. High throughput analysis of 21 perfluorinated compounds in drinking water, tap water, river water and plant effluent from southern China by supramolecular solvents-based microextraction coupled with HPLC-Orbitrap HRMS. Environ. Pollut. 0269-7491. 263, 114389. doi:10.1016/j.envpol.2020.114389.

Lirio, S., Liu, W.L., Lin, C.L., Lin, C.H., Huang, H.Y., 2016. Aluminum based metal-organic framework-polymer monolith in solid-phase microextraction of penicillins in river water and milk samples. J. Chromatogr. A. 1873-3778. 1428, 236–245. doi:10.1016/j.chroma.2015.05.043.

Liu, L., Meng, W.K., Zhou, Y.S., Wang, X., Xu, G.J., Wang, M.L., Lin, J.M., Zhao, R.S., 2019. β-Ketoenamine-linked covalent organic framework coating for ultra-high-performance solid-phase microextraction of polybrominated diphenyl ethers from environmental samples. Chem. Eng. J. 1385-8947. 356, 926–933. doi:10.1016/j.cej.2018.09.081.

Liu, R., Feng, F., Chen, G., Liu, Z., Xu, Z., 2016. Barbell-shaped stir bar sorptive extraction using dummy template molecularly imprinted polymer coatings for analysis of bisphenol A in water. Anal. Bioanal. Chem. 1618-2650. 408 (19), 5329–5335. doi:10.1007/s00216-016-9628-3.

Liu, S., Dasgupta, P.K., 1995. Liquid droplet. A renewable gas sampling interface. Anal. Chem. 1520-6882. 67 (13), 2042–2049. doi:10.1021/ac00109a023.

Liu, Y., Zhang, Y., Niu, J., Nie, L., Huang, S., Liu, H., Yuan, S., Zhou, Q., 2023. Selective fluorescent probe for sensitive determination of bisphenol A based on molecularly imprinted polymers decorated carbon dots derived from citric acid and ethylenediamine. Chemosphere. 0045-6535. 324, 138303. doi:10.1016/j.chemosphere.2023.138303.

López-Lorente, Á.I., Pena-Pereira, F., Pedersen-Bjergaard, S., Zuin, V.G., Ozkan, S.A., Psillakis, E., 2022. The ten principles of green sample preparation. TrAC. Trends Anal. Chem. 1879-3142. 148, 116530. doi:10.1016/j.trac.2022.116530.

Lorenzo-Parodi, N., Kaziur, W., Stojanović, N., Jochmann, M.A., Schmidt, T.C., 2019. Solventless microextraction techniques for water analysis. TrAC. Trends Anal. Chem. 1879-3142. 113, 321–331. doi:10.1016/j.trac.2018.11.013.

Madikizela, L.M., Pakade, V.E., Ncube, S., Tutu, H., Chimuka, L., 2020. Application of hollow fibre-liquid phase microextraction technique for isolation and pre-concentration of pharmaceuticals in water. Membranes. 2077-0375. 10 (11), 311. doi:10.3390/membranes10110311.

Mafra, G., Vieira, A.A., Merib, J., Anderson, J.L., Carasek, E., 2019. Single drop microextraction in a 96-well plate format: a step toward automated and high-throughput analysis. Anal. Chim. Acta. 1873-4324. 1063, 159–166. doi:10.1016/j.aca.2019.02.013.

Makoś, P., Przyjazny, A., Boczkaj, G., 2018. Hydrophobic deep eutectic solvents as "green" extraction media for polycyclic aromatic hydrocarbons in aqueous samples. J. Chromatogr. A. 0021-9673. 1570, 28–37. doi:10.1016/j.chroma.2018.07.070.

Manousi, N., Wojnowski, W., Płotka-Wasylka, J., Samanidou, V., 2023. Blue applicability grade index (BAGI) and software: a new tool for the evaluation of method practicality. Green. Chem. 1463-9262. 25 (19), 7598–7604. doi:10.1039/d3gc02347h.

Mansour, F.R., Bedair, A., Hamed, M., Magdy, G., Ali, I., Locatelli, M., 2024. Applications of (natural) deep eutectic solvents in liquid phase microextraction: a review. Microchem. J. 0026-265X. 198, 110178. doi:10.1016/j.microc.2024.110178.

Marín-san Román, S., Carot-Sierra, J.M., Sáenz de Urturi, I., Rubio-Bretón, P., Pérez-Álvarez, E.P., Garde-Cerdán, T., 2022. Optimization of stir bar sorptive extraction (SBSE) and multi-stir bar sorptive extraction (mSBSE) to improve must volatile compounds extraction. LWT. 0023-6438. 172, 114182. doi:10.1016/j.lwt.2022.114182.

Maszkowska, J., Synak, E., Fabiańska, A., Caban, M., Stolte, S., Husson, P., Costa Gomes, M.F., Synak, R., Stepnowski, P., 2015. Preliminary study on suitability of ionic liquids as potential passive-sampling media of polyaromatic-hydrocarbon (PAH) analyses in water. Anal. Bioanal. Chem. 1618-2642. 407 (12), 3531–3536. doi:10.1007/s00216-015-8545-1.

Mbayachi, V.B., Ndayiragije, E., Sammani, T., Taj, S., Mbuta, E.R., Khan, A.U., 2021. Graphene synthesis, characterization and its applications: a review. Results Chem. 2211-7156. 3, 100163. doi:10.1016/j.rechem.2021.100163.

Mlunguza, N.Y., Ncube, S., Mahlambi, P.N., Chimuka, L., Madikizela, L.M., 2020. Determination of selected antiretroviral drugs in wastewater, surface water and aquatic plants using hollow fibre liquid phase microextraction and liquid chromatography-tandem mass spectrometry. J. Hazard. Mater. 1873-3336. 382, 121067. doi:10.1016/j.jhazmat.2019.121067.

Naing, N.N., Li, S.F.Y., Lee, H.K., 2016. Application of porous membrane-protected chitosan microspheres to determine benzene, toluene, ethylbenzene, xylenes and styrene in water. J. Chromatogr. A. 1873-3778. 1448, 42–48. doi:10.1016/j.chroma.2016.04.062.

Nascimento, M.M., da Rocha, G.O., de Andrade, J.B., 2021. Customized dispersive micro-solid-phase extraction device combined with micro-desorption for the simultaneous determination of 39 multiclass pesticides in environmental water samples. J. Chromatogr. A. 1873-3778. 1639, 461781. doi:10.1016/j.chroma.2020.461781.

Negussie, B.T., Dube, S., Nindi, M.M., Tesfaw, A., 2025. Optimization of the ionic liquid-based dispersive liquid–liquid microextraction combined with high-performance liquid chromatography coupled with diode array detection for the determination of multiclass pesticide residues in water samples . J. Chromatogr. Open. 2772-3917. 8, 100231. doi:10.1016/j.jcoa.2025.100231.

Nowak, P.M., Wietecha-Posłuszny, R., Pawliszyn, J., 2021. White analytical chemistry: an approach to reconcile the principles of green analytical chemistry and functionality. TrAC. Trends Anal. Chem. 1879-3142. 138, 116223. doi:10.1016/j.trac.2021.116223.

Olasupo, A., Suah, F.B.M., 2022. Trends in hollow fibre liquid phase microextraction for the preconcentration of pharmaceutically active compounds in aqueous solution: a case for polymer inclusion membrane. J. Hazard. Mater. 1873-3336. 431, 128573. doi:10.1016/j.jhazmat.2022.128573.

Olcer, Y.A., Tascon, M., Eroglu, A.E., Boyacı, E., 2019. Thin film microextraction: towards faster and more sensitive microextraction. TrAC. Trends Anal. Chem. 1879-3142. 113, 93–101. doi:10.1016/j.trac.2019.01.022.

Ortega-Zamora, C., González-Sálamo, J., Rivero, D.S., Carrillo, R., Hernández-Borges, J., 2024. Tetrazine-based dynamic covalent polymers as degradable extraction materials in sample preparation. Anal. Chim. Acta. 0003-2670. 1318, 342925. doi:10.1016/j.aca.2024.342925.

Ortega-Zamora, C., Jiménez-Skrzypek, G., González-Sálamo, J., Mazzapioda, L., Navarra, M.A., Gentili, A., Hernández-Borges, J., 2022. Extraction of emerging contaminants from environmental waters and urine by dispersive liquid-liquid microextraction with solidification of the floating organic droplet using fenchol:acetic acid deep eutectic mixtures. ACS. Sustain. Chem. Eng. 2168-0485. 10 (48), 15714–15725. doi:10.1021/acssuschemeng.2c04044.

Pacheco-Fernández, I., González-Martín, R., E Silva, F.A., Freire, M.G., Pino, V., 2021. Insights into coacervative and dispersive liquid-phase microextraction strategies with hydrophilic media—a review. Anal. Chim. Acta. 1873-4324. 1143, 225–249. doi:10.1016/j.aca.2020.08.022.

Pang, J., Chen, H., Guo, H., Lin, K., Huang, S., Lin, B., Zhang, Y., 2024. High-sensitive determination of tetracycline antibiotics adsorbed on microplastics in mariculture water using pre-COF/monolith composite-based in-tube solid phase microextraction on-line coupled to HPLC-MS/MS. J. Hazard. Mater. 0304-3894. 469, 133768. doi:10.1016/j.jhazmat.2024.133768.

Pashayi Sarnaghi, S., Ayazi, Z., 2025. Synthesis of chitosan-based molecularly imprinted biopolymer for thin-film solid-phase microextraction of tramadol followed by its on-site detection using digital image colorimetry . Anal. Chim. Acta. 1873-4324. 1369, 344335. doi:10.1016/j.aca.2025.344335.

Pawliszyn, J., 2012. Comprehensive Sampling and Sample Preparation: Analytical Techniques for Scientists. Elsevier, Canada, pp. 1–2715, 978-0-12-381374-9.

Pedersen-Bjergaard, S., Rasmussen, K.E., 1999. Liquid-liquid-liquid microextraction for sample preparation of biological fluids prior to capillary electrophoresis. Anal. Chem. 0003-2700. 71 (14), 2650–2656. doi:10.1021/ac990055n.

Piri-Moghadam, H., Gionfriddo, E., Grandy, J.J., Alam, M.N., Pawliszyn, J., 2018. Development and validation of eco-friendly strategies based on thin film microextraction for water analysis. J. Chromatogr. A. 1873-3778. 1579, 20–30. doi:10.1016/j.chroma.2018.10.026.

Płotka-Wasylka, J., Szczepańska, N., de la Guardia, M., Namieśnik, J., 2015. Miniaturized solid-phase extraction techniques. TrAC. Trends Anal. Chem. 0165-9936. 73, 19–38. doi:10.1016/j.trac.2015.04.026.

Prasek, J., Drbohlavova, J., Chomoucka, J., Hubalek, J., Jasek, O., Adam, V., Kizek, R., 2011. Methods for carbon nanotubes synthesis-review. J. Mater. Chem. 1364-5501. 21 (40), 15872–15884. doi:10.1039/c1jm12254a.

Psillakis, E., Kalogerakis, N., 2003. Developments in liquid-phase microextraction. TrAC. Trends Anal. Chem. 0165-9936. 22 (9), 565–574. doi:10.1016/s0165-9936(03)01007-0.

Qiao, L., Sun, R., Yu, C., Tao, Y., Yan, Y., 2021. Novel hydrophobic deep eutectic solvents for ultrasound-assisted dispersive liquid-liquid microextraction of trace non-steroidal anti-inflammatory drugs in water and milk samples. Microchem. J. 0026-265X. 170, 106686. doi:10.1016/j.microc.2021.106686.

Quigley, A., Cummins, W., Connolly, D., 2016. Dispersive liquid-liquid microextraction in the analysis of milk and dairy products: a review. J. Chem. 2090-9063. 2016, 4040165. doi:10.1155/2016/4040165.

Rajendran, S., Hong Loh, S., Ariffin, M.M., Mohd Afiq Wan Mohd Khalik, W., 2022. Magnetic effervescent tablet-assisted ionic liquid dispersive liquid–liquid microextraction employing the response surface method for the preconcentration of basic pharmaceutical drugs: characterization, method development, and green profile assessment. J. Mol. Liq. 0167-7322. 367, 120411. doi:10.1016/j.molliq.2022.120411.

Ramos Payán, M., Santigosa, E., Fernández Torres, R., Bello López, M.Á., 2018. A new microchip design. A versatile combination of electromembrane extraction and liquid-phase microextraction in a single chip device. Anal. Chem. 1520-6882. 90 (17), 10417–10424. doi:10.1021/acs.analchem.8b02292.

Ramos-Payán, M., Villar-Navarro, M., Fernández-Torres, R., Callejón-Mochón, M., Bello-López, M.Á., 2013. Electromembrane extraction (EME)-an easy, novel and rapid extraction procedure for the HPLC determination of fluoroquinolones in wastewater samples. Anal. Bioanal. Chem. 1618-2650. 405 (8), 2575–2584. doi:10.1007/s00216-012-6664-5.

Rezaee, M., Assadi, Y., Milani Hosseini, M.R., Aghaee, E., Ahmadi, F., Berijani, S., 2006. Determination of organic compounds in water using dispersive liquid-liquid microextraction. J. Chromatogr. A. 0021-9673. 1116 (1–2), 1–9. doi:10.1016/j.chroma.2006.03.007.

Rocío-Bautista, P., Taima-Mancera, I., Pasán, J., Pino, V., 2019. Metal-organic frameworks in green analytical chemistry. Separations. 2297-8739. 6 (3), 33. doi:10.3390/separations6030033.

Rubio, S., 2020. Twenty years of supramolecular solvents in sample preparation for chromatography: achievements and challenges ahead . Anal. Bioanal. Chem. 1618-2650. 412 (24), 6037–6058. doi:10.1007/s00216-020-02559-y.

Rutkowska, M., Płotka-Wasylka, J., Sajid, M., Andruch, V., 2019. Liquid–phase microextraction: a review of reviews. Microchem. J. 0026-265X. 149, 103989. doi:10.1016/j.microc.2019.103989.

Sajid, M., Nazal, M.K., Ihsanullah, I., 2021. Novel materials for dispersive (micro) solid-phase extraction of polycyclic aromatic hydrocarbons in environmental water samples: a review. Anal. Chim. Acta. 1873-4324. 1141, 246–262. doi:10.1016/j.aca.2020.07.064.

Sánchez-Rojas, F., Bosch-Ojeda, C., Cano-Pavón, J.M., 2009. A review of stir bar sorptive extraction. Chromatographia. 1612-1112. 69 (1), 79–94. doi:10.1365/s10337-008-0687-2.

Seebunrueng, K., Phosiri, P., Apitanagotinon, R., Srijaranai, S., 2020. A new environment-friendly supramolecular solvent-based liquid phase microextraction coupled to high performance liquid chromatography for simultaneous determination of six phenoxy acid herbicides in water and rice samples. Microchem. J. 0026-265X. 152, 104418. doi:10.1016/j.microc.2019.104418.

Sereshti, H., Jazani, S.S., Nouri, N., Shams, G., 2020. Dispersive liquid–liquid microextraction based on hydrophobic deep eutectic solvents: application for tetracyclines monitoring in milk. Microchem. J. 0026-265X. 158, 105269. doi:10.1016/j.microc.2020.105269.

Shishov, A., Sviridov, I., Timofeeva, I., Chibisova, N., Moskvin, L., Bulatov, A., 2017. An effervescence tablet-assisted switchable solvent-based microextraction: on-site preconcentration of steroid hormones in water samples followed by HPLC-UV determination. J. Mol. Liq. 0167-7322. 247, 246–253. doi:10.1016/j.molliq.2017.09.120.

Smith, E.L., Abbott, A.P., Ryder, K.S., 2014. Deep eutectic solvents (DESs) and their applications. Chem. Rev. 1520-6890. 114 (21), 11060–11082. doi:10.1021/cr300162p.

Solomou, N., Bicchi, C., Sgorbini, B., Psillakis, E., 2020. Vacuum-assisted headspace sorptive extraction: theoretical considerations and proof-of-concept extraction of polycyclic aromatic hydrocarbons from water samples. Anal. Chim. Acta. 0003-2670. 1096, 100–107. doi:10.1016/j.aca.2019.10.050.

Song, C., Zheng, J., Zhang, Q., Yuan, H., Yu, A., Zhang, W., Zhang, S., Ouyang, G., 2023. Multifunctionalized covalent organic frameworks for broad-spectrum extraction and ultrasensitive analysis of per- and polyfluoroalkyl substances. Anal. Chem. 0003-2700. 95 (19), 7770–7778. doi:10.1021/acs.analchem.3c01137.

Sorouraddin, S.M., Farajzadeh, M.A., Dastoori, H., 2020. Development of a dispersive liquid-liquid microextraction method based on a ternary deep eutectic solvent as chelating agent and extraction solvent for preconcentration of heavy metals from milk samples. Talanta. 0039-9140. 208, 120485. doi:10.1016/j.talanta.2019.120485.

Sun, H., Zhu, J., 2024. Low-dimensional nanomaterials and their assembled structures and interfaces for energy applications. ACS. Appl. Nano. Mater. 2574-0970. 7 (24), 27797–27800. doi:10.1021/acsanm.4c06610.

Tang, S., Qi, T., Ansah, P.D., Nalouzebi Fouemina, J.C., Shen, W., Basheer, C., Lee, H.K., 2018. Single-drop microextraction. TrAC. Trends Anal. Chem. 1879-3142. 108, 306–313. doi:10.1016/j.trac.2018.09.016.

Trine, G.H., Halvorsen, G., Pedersen-Bjergaard, S., Rasmussen, K.E., 2001. Reduction of extraction times in liquid-phase microextraction. J. Chromatogr. B. 1570-0232. 760 (2), 219–226. doi:10.1016/S0378-4347(01)00272-9.

Trujillo-Rodríguez, M.J., Pacheco-Fernández, I., Taima-Mancera, I., Ayala, J.H., Pino, V., 2020. Evolution and current advances in sorbent-based microextraction configurations. J. Chromatogr. A. 1873-3778. 1634, 461670. doi:10.1016/j.chroma.2020.461670.

Trujillo-Rodríguez, M.J., Saura-Cayuela, M., Cabrera-González, S.D., Lara-Torres, S., Pino, V., 2025. Microextraction of organic and inorganic analytes for biological, medical, environmental, and forensics analyses. In: Comprehensive Sampling and Sample Preparation, vol. 3, second ed. Elsevier BV, pp. 730–758. doi:10.1016/b978-0-443-15978-7.00081-3.

Turiel, E., Díaz-Álvarez, M., Martín-Esteban, A., 2016. Supported liquid membrane-protected molecularly imprinted beads for the solid phase micro-extraction of triazines from environmental waters. J. Chromatogr. A. 0021-9673. 1432, 1–6. doi:10.1016/j.chroma.2015.11.086.

Ullah, N., Tuzen, M., 2023. A comprehensive review on recent developments and future perspectives of switchable solvents and their applications in sample preparation techniques. Green. Chem. 1463-9262. 25 (5), 1729–1748. doi:10.1039/d3gc00020f.

Villar-Navarro, M., Moreno-Carballo, M.D.C., Fernández-Torres, R., Callejón-Mochón, M., Bello-López, M.Á., 2016. Electromembrane extraction for the determination of parabens in water samples. Anal. Bioanal. Chem. 1618-2650. 408 (6), 1615–1621. doi:10.1007/s00216-015-9269-y.

Vosough, M., Hassanbeigi, Z., Salemi, A., 2018. Determination of ultraviolet filter compounds in environmental water samples using membrane-protected micro-solid-phase extraction. J. Sep. Sci. 1615-9306. 41 (11), 2401–2410. doi:10.1002/jssc.201701082.

Wang, G., Lei, Y., Song, H., 2015. Exploration of metal-organic framework MOF-177 coated fibers for headspace solid-phase microextraction of polychlorinated biphenyls and polycyclic aromatic hydrocarbons. Talanta. 0039-9140. 144, 369–374. doi:10.1016/j.talanta.2015.06.058.

Wang, H., Du, J., Zhen, Q., Zhang, R., Wang, X., Du, X., 2019. Selective solid-phase microextraction of ultraviolet filters in environmental water with oriented ZnO nanosheets coated nickel-titanium alloy fibers followed by high performance liquid chromatography with UV detection. Talanta. 0039-9140. 191, 193–201. doi:10.1016/j.talanta.2018.08.060.

Wang, H., Wang, T., Hong, M.X., Wang, Z., Jin, X., Wu, H., 2023. Direct solidification of switchable-hydrophilicity salicylic acid: a design for the on-site dispersive liquid-liquid microextraction of benzoylurea insecticides in water and honey samples. J. Chromatogr. A. 1873-3778. 1688, 463710. doi:10.1016/j.chroma.2022.463710.

Wang, S., Li, H., Huang, H., Cao, X., Chen, X., Cao, D., 2022. Porous organic polymers as a platform for sensing applications. Chem. Soc. Rev. 0306-0012. 51 (6), 2031–2080. doi:10.1039/d2cs00059h.

Wang, X., Yuan, K., Liu, H., Lin, L., Luan, T., 2014. Fully automatic exposed and in-syringe dynamic single-drop microextraction with online agitation for the determination of polycyclic musks in surface waters of the Pearl River Estuary and South China Sea. J. Sep. Sci. 1615-9306. 37 (14), 1842–1849. doi:10.1002/jssc.201400198.

Wang, Z., Jin, P., Zhou, S., Wang, X., Du, X., 2018. Controlled growth of a porous hydroxyapatite nanoparticle coating on a titanium fiber for rapid and efficient solid-phase microextraction of polar chlorophenols, triclosan and bisphenol A from environmental water. Anal. Methods. 1759-9660. 10 (26), 3237–3247. doi:10.1039/c8ay00570b.

Wang, Z., Yang, Y., Zhao, Z., Zhang, P., Zhang, Y., Liu, J., Ma, S., Cheng, P., Chen, Y., Zhang, Z., 2021. Green synthesis of olefin-linked covalent organic frameworks for hydrogen fuel cell applications. Nat. Commun. 2041-1723. 12, 1982. doi:10.1038/s41467-021-22288-9.

Wei, Y., Fang, Z., Zheng, L., Tan, L., Tsang, E.P., 2016. Green synthesis of Fe nanoparticles using Citrus maxima peels aqueous extracts. Mater. Lett. 1873-4979. 185, 384–386. doi:10.1016/j.matlet.2016.09.029.

Wojnowski, W., Tobiszewski, M., Pena-Pereira, F., Psillakis, E., 2022. AGREEprep-analytical greenness metric for sample preparation. TrAC. Trends Anal. Chem. 0165-9936. 149, 116553. doi:10.1016/j.trac.2022.116553.

Xiao, Z., He, M., Chen, B., Hu, B., 2016. Polydimethylsiloxane/metal-organic frameworks coated stir bar sorptive extraction coupled to gas chromatography-flame photometric detection for the determination of organophosphorus pesticides in environmental water samples. Talanta. 0039-9140. 156-157, 126–133. doi:10.1016/j.talanta.2016.05.001.

Xie, L., Liu, S., Han, Z., Jiang, R., Liu, H., Zhu, F., Zeng, F., Su, C., Ouyang, G., 2015. Preparation and characterization of metal-organic framework MIL-101(Cr)-coated solid-phase microextraction fiber. Anal. Chim. Acta. 0003-2670. 853 (1), 303–310. doi:10.1016/j.aca.2014.09.048.

Xu, S., Liu, H., Long, A., Li, H., Chen, C., Feng, S., Fan, J., 2022. Carbon dot-decorated graphite carbon nitride composites for enhanced solid-phase microextraction of chlorobenzenes from water. Nanomaterials. 2079-4991. 12 (3), 335. doi:10.3390/nano12030335.

Yamaguchi, A., Matsuda, K., Uehara, M., Honda, T., Saito, Y., 2016. A novel automated device for rapid nucleic acid extraction utilizing a zigzag motion of magnetic silica beads. Anal. Chim. Acta. 0003-2670. 906, 1–6. doi:10.1016/j.aca.2015.10.011.

Yang, Y., Guo, Y., Jia, X., Zhang, Q., Mao, J., Feng, Y., Yin, D., Zhao, W., Zhang, Y., Ouyang, G., Zhang, W., 2023. An ultrastable 2D covalent organic framework coating for headspace solid-phase microextraction of organochlorine pesticides in environmental water. J. Hazard. Mater. 0304-3894. 452, 131228. doi:10.1016/j.jhazmat.2023.131228.

Yang, Y., Rodriguez-Lafuente, A., Pawliszyn, J., 2014. Thermoelectric-based temperature-controlling system for in-tube solid-phase microextraction. J. Sep. Sci. 1615-9314. 37 (13), 1617–1621. doi:10.1002/jssc.201400041.

Yao, C., Li, T., Twu, P., Pitner, W.R., Anderson, J.L., 2011. Selective extraction of emerging contaminants from water samples by dispersive liquid-liquid microextraction using functionalized ionic liquids. J. Chromatogr. A. 0021-9673. 1218 (12), 1556–1566. doi:10.1016/j.chroma.2011.01.035.

Ye, P., Yu, L., Guo, J., Yao, M., Qiu, J., Chen, G., Xu, J., Zhu, F., Ouyang, G., 2024. Green determination of glucocorticoids in water environment based on novel polydopamine coated iron tetroxide via magnetic dispersive solid phase microextraction. Green. Anal. Chem. 2772-5774. 10, 100136. doi:10.1016/j.greeac.2024.100136.

Yıldırım, S., Cocovi-Solberg, D.J., Uslu, B., Solich, P., Horstkotte, B., 2022. Lab-In-Syringe automation of deep eutectic solvent-based direct immersion single drop microextraction coupled online to high-performance liquid chromatography for the determination of fluoroquinolones. Talanta. 0039-9140. 246, 123476. doi:10.1016/j.talanta.2022.123476.

Yu, H., Wang, Z., Wu, R., Chen, X., Chan, T.W.D., 2019. Water-dispersible pH/thermo dual-responsive microporous polymeric microspheres as adsorbent for dispersive solid-phase extraction of fluoroquinolones from environmental water samples and food samples. J. Chromatogr. A. 1873-3778. 1601, 27–34. doi:10.1016/j.chroma.2019.05.004.

Yu, J., Zhu, S., Pang, L., Chen, P., Zhu, G.T., 2018. Porphyrin-based magnetic nanocomposites for efficient extraction of polycyclic aromatic hydrocarbons from water samples. J. Chromatogr. A. 1873-3778. 1540, 1–10. doi:10.1016/j.chroma.2018.02.006.

Zgoła-Grześkowiak, A., Grześkowiak, T., 2011. Dispersive liquid-liquid microextraction. TrAC. Trends Anal. Chem. 0165-9936. 30 (9), 1382–1399. doi:10.1016/j.trac.2011.04.014.

Zhang, D., Zhang, L., Liu, T., 2018. A magnetic cellulose-based carbon fiber hybrid as a dispersive solid-phase extraction material for the simultaneous detection of six bisphenol analogs from environmental samples. Analyst. 0003-2654. 143 (13), 3100–3106. doi:10.1039/c8an00544c.

Zhang, Q., Yao, X., He, M., Chen, B., Zhu, N., Hu, B., 2023. Porous aromatic framework/polydimethylsiloxane coated stir bar sorptive extraction coupled with high performance liquid chromatography-diode array detection of trace polychlorinated biphenyls analysis in environmental waters. J. Chromatogr. A. 0021-9673. 1688, 463709. doi:10.1016/j.chroma.2022.463709.

Zhang, Y., Chen, M., Li, L., Lv, Y., Ma, Q., 2024. Recent advances in microextraction techniques using sustainable green solvents for mass spectrometry analysis. TrAC. Trends Anal. Chem. 0165-9936. 170, 117412. doi:10.1016/j.trac.2023.117412.

Zhou, W., Hu, K., Wang, Y., Jiang, R.W., Pawliszyn, J., 2024. Embedding mixed sorbents in binder: solid-phase microextraction coating with wide extraction coverage and its application in environmental water analysis. Environ. Sci. Technol. 1520-5851. 58 (1), 771–779. doi:10.1021/acs.est.3c07244.

Zuliani, A., Khiar, N., Carrillo-Carrión, C., 2023. Recent progress of metal–organic frameworks as sensors in (bio)analytical fields: towards real-world applications. Anal. Bioanal. Chem. 1618-2642. 415 (11), 2005–2023. doi:10.1007/s00216-022-04493-7.

Chapter 6

Sampling and extraction of solid samples in environmental analysis

Carolina Turnes Pasini Deolindo[a,b], Luana de Souza Futigami[a,b], Cristian Rafael Kleemann[a,b], Luan Valdemiro Alves De Oliveira[a], Milena Dutra Pierezan[a], and Rodrigo Barcellos Hoff[c]
[a]*Food Science, Federal University of Santa Catarina (UFSC), Florianópolis, Brazil*
[b]*Federal Laboratory of Animal and Plant Health and Inspection—LFDA/RS, Ministry of Agriculture and Livestock, São José, Brazil*
[c]*Federal Laboratory of Animal and Plant Health and Inspection, Ministry of Agriculture and Livestock, São José, Brazil*

1 Introduction

The increasing concern over contaminants of emerging concern (CECs) has led to substantial research focused on their identification, quantification, and monitoring in various environmental matrices (Alqarni, 2024; Maruya et al., 2014). CECs encompass a wide range of compounds—pharmaceuticals, personal care products (PPCPs), pesticides, flame retardants, plasticizers, and psychoactive substances—many of which are not yet regulated. Although some have been in use for decades, they are considered "emerging" due to their recent environmental detection, growing ecotoxicological evidence, and recognition of potential risks to ecosystems and human health (Schnoor, 2003).

Solid matrices such as soils, sediments, sludge, and biota are crucial for understanding CEC fate and transport, acting both as sinks and sources depending on environmental conditions (Bao et al., 2015). Hydrophobic and persistent compounds strongly adsorb to solids, highlighting the need for robust sampling and treatment strategies to produce reliable data on contaminant distribution and behavior. Advances in environmental chemistry have expanded knowledge beyond traditional pollutants to include less conventional CECs, which often present analytical challenges due to their low concentrations, instability, and the formation of complex transformation products through biotic and abiotic processes.

Contaminants of Emerging Concern in the Environment. https://doi.org/10.1016/B978-0-443-33821-2.00008-6

CECs have been detected in air, water, soils, sediments, and biota, with solid matrices providing critical insights into both current and historical contamination. Key groups detected include PPCPs, endocrine disruptors, PFAS, micro- and nanoplastics, flame retardants, engineered nanomaterials, and illicit drugs. Biota studies, especially involving fish and birds, are vital for assessing bioaccumulation and food chain transfer, while sediment cores help trace historical contamination (Bao et al., 2015; de Wit et al., 2006). Sludge and biosolids, originating from WWTPs, are particularly complex and contaminated, raising concerns about CEC transfer into soils and crops (Habimana and Sauvé, 2025).

Recent studies have reported high concentrations of antifungal agents, antibiotics, antidepressants, and plasticizers in sludge and soils. Factors such as hydrophobicity, soil pH, and organic carbon content influence CEC mobility and bioavailability. Land application of biosolids, manure, and treated effluents continues to introduce CECs into agricultural soils. Nguyen and Baduel (2023) demonstrated the effectiveness of modified QuEChERS methods for detecting a wide range of emerging pollutants in soils and sediments across multiple countries (Nguyen and Baduel, 2023).

Effective monitoring depends on matrix-appropriate sampling, extraction, and cleanup methods. Sample preparation typically involves drying, grinding, homogenization, and lyophilization to reduce variability (Dimpe and Nomngongo, 2016; Tadeo et al., 2012). Extraction has evolved toward more efficient and greener approaches, including pressurized liquid extraction (PLE), microwave-assisted extraction (MAE), QuEChERS, and dispersive solid-phase extraction (dSPE). Analytical determination relies on sensitive and selective techniques such as liquid chromatography coupled to tandem mass spectrometry (LC-MS/MS), gas chromatography coupled to tandem mass spectrometry (GC-MS/MS), and high-resolution mass spectrometry (Montes et al., 2022).

This chapter will cover theoretical foundations, sampling and preparation strategies, traditional and advanced extraction techniques, purification and preconcentration methods, and future trends in green and miniaturized technologies. By doing so, it aims to provide a comprehensive overview of analytical frameworks for solid environmental matrices, supporting environmental monitoring and risk mitigation efforts.

2 Fundamentals and instrumentation

This section aims to briefly revisit some fundamental concepts that are intrinsically connected to the main topics of this chapter—sampling and extraction of solid environmental samples for the determination of CECs. While these analytical fundamentals are explored in greater detail in other chapters of this book, a concise overview is presented here to contextualize the reader and enhance the understanding of the following sections.

2.1 Introduction to the fundamentals of analytical methods applied to the analysis of CECs in solid matrices

The determination of CECs in solid environmental matrices requires highly selective and sensitive analytical approaches due to sample complexity and low analyte concentrations. This section briefly reviews the analytical concepts underlying the methodologies employed for CEC determination, providing context for the sample preparation and instrumental techniques detailed throughout the chapter.

Modern workflows are predominantly based on chromatographic separation coupled with mass spectrometry detection. Although techniques such as capillary electrophoresis have been explored, chromatographic methods—especially liquid chromatography (LC)—remain the standard (Falandysz et al., 2024). LC is particularly versatile, accommodating the wide polarity range of CECs, from lipophilic flame retardants to hydrophilic pesticides and PPCPs. Gas chromatography (GC) is more restricted to volatile or derivatizable compounds, with notable applications in hormone and persistent pollutant analysis (Qin et al., 2023). In practice, the determination of CECs in solids primarily relies on these two instrumental components: chromatographic separation (LC or GC) and mass spectrometric detection. LC systems are favored for thermally labile and polar compounds, with advances such as ultrahigh pressure LC (UHPLC) enhancing resolution and speed (Ducrocq et al., 2024). GC remains essential for volatile analytes and is widely used for pollutants such as PAHs, PCBs, and brominated flame retardants (Pissinatti et al., 2015).

Mass spectrometry (MS) is the primary detection technique, especially in tandem (MS/MS) and high-resolution (HRMS) modes, offering the specificity and sensitivity needed for multiclass, multiresidue environmental analyses (Botella et al., 2024). Three major approaches are employed: **target analysis** (for known analytes using standards), **nontarget analysis** (for unknowns using accurate mass and fragmentation patterns), and **suspect screening** (identification based on predicted properties without full standards) (Schymanski et al., 2014). In quantitative workflows, MS/MS operated in multiple reaction monitoring (MRM) mode provides the required sensitivity and selectivity for trace analysis in complex solid matrices. High-resolution systems (Orbitrap, TOF) play a crucial role in nontarget and suspect screening, while conventional detectors such as UV-Vis, DAD, or fluorescence are generally inadequate for multiclass CEC determination in solids due to their limited selectivity and sensitivity (Ducrocq et al., 2024).

2.2 Basic principles of solid sample preparation and extraction

Accurate CEC determination in solids begins with appropriate sample preparation. Environmental solids—soils, sediments, sludge, dust, biota—are often heterogeneous and may have significant moisture content, especially in samples such as sewage sludge and aquatic organisms (Hoff et al., 2016).

Pretreatment steps such as drying, grinding, homogenization, lyophilization, and sieving are crucial to enhance extraction efficiency. Dust and dry soils may require minimal treatment, while biotic tissues often demand freeze-drying and cryogenic grinding (Botella et al., 2024; Radke et al., 2025).

In biological matrices, CECs can bind to proteins or lipids, requiring enzymatic or chemical hydrolysis prior to solvent extraction to recover parent compounds or metabolites (Shao et al., 2024). Analyte interactions vary by matrix type: soils and sediments typically feature adsorption onto particulates, while biota may involve metabolism or tissue binding. Thus understanding matrix composition and analyte behavior is essential for optimizing sample preparation, ensuring the efficiency and reliability of the analytical method.

3 Procedure

3.1 Sampling and sample preparation

3.1.1 Sampling strategies for environmental solid samples

Passive and grab sampling are widely used for monitoring CECs in environmental matrices, each offering distinct advantages depending on target compounds, environmental conditions, and analytical goals (Hawker et al., 2022).

Grab sampling, the most conventional method, involves collecting discrete samples at specific time points, providing a straightforward snapshot of contamination. However, it may not capture temporal variations, especially in dynamic environments such as rivers or effluents. In sediment or soil sampling, core or surface samples are selected based on study objectives, such as vertical profiling or surface concentration assessment. Typically, water is removed from sediments prior to extraction, with freeze-drying being widely adopted, despite some uncertainty regarding its effects on volatile compounds (Fernández-Sanjuan et al., 2009).

In the case of liquid samples, passive sampling, such as with polar organic chemical integrative samplers (POCIS), is available. The use of these devices offers time-integrated contaminant collection and enhanced sensitivity for trace-level detection and nontarget screening (NTS) workflows (Tadić et al., 2022). Despite longer deployment times and environmental dependencies such as flow and temperature, passive methods provide robust monitoring capabilities.

In aquatic environments, passive samplers are well-established and commercially available as described in Chapter 3. As expected, passive samplers are predominantly applied for liquid matrices. Notwithstanding, passive samplers can be configured with different receiving phases and diffusive gels to suit the specific requirements of solid samples, such as soil, to effectively capture organic pollutants. The diffusive gradients in thin films (DGT) technique, originally developed for aquatic environments, have been adapted to sample organic pollutants in soils. This adaptation, known as o-DGT, allows for the sampling of various organic compounds in solid matrices such as soils (Guibal et al., 2019).

A combined use of grab and passive strategies is recommended: grab sampling allows for immediate assessments, while passive sampling captures low-concentration and temporally variable contaminants.

For solid samples, passive sampling is still evolving, with exceptions in air and dust studies (Chapter 2 for more details). Devices using vacuum pumps or suction systems effectively trap particles for later analysis (Badry et al., 2022). Techniques such as multiratio equilibrium passive sampling (MR-EPS) enhance estimation of the bioaccessible fraction of semivolatile organic compounds (SVOCs) in indoor dust (Rusina et al., 2022). In situ sampling techniques have also proven valuable, especially for urban soil dust, where fine particle fractions are more sensitive to anthropogenic pollution, including trace metals. Complementary methods such as vacuum sampling—either via fresh collection or using dust from household vacuum cleaners—offer practical solutions for monitoring indoor environments, particularly for substances such as organophosphate esters (OPEs) (Fan et al., 2014). Electrostatic dust fall collectors and filter forensics, the latter involving HVAC filter dust analysis, add further dimensions to dust sampling strategies, enabling insights into airborne particle-bound contaminants (Haaland and Siegel, 2017). Each technique presents unique strengths depending on the target analytes and environmental context, and their combined use can enhance the comprehensiveness and comparability of CEC studies.

Regarding soil sampling, traditional techniques from agronomy and soil science are often employed. However, the stringent criteria of collecting soil by specific layers and depths are not consistently applied in studies focusing on emerging contaminants. In many environmental investigations, solid samples are often categorized as sediments or sludge, which differ from typical agricultural soils and may not necessitate the conventional sampling protocols established in soil science literature.

Soil sampling traditionally follows agronomic protocols, but strict layer-specific collection is often relaxed for CEC investigations, as solid samples frequently include sediments or sludge. No universal standards exist specifically for CECs in soils, though researchers adapt general guidelines, such as US EPA SW-846 methods, ISO 10381 series, and ASTM standards (Couch et al., 2000). Despite these resources, the unique properties of CECs—such as their low concentrations, diverse chemical nature, and potential for transformation—require tailored sampling strategies that are not always addressed in traditional protocols. As a result, many studies modify existing methods, for example by focusing on topsoil layers, using stainless-steel tools, and targeting specific particle size fractions (<2 mm or <63 μm) where CECs are more likely to accumulate.

There is growing demand for harmonized protocols that account for CECs' unique physicochemical behavior, particularly for emerging classes such as pharmaceuticals and plasticizers.

Sediment and soil sampling typically employ stainless-steel shovels or Van Veen grab samplers, depending on site conditions. Surface sediments from riverbanks can be collected with a shovel, whereas deeper aquatic zones require specialized equipment (Fig. 6.1).

Collected sediments, usually 50–500 g, are stored in refrigerated PET containers. Detailed characterization of sediment samples at the time of collection is crucial. Descriptions such as "anoxic", "sediment from the shoreline" or "shovel-collected riverbank sediment" are common in the literature and provide valuable context for interpreting analytical results. For instance, anoxic sediments are sediments collected from portions of aquatic habitats in which dissolved oxygen has been depleted: such zones are particularly relevant fot the fate and behavior of certain CECs. Therefore, documenting sample characteristics in a field notebook—or through audio recordings—and recording the precise GPS coordinates of each sampling site are highly recommended. Photographic records can also support later interpretation and traceability. Upon lab arrival, sediments are frozen, freeze-dried, and sieved (<250 μm) to obtain the fine fraction most suitable for chemical analysis.

Biota sampling for CEC determination is more heterogeneous. Rather than systematic collection, researchers often rely on opportunistic sampling of deceased or naturally available specimens. Park rangers or environmental staff frequently report findings for further laboratory analysis (Peris et al., 2023). Samples such as unhatched eggs are refrigerated, freeze-dried, homogenized, and stored frozen. Fish are the primary biota matrix in CEC studies, often collected under animal welfare guidelines or sourced from fisheries. Capture techniques may include electrofishing or other site-appropriate methods. Recording biometric data—weight, length, and age—is encouraged to improve exposure assessments. Biota samples are commonly dissected for specific organ analysis (e.g., muscle, liver, brain) (Manjarrés-López et al., 2023). For smaller species such as zooplankton and bivalves,

FIG. 6.1 General procedures for handling a Van Veen grab sampler during surface sediment collection.

multiple individuals are typically homogenized to create a pooled tissue sample, from which subsamples are extracted. Freeze-drying is commonly used to remove water before analysis.

Proper sample identification and transport are critical to ensuring the reliability and traceability of analytical results. All sample containers should be clearly labeled at the time of collection, including essential information such as sampling location (with GPS coordinates), date and time, and the name of the collector. When multiple containers are involved, a coding system is recommended. After collection, samples must be transported promptly to the analytical laboratories, ideally in one batch, using cooling boxes or dry ice to preserve their integrity until analysis.

3.1.2 Sample preparation

The preparation of solid environmental samples must follow a standardized sequence to ensure analytical accuracy, precision, and representativeness. For matrices such as soil, sediment, and sewage sludge, moisture content should be determined upon laboratory arrival, a critical step for normalizing results based on dry weight (Câmara et al., 2022). Following moisture assessment, samples are usually frozen and freeze-dried (lyophilized) to remove water without degrading thermolabile or volatile analytes. Freeze-drying preserves sample integrity and allows safe storage until analysis (Pena-Pereira et al., 2022; Qin et al., 2023; Chen et al., 2021). After drying, samples must be ground to reduce particle size and enhance homogeneity, improving analyte extraction efficiency. Care should be taken to avoid overheating, which can degrade sensitive compounds (Câmara et al., 2022; Chen et al., 2021). After grinding, the sample should be sieved—typically through a 2 mm mesh—to obtain a uniform particle size fraction. This improves the reproducibility of extraction procedures and standardizes sample granulometry, especially important for complex and heterogeneous environmental matrices (Câmara et al., 2022; Qin et al., 2023).

Beyond analytical performance, these steps are increasingly evaluated using tools such as AGREEprep, which assess sample preparation methods against green chemistry principles, considering factors such as energy use, waste generation, reagent toxicity, and operator safety (Pena-Pereira et al., 2022).

Preparation strategies must be tailored to the matrix. For indoor dust, often used in passive environmental exposure studies, samples are collected via vacuum cleaners with fine particulate filters (Moorchilot et al., 2025). After removing visible contaminants such as hairs or fibers, dust is typically sieved between 25 μm and 2 mm to obtain homogeneous fractions. Although freeze-drying is less common for dust, controlled drying is often necessary to eliminate residual moisture and prevent degradation of semivolatile compounds such as phthalates and PAHs. In some cases, samples are stored under inert conditions to preserve analytes until analysis (Orecchio et al., 2022).

Biota samples (e.g., fish, birds, insects) require specialized wet-based preparation methods, including tissue homogenization and solvent or enzymatic extraction. These procedures must maintain tissue integrity while enabling detection of low-concentration pollutants (Chen et al., 2014; Qin et al., 2023).

3.2 Extraction techniques

3.2.1 Solvent extraction

Solvent extraction techniques remain widely employed in environmental analysis due to their versatility, simplicity, and adaptability to different solid matrices and analytes, especially in nontarget or multiresidue screening approaches where broad recovery is prioritized (Sandoval et al., 2024). These methods are based on solid-liquid extraction, which may be performed in its simplest form—without the aid of external energy or specialized equipment—or through straightforward mechanical apparatus. Thus solvent extraction variants include Soxhlet extraction, accelerated solvent extraction, ultrasound-assisted extraction, and vortex-assisted extraction (Celeiro et al., 2022; Dürig et al., 2020; Havens et al., 2014; Song et al., 2023)—whose applications for CECs analysis are presented in Table 6.1.

The solvent choice plays a critical role in the efficiency and selectivity of CEC extraction. In most studies, the solvent extraction techniques involve acetonitrile (ACN)—often acidified with 0.1% formic acid—for polar and ionic analytes due to its ability to minimize coextraction of lipids and matrix interferences, especially in biota (Dürig et al., 2020). For non- and moderate polar compounds, solvent mixtures with methanol, dichloromethane, hexane, or acetone are commonly preferred due to their higher affinity and improved desorption from the matrix. In particular, combining acetone:hexane with elevated temperature and pressure or with ultrasound agitation enhances extraction by disrupting sorption interactions between analyte and matrix more effectively than heat alone (Havens et al., 2014). Overall, the combination of solvents with different polarity degrees is employed to cover a wide range of CECs physicochemical properties, and must be compatible with both the extraction technique and the instrumental analysis. Further cleanup steps are also commonly required (Song et al., 2023).

The simplest form of solvent-based extraction—commonly referred to as solid-liquid extraction—involves manual or mechanical homogenization of the solid sample with an organic solvent, without the application of external energy sources such as heat, pressure, or ultrasound. It stands out for its simplicity and broad coverage, especially in nontarget workflows. In biota (e.g. birds), acetonitrile acidified with 0.1% formic acid enabled satisfactory recoveries (82%) for over 200 CECs, including pharmaceuticals, drugs, stimulants, flame retardants, personal care products, plasticizers, and per- and polyfluoroalkyl classes (Dürig et al., 2020). Although matrix effects are higher without cleanup, its efficiency makes it ideal for large-scale screenings, and it requires minimal equipment with direct injection after centrifugation and filtration (Dürig et al., 2020).

Soxhlet extraction, on the contrary, allows repeated washing without the need for manual solvent replacement. It is a traditional yet exhaustive technique that consists of continuously cycling hot solvent through the solid sample in a specialized apparatus. It is particularly suitable for nonvolatile or thermally stable compounds, being applied for several classes of CECs (hormones, flame

TABLE 6.1 Examples of solvent extraction techniques applied to CECs analysis.

CECs	Matrix	Extraction technique	Extraction solvent	Instrumentation	Recovery (%)	LOD ($ng.g^{-1}$)	LOQ ($ng.g^{-1}$)	Reference
217 (pharmaceuticals, drugs, stimulants, flame retardants, personal care products, plasticizers, PFASs, among others)	Bird (0.5 g)	SLE	ACN + 0.1% FA	LC-HRMS	82 ± 32	0.05–120	0.2–1	(Dürig et al., 2020)
19 (hormones and their metabolites)	Sandy, silt loam, clay, high-organic matter soils (5 g)	Soxhlet	ACN:Hexane (1:1)	LC-MS/MS	58 ± 34	0.01–0.9	0.3–2.7	(Havens et al., 2014)
		UAE			71 ± 23			
		ASE			73 ± 16			
128 (organochlorine pesticides, pharmaceuticals, brominated flame retardants, personal care products, PFAS, plasticizers, UV filters, among others)	Indoor dust (0.1 g)	Soxhlet	DCM (GC)	GC-HRMS and LC-HRMS	63 ± 28[a]	0.47–933 (GC)	1.6–130 (GC)	(Song et al., 2023)
		UAE	MeOH (LC)		75 ± 16[a]	4.03-130 (LC)	13.4–435 (LC)	
		ASE			45 ± 19[a]			
173 (pharmaceuticals, flame retardants, PFASs, pesticides)	Bivalve mollusk, crab (0.5 g)	UAE	ACN + 0.1% FA	LC-HRMS	78 ± 31	14	0.01–22	(Manjarrés-López et al., 2023)

Continued

TABLE 6.1 Examples of solvent extraction techniques applied to CECs analysis—cont'd

CECs	Matrix	Extraction technique	Extraction solvent	Instrumentation	Recovery (%)	LOD ($ng.g^{-1}$)	LOQ ($ng.g^{-1}$)	Reference
60 (PCBs, flame retardants, pesticides)	Sediment (15 g)	UAE	Acetone, hexane, followed by acetone:hexane (1:1)	Open-column LC, GC-ECD-NPD	71–95	ND	ND	(de L Freitas et al., 2024)
164 (pharmaceuticals, pesticides, PFASs, parabens, personal care products, drugs, stimulants, among others)	Sludge (2 g)	UAE	ACN:water (1:1) followed by ACN:2-propanol:H_2O (3:3:4)	LC-MS/MS	ND	ND	0.2–8.5	(Golovko et al., 2021)
5 (UV filters)	Sediment (0.5 g)	VAE	Hexane:acetone (6:4)	LC-MS/MS	58–87	0.1–7	ND	(Tsui et al., 2017)
	Coral (0.02 g)					0.1–7.5		
30 (Flame retardants)	Dust (0.075 g)	VAE	Hexane:Acetone (3:1)	GC-MS	69–122	ND	0.04–370	(Van Den Eede et al., 2012)

PCBs, polychlorinated biphenyls; *PFASs*, per and polyfluoroalkyl substances; *UV*, ultraviolet; *UAE*, ultrasound-assisted extraction; *ASE*, accelerated solvent extraction; *SLE*, solid-liquid extraction; *VAE*, vortex-assisted extraction; *MeOH*, methanol; *ACN*, acetonitrile; *FA*, formic acid; *DCM*, dichloromethane; *GC*, gas chromatography; *LC*, liquid chromatography, *GC-MS*, gas chromatography coupled with mass spectrometry; *LC-MS/MS*, liquid chromatography coupled with tandem mass spectrometry; *LC-HRMS*, liquid chromatography coupled with high resolution mass spectrometer; *GC-HRMS*, gas chromatography coupled with high resolution mass spectrometry; *LOD*, limit of detection; *LOQ*, limit of quantitation.

[a] *Estimated values, based on the graphic provided by the authors; ND, not described.*

retardants, personal care products and endocrine-disruptors) in sediment, soil, dust, and sludge (Celeiro et al., 2022; Martín-Pozo et al., 2019). Soxhlet often requires high solvent volumes (100–500 mL), long extraction times (16–48 h), and despite its reliability and availability, the technique has low throughput and generates significant solvent waste (Celeiro et al., 2022). As an alternative, accelerated solvent extraction offers a faster automated extraction, using moderate volumes of pressurized solvent and varied temperature conditions to enhance analyte solubilization and mass transfer. Thus it has demonstrated high efficiency and reproducibility by extracting several hormones and their metabolites in many types of soils (sandy, silt loam, clay, and high-organic matter) using acetone:hexane (1:1) at 120°C and 1500 psi (Havens et al., 2014).

Apart from techniques that require thermal conditions, the ultrasound-assisted extraction has emerged as a milder and more versatile alternative. It is an environmentally friendly and lower energy consumption technology with several advantages such as a faster and yield improvement extraction. With this technique, cavitation bubbles are generated in the solvent, created due to the passage of ultrasound waves allowing for a greater penetration of the solvent within the material increasing the surface area (Celeiro et al., 2022). It has been widely applied to sediments, sludge, dust, and biota, showing excellent performance for non- and semivolatile contaminants (Llompart et al., 2019b; Celeiro et al., 2022; Manjarrés-López et al., 2023).

In a recent study with bivalve mollusks and crustaceans, several solvent combinations were tested for ultrasound-assisted technique extraction of a broad range of CECs, and ACN acidified with 0.1% formic acid provided acceptable recoveries for pharmaceuticals, pesticide, flame retardants, and per- and polyfluoroalkyl in biota samples (Manjarrés-López et al., 2023). On the contrary, Song et al. (2023) focused on nontarget screening of non- and semivolatile compounds in dust employing dichloromethane and methanol for GC-HRMS and LC-HRMS detection, respectively. The first enabled the extraction of less polar and semivolatile CECs such as triphenyl phosphate, 2,2',4,4'-Tetrabromodiphenyl ether (flame retardants), galaxolide, tonalide (personal care products), Di(2-ethylhexyl) phthalate, and Dibutyl phthalate (plasticizers), while the second was more efficient for nonvolatile and polar compounds, such as some pharmaceuticals and pesticides (e.g. carbamazepine, atrazine), thus covering a broad polarity range (Song et al., 2023). Ultrasound-assisted extraction consistently outperformed Soxhlet extraction and demonstrated similar or greater results than accelerated solvent extraction in recoveries and reproducibility across studies (Havens et al., 2014; Song et al., 2023).

Similarly, vortex-assisted extraction uses small volumes of solvent and vortex agitation and already demonstrated to be suitable for rapid extractions of UV filters, and flame retardants from sediments, coral and dust (Celeiro et al., 2022; Martín-Pozo et al., 2019). But even with generally acceptable recoveries, vortex-assisted extraction is more limited in scope and sensitivity than ultrasound-assisted or solid-liquid extraction (Celeiro et al., 2022; Van Den Eede et al., 2012).

Despite the effectiveness of organic solvents in the extraction of CECs, concerns regarding their volatility, toxicity, and environmental impact have spurred growing interest in greener alternatives. In this context, deep eutectic solvents (DESs) have emerged as promising candidates due to their low toxicity, biodegradability, and tunable polarity (Botella et al., 2024), although they are not yet widely applied in the conventional solvent extraction techniques discussed in this section. Therefore DESs will be further addressed in the section devoted to microextraction techniques.

3.2.2 *Microwave-assisted extraction*

Microwave-assisted extraction (MAE) is a green sample preparation technique that utilizes microwave energy to heat solvent-sample mixtures, facilitating rapid and efficient extraction of contaminants from solid environmental matrices such as soils, sediments, sewage sludge, and biota. This method is particularly relevant in the determination of CECs, including PPCPs, plasticizers, UV filters, and flame retardants. The advantages of MAE include reduced solvent use, shorter extraction times, and high extraction efficiency, aligning with the principles of green analytical chemistry (Llompart et al., 2019b; Celeiro et al., 2022).

MAE operates by exposing polar solvents and moist samples to microwave radiation (typically 2450 MHz), inducing dipole rotation and ionic conduction. These interactions cause rapid, uniform heating, breaking matrix-analyte bonds and enhancing desorption and diffusion of analytes into the solvent (Llompart et al., 2019b). Solvents such as MeOH, ACN, and their mixtures are commonly used. Extraction can be performed in closed systems (allowing elevated pressure and temperature) or open systems (atmospheric pressure). Critical parameters include solvent type and volume, extraction time and temperature, microwave power, and sample-to-solvent ratio (Panja, 2018).

Several operational factors critically affect MAE performance, as shown in Table 6.2:

These parameters, when systematically optimized, allow MAE to achieve efficient and reproducible extraction of CECs from solid environmental samples while reducing solvent use, time, and energy in accordance with green chemistry principles (Llompart et al., 2019b).

MAE significantly reduces solvent consumption and extraction time, allows for simultaneous processing of multiple samples, and is adaptable to various matrices and target analytes (Sanchez-Prado et al., 2015). It supports green chemistry principles. However, limitations include the need for specialized microwave equipment, potential degradation of thermolabile compounds, and occasional need for postextraction cleanup (Llompart et al., 2019b).

Recent studies highlight MAE's role in the extraction of multiclass CECs (Celeiro et al., 2022). Petrie et al. (2016)) demonstrated MAE for extracting up to 90 PPCPs from sludge with recoveries of 90%–100%. Azzouz and Ballesteros (2016) reported the use of MAE for retrieving parabens and phenolic preservatives from soil and sludge. In biota, MAE using acetone/heptane achieved recoveries >90% for UV filters from mussel tissue (Bachelot et al., 2012).

TABLE 6.2 Main operational parameters influencing microwave-assisted extraction (MAE) performance.

Parameter	Description
Solvent type	Polar solvents (e.g., methanol, ethanol, DES) for efficient microwave absorption.
Solvent volume	Sufficient to fully immerse the sample, avoiding excessive dilution.
Temperature	60°C–120°C, depending on analyte stability.
Extraction time	Typically 10–30 min; longer times may degrade thermolabile compounds.
Microwave power	Adjustable (200–1200 W) to control heating rate and prevent overheating.
Sample-to-solvent ratio	Commonly 1:10 to 1:20 (w/v) to ensure sufficient contact and efficient heating.
Clean-up step	SPE or filtration needed for complex matrices to remove interfering compounds.

Adapted from Llompart, M., Garcia-Jares, C., Celeiro, M., Dagnac, T., 2019a. Extraction | microwave-assisted extraction encyclopedia of analytical science. In: Reference Module in Chemistry, Molecular Sciences and Chemical Engineering. Elsevier, Spain, pp. 67–77. doi:10.1016/B978-0-12-409547-2.14442-7.

MAE is a powerful tool for the efficient, green, and reproducible extraction of CECs from solid environmental matrices. Its ability to work with various analytes and complex matrices makes it ideal for environmental monitoring applications. When combined with cleanup procedures and modern analytical techniques, MAE offers robust performance and contributes to sustainable analytical practice. Other recent applications of MAE for CECs determination in solid environmental samples are shown in Table 6.3.

3.2.3 Supercritical fluid extraction

Supercritical fluid extraction (SFE) is a method that utilizes supercritical fluids as solvents to isolate compounds from solid matrices. However, its application in the preparation of environmental solid samples for emerging contaminant analysis has been limited (Lindholm-Lehto et al., 2017). Only a small number of studies explored this technique, which employs solvents such as carbon dioxide and organic modifiers such as methanol, to analyze synthetic musks and related compounds in sewage sludge (Lee et al., 2003). Synthetic musk fragrances are widely used as additives in various consumer and personal-care products. Once these products reach the end of their life cycle, most of these compounds are ultimately discharged into municipal sewage systems.

TABLE 6.3 Applications of the microwave-assisted extraction (MAE) method for the determination of CECs in solid samples in environmental analysis.

Contaminant (CEC)	Matrix	Sample preparation	MAE conditions	Extract treatment	LOD/ LOQ	Determination method	Recovery (%)	Reference
Pharmaceuticals & Hormones/Veterinary Drugs								
15 Sex Hormones and Corticosteroids	Sewage Sludge	Lyophilization	10 mL MeOH; 500 W; 65°C; 4 min	Filtration, evaporation, and reconstitution in MeOH	1–7.9 ng/g	LC-MS/MS	>71	(Guedes-Alonso et al., 2016)
8 NSAIDs and 5 Estrogenic Hormones	Soil, Sediment, and Sludge	Drying, sieving, and homogenization	10 mL water (or mixture); 400 W; 115°C; 15 min	Evaporation, reconstitution, SPE, and derivatization	0.3–5.7 ng/g	GC-MS or LC-MS/MS	>50	(Kumirska et al., 2015)
Fluoroquinolones in Fish	Fish (0.5–10 g)	Lyophilization and tissue homogenization	5–8 mL ACN; 50–110°C; 5–20 min	SPE and/ or DLLME, depending on the method	0.05–0.1 ng/g	LC-MS/MS	70–96	(Huang et al., 2016)
7 Parabens and 2 Alkylphenols	Soil (2 g)	Drying, sieving, and homogenization	10 mL ACN; MAE at 80°C (10 min ramp plus 5 min hold)	Derivatization	0.4–1.1 ng/g	GC-MS	78–112	(Pérez et al., 2012)
5 Benzophenones and 2 Phototransformation Products	Sediment (4 g)	Drying	30 mL MeOH/ Acetone (1:1, v/v) with 5% formic acid; 800 W; ramp to 150°C; 10 min	Centrifugation, evaporation, dilution, SPE, and derivatization	0.1–1.4 ng/g	GC-MS	80–99	(Kotnik et al., 2014)
7 Benzotriazoles (BTRs)	Marine Sediment or Sludge (1 g)	–	2 mL ACN; 300 W; 5 min	Filtration and on-line SPE	0.053–0.1 ng/g	LC-MS/MS	50–87	(Montesdeoca-Esponda et al., 2013)

2 BTRs, 3 Benzothiazoles (BTs) and 2 Benzenesulfonamides (BSAs)	Soil	–	6 mL MeOH; MAE at 120°C; 10 min	Evaporation and reconstitution	LOQ: 100 ng/g	LC-UV	72–119	(Speltini et al., 2016)
33 Endocrine Disrupting Compounds (EDCs)	Wastewater and Sludge (0.5 g)	Filtration	25 mL MeOH/ H_2O (1:1, v/v); 110°C; 30 min	SPE	LOQ: 2 ng/g	LC-MS/MS	11–187	(Lopardo et al., 2019)
4 Polycyclic and Nitro Musks	Biota Fish (2 g)	Cleaning	4 mL MeOH + 15 mL Water with 4 g NaCl; MAE-HS-SPME; 80 W; 5 min	–	0.4–1.2 ng/g (LOQ)	GC-MS	80–92	(Wu et al., 2013)
2 Polycyclic Musks	Oysters (5 g)	Homogenization	10 mL Water + 3 g NaCl; MAE-HS-SPME; 180 W; 4 min	–	0.04 ng/g	GC-MS	80–89	(Wu et al., 2012)
Industrial Pollutants								
6 PBDEs	Sewage Sludge	Lyophilization and homogenization	MAE with 0.1 M HCl in MeOH	Addition of Tris-citrate buffer and extraction with iso-octane	<0.3 ng/g	GC-ICP-MS	ND	(Novak et al., 2016)
Plasticizers, Flame Retardants and Phenolic Compounds	Sediment or Soil	Drying and homogenization	MAE withorganic solvent mixture (Hexane/Acetone or MeOH) at elevated temperature (~120°C; 10–20 min)	Evaporation and purification (SPE)	ND	GC-MS/MS or LC-MS/MS	ND	(Cui et al., 2017)

3.2.4 QuEChERS extraction

Originally developed by Anastassiades in 2003 to address the limitations of multiresidue extraction methods, the QuEChERS technique is recognized for its speed, simplicity, robustness, and broad applicability to a wide range of analytes. The method typically involves extraction with ACN, partitioning by the addition of salts—most commonly magnesium sulfate and sodium chloride—and subsequent cleanup of the extract using dispersive solid-phase extraction (dSPE) (Anastassiades et al., 2003). A schematic overview of the main steps of the method is presented in Fig. 6.2.

Over time, several modifications have been proposed to improve the recovery of specific analyte groups or to expand the scope of the technique. Classical adaptations include the acetate-buffered QuEChERS, which employs ACN with 1% acetic acid and sodium acetate during the partitioning step to achieve buffering at pH 4.8; and the citrate-buffered QuEChERS, which uses trisodium citrate dihydrate and disodium hydrogen citrate sesquihydrate to buffer the system at pH 5.0–5.5. More recent developments include QuEChERSER (Lehotay, 2022) and p-QuEChERS (Kaufmann et al., 2022). QuEChERS and its adaptations have been widely used for the analysis of CECs in environmental solid samples, as presented in Table 6.4.

Modified versions of the QuEChERS method have been compared in terms of the recovery of 33 pharmaceutical compounds. It was observed that the acetate-buffered QuEChERS method did not enable the recovery of N-desmethyl-olanzapine, amitriptyline, fluoxetine, and norfluoxetine, whereas the citrate-buffered version allowed the recovery of all analytes evaluated (Miserli et al., 2023). The superior performance of the citrate-buffered

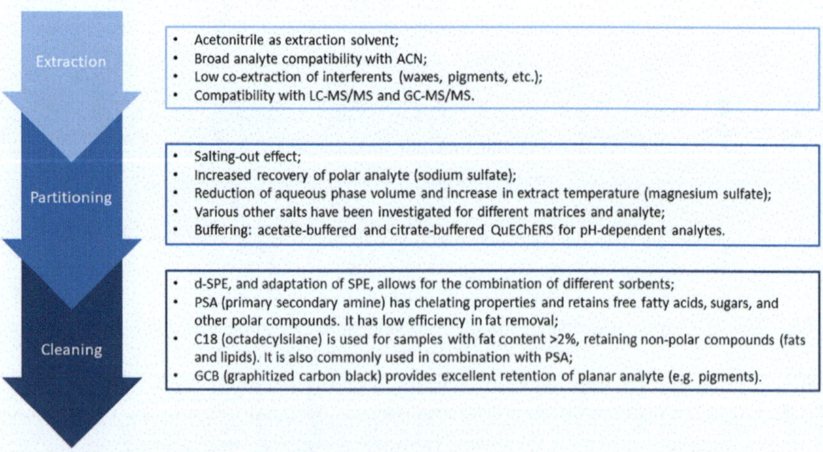

FIG. 6.2 Main steps of the QuEChERS method.

TABLE 6.4 Applications of the QuEChERS method for the determination of CECs in solid samples in environmental analysis.

CECs	Matrix (amount)	Extraction solvents	Salts	Clean-up step	Analytical technique	Recovery (%)	LOD ($ng.g^{-1}$)	LOQ ($ng.g^{-1}$)	Reference
1 (herbicide)	Paddy soil (5 g)	5 mL water and 10 mL ACN	2 g $MgSO_4$ and 1 g NaCl	150 mg of $MgSO_4$ and 100 mg PSA	UHPLC-(Q)- MS (ESI+)	96.7–106	1.3–3.6	4.89–12.48	(Liu et al., 2020)
33 pharmaceuticals and metabolites	Sewage sludge (1 g)	10 mL 0.1 M EDTA solution, 10 mL ACN (0.1%, v/v, acetic acid)	4 g $MgSO_4$, 1 g NaCl, 1 g tri-Na, 0.5 g di-Na	600 mg $MgSO_4$, 200 mg PSA, 75 mg of Z-Sep +	UHPLC- (Orbitrap)-MS (ESI+)	60-98	0.3-8.1	1.1–24.0	(Miserli et al., 2023)
8 (β-lactams antibiotics degradation product)	Wastewater sludge (0.5 g)	2 g EDTA-treated sand, 5 mL 0.1 M EDTA solution, 10 mL ACN	4 g $MgSO_4$, 1 g NaCl, 1 g tri-Na, 0.5 g di-Na	Strata X SPE cartridge	UHPLC LTQ/ Orbitrap (ESI+)	96–114	N.D.	1.9–17.4	(Guironnet et al., 2022)
17 (sulfonamides, fluoroquinolones, macrolides, tetracyclines, b-lactams, trimethoprim and thiamphenicol)	Sewage sludge (1 g)	10 mL 0.2 M Na_2EDTA, 8 mL acetonitrile, 2 mL methanol	Citrate buffer salt	200 mg $MgSO_4$ and 150 mg PSA sorbent	HPLC-MS/MS (ESI –/+)	4–125	0.003–120.0	0.01–365	(Ajibola et al., 2020)

Continued

TABLE 6.4 Applications of the QuEChERS method for the determination of CECs in solid samples in environmental analysis—cont'd

CECs	Matrix (amount)	Extraction solvents	Salts	Clean-up step	Analytical technique	Recovery (%)	LOD ($ng.g^{-1}$)	LOQ ($ng.g^{-1}$)	Reference
218 (pesticides)	Soil (10 g)	10 mL of ACN with 2.5% formic acid	6 g of MgSO4 and 1.5 g of CH_3COONa	no clean-up step	LC-MS/MS and GC-MS/MS	70–120	0.024–6.25	5.0	(Acosta-Dacal et al., 2021)
37 (pesticides, bisphenols, polycyclic musks and UV-filters)	Sediments (10 g)	10 mL of ultra-pure water plus 10 mL of ACN	2 g of magnesium sulphate and 0.5 g of sodium	EMR-Lipid (200 mg)	GC-EI-MS	71–112	0.25–2.5	1.25–200.00	(Cunha et al., 2022)
14 (PAHs)	Powder aerosol particles (10 mg)	0.4 mL ACN/ DCM (7:1, v/v)	20 mg Na_2SO_4 and 20 mg NaCl	8 mg PSA, 16 mg Na_2SO_4	HPLC-FLD	85–121	1.9–27.2	5.8–82.6	(Jing et al., 2023)
2 (Chlorinated paraffins)	Lake sediments (5 g)	5 mL water, 15 mL ACN	4 g $MgSO_4$, 1 g NaCl	1,200 mg $MgSO_4$, 400 mg PSA, 400 mg C_{18}, 45 mg	+2D-UHPLC(Orbitrap) MS (ESI-)	81-95	0,.41–60	1.38–7.0	(Wu et al., 2022)

QuEChERS for pharmaceutical compound recovery was also reported by Ajibola et al. (2020) and Herrero et al. (2014), whereas Peysson and Vulliet (2013) found better recoveries using the acetate-buffered version for both pharmaceuticals and hormones.

Citrate-buffered QuEChERS also yielded significantly higher recoveries of beta-lactams in sludge when compared to the acetate-buffered, original, and veterinary drug residue QuEChERS methods (the latter using sodium sulfate instead of magnesium sulfate) (Guironnet et al., 2022). According to the authors, this improved performance is attributed to the higher polarity of beta-lactams at pH 8, the buffering condition established by the citrate system.

With regard to the extraction solvent, Peysson and Vulliet (2013) evaluated several options for the analysis of pharmaceuticals and hormones in sludge. In general, methanol, ethyl acetate, and dichloromethane led to reduced recoveries, whereas tetrahydrofuran (THF) provided recoveries similar to those obtained with ACN for compounds such as famotidine (antihistamine), naproxen (anti-inflammatory), triclosan (antibacterial), fluoxetine (antidepressant), and several macrolide antibiotics (lincomycin, erythromycin, spiramycin, and tylosin). However, steroid hormone recoveries decreased by approximately 40% when THF was used instead of ACN (Peysson and Vulliet, 2013).

Under extraction conditions using 10 mL of water and 10 mL of ACN, neither tetracyclines nor fluoroquinolones were recovered. The authors suggested that the addition of EDTA and acid could improve extraction, as these analyte classes exhibit strong binding to matrix cations. Sample mass was also evaluated, and it was observed that reducing the amount of sample improved recoveries due to a decreased matrix effect. Penicillins (amoxicillin, ampicillin, cloxacillin) were more affected by matrix interference than cephalosporins.

The use of EDTA solution to chelate matrix metals (Ca^{2+} and Mg^{2+}) resulted in improved recovery of pharmaceutical compounds (Miserli et al., 2023). For beta-lactams, recoveries also increased with the use of EDTA solution or the addition of 0.1% formic acid. However, these additives were only effective when applied separately (Guironnet et al., 2022). The authors also noted that water absorption by sodium sulfate is a less exothermic reaction compared to magnesium sulfate, which reduces the likelihood of analyte degradation.

For the d-SPE step, magnesium sulfate and PSA were adopted as standard sorbents, and the effects of Z-Sep$^+$ and GCB on analyte recovery were evaluated. Z-Sep$^+$, at 18.75 mg mL^{-1}, provided higher analyte recoveries, whereas GCB, at any concentration, impaired recovery due to its strong adsorption capacity for the target analytes (Miserli et al., 2023). For beta-lactams, the use of d-SPE (tested with PSA and C18) was not feasible, as the analytes were retained by the sorbents. As an alternative for extract cleanup, Strata-X SPE cartridges were employed (Guironnet et al., 2022).

Li et al. (2020) evaluated different solvent and salt systems for the determination of the herbicide Glyamifop in rice, soil, and plant materials. Acetonitrile was the most effective extraction solvent, and salt composition had to be tailored to the matrix. Complex matrices such as rice straw required a combination of PSA, $MgSO_4$, and GCB for effective pigment removal.

Acosta-Dacal et al. (2021) developed a QuEChERS method for the simultaneous determination of 218 pesticides in clay loam soil, without the clean-up step. Despite this simplification, good recoveries were achieved using matrix-matched calibration. This emphasizes the robustness of QuEChERS even without d-SPE, provided appropriate calibration strategies are used.

For estuarine sediment analysis, Cunha et al. (2022) targeted a wide range of endocrine-disrupting compounds (EDCs), including pesticides, bisphenols, musks, and UV-filters. QuEChERS was combined with DLLME for preconcentration and derivatization. EMR-Lipid and Z-Sep^+/GCB sorbents were compared, with EMR-Lipid selected for its simplicity and high recovery. ACN alone was preferred over solvent mixtures to minimize extract contamination.

In the case of airborne particulates, Jing et al. (2023) optimized QuEChERS for the analysis of PAHs in aerosol particles. A binary solvent mixture of ACN:DCM (7:1) was superior to other solvents, enhancing analyte extraction from porous particles. Sodium sulfate and NaCl were preferred salts, and PSA was the most effective cleanup sorbent at low DCM content. This highlights how analyte polarity and matrix structure dictate solvent and sorbent choice.

Wu et al. (2022) developed a QuEChERS protocol for extracting long-chain chlorinated paraffins (LCCPs) and very-long-chain homologues (vLCCPs) from sediments. Sample amount and solvent volume were critical parameters: 5 g of sediment and 15 mL of MeCN provided better recoveries than the standard 10 g/10 mL combination. Among tested commercial d-SPE kits, the one containing $MgSO_4$, PSA, C18, and GCB showed the best performance in pigment removal and analyte recovery. Integration of a two-dimensional liquid chromatography (2D-LC) system postextraction enabled further online cleanup and eliminated matrix effects.

Taken together, these studies demonstrate that while the main principles of QuEChERS—such as the use of ACN and phase separation by salting-out—are broadly applicable, optimal performance requires case-by-case adjustment. Buffer systems (e.g., citrate vs. acetate), extraction solvent composition, salt ratios, and sorbent combinations must be tailored to the analyte classes and matrix properties. In complex matrices such as sediment, soil, or plant-based materials, additional cleanup or concentration steps (e.g., 2D-LC or DLLME) may be required. Furthermore, the strategic exclusion of cleanup, as seen in pesticide multiresidue determination, highlights the balance between throughput and selectivity. The accumulated evidence demonstrates the versatility and wide applicability of QuEChERS for extracting environmental solid samples.

3.2.5 Matrix solid-phase dispersion

Matrix solid-phase dispersion (MSPD) is a sample preparation technique that combines, in a single step, the extraction and purification of target analytes. It involves homogenizing the solid sample with a sorbent, such as Florisil or

C_{18}, using a mortar and pestle to form a dry and free-flowing mixture, which is then packed into cartridges and eluted with suitable solvents (Souza Futigami et al., 2024). The method was initially developed for biological samples but has increasingly been applied to the determination of CECs in solid environmental matrices such as soils, sediments, sewage sludge, and biota.

Several studies have demonstrated the effective use of MSPD for CEC extraction in environmental samples. For instance, the technique has been successfully applied to determine organochlorine pesticides in soils and emerging organic contaminants in aquatic plants and biota such as fish and insects. In the case of aquatic plants, combinations of MSPD with dispersing agents such as Florisil and $MgSO_4$ enabled the efficient extraction of a wide range of contaminants, including pharmaceuticals and personal care products, followed by GC-MS analysis (Aznar et al., 2017). Frequently, the technique is further associated with additional steps, such as ultrasound-assisted extraction, to improve recovery yields and minimize interferences (Souza Futigami et al., 2024).

Although MSPD offers advantages such as low solvent consumption, operational simplicity, and reduced costs, it also presents significant limitations: sample preparation is manual and operator-dependent, affecting reproducibility, and the technique's cleanup capability may not be sufficient for highly complex environmental matrices (Fontanals et al., 2021; Hoff and Pizzolato, 2018; Souza Futigami et al., 2024). Consequently, MSPD is rarely used alone for the determination of CECs in solid environmental samples and is often coupled with other techniques, such as pressurized liquid extraction (PLE) or additional solid-phase extraction (SPE) steps, to ensure adequate analytical performance and selectivity.

3.2.6 *Pressurized liquid extraction*

Pressurized liquid extraction (PLE), also known as accelerated solvent extraction (ASE), was commercially introduced in 1995 by Dionex Corporation, with its first documented scientific application published the following year (Richter et al., 1996). Since then, the technique has gained prominence for its effectiveness in extracting analytes from solid matrices and has been officially recognized by the United States Environmental Protection Agency (USEPA) as a method for the determination of persistent organic pollutants in environmental samples (Andreu and Picó, 2019).

PLE is an automated method that uses heated and pressurized solvents to enhance extraction efficiency. Under these conditions, solvents exhibit lower viscosity and surface tension, which facilitates their diffusion through the matrix and improves the solubilization of target compounds, reducing both solvent consumption and the time required for extraction compared to conventional techniques (Subedi et al., 2015).

The extraction is carried out in stainless steel cells (1–100 mL) capable of withstanding temperatures ranging from 50°C to 200°C and pressures from 500 to 3000 psi. Samples—whether fresh, dried, or lyophilized—are loaded with or without inert supports (such as diatomaceous earth or Florisil), using nylon or cellulose filters at the bottom to prevent clogging. Once assembled, the system performs automatic cycles of static extraction, rinse, and purge with an inert gas (N_2), collecting the extract in an appropriate container (Hoff and Pizzolato, 2018; Richter et al., 1996).

The use of sorbents in PLE allows the combination of extraction and purification in a single step, promoting the selective elution of analytes and the retention of interfering substances. This approach is known as selective PLE (SPLE) and has been referred to by terms such as EPLE (Enhanced PLE), "in-cell cleanup," or "in situ cleanup" (Fontanals et al., 2021; Subedi et al., 2015). Various natural materials have also been successfully tested as solid supports in PLE for the extraction of different analytes from diverse matrices, aiming to reduce environmental impact and promote sustainability (Souza Futigami et al., 2024).

Several parameters influence PLE efficiency, such as the type and composition of the solvent (commonly methanol, acetonitrile, acetone, dichloromethane), extraction temperature (typically between 40°C–120°C, considered the most critical variable due to its direct influence on solubility and selectivity), static time and number of cycles (which affect diffusion kinetics and phase equilibrium), rinse volume (essential for mobilizing the extract from inside the cell to the collection vessel), type of adsorbent (when applicable), sample/adsorbent mass ratio (when applicable), and the physicochemical characteristics of the matrix and target analytes. Proper selection of these parameters is essential to optimize analyte recovery and minimize the coextraction of interfering compounds (Hoff and Pizzolato, 2018; Subedi et al., 2015).

PLE is highly efficient, automated, and compatible with subsequent concentration or additional purification steps when necessary. It is often used as a preliminary stage for coupling with highly selective and sensitive analytical techniques, such as liquid or gas chromatography coupled to tandem mass spectrometry (LC-MS/MS, GC-MS/MS), particularly in studies focused on the detection of emerging contaminants at low levels (Fontanals et al., 2021; Hoff and Pizzolato, 2018).

Currently, other manufacturers also offer PLE systems, such as Fluid Management Systems, Inc. (FMS), Büchi, and CEM (EDGE). These systems present variations in heating methods, extraction times, and integration with purification steps, broadening the accessibility and versatility of the technique, which has been successfully applied in the recovery of CECs from environmental samples (Bosch-Orea et al., 2023; da Silva et al., 2025; Soriano et al., 2024).

Due to its characteristics, the technique has been widely studied and applied in laboratory routines and adapted to different types of complex samples, standing out as a promising method for the determination of CECs in solid environmental samples (Bonaccorso et al., 2024; da Silva et al., 2025; Martín-Pozo et al., 2019). Table 6.5 presents applications of the PLE method for the determination of CECs in solid samples in environmental analysis.

TABLE 6.5 Applications of the PLE method for the determination of CECs in solid samples in environmental analysis.

CECs	Matrix	Solvent/Sorbent	PLE conditions	LOD/LOQ	Analytical instrument	Recovery (%)	Reference
165 CECs(Pharmaceutical and illicit drugs	Sediment	MeOH:DCM (1:1 v/v)	100°C, 2 min, 1 cycle 150°C, 20 s, 1 cycle	LOQ: 1–5 ng/g	LC-MS/MS	Not reported	(da Silva et al., 2025)
34 PhCs pharmaceuticals (antibiotics, antidepressants, β-blockers, estrogens, etc.)	Three soil types: loam, silty-loam, and silty-clay-loam	Ultrapure water/ in-cell silica gel and glass beads packing	R-PHWE, 120°C, 110 bar, 7 static cycles, stainless-steel cell 50 × 5 mm	Not reported	LC–MS/MS	11–129. (80%–85% of target analytes showed *R*% ≥ 40%)	(Bonaccorso et al., 2024)
Marine biotoxins and CECs	Oysters and Mussels	H_2O : ACN, (80:20, *v* /v) with 0.1% de FA/ diatomaceous earth	MSPD-PLE I- 75°C ± 2°C, 10–15 s, 1 cycle, cell (refillable plastic capsules) MSPD-PLE II-80°C, 1 min, 1 cycle, aluminum cells	LOD > 5 μg/kg	LC-MS/MS	Not reported	(Bosch-Orea et al., 2023)

Continued

TABLE 6.5 Applications of the PLE method for the determination of CECs in solid samples in environmental analysis—cont'd

CECs	Matrix	Solvent/Sorbent	PLE conditions	LOD/LOQ	Analytical instrument	Recovery (%)	Reference
BTHs and UV filters	Sediment, and sludge	DCM/ACN (50:50, v/v)/ Na_2SO_4, silica and copper powder.	80°C–120°C, 5 min, 2–3 cycles, stainless-steel cells 40 mL	LOD: 0.01–1.77 ng/g (sediment), LOQ: 0.04–5.88 ng/g (sediment) LOD: 0.07–1.91 ng/g (sludge); LOQ: 0.22–6.61 ng/g (sludge)	GC-MS/MS	70–130	(Xiong et al., 2022)
Pesticides	Vineyard soils	• MeOH:ACN (70:30, v/v)MeOH:ACN: FA (65:30:5, v/v/v)/ diatomaceous earth	• 80°C, 5 min, 1 cycle.120°C, 5 min, 2 cycles.stain steel cells 11 mL, Flush volume 70%	LOQ: 0.2–13 ng/g	UPLC-ESI-QqQ-MS/MS	70–130	(Pérez-Mayán et al., 2020)
Carbazole and PHCs	Sediments	DCM/ diatomaceous earth	100°C, 5 min, 2 cycles, stain steel cells 32 mL	LOD: 0.003–0.22 ng/g LOQ: 0.01–0.75 ng/g	UHPLC-MS/MS	64.8–124.7	(Zhou et al., 2019)

OCPs	River sediments	DCM/*n*-hexane (4:3, v/v)/Silica gel	105°C, 103 bar, 4 cycles Clean-up sorbents: activated cooper and diatomaceous earth	LOD: 0.004–0.134 ng/mL LOQ: 0.012–0.360 ng/mL	GC-MS/MS	47.66–117.97	(Duodu et al., 2016)
EDCs (alkylphenols bisphenol A)	sediment and biota	MeOH/C18	150°C (sediment), 40°C (biota), and basic alumina as a clean-up sorbent	LOQ: 0.2-4.0 ng/g (sediment) 1.7–4.7 ng/g (biota)	LC-ESI-MS/MS	94–100 (sediment) 80–107(biota)	(Salgueiro-González et al., 2015)
BFRs, CFRs	Sewage sludge	DCM-Hexane, (1:1, v/v)/Mixed with Na_2SO_4	100°C, 3 cycles)—Acidic silica to remove lipids—Clean-up: multilayer silica column	LOD: 0.01–0.5 ng/g	HRGC/HRMS	92.6	(Zeng et al., 2014)
14 PPCPs	Soil and Biosolids	MeOH (70%)	80°C (biosolids), 100°C (soils), 5 min preheat, 5 min static, 1 cycle, 60% flush	LOQ: ng/g 0.1 a 15 ng/g (Soil) 0.3–27 (Biosolids)	LC-MS/MS	were over 50%, except diphenhydramine (~30%) in soils	(Wu et al., 2010)

BFRs, brominated flame retardants; *BSAs*, benzenesulfonamide; *BTHs*, benzothiazole; *BTRs*, benzotriazole; *CFRs*, chlorinated flame retardants; *EDCs*, endocrine disrupting chemicals; *HRGC/LRMS*, high-resolution gas chromatography/low resolution mass spectrometry; *OCPs*, organochlorine pesticides; *PhCs*, pharmaceuticals; *PPCPs*, pharmaceuticals and personal care products; *R-PHWE*, recirculated pressurized hot water extraction.

Typical and well-applied cases of PLE for the extraction of CECs have been reported in recent studies. Bonaccorso et al. (2024) evaluated 34 pharmaceutical compounds in soils by comparing two PLE methods (R-PHWE and D-PHWE), coupled with LC-MS/MS (Bonaccorso et al., 2024). Both methods used silica gel and glass beads mixed with soil as sorbents. The experimental design showed that the number of cycles significantly influenced recovery. Operating at 120°C and 110 bar, the main difference between methods was the extraction mode: R-PHWE employed automated hot water recirculation, while D-PHWE used a single-pass offline system. R-PHWE achieved statistically higher recoveries for 16 out of 34 pharmaceuticals (up to 103%), with greater efficiency and alignment with green chemistry principles.

Bosch-Orea et al. (2023) compared two PLE methods preceded by matrix solid-phase dispersion (MSPD-PLE I and MSPD-PLE II) with solid-liquid and ultrasound-assisted extraction for simultaneous determination of marine biotoxins and CECs in oysters and mussels. MSPD-PLE I used a coffee machine, while MSPD-PLE II employed a commercial automated system (EDGE), both using diatomaceous earth and extraction with H_2O:ACN (80:20) at 80°C. While MSPD-PLE II offered greater precision and robustness, MSPD-PLE I presented satisfactory results for most CECs despite its simplicity, highlighting PLE's versatility.

Another notable study (Pérez-Mayán et al., 2020) investigated pesticide residues in vineyard soils using PLE coupled to UPLC-ESI-QqQ-MS/MS. Lyophilized samples were extracted with sequential solvent mixtures under two sets of conditions (80°C and 120°C), achieving 70%–130% recovery for 44 out of 51 compounds, with LOQs between 0.2 and 13 ng/g. The method demonstrated high sensitivity and robustness across soils with varying properties, revealing the persistence of several fungicides and the insecticide imidacloprid through consecutive agricultural cycles.

Together, these studies demonstrate the adaptability, efficiency, and growing importance of PLE in the extraction and determination of CECs from complex environmental matrices.

3.2.7 *Ultrasound-assisted extraction*

Ultrasound-assisted extraction (UAE) has gained significant recognition as a viable and green technique for extracting contaminants of emerging concern (CECs) from environmental solid matrices such as soil, sediment, sludge, and biota. CECs—including pharmaceuticals, personal care products (PCPs), flame retardants, hormones, and heavy metals—are persistent pollutants that demand efficient, low-impact analytical methodologies. UAE operates on the principle of acoustic cavitation. High-frequency ultrasonic waves generate microbubbles in the extraction solvent, which collapse violently, causing microjets and localized high temperatures and pressures. This disrupts the structure of solid samples and enhances solvent penetration and analyte diffusion, increasing extraction efficiency (Albero et al., 2015; Medina-Torres et al., 2017). Two main types of instrumentation are employed: ultrasonic baths and ultrasonic probes. Ultrasonic baths are more accessible, allowing the simultaneous treatment of multiple samples with uniform energy distribution—making them suitable for routine laboratory operations. In contrast, ultrasonic probes deliver higher energy densities at localized points, often leading to more efficient extractions, especially for recalcitrant compounds, although they are more expensive and less suited for high-throughput workflows (Albero et al., 2015). Compared to conventional methods such as Soxhlet extraction and PLE, UAE presents several advantages: reduced solvent usage, shorter extraction times, lower energy requirements, and operation under mild conditions. These attributes make it highly compatible with the principles of green analytical chemistry (Mehta et al., 2022). Furthermore, UAE instrumentation is generally simpler and more cost-effective than alternatives such as MAE, while still achieving comparable recovery rates for various analytes (Albero et al., 2015). The versatility of UAE lies in its adaptability to a wide range of matrices and target compounds. Studies have demonstrated efficient extraction of antibiotics, hormones, fungicides, hydrocarbons, and polychlorinated biphenyls (PCBs) from soil and sludge with recoveries often exceeding 90% (Bezama et al., 2013; Kotowska et al., 2019; Merdassa et al., 2015). Additionally, UAE has been successfully combined with persulfate-based advanced oxidation processes, including activation with zero-valent iron, to enhance the degradation of persistent petroleum hydrocarbons in contaminated soils (Li et al., 2020). Despite its advantages, UAE is not free from limitations. The efficiency of extraction can vary based on several parameters, including solvent composition, ultrasound frequency, sonication duration, and matrix complexity. Prolonged sonication can lead to thermal degradation of sensitive analytes or coextraction of matrix interferences, especially in complex substrates such as sewage sludge. Therefore rigorous optimization and validation of the extraction conditions are critical for achieving reproducibility and accuracy (Albero et al., 2015; Leśniewska et al., 2016). Importantly, UAE is compatible with a broad range of solvents—from polar to less polar—enabling its application to diverse chemical classes. When appropriately configured, the technique minimizes environmental impact, positioning UAE as a sustainable and powerful tool in environmental analysis workflows. In conclusion, UAE stands out as a versatile, efficient, and eco-friendly technique for the extraction of emerging contaminants from environmental solid samples. Its adaptability to different instruments, sample types, and solvents, combined with its alignment to green chemistry, underscores its growing importance in modern environmental analytical practices.

3.2.8 Miniaturized and sustainable techniques for the extraction of solid samples

Headspace analysis is a widely used technique for extracting volatile compounds from solid samples. The sample is sealed in a vial, heated, and allowed to reach equilibrium, promoting the transfer of volatile analytes to the gas phase.

Only this vapor phase is introduced into the gas chromatograph, offering a clean and low-maintenance method (Sparkman et al., 2011). For the analysis of CECs, headspace is often coupled with solid-phase microextraction (HS-SPME), a solvent-free technique that uses a coated fiber exposed to the headspace to adsorb analytes without direct contact with the sample (Lancioni et al., 2022). This reduces matrix interferences and extends fiber lifespan. After equilibrium, analytes are thermally desorbed into the GC system. HS-SPME enables rapid, simplified, and in some cases in situ or in vivo sampling. However, it is limited to VOCs and SVOCs and does not allow for additional cleanup steps, which can lead to relatively high detection limits in complex matrices (Fig. 6.3).

A relevant application of HS-SPME for the analysis of CECs was demonstrated by Castro et al., who developed a method using HS-SPME Arrow coupled with GC–MS/MS to determine synthetic musk fragrances in fish samples (Castro et al., 2019). The technique was applied to lyophilized fish tissue, achieving detection limits as low as 0.5 ng g^{-1}. The use of PDMS-coated SPME Arrow significantly improved sensitivity—up to ten times greater than conventional fibers—due to the larger sorption phase. The optimized method proved effective for quantifying musk fragrances such as HHCB and AHTN in commercial fish.

According to Qiu et al., a novel in vivo solid-phase microextraction (SPME) method was developed using electrospun fibers composed of poly(lactic acid-co-caprolactone) (PLCL) doped with sulfonated γ-Al_2O_3 nanoparticles and coated with a polynorepinephrine (pNE) bioinspired sheath (Qiu et al., 2016b). This combination enhanced both the sorption capacity—via cation exchange interactions—and biocompatibility of the fiber. The SPME fibers were directly inserted into the dorsal-epaxial muscle of live tilapia and into the stems of okra

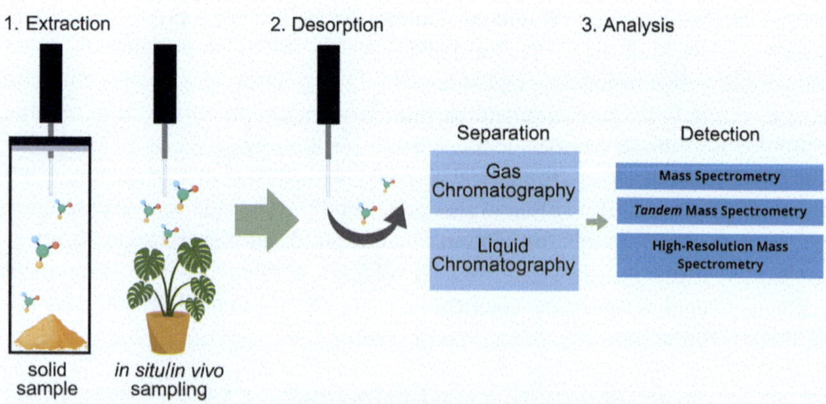

FIG. 6.3 Typical HS-SPME procedures with different separation and detection techniques applied in environmental solid samples. *(Adapted from Lancioni, C., Castells, C., Candal, R., Tascon, M., 2022. Headspace solid-phase microextraction: fundamentals and recent advances. Adv. Sample Prep. 2772-5820. 3, 100035. doi:10.1016/j.sampre.2022.100035. Licensed under Creative Commons Attribution 4.0 (CC BY 4.0).)*

plants previously exposed to pharmaceutical-spiked water (100 $\mu g \cdot L^{-1}$ for 5 days). After a 10 min sampling period, analytes were desorbed and quantified by LC-MS/MS.

The custom-made fibers showed superior performance over conventional polydimethylsiloxane (PDMS) coatings, with extraction efficiencies up to 85 times higher for sulfonamides. Limits of detection ranged from 0.16 to 5.35 $ng \cdot g^{-1}$ in fish and 0.02 to 8.02 $ng \cdot g^{-1}$ in vegetables. The method demonstrated high reproducibility (RSDs $\leq$ 15.6%) and sensitivity, with some compounds (e.g., sulfafurazole) detectable only by SPME. These results confirm the fiber's robustness and suitability for in vivo trace-level pharmaceutical monitoring in aquatic and plant matrices.

Focusing on a different matrix, Pastor-Belda et al. applied dispersive liquid-liquid microextraction (DLLME) for the determination of five new-generation pesticides in soil samples, coupled with LC-ESI-MS/MS (Pastor-Belda et al., 2015). This technique is based on the rapid dispersion of a high-density organic extraction solvent (carbon tetrachloride) into an aqueous medium using aACN as a disperser. The resulting cloudy solution significantly increases the surface contact between the phases, allowing fast and efficient analyte transfer into the extractant.

The method demonstrated excellent sensitivity, with detection limits as low as 0.0015–0.0090 $ng \cdot g^{-1}$ and enrichment factors ranging from 7 to 24. Recovery rates for spiked samples were within 87%–114%, and the precision (RSD) ranged between 5.5% and 14%. DLLME was shown to be effective for analyzing pesticide residues in complex solid matrices such as soils, with minimal solvent consumption and high extraction efficiency. Overall, the study highlights the technique's suitability for environmental monitoring of trace-level contaminants with high reliability and analytical performance.

The main microextraction techniques applied for solid environmental samples are briefly described in Table 6.6. Moreover, a summary of recent applications of miniaturized and sustainable techniques for the extraction of solid samples aiming for CECs determination is presented in Table 6.7.

3.2.9 *The concept of green methods*

The growing environmental awareness in science and industry led to the development of Green Chemistry in the 1990s, focused on reducing toxic solvents and waste. This movement gave rise to Green Analytical Chemistry (GAC), which promotes sustainable analytical practices aligned with environmental protection. In recent years, green extraction techniques have gained considerable attention as sustainable alternatives for the extraction of CECs from complex environmental matrices.

In the context of solid samples such as soils, sediments, sewage sludge, and biota, conventional extraction methods often involve high volumes of toxic organic solvents, long extraction times, and significant energy consumption. Green extraction aims to minimize these drawbacks by integrating the

TABLE 6.6 Microextraction techniques for solid environmental samples.

Microextraction technique	Working principle	Application matrix	Target contaminants	Advantages	Limitations	Key references
Solid-Phase Microextraction (SPME)	Adsorption of analytes on a solid fiber coated with stationary phase; solvent-free.	Soil, sediment, sewage sludge, biota	Pharmaceuticals, hormones, UV filters, musk fragrances	Portable, in vivo capable	Biofouling, coating fragility	(Vakondios et al., 2016; Vila et al., 2017; Wang et al., 2019)
Dispersive Micro Solid-Phase Extraction (d-μSPE)	Dispersion of a solid sorbent directly into the sample to increase contact efficiency.	Soil, dust, sludge	Pesticides, flame retardants, personal care products	Efficient, low solvent	Needs separation step	(Pastor-Belda et al., 2015)
Microextraction by Packed Sorbent (MEPS)	The sorbent is packed inside a 100–250 μL syringe, enabling sample handling, extraction, and elution in a single step	Sediment, wastewater, tap water	Plasticizers, parabens, drugs	Reusable, automatable	Sorbent saturation risk	(Amiri and Ghaemi, 2017)
Liquid-Phase Microextraction (LPME)	A small volume of solvent is used to extract and preconcentrate analytes from the sample in a single step.	Soil, sludge, biota	Pesticides, surfactants, pharmaceuticals	Selective, green solvent use	Limited to low polarity	(Elik et al., 2021)
Dispersive Liquid-Liquid Microextraction (DLLME)	Formation of fine extractant droplets dispersed in an aqueous phase using a disperser solvent to rapidly extract analytes	Soil	Pesticides, pharmaceuticals	High enrichment factors (EFs up to 24), Low solvent consumption, High sensitivity, Simple and rapid	Restricted to analytes with affinity to organic solvent, Limited extractant compatibility with LC mobile phases	(Pastor-Belda et al., 2015)
Single-Drop Microextraction (SDME)	A microdrop of solvent is suspended in the sample or headspace to extract analytes directly.	Wastewater, water	Volatile and semi-volatile organic compounds	Low-cost, simple	Drop instability, low repeatability	(Kiszkiel-Taudul and Starczewska, 2019)

TABLE 6.7 Summary of microextraction techniques applied to solid environmental matrices for the determination of contaminants of emerging concern (CECs).

Contaminant (CEC)	Matrix	Sample preparation	Instrumental determination	LOD	LOQ	Recovery (%)	Reference
Pharmaceuticals (12)	Fish and vegetable (biota)	In vivo-SPME-LC-MS/MS	LC-MS/MS	0.16–5.35 ng g^{-1}	0.55–17.8 ng g^{-1}	–	(Qiu et al., 2016b)
Acidic pharmaceuticals (10)	Fish (tilapia)	In vivo-SPME-LC-MS/MS	LC-MS/MS	0.13–8.44 ng g^{-1}	0.44–28.1 ng g^{-1}	–	(Qiu et al., 2016a)
Synthetic musk fragrances (9)	Fish	HS-SPME-GC-MS/MS	GC-MS/MS	0.5–2.5 ng g^{-1}	2.5–5 ng g^{-1}	–	(Castro et al., 2019)
UV filters (11)	Beach sand (sediment)	HS-SPME-GC-MS/MS	GC-MS/MS	0.001–0.087 ng g^{-1}	0.004–0.15 ng g^{-1}	71–119	(Vila et al., 2018)
Pharmaceuticals and PCPs (54)	Fish (biota)	MAE-hollow fiber-L/SME	UHPLC-MS/MS	0.01–0.50 ng g^{-1}	0.05–2.00 ng g^{-1}	56.3–119.9	(Zhang et al., 2017)
Pesticides (5)	Soil	SLE-DLLME-LC-ESI-MS/MS	LC-ESI-MS/MS	0.0015–0.0090 ng g^{-1}	0.005–0.030 ng g^{-1}	87–114	(Pastor-Belda et al., 2015)
Bisphenol A and replacements (6)	Dust	VALLME-LC-MS/MS	LC-MS/MS	0.5–10 ng g^{-1}	1–20 ng g^{-1}	69.0–108.1	(Dueñas-Mas et al., 2019)
Triazines (3)	Soil	LOC-electrochemical analysis	Electrochemical	0.36–0.55 nM	–	–	(Islam et al., 2015)
Methyl paraben, butyl paraben, diclofenac, triclosan	Mussels	LOV-BI-	LC-MS/MS	–	–	89–132	(Rosende et al., 2019)

SPME, solid-phase microextraction; *HS-SPME*, headspace solid-phase microextraction; *DI-SPME*, direct immersion solid-phase microextraction; *In vivo-SPME*, in vivo solid-phase microextraction; *MAE*, microwave-assisted extraction; *L/SME*, liquid/solid microextraction; *MEPS*: microextraction by packed sorbent; *DLLME*, dispersive liquid–liquid microextraction; *SLE*, solid–liquid extraction; *ETV*, electrothermal vaporization; *ICP-MS*, inductively coupled plasma mass spectrometry; *GC-MS*, gas chromatography–mass spectrometry; *GC-MS/MS*, gas chromatography–tandem mass spectrometry; *LC-MS*, liquid chromatography–mass spectrometry; *LC-MS/MS*, liquid chromatography–tandem mass spectrometry; *UHPLC-MS/MS*, ultra-high performance liquid chromatography–tandem mass spectrometry; *LC-DAD*, liquid chromatography with diode array detection; *VALLME*, vortex-assisted liquid–liquid microextraction; *LOC*, lab-on-a-chip; *PCPs*, personal care products.

principles of green chemistry, including the reduction of hazardous substances, lower energy use, and minimization of waste generation, while maintaining or improving analytical performance (Armenta et al., 2015; Gałuszka et al., 2013; Celeiro et al., 2022).

Among the available extraction techniques for solid environmental samples, supercritical fluid extraction (SFE), microwave-assisted extraction (MAE), miniaturized techniques, and QuEChERS stand out as green alternatives. SFE replaces toxic organic solvents with supercritical CO_2, thereby minimizing hazardous waste generation. MAE significantly reduces solvent use, extraction time, and energy consumption through efficient microwave heating. Miniaturized techniques contribute to greener practices by drastically reducing the amounts of solvents and samples required, while QuEChERS offers a simple, rapid, and effective procedure with minimal solvent consumption, aligning with several principles of green chemistry.

3.3 Purification and preconcentration techniques

3.3.1 *Purification using SPE (solid-phase extraction)*

With few exceptions, the determination of CECs in environmental samples deals with concentration levels ranging from parts per trillion and sub-ppb—i.e., in the order of nanograms per liter or nanograms per kilogram—up to approximately 1 to 10 or 20 parts per billion for the more concentrated analytes. Therefore it is virtually essential that this type of analysis includes analyte concentration techniques.

Although the main focus of this chapter is on sampling and extraction procedures, it is necessary to briefly address some of the cleanup and concentration techniques that arc subsequently applied. Undoubtedly, one of the techniques of choice is solid-phase extraction (SPE) using cartridges. This approach, long established in environmental chemistry—especially for emerging contaminants—remains the most practical and versatile method for concentrating and enriching analytes in environmental matrices. Various types of SPE cartridges are used for this purpose, mainly individually, but eventually in tandem (Hoff et al., 2016). A widely used option is a hydrophilic-lipophilic-balanced polymeric reversed-phase sorbent, such as Oasis HLB or Strata-X. Mixed-mode ion-exchange sorbents may also be employed when ionic selectivity is required.

Other commonly used options include cation-exchange or anion-exchange cartridges, often preceding an HLB cartridge. While some studies also employ Florisil or C18 cartridges, the vast majority use at least HLB cartridges, often as the final step.

Solid samples, once subjected to initial extraction procedures—be it solid-liquid extraction, pressurized liquid extraction, ultrasound-assisted extraction, etc.—almost invariably yield a liquid extract. This extract is then applied directly to the SPE. In some cases, the extract is diluted, pH-adjusted, or otherwise modified to ensure compatibility with the subsequent SPE step (Mastroianni et al., 2013; Andreu and Picó, 2019; Pico et al., 2019).

The extracts obtained from SPE can be concentrated directly via water bath evaporation under a nitrogen stream, or they may undergo additional purification and concentration steps. These may include low-temperature purification, complementary liquid-liquid partitioning, or simple centrifugation and filtration prior to direct injection. However, the most common workflow involves evaporation followed by reconstitution, centrifugation, and filtration, as will be discussed in the following section.

3.3.2 Filtration and centrifugation

Before injection of the extract into analytical systems, several final steps are typically crucial to prevent contamination, blockages, or other issues with the instruments. Simple techniques such as centrifugation and filtration play a key role. Centrifugation helps separate particulate matter from the extracts, especially in cases of highly turbid extractions, and also aids in precipitating fats and proteins that may remain in the extract even after cleanup procedures such as solid-phase extraction. This step is particularly important after applying partitioning techniques or low-temperature purification.

Centrifugation becomes especially crucial in certain types of matrices. For instance, extracts from lipid-rich samples such as fish, meat, or dairy products often retain residual fats and protein aggregates that can compromise chromatographic systems. Even when cold partitioning or solid-phase extraction is employed, these lipid components may not be fully removed, and centrifugation offers an effective way to pellet them out of the solution. Similarly, in environmental samples such as sediments, sludges, or airborne particulate matter, fine debris and colloidal particles frequently remain suspended in the extract. These can escape standard filtration and are best removed through centrifugation to prevent clogging and fouling of analytical equipment.

Plant-based matrices also benefit from this step. Extracts from leaves, seeds, or resinous plant material often carry chlorophyll, waxes, and resins, which are known to interfere with both chromatographic separation and mass spectrometric detection (Kurz et al., 2019). Centrifugation after initial extraction and cleanup facilitates the removal of these sticky or viscous compounds. Likewise, when working with biological fluids such as plasma, serum, or urine, residual macromolecules may persist even after protein precipitation. Centrifugation further clarifies the supernatant by sedimenting out remaining proteins, lipids, or extracellular vesicles, improving both recovery and instrument longevity.

This additional purification step is also important when extracts undergo liquid-liquid extraction or freeze-out partitioning. While temperature-driven separation techniques may precipitate some unwanted components, the resulting supernatants often remain cloudy. Centrifugation ensures that these precipitated residues are completely removed, yielding a cleaner, more stable sample for instrumental analysis.

Following centrifugation, a nearly mandatory step in all cases is the filtration of extracts using 0.22 or 0.45 μm membrane filters to eliminate any remaining particulate material that could clog or damage analytical instruments. The chemical nature of the membrane must be compatible with the solvent present in the extract. Depending on the solvent composition, nylon, PVDF, regenerated cellulose, hydrophilic PTFE, or conventional solvent-compatible PTFE membranes may be selected.

4 Case studies/application examples

4.1 Case study 1: Determination of pharmaceuticals and personal care products in sediment samples

A recent study conducted in the Lagoa da Conceição estuary in Florianópolis, Brazil, investigated the presence of 165 contaminants of emerging concern (CECs) in sediment, water, and biota as part of the REACQUA project (da Silva et al., 2025). Following an environmental disaster involving the release of treated effluent into the lagoon, sediment was prioritized as a key matrix for evaluating environmental contamination and potential ecological risks.

Sediment samples were manually collected from 18 shallow sites, corresponding to water sampling locations, using polypropylene jars. The surface layer (0–5 cm) was targeted, and after decanting excess water, samples were transferred to polypropylene tubes, transported on ice, and stored at −20°C. In the lab, samples were homogenized, dried at 60°C, ground with a ceramic pestle, and sieved through a 200-mesh screen. The prepared samples were kept frozen and protected from light until analysis.

Extraction was performed using an automated pressurized liquid extraction (PLE) system (EDGE, CEM Corporation). Dried sediment (1.0 ± 0.05 g) was extracted with a 1:1 mixture of methanol and dichloromethane. The system applied a two-step thermal cycle to obtain a 35 mL extract, which was then evaporated under nitrogen at 40°C. The residue was reconstituted in methanol with acid and ammonium acetate, and when necessary, a defatting step with hexane was applied before LC-MS/MS analysis.

Among the 35 CECs identified during initial screening, 12 were selected for quantitative analysis. Sediment samples showed the highest levels of contamination, with compounds such as caffeine, ciprofloxacin, clindamycin, diclofenac, and acetaminophen detected in significant concentrations. The highest values were observed near the wastewater treatment plant, with caffeine reaching up to 31,646 $ng \cdot kg^{-1}$ and ciprofloxacin up to 7543 $ng \cdot kg^{-1}$.

These results underscore the role of sediments as reservoirs for persistent contaminants. The study also demonstrated the effectiveness of automated PLE systems for extracting a wide range of analytes from solid environmental samples. Importantly, the findings point to the ecotoxicological risks posed by substances such as caffeine and diclofenac, reinforcing the need for continued monitoring and environmental protection in vulnerable estuarine systems.

4.2 Case study 2: Determination of polybrominated diphenyl ethers in sewage sludge

A novel analytical approach was developed for the detection of polybrominated diphenyl ethers (PBDEs) in sewage sludge using dispersive liquid-liquid microextraction (DLLME) with magnetic ionic liquids (MILs), combined with pyrolysis gas chromatography–mass spectrometry (Py-GC–MS) (Rodrigues et al., 2024). This configuration addressed the incompatibility of viscous MILs with direct GC injection by employing a pyrolyzer for sample introduction.

Samples were collected from an industrial sewage system and characterized for basic physicochemical properties. The MIL, composed of trihexyltetradecylphosphonium and manganese chloride, was synthesized and characterized by UV–Vis, FTIR, Raman, and thermal analysis. Extraction conditions were optimized via design of experiments, with the best results obtained using 10.0 mg MIL, 3.00 μL ACN, 8.50 mL of sample, and 60 s extraction—eliminating the need for centrifugation.

Pyrolysis conditions (220°C, 0.60 min, 9.00 μL) were selected to prevent MIL degradation while maximizing analyte transfer. The method showed good precision (RSD $\leq$ 12.5%) and low detection limits for all PBDE congeners tested. PBDE 28 was quantified at 800 $\mu g \cdot L^{-1}$, highlighting sewage sludge as a reservoir for persistent contaminants.

Compared to traditional extraction methods, this technique reduced both solvent usage and analysis time, proving to be more sustainable and efficient. It represents the first report coupling MIL-based DLLME with Py-GC–MS for PBDEs in sludge and offers a promising tool for environmental monitoring of emerging contaminants.

4.3 Case study 3: Determination of pharmaceuticals and personal care products in biota

This study focused on assessing the uptake of 55 pharmaceuticals and personal care products (PPCPs) by lettuce (from the variety *Cichorium intybus*) cultivated under real agricultural conditions using reclaimed water (Sunyer-Caldú and Diaz-Cruz, 2021). The experiment was conducted in two field plots located near a wastewater treatment plant in Girona, Spain, under varying combinations of irrigation systems (drip vs. sprinkling), soil textures (clayey vs. sandy), and water sources (treated effluent vs. water reclaimed via soil infiltration through reactive barriers).

At harvest, ten lettuces from each subplot (approx. 50 lettuces per subplot) were randomly collected. Samples were shaken to remove soil, stored frozen, and transported to the laboratory. There, the leaves were thawed, chopped, refrozen, lyophilized, and ground to a fine powder. Composite samples from each subplot were then prepared and stored frozen until extraction.

For the extraction of PPCPs, the researchers developed and validated a QuEChERS-based method adapted for lyophilized vegetable matrices. One gram of freeze-dried lettuce was hydrated with 9 mL of ultrapure water to mimic fresh lettuce moisture content. The sample was extracted with 10 mL of ACN, followed by the addition of citrate-buffered QuEChERS salts to promote analyte partitioning. After centrifugation, a cleanup step using PSA and C18 dispersive SPE sorbents was performed. The supernatant was acidified, evaporated under nitrogen, and reconstituted in methanol for LC-MS/MS analysis. Isotopically labeled internal standards were added for quality control and quantification.

The developed method demonstrated high sensitivity (method LODs as low as 0.01 ng/g dw), good linearity, and acceptable recovery rates (mostly within 80%–120%). The highest PPCP concentrations detected in lettuce were for the UV filter 4-hydroxybenzophenone (up to 84.1 ng/g dw), followed by salicylic acid, caffeine, diclofenac, and benzophenone-2. Several compounds, including transformation products of common pharmaceuticals (e.g., CBZ-E, N-desmethylvenlafaxine), were found in all samples, suggesting widespread occurrence and uptake.

This case highlights the effectiveness of a modified QuEChERS extraction method for analyzing complex vegetable matrices and underscores the environmental relevance of monitoring CECs in crops irrigated with reclaimed water under real agricultural practices.

5 Final considerations

The determination of CECs has exhibited a growing trend in recent years, gaining increasing attention within the scientific community. One of the main trends observed is the rise of integrative analyses. In these analyses, multiple matrices from the same environment are examined together, as commonly seen in aquatic environments, where sediment, water, and biota samples are simultaneously analyzed. The distributions of contaminants in each of these compartments are often assessed, and in some cases, risk analysis is also performed, especially if the biota includes fish or other animals that serve as direct food sources for humans.

The sampling phase still lacks proper harmonization and qualification. Recent literature does not significantly differ from the initial practices in this field. That is, sample collection from soil, sediment, sludge, or biota has primarily relied on grab sampling, where samples are collected at specific geographic and temporal points. Composite samples are rarely used for this type of collection, although some studies are striving for more detailed statistical planning and design in sample collection in areas of interest. Biota samples continue to be collected opportunistically, often involving small, medium, and large animals. Whenever possible, these animals, found dead or in biological parks and reserves, are collected. Collaborative agreements with universities and research centers allow conservation unit staff to collect materials such as egg remnants following predefined procedures, ensuring that the materials are refrigerated until repository personnel can perform the collection.

It is believed that the trend in this area will continue with this system, although there is room for improvement in solid sample collection. A future trend is the use of preextraction in the field, such as SPME, where fibers are used to collect the analytes directly from sludge or sediment and send the material to the laboratory. Passive samplers for this purpose are still in the early stages of development, but there is potential for growth.

Regarding sample extraction, there is considerable room for the expansion of green methods, especially those related to the use of pressure, such as SFE or PLE. These techniques have been successfully used in soil analysis but are less frequently applied in the analysis of sediments and sludge for the determination of CECs. Depending on the configurations, these methods can be considered green and have the potential for widespread use in this field.

Another approach involves the use of QuEChERS-like methods to broaden the scope of materials analyzed. However, this practice is not always successful, as the more analytes included, the greater the difficulty in assessing recovery and other validation parameters. It is believed that there will be a significant impact on the progress of this area with the growing use of alternative dispersive solid phases for precleaning and prepurification of extracts before instrumental analysis.

In summary, the determination of CECs in solid environmental matrices is not only complementary to aqueous investigations but essential for a comprehensive understanding of environmental contamination. Solid matrices such as soil, sludge, and biota exhibit distinct physicochemical characteristics that influence the retention, transformation, and bioavailability of CECs in ways that are fundamentally different from aquatic systems. Moreover, some contaminants are predominantly found in these matrices due to their hydrophobicity, strong sorption to organic matter, or biological accumulation—factors that often exclude them from water-based analyses. Solid matrices also act as reservoirs or long-term sinks, playing a critical role in the environmental fate and potential remobilization of these compounds under changing conditions. Finally, their integrative value in risk assessment—particularly when linked to trophic transfer or human exposure pathways—underscores the urgent need for continued innovation in sampling and extraction strategies tailored specifically for solids.

AI disclosure

During the preparation of this work the author(s) used ChatGPT for English-language revision. After using this tool/service, the authors reviewed and edited the content as needed and take full responsibility for the content of the publication.

References

Acosta-Dacal, A., Rial-Berriel, C., Díaz-Díaz, R., Bernal-Suárez, M.D.M., Luzardo, O.P., 2021. Optimization and validation of a QuEChERS-based method for the simultaneous environmental monitoring of 218 pesticide residues in clay loam soil. Sci. Total Environ. 00489697. 753, 142015. doi:10.1016/j.scitotenv.2020.142015.

Ajibola, A.S., Tisler, S., Zwiener, C., 2020. Simultaneous determination of multiclass antibiotics in sewage sludge based on QuEChERS extraction and liquid chromatography-tandem mass spectrometry. Anal. Methods. 17599679. 12 (4), 576–586. Available from: http://pubs.rsc.org/en/journals/journal/ay. doi:10.1039/c9ay02188d.

Albero, B., Sánchez-Brunete, C., García-Valcárcel, A.I., Pérez, R.A., Tadeo, J.L., 2015. Ultrasound-assisted extraction of emerging contaminants from environmental samples. TrAC Trends Anal. Chem. 18793142. 71, 110–118. Available from: www.elsevier.com/locate/trac. doi:10.1016/j.trac.2015.03.015.

Alqarni, A.M., 2024. Analytical methods for the determination of pharmaceuticals and personal care products in solid and liquid environmental matrices: a review. Molecules. 1420-3049. 29 (16), 3900. doi:10.3390/molecules29163900.

Amiri, A., Ghaemi, F., 2017. Microextraction in packed syringe by using a three-dimensional carbon nanotube/carbon nanofiber–graphene nanostructure coupled to dispersive liquid-liquid microextraction for the determination of phthalate esters in water samples. Microchim. Acta. 0026-3672. 184 (10), 3851–3858. doi:10.1007/s00604-017-2416-8.

Anastassiades, M., Lehotay, S.J., Štajnbaher, D., Schenck, F.J., 2003. Fast and easy multiresidue method employing acetonitrile extraction/partitioning and "dispersive solid-phase extraction" for the determination of pesticide residues in produce. J. AOAC Int. 10603271. 86 (2), 412–431. Available from: https://academic.oup.com/jaoac/issue. doi:10.1093/jaoac/86.2.412.

Andreu, V., Picó, Y., 2019. Pressurized liquid extraction of organic contaminants in environmental and food samples. TrAC Trends Anal. Chem. 01659936. 118, 709–721. doi:10.1016/j.trac.2019.06.038.

Armenta, S., Garrigues, S., de la Guardia, M., 2015. The role of green extraction techniques in green analytical chemistry. TrAC Trends Anal. Chem. 01659936. 71, 2–8. doi:10.1016/j.trac.2014.12.011.

Aznar, R., Albero, B., Sánchez-Brunete, C., Miguel, E., Martín-Girela, I., Tadeo, J.L., 2017. Simultaneous determination of multiclass emerging contaminants in aquatic plants by ultrasound-assisted matrix solid-phase dispersion and GC-MS. Environ. Sci. Pollut. Res. 16147499. 24 (9), 7911–7920. Available from: http://www.springerlink.com/content/0944-1344. doi:10.1007/s11356-016-6327-8.

Azzouz, A., Ballesteros, E., 2016. Determination of 13 endocrine disrupting chemicals in environmental solid samples using microwave-assisted solvent extraction and continuous solid-phase extraction followed by gas chromatography–mass spectrometry. Anal. Bioanal. Chem. 408 (1), 231–241. doi:10.1007/s00216-015-9096-1.

Bachelot, M., Li, Z., Munaron, D., Le Gall, P., Casellas, C., Fenet, H., Gomez, E., 2012. Organic UV filter concentrations in marine mussels from French coastal regions. Sci. Total Environ. 420, 273–279. doi:10.1016/j.scitotenv.2011.12.051.

Badry, A., Slobodnik, J., Alygizakis, N., Bunke, D., Cincinelli, A., Claßen, D., Dekker, R.W.R.J., Duke, G., Dulio, V., Göckener, B., Gkotsis, G., Hanke, G., Jartun, M., Movalli, P., Nika, M.C., Rüdel, H., Thomaidis, N.S., Tarazona, J.V., Tornero, V., Treu, G., Vorkamp, K., Walker, L.A., Koschorreck, J., 2022. Using environmental monitoring data from apex predators for chemicals management: towards harmonised sampling and processing of archived wildlife samples to increase the regulatory uptake of monitoring data in chemicals management. Environ. Sci. Eur. 21904715. 34 (1). Available from: http://www.springer.com/environment/journal/12302. doi:10.1186/s12302-022-00664-6.

Bao, L.J., Wei, Y.L., Yao, Y., Ruan, Q.Q., Zeng, E.Y., 2015. Global trends of research on emerging contaminants in the environment and humans: a literature assimilation. Environ. Sci. Pollut. Res. 16147499. 22 (3), 1635–1643. Available from: https://link.springer.com/journal/11356. doi:10.1007/s11356-014-3404-8.

Bezama, A., Flores, A., Araneda, A., Barra, R., Pereira, E., Hernández, V., Moya, H., Konrad, O., Quiroz, R., 2013. Assessment and optimization of an ultrasound-assisted washing process using organic solvents for polychlorinated biphenyl-contaminated soil. Waste Manag. Res. 0734-242X. 31 (10), 969–978. doi:10.1177/0734242x13492002.

Bonaccorso, G., Renai, L., Checchini, L., Del Bubba, M., 2024. A novel apparatus for the fully automated extraction and online liquid chromatographic analysis of solid environmental samples: application to the pressurized hot water extraction of pharmaceuticals in soil. Anal. Chem. 0003-2700. 96 (28), 11390–11396. doi:10.1021/acs.analchem.4c01479.

Bosch-Orea, C., Kleemann, C.R., Deolindo, C.T.P., Molognoni, L., Dallegrave, A., Daguer, H., de Oliveira Costa, A.C., Hoff, R.B., 2023. Integrated analysis of marine biotoxins and contaminants of emerging concern in bivalve mollusks from Santa Catarina, Brazil. Sci. Total Environ. 18791026. 905. Available from: www.elsevier.com/locate/scitotenv. doi:10.1016/j.scitotenv.2023.167254.

Botella, M.B., Lemos, A.A., Lujan, C.E., Wuilloud, R.G., Quintas, P.Y., 2024. Recent advances of extraction and separation of emerging organic contaminants through the application of natural deep eutectic solvents. TrAC Trends Anal. Chem. 18793142. 171. Available from: https://www.sciencedirect.com/science/journal/01659936. doi:10.1016/j.trac.2023.117518.

Câmara, J.S., Perestrelo, R., Berenguer, C.V., Andrade, C.F.P., Gomes, T.M., Olayanju, B., Kabir, A., Rocha, C.M.R., Teixeira, J.A., Pereira, J.A.M., 2022. Green extraction techniques as advanced sample preparation approaches in biological, food, and environmental matrices: a review. Molecules. 1420-3049. 27 (9), 2953. doi:10.3390/molecules27092953.

Castro, Ó., Trabalón, L., Schilling, B., Borrull, F., Pocurull, E., 2019. Solid phase microextraction arrow for the determination of synthetic musk fragrances in fish samples. J. Chromatogr. A. 00219673. 1591, 55–61. doi:10.1016/j.chroma.2019.01.032.

Celeiro, M., Llompart, M., Dagnac, T., 2022. Green analytical methodologies to determine personal care products in solid environmental matrices: soils, sediments, sludge and biota—a review. Adv. Sample Prep. 27725820. 2, 100013. doi:10.1016/j.sampre.2022.100013.

Chen, X.H., Zhao, Y.G., Shen, H.Y., Zhou, L.X., Pan, S.D., Jin, M.C., 2014. Fast determination of seven synthetic pigments from wine and soft drinks using magnetic dispersive solid-phase extraction followed by liquid chromatography-tandem mass spectrometry. J. Chromatogr. A 18733778. 1346, 123–128. Available from: www.elsevier.com/locate/chroma. doi:10.1016/j.chroma.2014.04.060.

Chen, X., Wu, X., Luan, T., Jiang, R., Ouyang, G., 2021. Sample preparation and instrumental methods for illicit drugs in environmental and biological samples: a review. J. Chromatogr. A. 00219673. 1640, 461961. doi:10.1016/j.chroma.2021.461961.

Couch, M.W., Schmidt, C.J., Wasdo, S.C., 2000. A comparison of sampling techniques for VOCs in soil. Adv. Environ. Res. 10930191. 4 (1), 91–96. doi:10.1016/S1093-0191(00)00007-1.

Cui, K., Wen, J., Zeng, F., Zhou, X., Li, S., Zeng, Z., 2017. Determination of organophosphate ester flame retardants and plasticizers in soil samples by microwave-assisted extraction coupled with silica gel/alumina multilayer solid-phase extraction and gas chromatography-mass spectrometry. Anal. Methods. 1759-9660. 9 (6), 986–993. doi:10.1039/c6ay02837c.

Cunha, S.C., Ferreira, R., Marmelo, I., Vieira, L.R., Anacleto, P., Maulvault, A., Marques, A., Guilhermino, L., Fernandes, J.O., 2022. Occurrence and seasonal variation of several endocrine disruptor compounds (pesticides, bisphenols, musks and UV-filters) in water and sediments from the estuaries of Tagus and Douro Rivers (NE Atlantic Ocean coast). Sci. Total Environ. 18791026. 838. Available from: www.elsevier.com/locate/scitotenv. doi:10.1016/j.scitotenv.2022.155814.

da Silva, A.C., Alves de Oliveira, L.V., Amaral Alexandre, L., Rocha Ribas, M., Lemos Dal Pizzol, J., Rocha, G., Kasuko Palmeiro, J., Perin, M., Hoff, R., Verruck, S., 2025. Suspect screening and quantitative analysis of 165 contaminants of emerging concern in water, sediments, and biota using LC-MS/MS: ecotoxicological and human health risk assessment. Sci. Total Environ. 18791026. 963. Available from: https://www.sciencedirect.com/science/journal/00489697. doi:10.1016/j.scitotenv.2025.178434.

de L Freitas, F., Costa, A.B., de Moraes, A.S.B., Lima, A.D.F., Santos, R.P., Silva, V.A.D., Pereira, N.S., Cavalcante, R.M., 2024. Contaminants of Emerging Concern (CECs): assessment of health and dietary risk in the consumption of Plagioscion squamosissimus in one of the largest rivers in a semi-arid region (Rio São Francisco, Brazil). Chemosphere. 363. doi:10.1016/j.chemosphere.2024.142889

de Wit, C.A., Alaee, M., Muir, D.C.G., 2006. Levels and trends of brominated flame retardants in the Arctic. Chemosphere. 00456535. 64 (2), 209–233. Available from: www.elsevier.com/locate/chemosphere. doi:10.1016/j.chemosphere.2005.12.029.

Dimpe, K.M., Nomngongo, P.N., 2016. Current sample preparation methodologies for analysis of emerging pollutants in different environmental matrices. TrAC Trends Anal. Chem. 18793142. 82, 199–207. Available from: www.elsevier.com/locate/trac. doi:10.1016/j.trac.2016.05.023.

Ducrocq, T., Merel, S., Miège, C., 2024. Review on analytical methods and occurrence of organic contaminants in continental water sediments. Chemosphere. 00456535. 365, 143275. doi:10.1016/j.chemosphere.2024.143275.

Dueñas-Mas, M.J., Ballesteros-Gómez, A., Rubio, S., 2019. Supramolecular solvent-based microextraction of emerging bisphenol A replacements (colour developers) in indoor dust from public environments. Chemosphere. 18791298. 222, 22–28. Available from: www.elsevier.com/locate/chemosphere. doi:10.1016/j.chemosphere.2019.01.095.

Duodu, G.O., Goonetilleke, A., Ayoko, G.A., 2016. Optimization of in-cell accelerated solvent extraction technique for the determination of organochlorine pesticides in river sediments. Talanta. 00399140. 150, 278–285. Available from: https://www.journals.elsevier.com/talanta. doi:10.1016/j.talanta.2015.12.049.

Dürig, W., Kintzi, A., Golovko, O., Wiberg, K., Ahrens, L., 2020. New extraction method prior to screening of organic micropollutants in various biota matrices using liquid chromatography coupled to high-resolution time-of-flight mass spectrometry. Talanta. 00399140. 219, 121294. doi:10.1016/j.talanta.2020.121294.

Dürig, W., Kintzi, A., Golovko, O., Wiberg, K., Ahrens, L., 2020. New extraction method prior to screening of organic micropollutants in various biota matrices using liquid chromatography coupled to high-resolution time-of-flight mass spectrometry. Talanta, 219, 121294. doi:10.1016/j.talanta.2020.121294.

Elik, A., Bingöl, D., Altunay, N., 2021. Ionic hydrophobic deep eutectic solvents in developing air-assisted liquid-phase microextraction based on experimental design: application to flame atomic absorption spectrometry determination of cobalt in liquid and solid samples. Food Chem. 03088146. 350, 129237. doi:10.1016/j.foodchem.2021.129237.

Falandysz, J., Liu, G., Rutkowska, M., 2024. Analytical progress on emerging pollutants in the environment: an overview of the topics. TrAC Trends Anal. Chem. 01659936. 175, 117719. doi:10.1016/j.trac.2024.117719.

Fan, X., Kubwabo, C., Rasmussen, P.E., Wu, F., 2014. Simultaneous determination of thirteen organophosphate esters in settled indoor house dust and a comparison between two sampling techniques. Sci. Total Environ. 00489697. 491–492, 80–86. doi:10.1016/j.scitotenv.2013.12.127.

Fernández-Sanjuan, M., Rigol, A., Sahuquillo, A., Rodríguez-Cruz, S., Lacorte, S., 2009. Determination of alkylphenols and alkylphenol ethoxylates in sewage sludge: effect of sample pre-treatment. Anal. Bioanal. Chem. 1618-2642. 394 (6), 1525–1533. doi:10.1007/s00216-009-2747-3.

Fontanals, N., Pocurull, E., Borrull, F., Marcé, R.M., 2021. Clean-up techniques in the pressurized liquid extraction of abiotic environmental solid samples. Trends Environ. Anal. Chem. 22141588. 29, e00111. doi:10.1016/j.teac.2020.e00111.

Gałuszka, A., Migaszewski, Z., Namieśnik, J., 2013. The 12 principles of green analytical chemistry and the SIGNIFICANCE mnemonic of green analytical practices. TrAC Trends Anal. Chem. 01659936. 50, 78–84. doi:10.1016/j.trac.2013.04.010.

Golovko, O., Örn, S., Sörengård, M., Frieberg, K., Nassazzi, W., Lai, F.Y., Ahrens, L., 2021. Occurrence and removal of chemicals of emerging concern in wastewater treatment plants and their impact on receiving water systems. Sci. Total Environ. 00489697. 754, 142122. doi:10.1016/j.scitotenv.2020.142122.

Guedes-Alonso, R., Santana-Viera, S., Montesdeoca-Esponda, S., Afonso-Olivares, C., Sosa-Ferrera, Z., Santana-Rodríguez, J.J., 2016. Application of microwave-assisted extraction and ultra-high performance liquid chromatography–tandem mass spectrometry for the analysis of sex hormones and corticosteroids in sewage sludge samples. Anal. Bioanal. Chem. 16182650. 408 (24), 6833–6844. Available from: http://link.springer.de/link/service/journals/00216/index.htm. doi:10.1007/s00216-016-9810-7.

Guibal, R., Buzier, R., Lissalde, S., Guibaud, G., 2019. Adaptation of diffusive gradients in thin films technique to sample organic pollutants in the environment: an overview of o-DGT passive samplers. Sci. Total Environ. 00489697. 693, 133537. doi:10.1016/j.scitotenv.2019.07.343.

Guironnet, A., Wiest, L., Vulliet, E., 2022. Improvement of the QuEChERS extraction step by matrix-dispersion effect and application on beta-lactams analysis in wastewater sludge by LC-MS/MS. Talanta. 00399140. 237, 122923. doi:10.1016/j.talanta.2021.122923.

Haaland, D., Siegel, J.A., 2017. Quantitative filter forensics for indoor particle sampling. Indoor Air. 09056947. 27 (2), 364–376. doi:10.1111/ina.12319.

Habimana, E., Sauvé, S., 2025. A review of properties, occurrence, fate, and transportation mechanisms of contaminants of emerging concern in sewage sludge, biosolids, and soils: recent advances and future trends. Front. Environ. Chem. 2673-4486. 6. doi:10.3389/fenvc.2025.1547596.

Havens, S.M., Hedman, C.J., Hemming, J.D.C., Mieritz, M.G., Shafer, M.M., Schauer, J.J., 2014. Comparison of accelerated solvent extraction, soxhlet and sonication techniques for the extraction of estrogens, androgens and progestogens from soils. J. Agric. Chem. Environ. 2325-7458. 03 (03), 103–120. doi:10.4236/jacen.2014.33013.

Hawker, D.W., Clokey, J., Gorji, S.G., Verhagen, R., Kaserzon, S.L., 2022. Monitoring techniques-Grab and passive sampling. In: Dalu, T., Tavengwa, N.T. (Eds.), Emerging Freshwater Pollutants: Analysis, Fate and Regulations. Elsevier, Australia, pp. 25–48. Available from: https://www.sciencedirect.com/book/9780128228500. doi:10.1016/B978-0-12-822850-0.00014-4.

Herrero, P., Borrull, F., Pocurull, E., Marcé, R.M., 2014. A quick, easy, cheap, effective, rugged and safe extraction method followed by liquid chromatography-(Orbitrap) high resolution mass spectrometry to determine benzotriazole, benzothiazole and benzenesulfonamide derivates in sewage sludge. J. Chromatogr. A 18733778. 1339, 34–41. Available from: www.elsevier.com/locate/chroma. doi:10.1016/j.chroma.2014.02.081.

Hoff, R.B., Pizzolato, T.M., 2018. Combining extraction and purification steps in sample preparation for environmental matrices: a review of matrix solid phase dispersion (MSPD) and pressurized liquid extraction (PLE) applications. TrAC Trends Anal. Chem. 18793142. 109, 83–96. Available from: www.elsevier.com/locate/trac. doi:10.1016/j.trac.2018.10.002.

Hoff, R., Pizzolato, T.M., Diaz-Cruz, M.S., 2016. Trends in sulfonamides and their by-products analysis in environmental samples using mass spectrometry techniques. Trends Environ. Anal. Chem. 22141588. 9, 24–36. Available from: http://www.journals.elsevier.com/trends-in-environmental-analytical-chemistry/. doi:10.1016/j.teac.2016.02.002.

Huang, P., Zhao, P., Dai, X., Hou, X., Zhao, L., Liang, N., 2016. Trace determination of antibacterial pharmaceuticals in fishes by microwave-assisted extraction and solid-phase purification combined with dispersive liquid–liquid microextraction followed by ultra-high performance liquid chromatography-tandem mass spectrometry. J. Chromatogr. B. 15700232. 1011, 136–144. doi:10.1016/j.jchromb.2015.12.059.

Islam, K., Chand, R., Han, D., Kim, Y.S., 2015. Microchip capillary electrophoresis based electroanalysis of triazine herbicides. Bull. Environ. Contam. Toxicol. 14320800. 94 (1), 41–45. Available from: http://link.springer.de/link/service/journals/00128/index.htm. doi:10.1007/s00128-014-1378-3.

Jing, W., Nakano, K., Shen, Z., Okuda, T., 2023. Optimization of the QuEChERS extraction method to determine polycyclic aromatic hydrocarbons (PAHs) in powder aerosol particles collected by cyclone. Environ. Technol. Innov. 23521864. 31, 103141. doi:10.1016/j.eti.2023.103141.

Kaufmann, A., Butcher, P., Maden, K., Walker, S., Widmer, M., 2022. Improving the QuEChERS liquid/liquid extraction of analytes with widely varying physicochemical properties: example of 201 veterinary drugs in milk. J. AOAC Int. 1060-3271. 105 (4), 1030–1042. doi:10.1093/jaoacint/qsac013.

Kiszkiel-Taudul, I., Starczewska, B., 2019. Single drop microextraction coupled with liquid chromatography-tandem mass spectrometry (SDME-LC-MS/MS) for determination of ranitidine in water samples. Microchem. J. 0026265X. 145, 936–941. doi:10.1016/j.microc.2018.12.015.

Kotnik, K., Kosjek, T., Krajnc, U., Heath, E., 2014. Trace analysis of benzophenone-derived compounds in surface waters and sediments using solid-phase extraction and microwave-assisted extraction followed by gas chromatography–mass spectrometry. Anal. Bioanal. Chem. 1618-2642. 406 (13), 3179–3190. doi:10.1007/s00216-014-7749-0.

Kotowska, U., Kapelewska, J., Kotowski, A., Pietuszewska, E., 2019. Rapid and sensitive analysis of hormones and other emerging contaminants in groundwater using ultrasound-assisted emulsification microextraction with solidification of floating organic droplet followed by GC-MS detection. Water. 2073-4441. 11 (8), 1638. doi:10.3390/w11081638.

Kumirska, J., Migowska, N., Caban, M., Łukaszewicz, P., Stepnowski, P., 2015. Simultaneous determination of non-steroidal anti-inflammatory drugs and oestrogenic hormones in environmental solid samples. Sci. Total Environ. 00489697. 508, 498–505. doi:10.1016/j.scitotenv.2014.12.020.

Kurz, M.H.S., Juliana, J.L., de Oliveira, L.G., Hoff, R., Martins, M.L., Gonçalves, F.F., 2019. Clean-up procedure development and method validation for pesticide residues analysis in carrots. Food Anal. Methods. 1936976X. 12 (1), 282–292. Available from: http://www.springer.com/life+sci/food+science/journal/12161. doi:10.1007/s12161-018-1359-y.

Lancioni, C., Castells, C., Candal, R., Tascon, M., 2022. Headspace solid-phase microextraction: fundamentals and recent advances. Adv. Sample Prep. 27725820. 3, 100035. doi:10.1016/j.sampre.2022.100035.

Lee, H., Peart, T.E., Sarafin, K., 2003. Occurrence of polycyclic and nitro musk compounds in canadian sludge and wastewater samples. Water Qual. Res. J. 38 (4), 683–702. . doi:10.2166/wqrj.2003.043.

Lehotay, S.J., 2022. The QuEChERSER mega-method. LCGC N. Am. 1939-1889. 40 (1), 13–19. doi:10.56530/lcgc.na.px2173m3.

Leśniewska, B., Krymska, M., Świerad, E., Wiater, J., Godlewska-Żyłkiewicz, B., 2016. An ultrasound-assisted procedure for fast screening of mobile fractions of Cd, Pb and Ni in soil. Insight into method optimization and validation. Environ. Sci. Pollut. Res. 0944-1344. 23 (24), 25093–25104. doi:10.1007/s11356-016-7745-3.

Li, Y.T., Li, D., Lai, L.J., Li, Y.H., 2020. Remediation of petroleum hydrocarbon contaminated soil by using activated persulfate with ultrasound and ultrasound/Fe. Chemosphere. 18791298. 238. Available from: www.elsevier.com/locate/chemosphere. doi:10.1016/j.chemosphere.2019.124657.

Lindholm-Lehto P.C., Ahkola H.S. and Knuutinen J.S., 2017. Procedures of determining organic trace compounds in municipal sewage sludge—a review. Environ. Sci. Pollut. Res. 24 (5), 4383–4412. doi:10.1007/s11356-016-8202-z.

Liu, L., Rao, L., Li, W., Zhou, W., Li, B., Tang, L., 2020. Detection of Glyamifop residues in rice and its environment by the QuEChERS method combined with HPLC–MS. Microchem. J. 158. doi:10.1016/j.microc.2020.105157

Llompart, M., Celeiro, M., Dagnac, T., 2019b. Microwave-assisted extraction of pharmaceuticals, personal care products and industrial contaminants in the environment. TrAC Trends Anal. Chem. 116, 136–150. doi:10.1016/j.trac.2019.04.029.

Llompart, M., Garcia-Jares, C., Celeiro, M., Dagnac, T., 2019a. Extraction | microwave-assisted extraction encyclopedia of analytical science. In: Worsfold, P., Townshend, A., Poole, C., Miró, M. (Eds.), Reference Module in Chemistry, Molecular Sciences and Chemical Engineering. Elsevier, Spain, pp. 67–77. doi:10.1016/B978-0-12-409547-2.14442-7.

Lopardo, L., Rydevik, A., Kasprzyk-Hordern, B., 2019. A new analytical framework for multiresidue analysis of chemically diverse endocrine disruptors in complex environmental matrices utilising ultra-performance liquid chromatography coupled with high-resolution tandem quadrupole time-of-flight mass spectrometry. Anal. Bioanal. Chem. 1618-2642. 411 (3), 689–704. doi:10.1007/s00216-018-1483-y.

Manjarrés-López, D.P., Vitale, D., Callejas-Martos, S., Usuriaga, M., Picó, Y., Pérez, S., Montemurro, N., 2023. An effective method for the simultaneous extraction of 173 contaminants of emerging concern in freshwater invasive species and its application. Anal. Bioanal. Chem. 16182650. 415 (29-30), 7085–7101. Available from: https://www.springer.com/journal/216. doi:10.1007/s00216-023-04974-3.

Martín-Pozo, L., de Alarcón-gómez, B., Rodríguez-Gómez, R., García-Córcoles, M.T., Çipa, M., Zafra-Gómez, A., 2019. Analytical methods for the determination of emerging contaminants in sewage sludge samples. A review. Talanta. 00399140. 192, 508–533. Available from: https://www.journals.elsevier.com/talanta. doi:10.1016/j.talanta.2018.09.056.

Maruya, K.A., Dodder, N.G., Schaffner, R.A., Weisberg, S.B., Gregorio, D., Klosterhaus, S., Alvarez, D.A., Furlong, E.T., Kimbrough, K.L., Lauenstein, G.G., Christensen, J.D., 2014. Refocusing mussel watch on contaminants of emerging concern (CECs): the California pilot study (2009–10). Mar. Pollut. Bull. 18793363. 81 (2), 334–339. Available from: www.elsevier.com/locate/marpolbul. doi:10.1016/j.marpolbul.2013.04.027.

Mastroianni, N., Postigo, C., De Alda, M.L., Barcelo, D., 2013. Illicit and abused drugs in sewage sludge: method optimization and occurrence. J. Chromatogr. A. 18733778. 1322, 29–37. doi:10.1016/j.chroma.2013.10.078.

Medina-Torres, N., Ayora-Talavera, T., Espinosa-Andrews, H., Sánchez-Contreras, A., Pacheco, N., 2017. Ultrasound assisted extraction for the recovery of phenolic compounds from vegetable sources. Agronomy. 2073-4395. 7 (3), 47. doi:10.3390/agronomy7030047.

Mehta, N., S, J., Kumar, P., Verma, A.K., Umaraw, P., Khatkar, S.K., Khatkar, A.B., Pathak, D., Kaka, U., Sazili, A.Q., 2022. Ultrasound-assisted extraction and the encapsulation of bioactive components for food applications. Foods. 23048158. 11 (19). Available from: www.mdpi.com/journal/foods. doi:10.3390/foods11192973.

Merdassa, Y., Liu, J.F., Megersa, N., 2015. Development and validation of one-step ultrasound-assisted extraction for simultaneous determination of multiclass fungicides in soils. J. AOAC Int. 10603271. 98 (1), 192–200. Available from: http://docserver.ingentaconnect.com/deliver/connect/aoac/10603271/v98n1/s27.pdf?expires=1470916493&id=88370670&titleid=41000050&accname=Elsevier+BV&checksum=ED97CEB2558F64B8BE781AB96536BEF7. doi:10.5740/jaoacint.14-033.

Miserli, K., Kosma, C., Konstantinou, I., 2023. Determination of pharmaceuticals and metabolites in sludge and hydrochar after hydrothermal carbonization using sonication—QuEChERS extraction method and UHPLC LTQ/Orbitrap MS. Environ. Sci. Pollut. Res. 0944-1344. 30 (1), 1686–1703. doi:10.1007/s11356-022-22215-5.

Montes, R., Méndez, S., Carro, N., Cobas, J., Alves, N., Neuparth, T., Santos, M.M., Quintana, J.B., Rodil, R., 2022. Screening of contaminants of emerging concern in surface water and wastewater effluents, assisted by the persistency-mobility-toxicity criteria. Molecules. 14203049. 27 (12). Available from: https://www.mdpi.com/1420-3049/27/12/3915/pdf?version=1655545448. doi:10.3390/molecules27123915.

Montesdeoca-Esponda, S., Sosa-Ferrera, Z., Santana-Rodríguez, J.J., 2013. Microwave-assisted extraction combined with on-line solid phase extraction followed by ultra-high-performance liquid chromatography with tandem mass spectrometric determination of benzotriazole UV stabilizers in marine sediments and sewage sludges. J. Sep. Sci. 16159314. 36 (4), 781–788. doi:10.1002/jssc.201200664.

Moorchilot, V.S., Louis, H., Haridas, A., Praveena, P., Arya, S., Nair, A.S., Aravind, U.K., Aravindakumar, C., 2025. Bisphenols in indoor dust: A comprehensive review of global distribution, exposure risks, transformation, and biomonitoring. Chemosphere, 370, 143798. doi:10.1016/j.chemosphere.2024.143798.

Nguyen, T.T.N., Baduel, C., 2023. Optimization and validation of an extraction method for the analysis of multi-class emerging contaminants in soil and sediment. J. Chromatogr. A 18733778. 1710. Available from: www.elsevier.com/locate/chroma. doi:10.1016/j.chroma.2023.464287.

Novak, P., Zuliani, T., Milačič, R., Ščančar, J., 2016. Development of an analytical method for the determination of polybrominated diphenyl ethers in sewage sludge by the use of gas chromatography coupled to inductively coupled plasma mass spectrometry. Anal. Chim. Acta. 00032670. 915, 27–35. doi:10.1016/j.aca.2016.02.022.

Orecchio, S., Amorello, D., Indelicato, R., Barreca, S., Orecchio, S., 2022. A short review of simple analytical methods for the evaluation of PAHs and PAEs as indoor pollutants in house dust samples. Atmosphere. 2073-4433. 13 (11), 1799. doi:10.3390/atmos13111799.

Panja, P., 2018. Green extraction methods of food polyphenols from vegetable materials. Curr. Opin. Food Sci. 22147993. 23, 173–182. doi:10.1016/j.cofs.2017.11.012.

Pastor-Belda, M., Garrido, I., Campillo, N., Viñas, P., Hellín, P., Flores, P., Fenoll, J., 2015. Dispersive liquid–liquid microextraction for the determination of new generation pesticides in soils by liquid chromatography and tandem mass spectrometry. J. Chromatogr. A. 00219673. 1394, 1–8. doi:10.1016/j.chroma.2015.03.032.

Pena-Pereira, F., Tobiszewski, M., Wojnowski, W., Psillakis, E., 2022. A tutorial on AGREEprep an analytical greenness metric for sample preparation. Adv. Sample Prep. 27725820. 3, 100025. doi:10.1016/j.sampre.2022.100025.

Pérez-Mayán, L., Ramil, M., Cela, R., Rodríguez, I., 2020. Multiresidue procedure to assess the occurrence and dissipation of fungicides and insecticides in vineyard soils from Northwest Spain. Chemosphere. 00456535. 261, 127696. doi:10.1016/j.chemosphere.2020.127696.

Pérez, R.A., Albero, B., Miguel, E., Sánchez-Brunete, C., 2012. Determination of parabens and endocrine-disrupting alkylphenols in soil by gas chromatography–mass spectrometry following matrix solid-phase dispersion or in-column microwave-assisted extraction: a comparative study. Anal. Bioanal. Chem. 1618-2642. 402 (7), 2347–2357. doi:10.1007/s00216-011-5248-0.

Peris, A., Baos, R., Martínez, A., Sergio, F., Hiraldo, F., Eljarrat, E., 2023. Pesticide contamination of bird species from Doñana National Park (southwestern Spain): temporal trends (1999–2021) and reproductive impacts. Environ. Pollut. 02697491. 323, 121240. doi:10.1016/j.envpol.2023.121240.

Petrie, B., Youdan, J., Barden, R., Kasprzyk-Hordern, B., 2016. Multi-residue analysis of 90 emerging contaminants in liquid and solid environmental matrices by ultra-high-performance liquid chromatography tandem mass spectrometry. J. Chromatogr. A 1431, 64–78. doi:10.1016/j.chroma.2015.12.036.

Peysson, W., Vulliet, E., 2013. Determination of 136 pharmaceuticals and hormones in sewage sludge using quick, easy, cheap, effective, rugged and safe extraction followed by analysis with liquid chromatography–time-of-flight-mass spectrometry. J. Chromatogr. A. 00219673. 1290, 46–61. doi:10.1016/j.chroma.2013.03.057.

Pico, Y., Belenguer, V., Corcellas, C., Diaz-Cruz, M.S., Eljarrat, E., Farré, M., Gago-Ferrero, P., Huerta, B., Navarro-Ortega, A., Petrovic, M., Rodríguez-Mozaz, S., Sabater, L., Santín, G., Barcelo, D., 2019. Contaminants of emerging concern in freshwater fish from four Spanish Rivers. Sci. Total Environ. 00489697. 659, 1186–1198. doi:10.1016/j.scitotenv.2018.12.366.

Pissinatti, R., Nunes, C.M., de Souzade Souza, A.G., Junqueira, R.G., de Souzade Souza, S.V.C., 2015. Simultaneous analysis of 10 polycyclic aromatic hydrocarbons in roasted coffee by isotope dilution gas chromatography-mass spectrometry: optimization, in-house method validation and application to an exploratory study. Food Control. 09567135. 51, 140–148. doi:10.1016/j.foodcont.2014.11.003.

Qin, H., Liu, H., Liu, Y., Di, S., Bao, Y., Zhai, Y., Zhu, S., 2023. Recent advances in sample preparation and chromatographic analysis of pharmaceuticals and personal care products in environment. TrAC Trends Anal. Chem. 01659936. 164, 117112. doi:10.1016/j.trac.2023.117112.

Qiu, J., Chen, G., Liu, S., Zhang, T., Wu, J., Wang, F., Xu, J., Liu, Y., Zhu, F., Ouyang, G., 2016a. Bioinspired polyelectrolyte-assembled graphene-oxide-coated C18 composite solid-phase microextraction fibers for in vivo monitoring of acidic pharmaceuticals in fish. Anal. Chem. 0003-2700. 88 (11), 5841–5848. doi:10.1021/acs.analchem.6b00417.

Qiu, J., Chen, G., Zhu, F., Ouyang, G., 2016b. Sulfonated nanoparticles doped electrospun fibers with bioinspired polynorepinephrine sheath for in vivo solid-phase microextraction of pharmaceuticals in fish and vegetable. J. Chromatogr. A. 00219673. 1455, 20–27. doi:10.1016/j.chroma.2016.05.082.

Radke, M.J., Cresswell, S.L., Leusch, F.D.L., 2025. Combining non-targeted high resolution mass spectrometry with effect-directed analysis to identify contaminants of emerging concern in the field of ecotoxicology: a systematic quantitative literature review. Sci. Total Environ. 18791026. 972. Available from: https://www.sciencedirect.com/science/journal/00489697. doi:10.1016/j.scitotenv.2025.179122.

Richter, B.E., Jones, B.A., Ezzell, J.L., Porter, N.L., Avdalovic, N., Pohl, C., 1996. Accelerated solvent extraction: a technique for sample preparation. Anal. Chem. 00032700. 68 (6), 1033–1039. Available from: http://pubs.acs.org/journal/ancham. doi:10.1021/ac9508199.

Rodrigues, T., Ferreira, K.C., Isquibola, G., Franco, D.F., Anderson, J.L., Merib, J.de., Lima Gomes, P.C., 2024. Investigating a new approach for magnetic ionic liquids: Dispersive liquid-liquid microextraction coupled to pyrolysis gas-chromatography-mass spectrometry to determine flame retardants in sewage sludge samples. J. Chromatogr. A 1730, 465038. doi:10.1016/j.chroma.2024.465038.

Rosende, M., Prieto, A., Etxebarria, N., Martorell, G., Miró, M., 2019. Automatic mesofluidic system combining dynamic gastrointestinal bioaccessibility with lab-on-valve-based sorptive microextraction for risk exposure of organic emerging contaminants in filter-feeding organisms. Anal. Chem. 0003-2700. 91 (9), 5739–5746. doi:10.1021/acs.analchem.8b05870.

Rusina, T.P., Jílková, S.R., Melymuk, L., Vrana, B., Smedes, F., 2022. Accessibility investigation of semi-volatile organic compounds in indoor dust estimated by multi-ratio equilibrium passive sampling. SSRN. 15565068. Available from: https://www.ssrn.com/index.cfm/en/. doi:10.2139/ssrn.4223453.

Salgueiro-González, N., Turnes-Carou, I., Besada, V., Muniategui-Lorenzo, S., López-Mahía, P., Prada-Rodríguez, D., 2015. Occurrence, distribution and bioaccumulation of endocrine disrupting compounds in water, sediment and biota samples from a European River Basin. Sci. Total Environ. 00489697. 529, 121–130. doi:10.1016/j.scitotenv.2015.05.048.

Sanchez-Prado, L., Garcia-Jares, C., Dagnac, T., Llompart, M., 2015. Microwave-assisted extraction of emerging pollutants in environmental and biological samples before chromatographic determination. TrAC Trends Anal. Chem. 01659936. 71, 119–143. doi:10.1016/j.trac.2015.03.014.

Sandoval, M.A., Calzadilla, W., Vidal, J., Brillas, E., Salazar-González, R., 2024. Contaminants of emerging concern: occurrence, analytical techniques, and removal with electrochemical advanced oxidation processes with special emphasis in Latin America. Environ. Pollut. 02697491. 345, 123397. doi:10.1016/j.envpol.2024.123397.

Schnoor, J.L., 2003. Emerging chemical contaminants. Environ. Sci. Technol. 0013936X. 37 (21), 375. Available from: http://pubs.acs.org/journal/esthag. doi:10.1021/es032604j.

Schymanski, E.L., Jeon, J., Gulde, R., Fenner, K., Ruff, M., Singer, H.P., Hollender, J., 2014. Identifying small molecules via high resolution mass spectrometry: communicating confidence. Environ. Sci. Technol. 15205851. 48 (4), 2097–2098. doi:10.1021/es5002105.

Shao, Y., Li, S., Li, T., Wei, X., Tian, Y., Yang, Z., Li, X., 2024. Degradation of emerging contaminants in synthetic hydrolyzed urine by UV/peracetic acid: free radical chemistry, and toxicity analysis. Environ. Pollut. 02697491. 359, 124557. doi:10.1016/j.envpol.2024.124557.

Song, Z., Shi, M., Ren, X., Wang, L., Wu, Y., Fan, Y., Zhang, Y., Xu, Y., 2023. An integrated non-targeted and targeted analysis approach for identification of semi-volatile organic compounds in indoor dust. J. Hazard. Mater. 03043894. 459, 132202. doi:10.1016/j.jhazmat.2023.132202.

Soriano, Y., Andreu, V., Picó, Y., 2024. Pressurized liquid extraction of organic contaminants in environmental and food samples. TrAC Trends Anal. Chem. 01659936. 173, 117624. doi:10.1016/j.trac.2024.117624.

Souza Futigami, L., Barcellos Hoff, R., Turnes Pasini Deolindo, C., Kleemann, C.R., Alves de Oliveira, L.V., de Francisco de Casas, A., Burin, V.M., 2024. Search for new green natural solid phases for sample preparation for PAHs determination in seafood samples followed by LC and GC–MS/MS analysis. Food Res. Int. 18737145. 183. Available from: https://www.sciencedirect.com/science/journal/09639969. doi:10.1016/j.foodres.2024.114240.

Sparkman, O.D., Penton, Z.E., Kitson, F.G., 2011. Gas chromatography. In: Kitson F.G., Larsen, B.S., McEwen, C.N., (Eds.), Gas Chromatography and Mass Spectrometry: A Practical Guide. Elsevier BV, pp. 15–83. doi:10.1016/b978-0-12-373628-4.00002-2.

Speltini, A., Sturini, M., Maraschi, F., Porta, A., Profumo, A., 2016. Fast low-pressurized microwave-assisted extraction of benzotriazole, benzothiazole and benezenesulfonamide compounds from soil samples. Talanta. 00399140. 147, 322–327. doi:10.1016/j.talanta.2015.09.074.

Subedi, B., Aguilar, L., Robinson, E.M., Hageman, K.J., Björklund, E., Sheesley, R.J., Usenko, S., 2015. Selective pressurized liquid extraction as a sample-preparation technique for persistent organic pollutants and contaminants of emerging concern. TrAC Trends Anal. Chem. 18793142. 68, 119–132. Available from: www.elsevier.com/locate/trac. doi:10.1016/j.trac.2015.02.011.

Sunyer-Caldú, A., Diaz-Cruz, M.S., 2021. Development of a QuEChERS-based method for the analysis of pharmaceuticals and personal care products in lettuces grown in field-scale agricultural plots irrigated with reclaimed water. Talanta. 00399140. 230, 122302. doi:10.1016/j.talanta.2021.122302.

Tadeo, J., Sánchez-Brunete, C., Albero, B., García-Valcárcel, A., Pérez, R., 2012. Analysis of emerging organic contaminants in environmental solid samples. Open Chem. 2391-5420. 10 (3), 480–520. doi:10.2478/s11532-011-0157-9.

Tadić, Đ., Manasfi, R., Bertrand, M., Sauvêtre, A., Chiron, S., 2022. Use of passive and Grab sampling and high-resolution mass spectrometry for non-targeted analysis of emerging contaminants and their semi-quantification in water. Molecules. 1420-3049. 27 (10), 3167. doi:10.3390/molecules27103167.

Tsui, M.M.P., Lam, J.C.W., Ng, T.Y., Ang, P.O., Murphy, M.B., Lam, P.K.S., 2017. Occurrence, distribution, and fate of organic UV filters in coral communities. Environ. Sci. Technol. 15205851. 51 (8), 4182–4190. Available from: http://pubs.acs.org/journal/esthag. doi:10.1021/acs.est.6b05211.

Vakondios, N., Mazioti, A.A., Koukouraki, E.E., Diamadopoulos, E., 2016. An analytical method for measuring specific endocrine disruptors in activated sludge (biosolids) using solid phase microextraction-gas chromatography. J. Environ. Chem. Eng. 22133437. 4 (2), 1910–1917. Available from: http://www.journals.elsevier.com/journal-of-environmental-chemical-engineering/. doi:10.1016/j.jece.2016.03.018.

Van Den Eede, N., Dirtu, A.C., Ali, N., Neels, H., Covaci, A., 2012. Multi-residue method for the determination of brominated and organophosphate flame retardants in indoor dust. Talanta. 00399140. 89, 292–300. Available from: https://www.journals.elsevier.com/talanta. doi:10.1016/j.talanta.2011.12.031.

Vila, M., Celeiro, M., Lamas, J.P., Garcia-Jares, C., Dagnac, T., Llompart, M., 2017. Simultaneous in-vial acetylation solid-phase microextraction followed by gas chromatography tandem mass spectrometry for the analysis of multiclass organic UV filters in water. J. Hazard. Mater. 18733336. 323, 45–55. Available from: www.elsevier.com/locate/jhazmat. doi:10.1016/j.jhazmat.2016.06.056.

Vila, M., Llompart, M., Garcia-Jares, C., Homem, V., Dagnac, T., 2018. Development and optimization of a solid-phase microextraction gas chromatography–tandem mass spectrometry methodology to analyse ultraviolet filters in beach sand. J. Chromatogr. A. 00219673. 1564, 59–68. doi:10.1016/j.chroma.2018.06.016.

Wang, X., Huang, P., Ma, X., Du, X., Lu, X., 2019. Enhanced in-out-tube solid-phase microextraction by molecularly imprinted polymers-coated capillary followed by HPLC for endocrine disrupting chemicals analysis. Talanta. 00399140. 194, 7–13. doi:10.1016/j.talanta.2018.10.027.

Wu, C., Spongberg, A.L., Witter, J.D., Fang, M., Ames, A., Czajkowski, K.P., 2010. Detection of pharmaceuticals and personal care products in agricultural soils receiving biosolids application. Clean Soil Air Water. 18630669. 38 (3), 230–237. Available from: http://www3.interscience.wiley.com/cgi-bin/fulltext/123327054/PDFSTART. doi: 10.1002/clen.200900263.

Wu, M.W., Yeh, P.C., Chen, H.C., Liu, L.L., Ding, W.H., 2013. A microwave-assisted headspace solid-phase microextraction for rapid determination of synthetic polycyclic and nitro-aromatic musks in fish samples. J. Chin. Chem. Soc. 21926549. 60 (9), 1169–1174. doi:10.1002/jccs.201300029.

Wu, S.F., Liu, L.L., Ding, W.H., 2012. One-step microwave-assisted headspace solid-phase microextraction for the rapid determination of synthetic polycyclic musks in oyster by gas chromatography-mass spectrometry. Food Chem. 18737072. 133 (2), 513–517. doi:10.1016/j.foodchem.2012.01.017.

Wu, Y., Gao, S., Cui, J., Zhang, B., Zhu, Z., Song, Q., Zeng, X., Liang, Y., Yu, Z., 2022. QuEChERS-based extraction and two-dimensional liquid chromatography-high resolution mass spectrometry for the determination of long chain chlorinated paraffins in sediments. J. Chromatogr. A. 00219673. 1684, 463585. doi:10.1016/j.chroma.2022.463585.

Xiong, Q., Wu, D., Cheng, Y.X., Hu, L.X., Chen, Q.L., Wu, H.Y., Sun, Y.H., Liu, Y.S., Ying, G.G., 2022. Development and validation of a simultaneous method for the analysis of benzothiazoles and organic ultraviolet filters in various environmental matrices by GC–MS/MS. Anal. Bioanal. Chem. 16182650. 414 (22), 6541–6555. Available from: https://link.springer.com/journal/216/volumes-and-issues. doi:10.1007/s00216-022-04212-2.

Zeng, L., Yang, R., Zhang, Q., Zhang, H., Xiao, K., Zhang, H., Wang, Y., Lam, P.K.S., Jiang, G., 2014. Current levels and composition profiles of emerging halogenated flame retardants and dehalogenated products in sewage sludge from municipal wastewater treatment plants in China. Environ. Sci. Technol. 15205851. 48 (21), 12586–12594. Available from: http://pubs.acs.org/journal/esthag. doi:10.1021/es503510q.

Zhang, Y., Guo, W., Yue, Z., Lin, L., Zhao, F., Chen, P., Wu, W., Zhu, H., Yang, B., Kuang, Y., Wang, J., 2017. Rapid determination of 54 pharmaceutical and personal care products in fish samples using microwave-assisted extraction—hollow fiber—liquid/solid phase microextraction. J. Chromatogr. B. 15700232. 1051, 41–53. doi:10.1016/j.jchromb.2017.01.026.

Zhou, W., Huang, X., Lin, K., 2019. Analysis of polyhalogenated carbazoles in sediment using liquid chromatography–tandem mass spectrometry. Ecotoxicol. Environ. Saf. 01476513. 170, 148–155. doi:10.1016/j.ecoenv.2018.11.131.

Part 3

Determination of emerging organic contaminants from environmental samples

Chapter 7

Gas chromatography for environmental analysis

Carles Moreu-Romero[a,b], Naroa Lopez-Herguedas[a,b,c], Iker Alvarez-Mora[a,b], Mireia Irazola[a,b], Belén González-Gaya[a,b,d], Mikel Musatadi[a,b,e], and Juan F. Ayala-Cabrera[a,b]

[a]*Faculty of Science and Technology, Department of Analytical Chemistry, University of the Basque Country (UPV/EHU), Leioa, Basque Country, Spain*

[b]*Research Centre for Experimental Marine Biology and Biotechnology (Plentzia Marine Station; PiE-UPV/EHU), University of the Basque Country (UPV/EHU), Plentzia, Basque Country, Spain*

[c]*Department of Cell Toxicology, Helmholtz Centre for Environmental Research—UFZ, Leipzig, Germany*

[d]*IKERBASQUE, Basque Foundation for Science, Bilbao, Spain*

[e]*Toxicological Centre, University of Antwerp, Wilrijk, Belgium*

Abbreviations

AED	plasma atomic emission
APCI	atmospheric pressure chemical ionization
API	atmospheric pressure ionization
APLI	atmospheric pressure laser ionization
APPI	atmospheric pressure photoionization
AP-Plasma	atmospheric pressure plasma ionization
BFRs	brominated flame retardants
BSC	Benzenesulfonyl chloride
BSTFA	N,O-Bis(trimethylsilyl)trifluoroacetamide
BTF	benzotrifluoride
CAR	carboxen
CEC	contaminants of emerging concern
CI	chemical ionization
cIM	cyclic ion mobility
CIS	cooled injection system
CMF-ID	competitive fragmentation modeling for metabolite identification
DCM	dichloromethane
DLLME	dispersive liquid–liquid microextraction
DP	dechlorane plus

Contaminants of Emerging Concern in the Environment. https://doi.org/10.1016/B978-0-443-33821-2.00004-9

DVB	divinylbenzene
ECD	electron capture detector
ECF	ethyl chloroformate
EDAC	N-(3-dimethylaminopropyl)-N'-ethylcarbodiimide
EDCs	endocrine disrupting chemicals
EI	electron ionization
ELCD	electrolytic conductivity detector
EtOAc	ethyl acetate
FID	flame ionization detector
FPD	flame photometric detector
FTOHs	fluorotelomer alcohols
GC	gas chromatography
GC × GC	comprehensive two-dimensional gas chromatography
GCB	graphitized carbon black
GC-GC	heart-cutting multidimensional gas chromatography
GLC	gas–liquid chromatography
GPC	gel permeation chromatography
GSC	gas–solid chromatography
Hex	hexane
HF-LPME	hollow fiber liquid-phase microextraction
HFRs	halogenated flame retardants
HRMS	high-resolution mass spectrometry
HS-SPME	headspace-solid-phase microextraction
ICP	inductively coupled plasma
iLODs	instrumental limits of detection
IT	ion trap
IUPAC	International Union of Pure and Applied Chemistry
LC	liquid chromatography
LDS	low-density solvent
LLE	liquid–liquid extraction
LLME	liquid–liquid microextraction
LOD	limit of detection
LOI	limit of identification
LPME	liquid-phase microextraction
LVI	large-volume injection
MAHs	monoaromatic hydrocarbons
MDGC	multidimensional gas chromatography
MEPS	microextraction by packed sorbent
mLODs	method limits of detection
mLOQs	method limits of quantification
MS	mass spectrometry
MS/MS	tandem mass spectrometry
MSPD	matrix solid phase dispersion
MSTFA	N-methyl-N-(trimethylsilyl)trifluoroacetamide)

MTBE	methyl tert-butyl ether
MTBSTFA	N-tert-Butyldimethylsilyl-N-methyltrifluoroacetamide
MTPA-Cl	(S)-(+)-α-methoxy-α-trifluoromethylphenylacetyl chloride
NICI	negative ion chemical ionization
NPD	nitrogen–phosphorus detector
NSAIDs	nonsteroidal antiinflammatory drugs
NTS	nontarget screening
OCPs	organochlorine pesticides
OPAHs	oxygenated polycyclic aromatic hydrocarbons
OPFRs	organophosphorus flame retardants
OPPs	organophosphorus pesticides
PAEs	phthalate esters
PAHs	polycyclic aromatic hydrocarbons
PBDEs	polybrominated diphenyl ethers
PCBs	polychlorinated biphenyls
PDMS	polydimethylsiloxane
PEG	polyethylene glycol
PFAS	per- and polyfluoroalkyl substances
PFBBr	pentafluorobenzyl bromide
PFCAs	perfluorocarboxylic acids
PFP	pentafluoropropionyl
PFPA	pentafluoropropionic anhydride
PFPOH	2,2,3,3,3-pentafluoro-1-propanol
PFSAs	perfluorosulfonic acids
PID	photoionization detector
PLE	pressurized liquid extraction
PLOT	porous layer open tubular
PPCPs	pharmaceuticals and personal-care products
PSA	ethylenediamine-N-propyl
PTV	programmable temperature vaporizer
PUF	polyurethane foam
Q	single quadrupole
QqQ	triple quadrupole
qTOF	quadrupole time of flight
QuEChERS	quick, easy, cheap, effective, rugged, and safe
SALLE	salting-out assisted liquid–liquid extraction
SBSE	stir bar sorptive extraction
SCCPs	short-chain chlorinated paraffins
SCOT	support-coated open tubular
SEV-DLLME	silylated vessel-dispersive liquid–liquid microextraction
SFE	supercritical fluid extraction
SLE	solid–liquid extraction
SPE	solid-phase extraction
SPME	solid-phase microextraction

SS	suspect screening
SSL	split/splitless
SVOCs	semi-volatile organic compounds
TBDMSCl	tert-Butyldimethylsilyl chloride
TCD	thermal conductivity detector
TDU	thermal desorption unit
TFEA	2,2,2-trifluoroethylamine
TMCS	trimethylchlorosilane
TMS	trimethylsilyl
TOF	time-of-flight
TPI	tube plasma ionization
UA-DSPE-LDS-DLLME	ultrasonic-assisted dispersive solid-phase extraction and low-density solvent dispersive liquid–liquid microextraction
UAE	ultrasound assisted extraction
VOCs	volatile organic compounds
WCOT	wall-coated open tubular

1 Introduction and fundamentals

According to IUPAC, "chromatography is a physical method of separation in which the components to be separated are distributed between two phases, one of which is stationary (stationary phase) while the other (the mobile phase) moves in a definite direction"(Ettre, 1993).

Gas chromatography (GC) refers to the specific type of chromatographic technique in which the mobile phase is a gas. GC was the first chromatographic technique developed aiming to separate and determine volatile compounds. In GC, the sample is vaporized and carried by the mobile gas phase (the *carrier gas*) through a column (stationary phase). In most analyses, samples partition (equilibrate) between the stationary liquid phase and the mobile phase, based on their solubility in the stationary phase at the given temperature. The components of the sample (called solutes or analytes) separate from one another based on their *relative* vapor pressures and affinities for the stationary phase. Among the GC techniques, a subclassification can be made according to the state of the stationary phase in: (i) gas–solid chromatography (GSC) and the most common (ii) gas–liquid chromatography (GLC).

GC is a suitable separation technique to carry out the determination of analytes with significant volatility and thermal stability at temperatures below 350°C. These properties can be predicted based on the chemical structure and the molecular mass of the compound. As a rule of thumb, the highest the molecular mass or polarity, the lowest the volatility. For instance, while hydrocarbons up to approximately 500 Da can be analyzed using conventional GC, those exceeding 1400 Da may require specialized conditions. Similarly, compounds containing polar functional groups typically exhibit reduced volatility, and are therefore not suitable for GC analysis unless a derivatization step is applied to increase their volatility. Due to these capabilities, GC is a powerful tool not only for legacy pollutants but also for the determinations of contaminants of emerging concern (CECs) in complex matrices such as those coming from environmental samples.

2 Instrumentation

Fig. 7.1 shows a general scheme of a gas chromatograph. First, the mobile phase, also called carrier gas, transports the analytes from the column to the detector. The carrier gas is inert and, consequently, it does not interact with the analytes or the stationary phase. Helium, nitrogen, and hydrogen are among the most common carrier gases, although argon and carbon dioxide can occasionally be used. All the gases used for GC must be of high purity (>99.995%) and the moisture, oxygen, and hydrocarbons must be removed from the mobile phase, generally using adsorbent cartridges, before entering the instrument. The main criterion for selecting the carrier gas is efficiency. According to the Van Deemter's equation, helium and hydrogen are often used, as they allow high efficiency at higher flow rates which results in lower analysis times. Among them, helium is usually preferred since hydrogen is a flammable gas which requires strong safety conditions. Other variables that can significantly influence GC distributions are the gas flow and the pressure at the head of the column, as they limit the distribution speed.

The injector allows the introduction of a precise sample volume as a narrow band onto the column. In GC, the sample is typically injected as a liquid

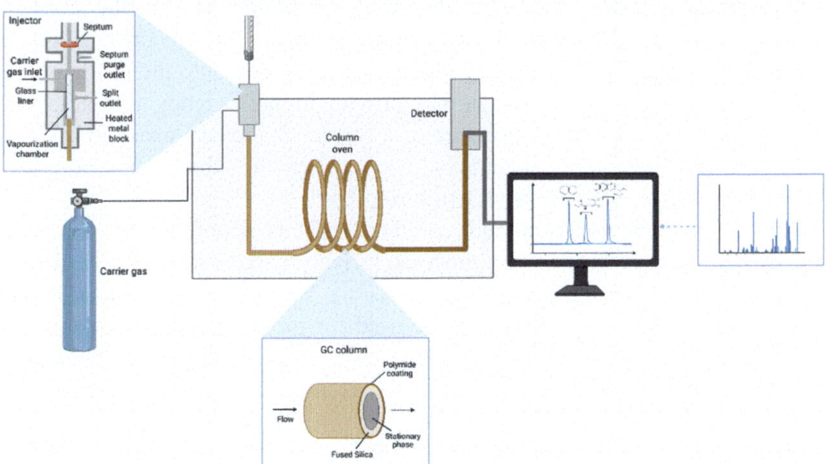

FIG. 7.1 General scheme of gas chromatography (GC) instrumentation. This figure illustrates the main components of a classical gas chromatography (GC) instrument, including the carrier gas, injector, oven with the analytical column, and detector. *(Created with BioRender.com.)*

using a syringe. Once the syringe has pierced the rubber disc called septum, it reaches a silanized glass tube (liner) where the sample evaporates quickly at high temperatures and the carrier gas guides the sample toward the head of the column. The injection temperature must be high enough to ensure the simultaneous evaporation of all analytes while preventing potential thermal degradation. Typical injection volumes in GC range from 0.1 to 2 μL. The split/splitless (SSL) injector is the most used type due to its versatility, allowing it to operate in both split and splitless modes depending on the goals of the analysis. During the splitless injection, the split valve is closed, which directs the whole sample volume into the head of the column, maximizing sensitivity for trace analysis. In contrast, during split injection, the split valve is opened, allowing only a fraction of the sample to enter the column. This mode is useful when analyte concentrations are high enough to allow partial transfer into the column, resulting in sharper and narrower peaks while minimizing interferences. In addition, other injectors such as the programmable temperature vaporizer (PTV) are also widely used due to their suitability for the analysis of thermally labile compounds as well as large volume (up to 100 μL) injections.

The column is the key component of the chromatographic system, facilitating the separation of compounds through interactions between the analytes and the stationary phase. Today, the most popular columns are made of fused silica and are open tubes with capillary-scale dimensions, typically ranging from 10 to 100 m in length and 0.1 to 0.53 mm in inner diameter. The stationary liquid phase is coated on the inside surface of the capillary wall, with typical thicknesses of 0.1–5 μm. GC columns can be classified into three types (i) porous layer open tubular (PLOT), (ii) support-coated open tubular (SCOT), and (iii) wall-coated open tubular (WCOT). PLOT columns have a porous and solid sorbent material on the inner walls, and they are mainly used for the analysis of permanent gases. However, for the determination of CECs, WCOT columns, which consist of a liquid stationary phase coated directly onto the inside of the column, are generally preferred as they provide high efficiency, inertness and fast analysis. The most used polymer as a bonded stationary phase in WCOT columns is substituted polysiloxane (silicone), where the substituent groups can be methyl, phenyl, cyano-propyl, trifluoro propyl, etc. Depending on the level of substitution the main interactions could differ. Thus, in polydimethylsiloxane (PDMS) columns, retention is almost exclusively based on dispersive forces while diphenyl-PDMS column interactions are based on dispersive, π-π and dipole forces. These columns could be used for more polar analytes such as aromatic hydrocarbons. When the polarity of the analytes increases, cyanopropyl-PDMS columns could be an alternative, as they show an increased interaction with O, N, OH, and COOH groups due to dipole–dipole and hydrogen bonding forces. Besides polysiloxane-based columns, another common type of stationary phase is polyethylene glycol (PEG), which provides an appropriate separation for polar compounds containing functional groups such as COOH, OH, NH_2, and C(O)OR due to dipole–dipole interactions. Ionic liquids are also suitable for the separation of polar compounds, as they combine unique selectivity and high thermal stability.

The separation of the analytes in GC can be modulated by programming the temperature of the oven. Thus chromatographic separation can be made in isothermal or gradient conditions. In isothermal mode, the temperature remains constant during the whole analysis, being suitable to separate compounds with similar vapor pressures/boiling points. On the other hand, gradient elution is generally applied for the separation of complex samples (e.g., environmental samples) allowing the progressive increase of the temperature during the run. Finally, the detector allows the monitoring of the compounds by means of a signal/response which can be correlated with the concentration. Table 7.1 summarizes the main detectors used in GC analysis.

TABLE 7.1 Main detectors used for GC separations.

Detector	Type	Selective for	iLODs[a]
Flame ionization detector (FID)	Ionization	Carbon-containing compounds (universal)	1 pg
Thermal conductivity detector (TCD)	Ionization	Universal	1 ng
Electron capture detector (ECD)	Ionization	Highly electronegative compounds (i.e., halogens)	1–100 fg
Nitrogen-phosphorous detector (NPD)	Ionization	N- and P-containing compounds	0.01–0.1 pg
Photoionization detector (PID)	Ionization	Aromatic compounds and olefins	2–200 fg
Flame photometric detector (FPD)	Emission	S- and P-containing compounds	10–100 pg
Plasma atomic emission (AED)	Emission	Metals, halogens, C, and O	0.1 pg
Electrolytic conductivity detector (ELCD)	Electro-chemical	Halogens, S-, and N-containing compounds	5–20 pg
Chemiluminescent detector	Others	S- and N-containing compounds	0.5–3 pg
Mass spectrometry (MS)	Others	Ionizable substances	0.1–10 pg

[a]*iLODs, instrumental limits of detection.*

3 Classical approaches

Some of the most relevant CECs involve pharmaceuticals and personal-care products (PPCPs), illicit drugs and novel psychoactive substances, industrial chemicals such as plasticizers, flame retardants or per- and polyfluoroalkyl substances (PFAS), or pesticides, which are the most widely determined compound families by GC, as their extensive and continuous use in agriculture has led to elevated concentrations in various environmental compartments (da Silva Sousa et al., 2021). Thus GC classical techniques have played a crucial role in the identification and quantification of CECs, particularly for the volatile and semivolatile nonpolar compounds. Despite advances in modern analytical techniques, these classical approaches remain highly relevant, providing robust and cost-effective methods for CEC determination. This section discusses the key components involved in the classical GC analysis of CECs highlighting their strengths, limitations, and continued importance of them in the research. Table 7.2 summarizes some of the main applications for the determination of CECs using GC classical approaches.

3.1 Classical injection modes

As mentioned above, three injection modes can be chosen based on factors such as sample type, analyte concentration, physicochemical properties, and required sensitivity. For CECs analysis, the SSL injection mode is preferred as shown in Table 7.2. As CECs are at trace levels in environmental samples, spitless injection mode is usually carried out. Besides their advantages for trace analysis, this injection mode is also beneficial when analytes tend to get adsorbed or transformed due to the slow flow rate during splitless injection, which can result in broader peaks, it can happen with some industrial chemicals and pesticides (Bi et al., 2015; Kadhum et al., 2024). In addition, the incorporation of a PTV allows for large-volume injection (LVI). This means that, unlike conventional SSL injection, which typically uses 1–2 μL, PTV can handle significantly larger volumes (20–100 μL), as shown for pesticides (Carro et al., 2012) and PPCPs (Guitart and Readman, 2010) leading to pretty low limits of detection (up to fg L^{-1}). Apart from that, unlike standard injectors that operate at a constant temperature, PTV enables dynamic temperature control. This allows to start at a low temperature, remove the solvent, before rapidly increasing the temperature to evaporate the analytes. This approach is effective when there is a significant temperature difference between the analytes and the solvent (Gustavsson et al., 2017), as minimal differences may lead to the loss of certain analytes during the solvent removal process. The combination of these factors allows to reduce the interferences and achieve better sensitivity as in the study conducted by Carro et al., where a PTV was employed for LVI (50 μL) after a dispersive liquid–liquid microextraction (DLLME) for the determination of seventeen pesticides and related compounds in environmental and wastewater samples. This approach enabled significantly lower method LODs (0.5–18 ng/L) compared to

TABLE 7.2 Main applications for the determination of CECs using GC classical approaches.

Analytes	Matrix	Sample treatment	Analytical method	Injection	Column	Carrier gas	LODs/LOQs	Ref.
Pesticides								
OCPs, OPPs, pyrethroids, carbamates and triazoles ($n = 58$)	Soil	QuEChERS	GC-EI-MS/MS (QqQ)	Splitless (1 μL)	HP-5 MS UI (30 m x 0.25 mm, 0.25 μm)	He (1 mL min^{-1})	mLOQs: 0.1–5 μg kg^{-1}	(Yu et al., 2016)
Fungicides ($n = 3$)	Soil	• UAE (MeOH:H_2O, 1:1 v/v) • SPE (PSA)	GC-NPD	Splitless (1 μL)	Rtx-5 (30 m x 0.25 mm, 0.25 μm)	N_2 (1.2 mL min^{-1})	iLODs: 3–32 μg kg^{-1}	(Deng et al., 2010)
Spiromesifen	Soil	QuEChERS	GC-FID	Split (n.r.)[a]	Rti-5Sil MS (60 m x 0.25 mm, 0.25 μm)	N_2 (40 cm s^{-1})	mLOQ: 6 μg L^{-1}	(Kadhum et al., 2024)
Acaricides, insecticides, herbicides and fungicides ($n = 150$)	Soil	MSPD (Florisil)	GC-ECD	n.r.	DB-35 (30 m x 0.32 mm, 0.5 μm)	He (1.9 mL min^{-1})	mLODs: 1–20 μg kg^{-1}	(Łozowicka et al., 2012)
OPPs ($n = 6$)	Soil	UA-DSPE-LDS-DLLME (Acetone)	GC-FID	Splitless (1 μL)	Rtx-1701 (30 m x 0.25 mm, 0.25 μm)	N_2 (1 mL min^{-1})	mLOQs: 0.5–1.2 μg kg^{-1}	(Wang et al., 2014)
OCPs ($n = 14$)	Sediment	• UAE and vortex (Hex:acetone, 1:1 v/v)SPE (Florisil)	GC-ECD	Splitless (2 μL)	HP-5 (30 m x 0.25 mm, 0.25 μm)	He (1 mL min^{-1})	mLOQs: 4.5–15 μg L^{-1}	(García et al., 2020)
			GC-NICI-MS (Q)		DB-35 (60 m x 0.25 mm, 0.25 μm)		mLOQs: 0.1–5 μg L^{-1}	

Continued

TABLE 7.2 Main applications for the determination of CECs using GC classical approaches—cont'd

Analytes	Matrix	Sample treatment	Analytical method	Injection	Column	Carrier gas	LODs/LOQs	Ref.
OPP, OCP pyrethroid, carbamate and oxadiazine ($n=7$)	Sediment	• PLE (DCM:acetone, 1:1 v/v)SPE (GCB/PSA)	GC-ECD	Splitless (2 μL)	HP-5MS (30 m x 0.25 mm, 0.25 μm)	He (1.8 mL min^{-1})	mLODs: 0.23–1.8 μg kg^{-1}	(Wang et al., 2010)
			GC-NPD		DB-608 (30 m x 0.32 mm, 0.5 μm)	He (3.8 mL min^{-1})		
Triazoles ($n=6$)	Tap water	SEV-DLLME (MeOH)	GC-FID	Splitless (n.r.)[a]	PT-5 (30 m x 0.25 mm, 0.25 μm)	He (30 cm s^{-1})	mLOQs: 0.3–2.6 μg L^{-1}	(Farajzadeh et al., 2011)
Multiclass ($n=17$)	Surface water	DLLME (ACN)	GC-EI-MS/MS (QqQ)	PTV (50 μL)	VF-5MS (30 m x 0.25 mm, 0.25 μm)	He (1 mL min^{-1})	mLOQs: 0.5–18 ng L^{-1}	(Carro et al., 2012)
Pharmaceuticals and personal-care products								
NSAIDs and estrogenic hormones ($n=13$)	Sewage water	• SPE (Oasis HLB) • BSTFA + TCMS (1%)	GC-EI-MS (Q)	Splitless (1 μL)	Rtx-5 (30 m x 0.25 mm, 0.25 μm)	He (2 mL min^{-1})	mLOQs: 10–114 ng L^{-1}	(Migowska et al., 2012)
		• SPE (Oasis HLB) • PFPOH + PFPA	GC-ECD			Ar (2 mL min^{-1})	mLOQ: 2–16 ng L^{-1}	
Acidic pharmaceuticals and phenolic EDCs ($n=9$)	Sewage water	• SPE (C18) • BSTFA + TCMS (1%)	GC-EI-MS (Q)	Splitless (1 μL)	DB-5MS (60 m x 0.32 mm, 0.25 μm)	He (0.9 mL min^{-1})	mLOQs: 2.7–48 ng L^{-1}	(Samaras et al., 2011)
	Sludge	• UAE (MeOH:H_2O, 1:1 v/v) • SPE (C18) • BSTFA + TCMS (1%)					mLOQs: 59–108 ng L^{-1}	

Antiinflammatory, antidepressant, hormones, and lipid regulator drugs ($n = 9$)	Sewage water	In situ derivatization (ECF) in SPME (PDMS/DVB)	GC-EI-MS (Q)	Splitless (-)	Rtx-5MS (15 m x 0.25 mm, 0.25 μm)	He (1.3 mL min^{-1})	mLODs: 95–160 ng L^{-1}	(De Lima Gomes et al., 2011)
Pharmaceuticals, personal-care products, phenolic endocrine disruptors and sewage molecular markers	Sewage water	• SPE (Oasis HLB) • MTBSTFA + TBDMSCl[a] (1%)	GC-EI-MS (Q)	PTV (20 μL)	HP-5MS (30 m x 0.25 mm, 0.25 μm)	He (1 mL min^{-1})	iLODs: 0.1–19.5 ng L^{-1}	(Guitart and Readman, 2010)
NSAIDs ($n = 5$)	Sewage, surface and tap water	In situ derivatization (EDC + TFEA) in SPME (MEPS)	GC-EI-MS (Q)	PTV (-)	HP-5MS (30 m x 0.25 mm, 0.25 μm)	He (n.r.)[a]	mLODs: 4–540 ng L^{-1}	(Noche et al., 2011)
Triclosan, bisphenol A and UV filters ($n = 5$)	Dust	• PLE (acetone) • SPE (Oasis HLB) • BSTFA + TCMS (1%)	GC-EI-MS/MS (QqQ)	Splitless (1 μL)	Rtx-5 (30 m x 0.25 mm, 0.25 μm)	He (1.2 mL min^{-1})	mLOQs: 0.5–12,100 ng kg^{-1}	(Ao et al., 2017)
Human pharmaceuticals ($n = 12$)	Particulate matter	• UAE (DCM: acetone:MeOH, 5:3:2 v/v/v) • Silica fractionation MTBSTFA	GC-EI-MS (Q)	On-column (n.r.)[a]	DB-1701 (50 m x 0.32 mm, 0.25 μm)	n.r.[a]	iLOQs: 9–140 μg L^{-1}	(Cecinato et al., 2017)

Continued

TABLE 7.2 Main applications for the determination of CECs using GC classical approaches—cont'd

Analytes	Matrix	Sample treatment	Analytical method	Injection	Column	Carrier gas	LODs/LOQs	Ref.
Industrial chemicals								
PFCAs ($n = 4$)	Soil	*In situ* derivatization SFE (n-butanol) HS-SPME	GC-NICI-MS/MS (QqQ)	Splitless (-)	DB-624 (60 m x 0.25 mm, 1.4 μm)	He (1.2 mL min^{-1})	iLOQs: 1.3–1.8 μg kg^{-1}	(Liu et al., 2011)
BFRs and PFAS ($n = 53$)	Sediment	PLE (Hex and DCM)	GC-EI-MS/MS (QqQ)	PTV (2 μL)	Restek 1614 (15 m x 0.25 mm, 0.1 μm)	He (2.5 mL min^{-1})	mLOQs: 4–4252 ng kg^{-1}	(Sühring et al., 2015)
DP and related compounds ($n = 6$)	Sediment	PLE (Hex:DCM, 1:1 v/v)	GC-NICI-MS/MS (QqQ)	Splitless (n.r.)[a]	DB-5MS (15 m x 0.10 mm, 0.1 μm)	He (2.25 mL min^{-1})	mLOQs: 0.1–1.3 ng kg^{-1}	(Barón et al., 2012)
	Sludge	• PLE (Hex:DCM, 1:1 v/v) • SPE (Silica) • SPE (Alumina)					mLOQs: 1–2.9 ng kg^{-1}	
	Biota	• PLE (Hex:DCM, 1:1 v/v) • SPE (Alumina)					mLOQs: 2–21 ng kg^{-1}	
PAEs ($n = 10$)	Sludge	UAE (Hex:acetone, 1:1 v/v)	GC-EI-MS (Q)	Splitless (n.r.)[a]	HP-5MS (30 m x 0.25 mm, 0.25 μm)	He (1 mL min^{-1})	mLODs: 3–46 μg kg^{-1}	(Dong et al., 2020)
Volatile PFAS ($n = 65$)	Sewage water	SPE (Oasis HLB)	GC-EI-MS (Q)	Splitless (1 μL)	VMS (30 m x 0.25 mm, 3 μm)	He (1.8 mL min^{-1})	mLODs: 10 ng L^{-1}	(Dimzon et al., 2017)
Neutral PFAS ($n = 16$)	Surface water	SPE (Oasis HLB)	GC-EI-MS (Q)	Split (1 μL, 1:15)	DB-624 (60 m x 0.25 mm, 1.4 μm)	He (1 mL min^{-1})	iLODs: 0.2–17 μg L^{-1}	(Ayala-Cabrera et al., 2020b)
			GC-CI-MS (Q)				iLODs: 0.06–430 μg L^{-1}	

OPFRs($n = 14$)	Surface water	SPE (Oasis HLB)	GC-EI-MS (Q)	Splitless (n.r.)[a]	Rxi-17Sil MS (60 m x 0.25 mm, 0.25 μm)	He (1.5 mL min^{-1})	iLOQs: 0.12–5.03 ng L^{-1}	(Loos et al., 2017)
PBDEs, HFRs, and OPFRs ($n = 102$)	Surface water	SPE (Alumina)	GC-EI-MS/MS (QqQ)	PTV (2 μL)	DB-5MS (15 m x 0.25 mm, 0.1 μm)	n.r.[a]	iLODs: 1–8336 μg L^{-1}	(Gustavsson et al., 2017)
			GC-EI-MS (Q)				iLODs: 12–1057 μg L^{-1}	
			GC-NICI-MS (Q)				iLODs: n.d.[a]-5089 μg L^{-1}	
PBDEs, BFRs, and DPs ($n = 14$)	Surface water	SBSE (PDMS)	GC-NICI-MS (Q)	Splitless (1 μL)	DB-5HT (15 m x 0.25 mm, 0.1 μm)	He (1 mL min^{-1})	mLODs: 0.4–10 ng L^{-1}	(Valls-Cantenys et al., 2013)
Disinfection by products ($n = 26$)	Surface water	–	GC-ECD	On-column (2 μL)	ZB-624 (60 m x 0.32 mm, 1.8 μm)	He (2.2 mL min^{-1})	mLODs: 0.03–5.2 μg L^{-1}	(Tobiszewski and Namieśnik, 2011)
Benzotrifluoride and chlorinated benzotrifluorides ($n = 4$)	Ground-water	Purge-and-trap	GC-EI-MS (Q)	Splitless (1 μL)	Rtx-624 (30 m x 0.25 mm, 0.25 μm)	He (1 mL min^{-1})	iLOQs: 5–18 ng L^{-1}	(Lava et al., 2013)
Amino- and nitro-benzotrifluoride derivatives ($n = 5$)		SPE (Oasis HLB)			VF-5MS (30 m x 0.25 mm, 0.25 μm)			

Continued

TABLE 7.2 Main applications for the determination of CECs using GC classical approaches—cont'd

Analytes	Matrix	Sample treatment	Analytical method	Injection	Column	Carrier gas	LODs/LOQs	Ref.
OPFRs ($n = 8$)	Dust	• UAE (Hex:acetone, 3:1 v/v) • SPE (Florisil ENVI)	GC-EI-MS (Q)	PTV (1 μL)	HT-8 (25 m x 0.22 mm, 0.25 μm)	He (1–1.5 mL min^{-1})	mLOQs: 6–90 μg kg^{-1}	(Christia et al., 2018)
PAEs ($n = 14$)	Dust	UAE (EtOAc)	GC-EI-MS (Q)	Split (n.r.[a], 1:5)	SHRXI-5MS (30 m x 0.25 mm, 0.25 μm)	He (1 mL min^{-1})	mLOQs: 73–359 μg kg^{-1}	(Bi et al., 2015)
BFRs ($n = 21$)	Dust	• UAE (Hex:acetone, 1:1 v/v) • SPE (Alumina)	GC-NICI-MS (Q)	n.r.[a]	DB-5MS (15 m x 0.25 mm, 0.1 μm)	He (1.1 mL min^{-1})	mLODs: 0.83–8 μg kg^{-1}	(Genisoglu et al., 2019)
	Particulate matter						mLODs: 4.5–8 pg m^{-3}	
	Air	• Soxhlet (Hex:acetone, 1:1 v/v) • SPE (Alumina)					mLODs: 4–42 pg m^{-3}	
Cl-PAHs ($n = 24$)	Particulate matter	• PLE (DCM) • SPE (Silica)	GC-EI-MS (Q)	Splitless (1 μL)	InertCap 5 MS/NP (30 m x 0.25 mm, 0.25 μm)	He (1 mL min^{-1})	mLODs: 0.07–3.3 pg	(Kamiya et al., 2015)
			GC-NICI-MS (Q)				mLODs: n.d.[a]-4.4 pg	
PBDEs, HFRs, and DPs ($n = 17$)	Particulate matter	• UAE (DCM) • Florisil and silica fractionation	GC-NICI-MS (Q)	Splitless (1 μL)	DB-5MS (15 m x 0.25 mm, 0.1 μm)	He (1.5 mL min^{-1})	iLODs: 0.003–5.44 pg	(Chen et al., 2019)
OPFRs ($n = 12$)			GC-EI-MS/MS (QqQ)		TG-5MS (30 m x 0.25 mm, 0.25 μm)			

Volatile PFAS ($n = 65$)	Air	SPE (Oasis HLB)	GC-EI-MS (Q)	Splitless (1 μL)	VMS (30 m x 0.25 mm, 3 μm)	He (1.8 mL min^{-1})	mLODs: 1–40 ng m^{-3}	(Dimzon et al., 2017)
Neutral PFAS ($n = 16$)	Air	Cold column extraction (MTBE:acetone, 1:1 v/v)	GC-CI-MS (Q)	Splitless (2 μL)	Supelcowax10 (60 m x 0.25 mm, 1.4 μm)	He (1.5 mL min^{-1})	iLOQs: 0.2–8.2 pg L^{-1}	(Weinberg et al., 2011)
Illicit drugs and novel psychoactive substances								
Amphetamine-type stimulants, synthetic cathinones, and plant-based alkaloids ($n = 11$)	Sewage water	LPME (Toluene:DCM: EtOAc, 1:1:1 v/v/v)	GC-EI-MS (Q)	Splitless (1 μL)	Rtx-5MS (30 m x 0.25 mm, 0.25 μm)	He (1 mL min^{-1})	mLOQs: 35–73 ng L^{-1}	(Nascimento et al., 2023)
Multiresidue								
PAEs, PAHs, adipates vulcanization additives and antioxidants ($n = 40$)	Runoff water	HS-SPME (PDMS/DVB)	GC-EI-MS (Q)	Splitless (-)	ZB-Semivolatiles (30 m x 0.25 mm, 0.25 μm)	He (1 mL min^{-1})	iLODs: 0.11–1000 μg L^{-1}	(Celeiro et al., 2018)
PAEs, UV filters, musks, antioxidants and fragrance allergens ($n = 37$)	Surface water	HS-SPME (PDMS/DVB)	GC-EI-MS/MS (QqQ)	Splitless (-)	ZB-Semivolatiles (30 m x 0.25 mm, 0.25 μm)	He (1 mL min^{-1})	mLOD: 0.02–5.7 ng L^{-1}	(Homem et al., 2022)
Pesticides, pharmaceuticals, and hormones ($n = 35$)	Surface waters	SPE (Oasis HLB)	GC-EI-MS/MS (QqQ)	n.r.[a] (1 μL)	DB-5HT (15 m x 0.25 mm, 0.1 μm)	He (1 mL min^{-1})	mLODs: 0.4–4.5 ng L^{-1}	(He and Aga, 2019)

Continued

TABLE 7.2 Main applications for the determination of CECs using GC classical approaches—cont'd

Analytes	Matrix	Sample treatment	Analytical method	Injection	Column	Carrier gas	LODs/LOQs	Ref.
VOCs, EDCs, odor compounds, fragrance allergens and pesticides ($n=76$)	Surface water	HS-SPME (PDMS/DVB)	GC-EI-MS (Q)	Splitless (-)	TRB-5MS (60 m x 0.32 mm, 1 μm)	He (1 mL min^{-1})	mLOQs: 2–333 ng L^{-1}	(Martínez et al., 2013)
OCPs, OPPs, PAHs, PBDEs, PAEs, musks, and pyrethroids ($n=55$)	Dust	• PLE (DCM)SPE (Amino sorbent)	GC-EI-MS (Q)	Splitless (1 μL)	Rtx-PCB (60 m x 0.25 mm, 0.25 μm)	He (2 mL min^{-1})	mLOQs: 26–197 μg kg^{-1}	(Mercier et al., 2014)
OCPs and PBDEs ($n=22$)	Particulate matter	UAE (DCM)	GC-NICI-MS (Q)	Splitless (2 μL)	DB-35MS (60 m x 0.25 mm, 0.25 μm)	He (1.2 mL min^{-1})	mLOQs: 1.7–9.0 pg m^{-3}	(Beristain-Montiel et al., 2016)
PAEs, PAHs, OPFRs, musks and bisphenols ($n=50$)	Particulate matter	UAE and vortex (Hex:acetone, 1:1 v/v)	GC-EI-MS/MS (QqQ)	PTV (8 μL)	DB-XLB (60 m x 0.25 mm, 0.25 μm)	He (1.2 mL min^{-1})	mLOQs: 1.2–1790 pg m^{-3}	(Sánchez-Piñero et al., 2021)
Plasticizers (PAEs and nonPAEs) and OPFRs ($n=48$)	Air	UAE (acetone)	GC-EI-MS (Q)	Splitless (2 μL)	DB-5MS (30 m x 0.25 mm, 0.25 μm)	He (n.r.)[a]	mLOQs; 0.1–30 ng m^{-3}	(Takeuchi et al., 2014)

[a]*not reported.*

other studies (Carro et al., 2012). In addition, this injector has also been applied for the determination of organophosphate flame retardants (OPFRs), polycyclic aromatic hydrocarbons (PAHs), and phthalate esters (PAEs) in particulate matter and dust (Christia et al., 2018; Sánchez-Piñero et al., 2021). Furthermore, a PTV has also been employed to inject larger volumes using a solvent-vent mode with a liquid CO_2 cooling system. Grueiro Noche et al. injected nonsteroidal antiinflammatory drugs (NSAIDs) at a rate of 25 $\mu L\ s^{-1}$ for 2.5 min (Noche et al., 2011), while Guitart and Readman tested injection volumes from 20 to 250 μL for PPCPs, endocrine disrupting chemicals (EDCs), and sewage molecular markers (Guitart and Readman, 2010). Nevertheless, they noted the limited applicability of the PTV inlet for environmental applications, affording only a modest improvement in chromatographic signal-to-noise ratios (Guitart and Readman, 2010). In contrast, Ahmed et al. observed that LVI up to 300 μL improved sensitivity for detecting oxygenated PAHs (OPAHs) in particulate matter samples using a hyphenated LC-GC/MS system which allows a clean-up in a normal phase LC column to remove the alkanes followed by the transfer of the OPAHs fraction to a PTV injector for the GC-MS determination (Ahmed et al., 2015).

On-line extraction can be also programmed in the injector port. One of the most common on-line approaches to extract CECs is the solid-phase microextraction (SPME). It involves a fiber coated with a film of a polymeric material that adsorbs analytes. This polymeric material can vary depending on the properties of the analyte. Typically, the fiber is placed in the headspace of the vial (HS-SPME), allowing a solvent-free approach that reduces the matrix effect while simultaneously preconcentrating the analytes. Once the extraction is performed, the fiber is removed and desorbed into the injector. This technique is particularly useful for the more volatile fraction of CECs such as EDCs, fragrance allergens, phthalates, and some pesticides (Homem et al., 2022; Martínez et al., 2013). When combined with other extraction techniques, such as the supercritical fluid extraction, it can reduce the sample management minimizing the analyte loss (Liu et al., 2011).

Finally, the sample could also be injected directed on column without a liner, called on-column injection. This is interesting for high-boiling components or active analytes, such as pharmaceutical products, that would not be quantitatively transferred to the column with the other injectors, providing more intense and symmetrical peaks than conventional SSL and PTV injectors (Cecinato et al., 2017). Nevertheless, the injected sample should be free of any residues, as these can be transferred to the column, thereby reducing its lifespan.

3.2 Classical GC stationary phases

The successful separation of analytes from matrix interferences relies heavily on the appropriate selection of the chromatographic column. The ideal column depends on both the sample matrix and the properties of the analytes. For the CECs determination in GC the most used columns are the low polar polysiloxane-base capillary columns, as shown in Table 7.2. Among them, the stationary phase composed of 5% phenyl-95% PDMS is the gold standard column as it allows the interaction with the analytes by means of π–π and dipole–dipole interactions and dispersion forces. These columns have demonstrated excellent performance in the analysis of diverse industrial chemicals in complex matrices, including indoor air (Takeuchi et al., 2014), particulate matter (Chen et al., 2022b), sludge (Dong et al., 2020), wastewaters (Huang et al., 2023), or surface waters (Minho et al., 2022) among others. Also, it has been proven to be the most used for illicit drugs, pesticides, and PPCPs. By changing the functional group or increasing the level of substitution, the polarity of the stationary phase can be increased, making it suitable for the retention and separation of semi-polar and moderate polar compounds. In addition, these columns are well-suited for derivatized compounds that require high temperatures for vaporization. This can be seen in several practical applications, for instance, a (14% cyanopropyl-phenyl)-86% PDMS column for derivatized pharmaceuticals (Cecinato et al., 2017), 35% phenyl- 65% PDMS phases for organochlorine pesticides (OCPs) (García et al., 2020; Beristain-Montiel et al., 2016), a 14% cyanopropyl- 86% PDMS for organophosphorus pesticides (OPPs) (Wang et al., 2014), or 50% trifluoropropyl- 50% PDMS for a broader range of compounds, including halogenated pesticides (Wang et al., 2010) among others. For highly polar compounds, PEG-based columns are generally used although they have a lower thermal stability, with a maximum temperature limit of 260°C–270°C and, in some cases, these compounds can also be analyzed using a liquid chromatograph. On the other hand, nonpolar columns with a 100% PDMS can also be used to add an extra confirmatory dimension in the analysis as in the case of Migowska et al. that used this kind of column to confirm the detection of derivatized NSAIDs and estrogenic hormones in water and wastewater samples (Migowska et al., 2012). Finally, some columns have stationary phases specifically designed for the separation of only one family or one specific group of compounds, such as polychlorinated biphenyls (PCBs) (Wang et al., 2010; Mercier et al., 2014; Van den Eede et al., 2011), pesticides (Wang et al., 2010, 2014; Cecinato et al., 2017), PAEs (Demirtepe et al., 2021), polybrominated diphenyl ethers (PBDE) (Sühring et al., 2015), volatile organic compounds (VOCs) (Liu et al., 2011; Dimzon et al., 2017; Tobiszewski and Namieśnik, 2011; Lava et al., 2013), or semivolatile organic compounds (SVOCs) (Celeiro et al., 2018; Homem et al., 2022).

Besides the stationary phase, column length is another key element for the separation of the analytes. The standard column length for CECs separation is typically 30 m, as shown in Table 7.2. Nevertheless, if the matrix contains many interferences that coelute with the analytes, or if the analytes themselves are very similar, a higher degree of separation is required. To achieve that, larger columns (50–60 m) are used to resolve complex mixtures in environmental matrices such as atmospheric samples (Cecinato et al., 2017), as well as for wastewater and sewage (Samaras et al., 2011). However, this increase in the length leads to longer analysis times, higher back pressure and column cost. On the other hand, shorter columns reduce analysis time and increase sensitivity, making them particularly useful for compounds that can get thermally degraded, such as brominated flame retardants (BFRs) although the resolution is reduced (Deboer and Wells, 2006).

Another important parameter influencing chromatographic separation is the internal diameter of the column. While the internal diameter is usually 0.25 mm, it can vary depending on the application. A smaller internal diameter increases column efficiency and produces narrower peaks, but it also results in higher internal pressure. On the contrary, a larger internal diameter reduces efficiency but increases sample capacity.

Finally, the film thickness can also vary between columns. While it is typically 0.25 μm, it can be adjusted depending on the nature of the analytes of interest. For highly volatile compounds such as solvents (Tobiszewski and Namieśnik, 2011) or volatile PFAS (Dimzon et al., 2017) a thicker film is required as it increases the interaction of the analytes with the stationary phase. Also, midpolar columns often use larger internal diameters (0.32–0.53 mm) and thicker film coatings (0.5–1 μm) to ensure effective analyte retention (Table 7.2). However, the drawbacks are the reduction of the resolution and the increase of column bleed. On the other hand, for analytes with a higher boiling point, such as BFRs, a thinner film is preferred (Gustavsson et al., 2017; Barón et al., 2012; Chen et al., 2019). This allows for shorter retention time, minimizing the risk of thermal degradation. In addition, a thinner film reduces column bleed and produces sharper peaks, although it may affect the separation efficiency.

3.3 GC coupling to classical detectors

Once the chromatographic separation has been achieved, the CECs have to be detected. Some detectors, such as flame ionization detectors (FID), are considered universal as they can detect virtually all organic compounds, including the organic CECs. However, a very specific clean-up is required, as the main drawback of this technique is its lack of selectivity since all organic compounds give a response and coeluting compounds in a sample will interfere with the analysis. With the adequate cleaning step, low limits of detection and quantification can be achieved for various analytes, such as pesticides in tap water (mLOQs: 0.3–2.6 μg L^{-1} (Farajzadeh et al., 2011) or soil (mLOQs: 0.5–1.2 μg kg^{-1}) (Wang et al., 2014).

To reduce potential interferences, other similar detectors are used for specific compounds. For instance, the flame photometric detector (FPD) only detects compounds containing sulfur and phosphorus, making it suitable for the analysis of OPPs providing a high selectivity (Wang et al., 2014). Similarly, the nitrogen phosphorus detector (NPD) is commonly employed for pesticide detection, offering enhanced sensitivity for nitrogen- and phosphorus-containing compounds (Deng et al., 2010; Łozowicka et al., 2012; Wang et al., 2010) as shown by Deng et al. for the simultaneous determination of triazole fungicides (hexaconazole, myclobutanil, and tebuconazole) in soil. This method proved to be simpler and more cost-effective than GC coupled to tandem mass spectrometry (GC–MS/MS) (Deng et al., 2010).

Another powerful detector for CECs is the electron capture detector (ECD). This detector is very selective and extremely sensitive as it requires that the substances subjected to the analysis get ionized by the electrons emitted after the ionization of the carrier gas by β-particles. Therefore only CECs with electronegative groups can be analyzed this way, typically the halogens bromine and chlorine as shown for the detection of organochlorine pesticides (Łozowicka et al., 2012; García et al., 2020; Wang et al., 2010) or chlorinated by-products (Tobiszewski and Namieśnik, 2011).

Mass spectrometry (MS) is among the most used techniques for the analysis of CECs. Its main advantage lies in its ability to confirm analytes through their unique mass spectra, its high selectivity and sensitivity. The ionization source and the mass analyzer used in GC-MS play a crucial role in determining the sensitivity, selectivity, and fragmentation patterns of analytes. Among the different ionization techniques (Fig. 7.2), electron ionization (EI) is the predominant ionization technique, as shown in Table 7.2. It relies on high-energy electrons to ionize molecules, generating extensive fragmentation that aids

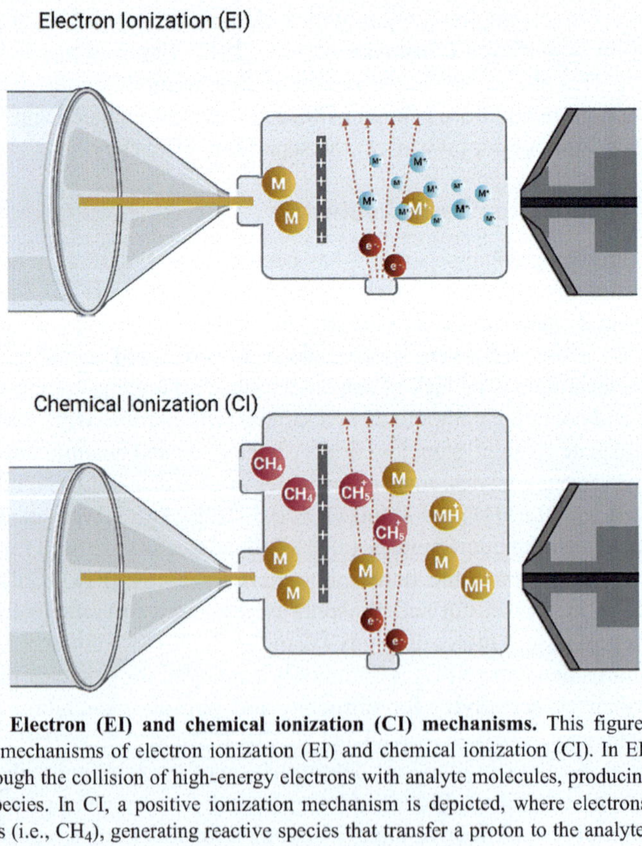

FIG. 7.2 Electron (EI) and chemical ionization (CI) mechanisms. This figure shows the ionization mechanisms of electron ionization (EI) and chemical ionization (CI). In EI, ionization occurs through the collision of high-energy electrons with analyte molecules, producing positively charged species. In CI, a positive ionization mechanism is depicted, where electrons ionize the reagent gas (i.e., CH_4), generating reactive species that transfer a proton to the analyte molecules, forming positively charged ions. *(Created with BioRender.com.)*

in structural analysis. The ionization energy is typically 70 eV but, in some cases, it can be adjusted to reduce fragmentation, for example, in the analysis of PBDEs (Gripp et al., 2022). Moreover, chemical ionization is a widely used soft ionization method that can operate in either positive (CI) or negative (NICI) mode. As a soft ionization method, it induces less fragmentation, which allows for better detection of molecular or *quasi*-molecular ions, offering higher sensitivity and selectivity. CI is typically used for analytes that readily undergo protonation or cation attachment as it can be the case of some PFAS and PBDEs (Weinberg et al., 2011). For example, in the determination of fluorotelomer alcohols (FTOHs) in river waters, CI yielded significantly lower instrumental LODs (iLODs) compared to EI, making it more suitable for trace level analysis of FTOHs in complex environmental samples (Ayala-Cabrera et al., 2020a). NICI is particularly effective for analytes containing electronegative functional groups. This includes perfluorocarboxylic acids (PFCAs) (Liu et al., 2011), chlorinated (Barón et al., 2012) and BFRs (Valls-Cantenys et al., 2013; Genisoglu et al., 2019; Chen et al., 2019; Bendig and Vetter, 2013), OCPs (Beristain-Montiel et al., 2016), chlorinated PAHs (Kamiya et al., 2015), and other polybrominated and polychlorinated compounds (Beristain-Montiel et al., 2016). In addition, in a comparative study by García et al., ECD and NICI-MS were evaluated for OCP analysis in sediments. Both methods provided satisfactory performance, though NICI-MS demonstrated superior linearity, sensitivity, and accuracy (García et al., 2020). For the mass analyzer, classical GC analysis relies on two main available options, the single quadrupole (Q) and triple quadrupole (QqQ). The single quadrupole consists of a single mass analyzer allowing it to measure only precursor ions. In contrast, the triple quadrupole features two quadrupole mass analyzers separated by a collision cell, enabling fragmentation experiments. These differences make the triple quadrupole more suitable for complex matrices or mixtures while also offering higher sensitivity for some compounds (Gustavsson et al., 2017; Barón et al., 2012).

3.4 Classical GC derivatization approaches

The analysis of polar and semi-polar emerging contaminants by GC is not feasible without a prior derivatization step. This chemical transformation enables analytes to be compatible with GC separation. Additionally, an appropriate derivatization reagent can help to reduce analyte adsorption within the GC system, enhance detector response, and improve chromatographic resolution and peak symmetry. Usually, derivatization takes place in a nonwater miscible solvent and, thus, liquid injection is typically carried out in the splitless mode, injecting 1–2 μL (see Table 7.2).

In the analysis of environmental samples, three major families of compounds have been the focus of GC derivatization methods: pharmaceuticals, industrial chemicals, and illicit drugs. For pharmaceuticals, the most widely used derivatization reaction is silylation with reagents such as N,O-Bis(trimethylsilyl) trifluoroacetamide/N-tert-Butyldimethylsilyl-N-methyltrifluoroacetamide(BSTFA/MTBSTFA) + 1% Trimethylchlorosilane/ tert-Butyldimethylsilyl chloride (TMCS/TBDMSCl). This approach has demonstrated strong performance for certain NSAIDs and EDCs, such as estrogenic hormones, achieving method LODs (mLODs) in the low ng/L range. Ademoyegun et al. compared silylation using BSTFA with 1% TMCS for the acetylation of various pharmaceuticals and found that, in addition to providing higher accuracy and narrower peaks, silylation enabled the detection of several antibiotics that were not detectable through acetylation (Ademoyegun et al., 2020).

The conversion of analytes into derivatives containing multiple halogens, such as in the study by Migowska et al., where a mixture of 2,2,3,3,3-pentafluoro-1-propanol (PFPOH) and pentafluoropropionic anhydride (PFPA) was used for NSAIDs and estrogenic hormone (Migowska et al., 2012), is often linked to the chosen detector. Introducing halogens into molecules significantly enhances sensitivity when an ECD is used, as in that study (Migowska et al., 2012). In fact, the determination of pentafluoropropionyl (PFP) derivatives by GC-ECD proved more sensitive results than that of trimethylsilyl (TMS) derivatives by GC-MS in the same study (mLOQs 2.2–16.3 ng/L and 9.8–114.2 ng/L, respectively), identifying more pharmaceuticals and hormones with the former in the same water sample. However, the synthesis of PFP derivatives was more labor-intensive, and ECD solely relies on retention time identification. To minimize misidentification, the use of a second column with a different stationary phase polarity is recommended (Migowska et al., 2012). Overall, analytical methods incorporating derivatization, besides being time-consuming, tend to be tailored to a limited set of compounds rather than targeting a broad range of contaminants. Most of these methods focus on determining at most one or two dozen compounds, while those with broader coverage can detect around 50 chemicals.

4 Updated methodologies

Although the classical GC approaches are widely used, some of their limitations have led to the application of novel analytical strategies or new instrumentation aiming to improve the chromatographic resolution, the sensitivity and the selectivity, to extend the chemical space and to increase the throughput of the GC determinations. On the one hand, some of these approaches are not necessarily new (i.e., derivatization or multidimensional chromatography) but their application to the analysis of CECs is still emerging. On the other hand, other methodologies rely on new instrumentation developments (i.e., new ionization sources for GC-MS, coupling to high-resolution mass spectrometry—HRMS—or portable GC-MS) or novel approaches such as multitarget, suspect—SS—or nontarget screening—NTS—). This section explores the different approaches to determine CECs using cutting-edge GC techniques. Table 7.3 summarizes some of the main applications for the determination of CECs using GC updated approaches.

TABLE 7.3 Main applications for the determination of CECs using GC updated approaches (novel aspect highlighted in bold).

Analytes	Matrix	Sample treatment	Analytical method	Injection	Column	Carrier gas	LODs/LOQs	Ref.
Illicit drugs								
Psychoactive substances and B-blockers ($n = 20$)	Sewage water	• SPE (Oasis MCX) • **Derivatization** (Enantiomerically pure derivatizing reagent = (R)-MTPA-Cl + MSTFA with 1% TMCS)	GC-EI-MS (IT)	Splitless (1 μL)	Rtx-5 (30 m x 0.25 mm, 0.25 μm)	He (1 mL min^{-1})	mLOQ: 0.15–104 ng L^{-1}	(Gonçalves et al., 2019)
Psychoactive substances ($n = 14$)	Sewage water	• SPE (Oasis HLB) • **Derivatization** (MSTFA)	GC-EI-MS/MS (QqQ)	Splitless (2 μL)	HP-5 MS (30 m x 0.25 mm, 0.25 μm)	He (1.3 mL min^{-1})	mLOD: 0.6–18 ng L^{-1}	(González-Mariño et al., 2010)
Pesticides								
OCPs, OPPs ($n = 100$)	Water	SPE (C18 & HLB)	**GC × GC**(cryogenic)-**EI-HRMS** (TOF)	Splitless (2 μL)	Rxi-5Sil MS (30 m x 0.25 mm, 0.25 μm) x Rtx-200 (2 m x 0.15 mm, 0.15 μm)	n.r.[a]	mLOQ 0.1 ug kg^{-1}	(Wang et al., 2024)
OCPs, OPPs ($n = 100$)	Soil	QuEChERS	**GC × GC**(cryogenic)-**EI-HRMS** (TOF)	Splitless (2 μL)	Rxi-5Sil MS (30 m x 0.25 mm, 0.25 μm) x Rtx-200 (2 m x 0.15 mm, 0.15 μm)	n.r.[a]	mLOQ 10 ug kg^{-1}	(Wang et al., 2024)

Continued

TABLE 7.3 Main applications for the determination of CECs using GC updated approaches (novel aspect highlighted in bold)—cont'd

Analytes	Matrix	Sample treatment	Analytical method	Injection	Column	Carrier gas	LODs/LOQs	Ref.
Nonchlorinated and chlorinated pesticides ($n = 53$)	Surface water and groundwater	SPE (Oasis HLB)	**GC × GC**(cryogenic)-**EI-HRMS** (TOF)	Split (2 μL, 1:10)	Rxi-5Sil MS (30 m x 0.25 mm, 0.25 μm) x Rxi-17Sil MS (2 m x 0.25 mm, 0.25 μm)	He (1.3 mL min^{-1})	mLODs: 0.3–2.2 ng L^{-1}	(Romagnoli et al., 2023)
OCPs ($n = 21$)	Surface water	SPE (Oasis HLB)	**GC-NICI-HRMS** (qTOF)	Splitless (2.5 μL)	HP-5 MS (30 m x 0.25 mm, 0.25 μm)	He (0.8 mL min^{-1})	n.r.[a]	(Moschet et al., 2017)
OCPs ($n = 15$)	Surface water	SPE (PEP)	**GC-APCI-HRMS** (qTOF)	n.r.[a]	HP-5 MS (30 m x 0.25 mm, 0.25 μm)	He (2 mL min^{-1})	iLOQs: 0.9–10 μg L^{-1}	(Cheng et al., 2016)
	Soil	• SLE (Hex:Ace, 9:1 v/v)SPE (Florisil)						
OPPs and pyrethroids ($n = 19$)	Surface water	LLE (Hexane)	**GC-APCI-MS/MS** (QqQ)	Splitless (5 μL)	Rtx-5MS (30 m x 0.25 mm, 0.25 μm)	He (3 mL min^{-1})	mLOQs: 12.5–125 pg L^{-1}	(Rösch et al., 2019)
Pharmaceuticals and personal-care products								
Pharmaceuticals and PCPs ($n = 30$)	Sediments	• PLE (MeOH, sediments) • SBSE (PDMS) • Liquid desorption (EtOAc) • **Derivatization** (MTBSTFA)	GC-EI-MS (Q)	Splitless (1 μL)	HP-5 MS (30 m × 0.25 mm, 0.25 μm)	He (1 mL min^{-1})	mLODs: 0.01–5 ng g^{-1}	(Pintado-Herrera et al., 2013)
	Wastewater and surface water						mLODs: 0.2–853 ng L^{-1}	

NSAIDs and estrogenic hormones ($n = 13$)	Sewage water	• SPE (Oasis HLB)BSTFA + TCMS (1%)	GC-EI-MS (Q)	Splitless (1 μL)	Rtx-5 (30 m x 0.25 mm, 0.25 μm)	He (2 mL min^{-1})	mLOQs: 10–114 ng L^{-1}	(Migowska et al., 2012)
		• SPE (Oasis HLB) • **PFPOH + PFPA**	GC-ECD			Ar (2 mL min^{-1})	mLOQ: 2–16 ng L^{-1}	
Pharmaceuticals and antiseptics ($n = 18$)	Sewage water	• SPE (Strata X) • **Derivatization** (PFBBr)	GC-EI-MS (Q)	Splitless (2 μL)	DB-5 MS (30 m x 0.25 mm, 0.25 μm)	n.r.[a]	mLODs: 1–17 ng L^{-1}	(Yu et al., 2012)
Antiinflammatory, antidepressant, hormones, and lipid regulator drugs ($n = 9$)	Sewage water	**In situ derivatization** (ECF) **in SPME** (PDMS/DVB)[a]	GC-EI-MS (Q)	Splitless (-)	Rtx-5MS (15 m x 0.25 mm, 0.25 μm)	He (1.3 mL min^{-1})	mLODs: 95–160 ng L^{-1}	(De Lima Gomes et al., 2011)
Industrial chemicals								
DPs (**NTS**)	Biota	• UAE (DCM)GPCSPE (Silica)	**GC-APCI-HRMS** (qTOF)	Splitless (-)	DB-5HT (15 m x 0.25 mm, 0.1 μm)	n.r.[a]	n.r.[a]	(Liu et al., 2019)
DPs ($n = 9$)	Biota	• PLE (Hex:DCM, 1:1 v/v)SPE (Florisil)	**GC-APPI-HRMS** (qOrbitrap)	Splitless (1 μL)	TG-5MS (15 m x 0.25 mm, 0.25 μm)	He (1.2 mL min^{-1})	mLODs: 0.1–3.5 pg g^{-1}	(Ayala-Cabrera et al., 2021a)
SCCPs ($n \leq 1000$)	Biota	• PLE (Hex:DCM, 1:1 v/v)SPE (Florisil)	**GC-APPI-HRMS** (qOrbitrap)	Splitless (1 μL)	TG-5MS (15 m x 0.25 mm, 0.25 μm)	He (1 mL min^{-1})	mLODs: 17–34 pg g^{-1}	(Ayala-Cabrera et al., 2021b)
PFCAs ($n = 9$)	Tap water	• SBSE (PDMS) • **Derivatization** (Isobutyl alcohol, isobutyl chloroformate)	GC-EI-MS (Q)	Splitless (CIS/TDU)	HP-5 MS UI (30 m × 0.25 mm, 0.25 μm)	He (1.2 mL min^{-1})	mLODs: 294–1459 ng L^{-1}	(Habib et al., 2024)

Continued

TABLE 7.3 Main applications for the determination of CECs using GC updated approaches (novel aspect highlighted in bold)—cont'd

Analytes	Matrix	Sample treatment	Analytical method	Injection	Column	Carrier gas	LODs/LOQs	Ref.
Neutral PFAS ($n = 12$)	Surface water	SPME (DVB/CAR/PDMS)	**GC-APPI-HRMS** (qOrbitrap)	Splitless (-)	DB-624 (60 m x 0.25 mm, 1.4 μm)	He (1.3 mL min^{-1})	mLOQs: 0.08–50 ng L^{-1}	(Ayala-Cabrera et al., 2020a)
PAEs ($n = 16$)	Drinking water	SPME (DVB/PDMS)	**GC-TPI-HRMS** (qTOF)	Splitless (-)	DB-5MS (30 m x 0.25 mm, 0.25 μm)	He (1.5 mL min^{-1})	mLODs: 1–164 ng L^{-1}	(Löbbecke et al., 2025)
Halogenated FRs ($n = 65$)	Dust	UAE (n.r.[a])	**GC-APCI-HRMS** (qTOF)	n.r.[a] (1 μL)	DB-5 HT (15 m × 0.25 mm, 0.1 μm)	n.r.[a]	iLOD: 0.0123–2.5 ug L^{-1}	(Megson et al., 2016)
PAEs & alternative plasticizers ($n = 13$)	Dust	• UAE (Hex:Ace)SPE (silica)	**GC-EI-HRMS** (qOrbitrap)	Splitless (1 μL)	Rxi5-SIL-MS (30 m x 0.25 mm, 0.25 um)	n.r.[a]	mLOQ 0.001–105.5 ug g^{-1}	(Demirtepe et al., 2021)
OPAHs & PAHs ($n = 29$)	Dust	• PLE (Tol:MeOH)SPE (silica)LC clean-up (Nucleosil 100-5NO_2)	GC-EI-MS (Q)	PTV (n.r.[a])	DB17-MS (60 m × 0.25 mm, 0.15 μm)	He (1 mL min^{-1})	iLODs: 0.06–0.8 pg	(Ahmed et al., 2015)
Amines ($n = 14$)	Gaseous and particulate matter	• UAE (H_2O) • **Derivatization** (BSC)	GC-EI-MS (Q)	Splitless (1 μL)	HP-5MS (30 m x 0.25 mm, 0.25 μm)	He (1 mL min^{-1})	iLODs: 0.04–0.4 μg mL^{-1}	(Chen et al., 2022b)
Nonphthalate plasticizers ($n = 28$)	Soil	SLE (H_2O, EtOAc)	**GC-EI-HRMS** (qOrbitrap)	Splitless (1 μL)	BP5-MS (30 x 0.25 mm, 0.25 um)	He (1 mL min^{-1})	mLOQ 10–200 μg kg^{-1}	(Capilla-Flores et al., 2024)
Multiresidue—suspect and nontarget screening								
Rubber compounds (**SS & NTS**)	Drinking water	SPE (Oasis HLB)	GC-EI-MS (Q)	Splitless (1 μL)	ZB-5 (60 m x 0.25 mm, 0.25 um)	He (2 mL min^{-1})	n.r.[a]	(Diera et al., 2023)

CECs (**SS**)	Ground, surface, and effluent wastewater	SPE (Oasis HLB)	**GC-APCI-HRMS** (qTOF)	Splitless (1 μL)	DB-5 MS (30 m x 0.25 mm, 0.25 μm)	He (2 mL min^{-1})	mLOIs: 0.02–0.5 μg L^{-1}	(Hernández et al., 2015)
Legacy and emerging contaminants: (**multitarget *n* = 102 & SS**)	Ground, river, sea, and effluent waters	• **In situ derivatization** (acetic anhydride) • SBSE (PDMS, Liquid desorption: EtOAc)	**GC-APCI-HRMS** (TOF)	Splitless (1 μL)	HP-5 MS (30 m x 0.25 mm, 0.25 μm)	He (1 mL min^{-1})	mLODs: 0.01–18 ng L^{-1}	(Pintado-Herrera et al., 2014)
PAEs degradation products and CECs (**SS**)	Surface water	• HF-LPME (PP with cellulose nanofibers) • Derivatization (BSTFA and TMS)	**GC** × **GC** (cryogenic)-**EI-HRMS** (qTOF)	Splitless (0.2 μL)	DB-5 MS (30 m x 0.25 mm, 0.25 μm) Rtx-200 (2 m x 0.15 mm, 0.15 μm)	He (1.3 mL min^{-1})	n.r.[a]	(Minho et al., 2022)
Water contaminants (**SS**, *n* > 5000)	Surface water	SPE (HLB)	**GC-EI-HRMS** (qTOF)	Splitless (2.5 μL)	HP-5 MS (30 m x 0.25 mm, 0.25 μm)	He (1.35 mL min^{-1})	n.r.[a]	(Moschet et al., 2017)
CECs (**NTS**)	Surface water	• Passive sampling (PDMS)TD	**GC** × **GC** (cryogenic)-**EI-HRMS** (TOF)	Splitless (TD)	Rxi 5Sil MS (30 m × 0.25 mm, 0.25 μm) Rxi17Sil MS (1 m × 0.25 mm, 0.25 μm)	He (2 mL min^{-1})	n.r.[a]	(Naudé et al., 2015)
EDCs and antiretroviral compounds (multitarget *n* = 11 & **NTS**)	Surface water	• SBSE (PDMS)TD	**GC** × **GC** (cryogenic)-**EI-HRMS** (TOF)	Splitless (TD)	Rtx-CLPesticides II (30 m x 0.25 mm, 0.2 μm) Rxi-17Sil MS (1 m x 0.25 mm, 0.25 μm)	He (1.4 mL min^{-1})	mLODs: 1–98 pg L^{-1}	(Wooding et al., 2017)

Continued

TABLE 7.3 Main applications for the determination of CECs using GC updated approaches (novel aspect highlighted in bold)—cont'd

Analytes	Matrix	Sample treatment	Analytical method	Injection	Column	Carrier gas	LODs/LOQs	Ref.
PAHs, MAHs, phenols, plasticizers, pesticides (**multi-target** $n = 70$)	Surface water	LLE (DCM)	GC-EI-MS (IT)	Splitless (2 μL)	MXT-5 (5 m x 0.1 mm, 0.4 μm) J&W DB-5MS (30 m x 0.25 mm, 0.25 μm)	n.r.[a]	mLOD: 200–1600 ng L^{-1}	(Duff et al., 2023)
PFAS (**SS**)	Sewage water	LLE (DCM)	**GC-cIM-APCI-HRMS** (qTOF)	Splitless (1 μL)	Rtx-5 (15 m x 0.25 mm, 0.1 μm)	He (3 mL min^{-1})	mLODs: 0.2–7.9 ng L^{-1}	(Steeves et al., 2023)
SVOCs (**NTS**)	Sewage water	LLE (DCM)	**GC × GC-EI-HRMS** (qTOF)	Splitless (1 μL)	Rxi-5Sil MS (60 m × 0.25 mm, 0.25 μm) Rtx-200 (1.2 m × 0.25 mm, 0.25 μm)	He (2 mL min^{-1})	n.r.[a]	(Castillo Meza et al., 2020)
Plasticizers (Target $n = 10$ & **SS**)	Sewage water	LLE (DCM)	**GC × GC** (solid state modulator)-**EI-HRMS** (qTOF)	n.r.[a]	DB-5 MS (30 m x 0.25 mm; 0.25 μm) DB-17 MS (1 m x 0.18 mm; 0.18 μm)	He (1 mL min^{-1})	LOQs: 100 ng L^{-1}	(Huang et al., 2023)
VOCs, pesticides, plasticizers and bisphenols (multitarget $n = 86$ & **SS & NTS**)	Dust	UAE (Hex:Ace)	**GC-EI-HRMS** (qOrbitrap)	Splitless (1 μL)	DB-5 MS (30 m x 0.25 mm, 0.25 μm)	n.r.[a]	iLOQs: 0.008–140 ng g^{-1}	(Dubocq et al., 2021)

PFAS (**NTS**)	Dust	UAE (Hex:Ace:MeOH:ACN)	**GC-APCI-MS**	Splitless (1 μL)	Rtx-5 (30 m x 0.25 mm, 0.25 μm)	He (1.4 mL min^{-1})	n.r.[a]	(Macneil et al., 2022)
CECs (**SS & NTS**)	Dust	PLE (Hex:DCM)	**GC-APCI-HRMS** (qTOF)	Splitless (1 μL)	DB-5 HT (15 m x 0.25 mm, 0.1 μm)	He (3 mL min^{-1})	n.r.[a]	(Schreckenbach et al., 2021)
PAHs, PCBs, PBDEs, pesticides (target $n = 118$ & **NTS**)	Soil	PLE (DCM)	**GC-APCI-HRMS** (qTOF)	Splitless (1 μL)	BR-5MS (30 m x 0.25 mm, 0.25 μm)	He (1 mL min^{-1})	iLOQs: 0.02–22.7 ng g^{-1}	(Costa et al., 2024)
Neutral PFAS (**NTS**)	Soil	SLE (nanopure H_2O + MTBE)	**GC-CI-HRMS** (qTOF)	Splitless (3 μL)	Rtx-1701 (30 m x 0.25 mm, 0.25 um)	He (1.1 mL min^{-1})	iLOD 0.5–0.97 ng g^{-1}	(Henderson et al., 2024)
PFAS (Target $n = 13$ & **SS**)	Sediments	• Shake (DCM)SPE (Envi-Carb)	GC-EI-MS (QqQ) **GC-EI-HRMS** (qTOF)	Splitless (2 μL)	Target: DB-WAX UI (30 m x 0.25 mm, 0.25 μm) NTS: DB-5 MS (30 m x 0.25 mm, 0.25 μm)	He (1 mL min^{-1}) He (1.2 mL min^{-1})	mLOQs: 0.07–0.76 ng g^{-1}	(Mok et al., 2023)
	Sewage water	• PLE (EtOAc)SPE (Envi-Carb)					mLOQs: 11–68 ng L^{-1}	
	Air						mLOQs (QqQ): 0.3–3.8 pg m^{-3}	
CECs (**SS** & NTS)	Air	• Quartz filter (particles) & PUF/XAD-2 (gas phase compounds)Soxhlet (Hex:Ace)	**GC-EI-HRMS** (qOrbitrap)	n.r.[a]	TG-5SIL MS (30 m x 0.25 mm, 0.25 μm)	n.r.[a]	n.r.[a]	(Zhu et al., 2024)

[a]*not reported.*

4.1 Novel derivatization strategies

One of the main limitations of classical derivatization approaches is that derivatizing reagents are not compatible with water, making them unstable to humidity and unsuitable for direct application in aqueous matrices, where these hydrophilic compounds are most expected. However, some water-soluble alternatives are compatible with this matrix, such as the combination of N-(3-dimethylaminopropyl)-N′-ethylcarbodiimide (EDAC) (Noche et al., 2011; De Lima Gomes et al., 2011) with 2,2,2-trifluoroethylamine (TFEA) to form amide derivatives, ethyl chloroformate (ECF) to form esters, or pentafluorobenzyl bromide (PFBBr) to substitute proton donor groups with pentafluorobenzyl (Yu et al., 2012).

Most of the GC methods for industrial chemicals, also pesticides, do not require a derivatization step (see Tables 7.2 and 7.3). However, in specific cases, derivatization agents have been tested to improve the volatility and stability of specific classes of compounds. For example, benzenesulfonyl chloride (BSC) has been used for the determination of atmospheric amines in gaseous and particulate phases (Chen et al., 2022b). Moreover, although PFAS are widely analyzed by LC-MS, there is evidence suggesting that LC systems may contain fluorotelomer-based chemicals, leading to background contamination. These challenges highlight GC as a valuable alternative, especially for volatile and semi-volatile PFAS analysis. In fact, GC has been successfully applied for PFAS analysis in a wide range of environmental matrices, including sewage (Mok et al., 2023) and tap waters, soils (Henderson et al., 2024), sediments (Mok et al., 2023), dust (Macneil et al., 2022), or gaseous (Mok et al., 2023) samples. For volatile PFAS, derivatization is not required (Henderson et al., 2024), whereas for the nonvolatile PFAS, derivatization is an essential step (Habib et al., 2024; Liu et al., 2011). Derivatization agents like n-butanol and isobutyl alcohol have been extensively used for the analysis of PFCAs in sediments (Liu et al., 2011) and aqueous samples (Habib et al., 2024), respectively. However, to the best of our knowledge, the derivatization of other relevant family of PFAS such as perfluorosulfonic acids (PFSAs) remains a challenge.

In the case of illicit drugs, although LC-MS is also the preferred technique, GC-MS remains a cost-effective alternative, offering robust compound identification without ionization matrix effects. Their polar nature requires a derivatization step that has been commonly done using N-Methyl-N-(trimethylsilyl)trifluoroacetamide) (MSTFA) as derivatization reagent as shown by González-Mariño et al. for cocaine and amphetamines, achieving LODs of 0.6–18 ng/L (González-Mariño et al., 2010). For further sensitivity improvements, Gonçalves et al. (2019), combined MSTFA with (S)-(+)-α-methoxy-α-trifluoromethylphenylacetyl chloride (MTPA-Cl) for derivatization, improving mLODs to 0.03–26.0 ng/L for 20 contaminants. Particularly, applications of GC for the determination of illicit drugs have been reported for environmental waters including sewage and surface waters (see Table 7.3).

In addition, several studies have focused on increasing the throughput of the derivatization step by implementing in situ derivatization within the extractant phase. The most common strategy in this context is the simultaneous derivatization and extraction using SPME. For example, Díaz and Peña-Alvarez (2017) carried out on-fiber derivatization of PPCPs in headspace mode, whereas De Lima Gomes et al. (2011) also combined in situ derivatization with SPME extraction, but directly in the aqueous phase and specifically for pharmaceuticals.

4.2 Multidimensional gas chromatography

Due to the high complexity of environmental samples, one-dimensional GC is often insufficient to achieve adequate separation. This frequently results in a high number of coelutions that hinders both qualitative identification and quantitative analysis. This is especially critical for GC systems interfaced with the EI source as the hard ionization may lead to complex overlapped mass spectra that could be difficult to deconvolute. In this sense, multidimensional gas chromatography (MDGC) has become a powerful analytical instrumentation to overcome these limitations by adding a short second column with a different stationary phase that improves the resolution and the selectivity of the separation (Semard et al., 2009; Tranchida et al., 2022). MDGC was first described by Simmons and Sanders in 1958 for the separation of hydrocarbons (Tranchida et al., 2022) and it has significantly evolved over the years. Depending on the goal, two different types of MDGC can be found: (i) heart-cutting MDGC (GC-GC) and (ii) comprehensive two-dimensional GC (GC × GC). On the one hand, GC-GC is generally employed in those cases where specific fractions of the first dimension (^{1}D) separation need additional separation to isolate the analyte from interferences in a way that the ^{1}D separation can be seen as a sample preparation step. On the other hand, GC × GC is usually applied when the whole sample requires further separation becoming a relevant strategy for multitarget and nontarget analysis of complex samples, such as environmental matrices (Fig. 7.3).

In this last approach, one of the most relevant variables to increase the peak capacity (number of potentially resolved peaks) lies in the use of complementary ^{1}D and second dimension (^{2}D) separation mechanisms (high orthogonality). Generally, combination of ^{1}D nonpolar columns (i.e., 5% phenyl-) with semi-polar or polar stationary phases (i.e., WAX) in the second dimension are preferred to enhance the separation of environmental samples. Another key factor is the modulator which allows the fast and reliable transference of the analytes from the first to the second dimension. In most cases, thermal modulators that use cryogenic solvents (i.e., liquid N_2) and hot jets to modulate the ^{1}D eluate are generally employed. However, valve-based modulators which collect ^{1}D fraction in a collector channel and fastly transfer then onto the ^{2}D column by applying a high flow (around 60 mL/min) are gaining more interest due to their simplicity, easier automation and reduced costs compared to thermal modulators (Tranchida, 2018; Mondello et al., 2025). Although one of the bottlenecks of GC × GC is the data analysis, the last advances and software development are intended to simplify this step,

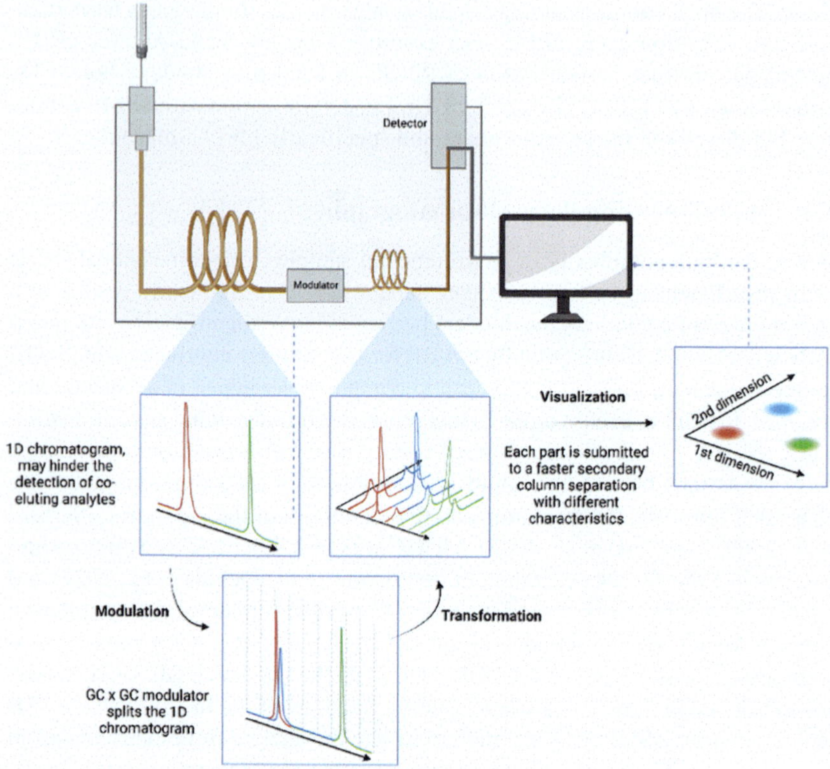

FIG. 7.3 General scheme of GC × GC instrumentation. This figure represents a comprehensive two-dimensional gas chromatography (GC × GC) system. The key differences lie in the oven section, where two columns with orthogonal polarities are connected via a modulator. The modulator periodically refocuses and reinjects fractions of the first-dimension (1D) effluent into the second-dimension (2D) column, enhancing separation capacity. *Created with BioRender.com.*

making this technology a strong approach to enhance the performance of GC separations (Vial et al., 2011; Favela et al., 2021).

In recent years, some studies have implemented this separation technique, especially in the determination of industrial chemicals and pesticides (Table 7.3). As observed, 5% phenyl- 95% PDMS stationary phases are used as ^{1}D column whereas 50% phenyl- 50% PDMS and trifluoropropyl-PDMS stationary phases are generally chosen in the second dimension. Moreover, almost all methods employed a thermal modulator. For instance, GC × GC has been used to improve the resolving power in the analysis of OCPs and OPPs, using a 30 m long 5% phenyl-95%PDMS column in the ^{1}D, followed by 2 m long polar ^{2}D columns, such as 50% phenyl-50% PDMS (Romagnoli et al., 2023) or 100% trifluoropropyl-PDMS (Wang et al., 2024). GC × GC has been extensively used in surface (Romagnoli et al., 2023; Minho et al., 2022; Naudé et al., 2015; Wooding et al., 2017; Ochiai et al., 2011), sewage (Castillo Meza et al., 2020; Huang et al.,

2023), and groundwater (Romagnoli et al., 2023). For example, Huang et al. used a solid-state modulator for the determination of phthalate, phosphate, and citrate esters in sewage water (Huang et al., 2023). The application of multidimensional chromatography has also been extended to LC-GC-MS system to facilitate the simultaneous determination of PAHs and OPAHs in dust samples by minimizing matrix interference and enhancing analyte detection (Ahmed et al., 2015).

4.3 Atmospheric pressure ionization for GC-MS couplings

As previously mentioned, most GC-MS approaches employ EI as gold standard ionization technique. EI provides a hard ionization that allows achieving structural information. This fact, together with the high stability of the mass spectra within different MS branches, has led to highly developed MS spectral libraries like NIST or MoNa, among others. However, one of the major drawbacks of EI lies in the need to find a compromise between selectivity and sensitivity due to the high fragmentation observed (Li et al., 2015). Therefore, in the last decades, the use of atmospheric pressure ionization (API) techniques to interface GC-MS instrumentation has emerged as an alternative to sort out some of these disadvantages (Ayala-Cabrera et al., 2023a, 2023b). In contrast to vacuum ionization techniques, API sources provide a soft ionization which largely preserves the $[M]^{+}$ and/or the protonated/deprotonated molecule. These ionization techniques provide several advantages over EI, such as the easy correlation of a feature/compound with an ion, the absence of a compromise between selectivity and sensitivity, as well as the easy coupling to low- and high-resolution MS instruments initially thought for LC applications. This could help to enhance the laboratory throughput by switching GC and LC separations in the same MS instrument (Li et al., 2015; Ayala-Cabrera et al., 2023a). API sources also present some disadvantages, such as the influence of the lab conditions (humidity, carryover, etc.) on the ionization efficiency or the lack of MS/HRMS spectral libraries and workflows for NTS (Ayala-Cabrera et al., 2023a). Fig. 7.4 shows the main API sources for GC-MS coupling.

Among all the API sources described in the literature, atmospheric pressure chemical ionization (APCI) is the most widely used mainly due to the vast number of instrumental advances in the last years and the commercialization by several MS companies (i.e., Waters, Sciex, Agilent Tech.) (Fang et al., 2020). GC-APCI-MS is gaining interest in the environmental field helping to overcome some of the limitations achieved for legacy and emerging pollutants like the low detection capability with common GC-MS instrumentation (i.e., PCDD/Fs, PCBs) or the low selectivity (i.e., PBDEs, PAEs, flame retardants, etc.) (Ayala-Cabrera et al., 2023b). For instance, it has been used to determine pesticides in soil and water (Cheng et al., 2016; Rösch et al., 2019), dechlorane plus (DP) and analogs in gull eggs and shark liver (Liu et al., 2019) or halogenated flame retardants in dust (Megson et al., 2016). Other API sources which are also commercially available are the atmospheric pressure photoionization

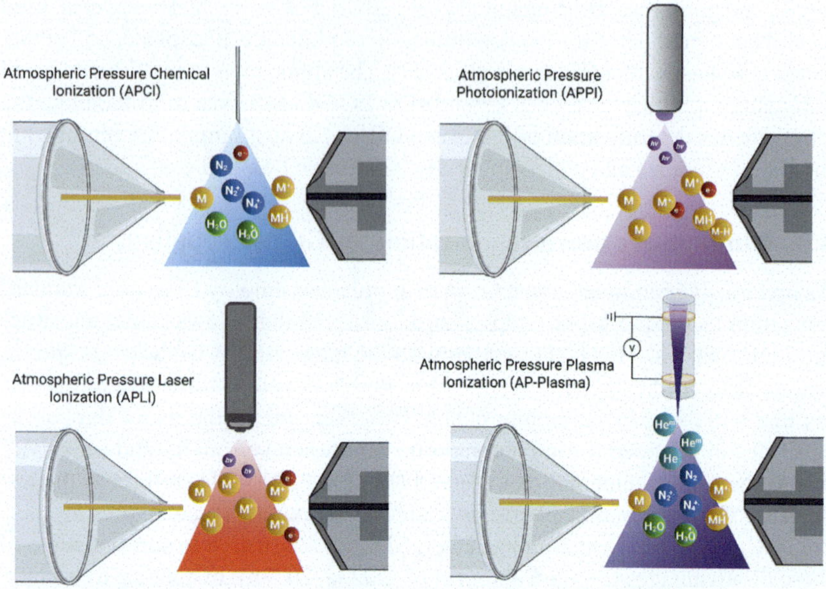

FIG. 7.4 Main API sources for GC-MS coupling. This figure summarizes the main atmospheric pressure ionization (API) sources, including atmospheric pressure chemical ionization (APCI), atmospheric pressure photoionization (APPI), atmospheric pressure laser ionization (APLI), and atmospheric pressure plasma ionization (AP-Plasma). These sources provide softer ionization compared to EI, preserving the molecular ion or quasi-molecular ion and reducing fragmentation. *(Created with BioRender.com.)*

(GC-APPI from MasCom Tech.), suitable for compounds with high aromaticity and electronegativity (i.e., PCDD/Fs, PFAS, chlorinated pollutants, etc.) (Ayala-Cabrera et al., 2023a, 2023b; Di Lorenzo et al., 2019) or atmospheric pressure plasma ionization (AP-Plasma) (SICRIT from Plasmion) which has shown the possibility to extend the chemical coverage leading to an efficient ionization for nonpolar and moderate polar compounds. For instance, tube plasma ionization (TPI), an AP-plasma source configuration, has been used to selectively determine PAEs in drinking water (Löbbecke et al., 2025) achieving comparable results than those obtained by EI. Moreover, all these API sources allow the addition of modifiers/reagents to customize the ionization, thus helping to solve detection issues or to promote the ionization of specific families of compounds. These strategies have been applied to improve the determination of several families of pollutants in environmental samples (Ayala-Cabrera et al., 2023a). For instance, Portolés et al. broaden the range of pesticides combining direct analysis with the addition of water as modifier in GC-APCI-MS (Portolés et al., 2014) while Ayala-Cabrera et al. enhanced the detection capability of DP and analogs in gull eggs by means of dopant-assisted GC-APPI-HRMS (Ayala-Cabrera et al., 2021a). Those strategies can also be used to improve selectivity as shown in the determination of short-chain chlorinated paraffins (SCCPs) where the use of a dopant mixture of acetone/carbon tetrachloride in GC-APPI helped to promote the formation of characteristic $[M+Cl]^-$ ions for each group of SCCP congeners (Ayala-Cabrera et al., 2021b). Other sources such as atmospheric pressure laser ionization (APLI) have been used to a lesser extent for the determination of PAHs coastal and harbor water (Dohmann et al., 2019), dust, sediments, and soil (Große Brinkhaus et al., 2017) as they are not commercially available.

4.4 Expanding the analyte scope with HRMS coupling

Although target analysis remains the predominant approach, (see Tables 7.2 and 7.3) an increasing number of laboratories are incorporating GC systems coupled with modern HRMS instruments into their inventories. These instruments enable ion detection with resolutions exceeding a mass-to-mass difference ratio of ≥20,000, along with mass accuracy within 5 ppm. Further details on the types of analyzers, ionization sources, and analytical strategies (targeted or nontargeted) used in HRMS studies are provided in Table 7.3. In GC-(HR) MS, where EI is the predominant ionization method, the use of HRMS allows for the acquisition of high-resolution fragmentation spectra, significantly narrowing the number of candidate structures for a given spectrum. This approach, known as SS, enables compound identification by matching experimental mass spectra with those available in spectral libraries. While this strategy has also been applied in low-resolution analyses, supported by extensive libraries such as NIST, its main limitation, i.e., detecting multiple candidates with the same reference spectrum, is mitigated when spectra are acquired in high resolution. However, the number of high-resolution spectral databases available remains limited compared to low-resolution, preventing full exploitation of this approach. Among the few spectral libraries, most of which are homemade (Zhu et al., 2024), some have begun to be openly distributed, such as the RECETOX library (Price et al., 2020), while others remain available for use in vendor software only (Moschet et al., 2017). With few exceptions utilizing these high-resolution spectral libraries, most studies perform SS by matching spectra to low resolution libraries such as NIST. Therefore data acquired using MS and HRMS instruments are often processed following the same methods (Castillo Meza et al., 2020). As a result, HRMS does not provide a significant advantage for this approach, particularly when considering the cost difference between instruments. Some examples of the implementation of suspect and nontarget approaches in environmental analysis include the detection of nonphthalate plasticizers in soils (Capilla-Flores et al., 2024) and phthalates and their alternatives in dust (Demirtepe et al., 2021), surface waters (Minho et al., 2022), or sewage waters (Huang et al., 2023).

Another advantage of high-resolution analysis is the ability to extract structural information from the exact mass of the parent ion. However, this conflicts with the high energy of EI (70 eV), requiring the use of lower energies to preserve the parent ion. Beyond spectral library matching, this additional information enables NTS through various structural elucidation approaches, such as competitive fragmentation modeling for metabolite identification (CMF-ID) (Zhu et al., 2024; Wang et al., 2021a). As explained above, one of the key advantages of API sources is their ability to preserve the parent ion, which has led to the development of nontargeted analysis strategies following workflows similar to those used in LC-HRMS (Hernández et al., 2015; Schreckenbach et al., 2021; Costa et al., 2024). Some of these studies employ SS approaches based on accurate mass, followed by in silico fragmentation matching, typically using vendor software. Others rely on exact mass matching within ±3 ppm accuracy and isotopic pattern fitting, followed by MS/MS spectral analysis comparison using open-access libraries, such as MassBank (Costa et al., 2024). Also based on accurate mass detection by GC-HRMS, some nontargeted approaches use Kendrick mass defect calculation to find homologous series of fluorinated compounds to identify unknown PFAS (Macneil et al., 2022; Henderson et al., 2024). Despite the high capability of resolving close masses, compounds with the same exact mass that coelute may still not be properly identified using this approach. Therefore techniques such as GC × GC, previously described, can be a valuable addition to SS and NTS workflows (Wooding et al., 2017; Castillo Meza et al., 2020). Moreover, ion mobility (IM) can also add an extra separation capacity as well as increase the confidence of the identifications by means of collision cross-section comparison. Indeed, Steeves et al. (2023) already proposed a GC-cyclic IMS as a powerful instrumentation to carry out NTS of fluorotelomer ethoxylates in indoor dust and wastewater.

4.5 On-site GC analysis

The high workload faced by analytical laboratories remains one of the main bottlenecks for the analysis of CECs in complex environmental samples. Conventional benchtop methodologies are often time-consuming and costly, which delays the delivery of reliable results. In some cases, the qualitative identification of a pollutant is more relevant than its quantification, as it may be sufficient to determine whether the compound, if regulated, exceeds the established threshold. In this sense, in-field results, which do not need to be confirmatory, would help the field officer to manage the environmental protection of a site in a short period of time (Spikmans, 2019). Thus portable instruments that allow on-site analysis are gaining interest as they allow miniaturization and shorter analysis times even under uncontrolled environmental conditions with the minimal loss of performance (Lee et al., 2024; Chen et al., 2024). Nowadays, many portable GC-MS instruments are commercially available (Chen et al., 2024). For instance, Duff et al. carried out an on-site screening of organic pollutants in soil and water (Duff et al., 2023). Coupling these approaches with rapid, solvent-free extraction techniques such as SPME represents a smart strategy to enhance the analytical throughput in environmental as shown by Wang et al. in the determination trichloroethylene in soil vapor and groundwater (Wang et al., 2021b).

Sample extraction can be even integrated with the sampling to enable continuous monitoring of pollutants, as demonstrated by Chen et al. (2022b) through a drone-based SPME coupled to portable GC-MS for the detection of air pollutants. Such advances highlight the potential of portable GC-MS as a screening tool to enhance laboratory throughput.

5 Future trends and conclusions

The new advances and developments related to GC-based methodologies are helping to assess the determination of CECs in environmental samples. One the one hand, classical approaches remain widely used to sort out some of the challenges associated with CECs analysis. However, the new trends toward extending the chemical coverage of analytical methodologies by means of multiresidue or even suspect and nontarget approaches have led to new emerging technologies or the reuse of conventional instrumentation which has been scarcely used. In this way, the development of new derivatization strategies that are compatible with aqueous solutions has been emerging specially in the determination of PFAS, PPCPs, or illicit drugs, making GC methodologies suitable for the analysis of these more polar compounds. Also, in situ derivatization methodologies that help to speed up the sample treatment could also help to improve the throughput of GC analysis. Another way to extend the chemical space of GC-based methodologies is the use of multidimensional chromatography to enhance the separation capacity. This "old" technology is now emerging as a powerful separation approach especially for the analysis of complex samples. Besides, the introduction of flow modulators has also helped to overcome one of the biggest limitations of MDGC such as automatization. On the other hand, the coupling of GC to MS systems interfaced with soft API sources has also significantly grown during the last decades as it allows to efficiently preserve the molecular/*quasi*-molecular ion enhancing the sensitivity and the selectivity of GC methods. Thus sources such as APCI, APPI or atmospheric pressure plasma sources had helped to improve the performance on the determination of CECs that present a bad ionization efficiency with EI or CI techniques. Finally, without a doubt, future challenges include everything related to suspect and nontarget analysis, as these strategies are opening a new scenario toward nondiscriminative analytical analysis. Thus most of the upcoming challenges to face in GC analysis are related to data processing (especially in nontarget and multidimensional approaches) due to the huge amount of data that is being generated. In this way, the combination of EI and API sources could help to increase the confidence on the annotations proposed for CECs, although efforts must be made to develop high-resolution mass spectral libraries, especially for API sources. Besides that, well-known challenges such as the adaptation of GC to a world without using helium as carrier gas or the development of instruments to carry out fast GC analyses and on-site analysis are still needed to expand their use in environmental analysis.

Acknowledgments

The authors thank the financial support from the State Research Agency of the Ministry of Science and Innovation, Government of Spain (project PID2020-117686RB-C31) and the Basque Government as a consolidated group of the Basque Research System (IT-1446-22). Carles Moreu-Romero acknowledges his predoctoral grant from the University of the Basque Country (UPV/EHU). Naroa Lopez-Herguedas, Iker Alvarez-Mora and Mikel Musatadi acknowledge the Postdoctoral Program for the Advancement of Doctoral Research Personnel of the Basque Government. Juan F. Ayala-Cabrera also acknowledges his Maria Zambrano Postdoctoral Fellowship funded by UPV/EHU, the Spanish Ministry of Universities and European Union (Next Generation EU).

References

Ademoyegun, O.T., Okoh, O.O., Okoh, A.I., 2020. Method validation and investigation of the levels of pharmaceuticals and personal care products in sludge of wastewater treatment plants and soils of irrigated golf course. Molecules. 1420-3049. 25 (14), 3114. doi:10.3390/molecules25143114.

Ahmed, T.M., Bergvall, C., Åberg, M., Westerholm, R., 2015. Determination of oxygenated and native polycyclic aromatic hydrocarbons in urban dust and diesel particulate matter standard reference materials using pressurized liquid extraction and LC-GC/MS. Anal. Bioanal. Chem. 16182650. 407 (2), 427–438. Available from: http://link.springer.de/link/service/journals/00216/index.htm. doi:10.1007/s00216-014-8304-8.

Ao, J., Yuan, T., Ma, Y., Gao, L., Ni, N., Li, D., 2017. Identification, characteristics and human exposure assessments of triclosan, bisphenol-A, and four commonly used organic UV filters in indoor dust collected from Shanghai, China. Chemosphere. 18791298. 184, 575–583. Available from: www.elsevier.com/locate/chemosphere. doi:10.1016/j.chemosphere.2017.06.033.

Ayala-Cabrera, J.F., Contreras-Llin, A., Moyano, E., Santos, F.J., 2020a. A novel methodology for the determination of neutral perfluoroalkyl and polyfluoroalkyl substances in water by gas chromatography-atmospheric pressure photoionisation-high resolution mass spectrometry. Anal. Chim. Acta. 00032670. 1100, 97–106. doi:10.1016/j.aca.2019.12.004.

Ayala-Cabrera, J.F., Moyano, E., Santos, F.J., 2020b. Gas chromatography and liquid chromatography coupled to mass spectrometry for the determination of fluorotelomer olefins, fluorotelomer alcohols, perfluoroalkyl sulfonamides and sulfonamido-ethanols in water. J. Chromatogr. A. 00219673. 1609, 460463. doi:10.1016/j.chroma.2019.460463.

Ayala-Cabrera, J.F., Lacorte, S., Moyano, E., Santos, F.J., 2021a. Analysis of dechlorane plus and related compounds in gull eggs by GC-HRMS using a novel atmospheric pressure photoionization source. Anal. Bioanal. Chem. 16182650. 413 (13), 3421–3431. Available from: https://link.springer.com/journal/216/volumes-and-issues. doi:10.1007/s00216-021-03286-8.

Ayala-Cabrera, J.F., Galceran, M.T., Moyano, E., Santos, F.J., 2021b. Chloride-attachment atmospheric pressure photoionisation for the determination of short-chain chlorinated paraffins by gas chromatography-high-resolution mass spectrometry. Anal. Chim. Acta. 00032670. 1172, 338673. doi:10.1016/j.aca.2021.338673.

Ayala-Cabrera, J.F., Montero, L., Meckelmann, S.W., Uteschil, F., Schmitz, O.J., 2023a. Review on atmospheric pressure ionization sources for gas chromatography-mass spectrometry. Part I: current ion source developments and improvements in ionization strategies. Anal. Chim. Acta. 00032670. 1238, 340353. doi:10.1016/j.aca.2022.340353.

Ayala-Cabrera, J.F., Montero, L., Meckelmann, S.W., Uteschil, F., Schmitz, O.J., 2023b. Review on atmospheric pressure ionization sources for gas chromatography-mass spectrometry. Part II: current applications. Anal. Chim. Acta. 00032670. 1238, 340379. doi:10.1016/j.aca.2022.340379.

Barón, E., Eljarrat, E., Barceló, D., 2012. Analytical method for the determination of halogenated norbornene flame retardants in environmental and biota matrices by gas chromatography coupled to tandem mass spectrometry. J. Chromatogr. A. 00219673. 1248, 154–160. doi:10.1016/j.chroma.2012.05.079.

Bendig, P., Vetter, W., 2013. Improved GC/ECNI-MS sensitivity of decabromodiphenyl ether determination on 30 m columns by increasing the carrier gas flow after a modified gel permeation chromatographic cleanup protocol employing cyclohexane/ethyl acetate as eluent. Int. J. Environ. Anal. Chem. 0306-7319. 93 (9), 1019–1029. doi:10.1080/03067319.2012.690150.

Beristain-Montiel, E., Villalobos-Pietrini, R., Arias-Loaiza, G.E., Gómez-Arroyo, S.L., Amador-Muñoz, O., 2016. An innovative ultrasound assisted extraction micro-scale cell combined with gas chromatography/mass spectrometry in negative chemical ionization to determine persistent organic pollutants in air particulate matter. J. Chromatogr. A. 00219673. 1477, 100–107. doi:10.1016/j.chroma.2016.11.043.

Bi, X., Yuan, S., Pan, X., Winstead, C., Wang, Q., 2015. Comparison, association, and risk assessment of phthalates in floor dust at different indoor environments in Delaware, USA. J. Environ. Sci. Health A. 1093-4529. 50 (14), 1428–1439. doi:10.1080/10934529.2015.1074482.

Capilla-Flores, R., Egea-Castro, G., López-Ruíz, R., Romero-González, R., Garrido Frenich, A., 2024. Development of novel methods based on GC-HRMS and LC-HRMS for the determination of non-phthalate plasticizers in soil. Sci. Total Environ. 00489697. 917, 170150. doi:10.1016/j.scitotenv.2024.170150.

Carro, A.M., Fernández, S., Racamonde, I., García-Rodríguez, D., González, P., Lorenzo, R.A., 2012. Dispersive liquid-liquid microextraction coupled with programmed temperature vaporization-large volume injection-gas chromatography-tandem mass spectrometry for multiclass pesticides in water. J. Chromatogr. A. 18733778. 1253, 134–143. doi:10.1016/j.chroma.2012.06.089.

Castillo Meza, L., Piotrowski, P., Farnan, J., Tasker, T.L., Xiong, B., Weggler, B., Murrell, K., Dorman, F.L., Vanden Heuvel, J.P., Burgos, W.D., 2020. Detection and removal of biologically active organic micropollutants from hospital wastewater. Sci. Total Environ. 00489697. 700, 134469. doi:10.1016/j.scitotenv.2019.134469.

Cecinato, A., Romagnoli, P., Perilli, M., Balducci, C., 2017. Pharmaceutical substances in ambient particulates: a preliminary assessment. Chemosphere. 00456535. 183, 62–68. doi:10.1016/j.chemosphere.2017.05.100.

Celeiro, M., Dagnac, T., Llompart, M., 2018. Determination of priority and other hazardous substances in football fields of synthetic turf by gas chromatography-mass spectrometry: a health and environmental concern. Chemosphere. 00456535. 195, 201–211. doi:10.1016/j.chemosphere.2017.12.063.

Chen, J., Kuang, Y., Feng, X., Mao, C., Zheng, J., Ouyang, G., 2024. Solid phase microextraction coupled with portable gas chromatography-mass spectrometry for on-site analysis. Green Anal. Chem. 27725774. 11, 100164. doi:10.1016/j.greeac.2024.100164.

Chen, W., Zou, Y., Mo, W., Di, D., Wang, B., Wu, M., Huang, Z., Hu, B., 2022a. Onsite identification and spatial distribution of air pollutants using a drone-based solid-phase microextraction array coupled with portable gas chromatography-mass spectrometry via continuous-airflow sampling. Environ. Sci. Technol. 0013-936X. 56 (23), 17100–17107. doi:10.1021/acs.est.2c05259.

Chen, Y., Chen, Y.J., Zhang, Y., Li, R., Chen, W., Yan, S.C., Qi, Z., Chen, Z.F., Cai, Z., 2019. Determination of HFRs and OPFRs in PM2.5 by ultrasonic-assisted extraction combined with multi-segment column purification and GC-MS/MS. Talanta. 00399140. 194, 320–328. Available from: https://www.journals.elsevier.com/talanta. doi:10.1016/j.talanta.2018.10.025.

Chen, Y., Lin, Q., Li, G., An, T., 2022b. A new method of simultaneous determination of atmospheric amines in gaseous and particulate phases by gas chromatography-mass spectrometry. J. Environ. Sci. 10010742. 114, 401–411. doi:10.1016/j.jes.2021.09.027.

Cheng, Z., Dong, F., Xu, J., Liu, X., Wu, X., Chen, Z., Pan, X., Zheng, Y., 2016. Atmospheric pressure gas chromatography quadrupole-time-of-flight mass spectrometry for simultaneous determination of fifteen organochlorine pesticides in soil and water. J. Chromatogr. A. 00219673. 1435, 115–124. doi:10.1016/j.chroma.2016.01.025.

Christia, C., Poma, G., Besis, A., Samara, C., Covaci, A., 2018. Legacy and emerging organophosphorus flame retardants in car dust from Greece: implications for human exposure. Chemosphere. 00456535. 196, 231–239. doi:10.1016/j.chemosphere.2017.12.132.

Costa, L.R., Tovar Salvador, M.D.L.L., Pintado-Herrera, M.G., Albergaria-Barbosa, A.C.R., Martins, C.C., Lourenço, R.A., Combi, T., 2024. Legacy and novel contaminants in surface sediments of Admiralty Bay, Antarctica Peninsula. Sci. Total Environ. 18791026. 951. Available from: https://www.sciencedirect.com/science/journal/00489697. doi:10.1016/j.scitotenv.2024.175551.

da Silva Sousa, J., Do Nascimentodo Nascimento, H.O., de Oliveira Gomes, H., Do Nascimentodo Nascimento, R.F., 2021. Pesticide residues in groundwater and surface water: recent advances in solid-phase extraction and solid-phase microextraction sample preparation methods for multiclass analysis by gas chromatography-mass spectrometry. Microchem. J. 0026265X. 168, 106359. doi:10.1016/j.microc.2021.106359.

De Lima Gomes, P.C.F., Barletta, J.Y., Nazario, C.E.D., Santos-Neto, A.J., Von Wolff, M.A., Coneglian, C.M.R., Umbuzeiro, G.A., Lancas, F.M., 2011. Optimization of in situ derivatization SPME by experimental design for GC-MS multi-residue analysis of pharmaceutical drugs in wastewater. J. Sep. Sci. 16159314. 34 (4), 436–445. doi:10.1002/jssc.201000708.

Deboer, J., Wells, D., 2006. Pitfalls in the analysis of brominated flame retardants in environmental, human and food samples—including results of three international interlaboratory studies. TrAC Trends Anal. Chem. 01659936. 25 (4), 364–372. doi:10.1016/j.trac.2006.01.008.

Demirtepe, H., Melymuk, L., Codling, G., Murínová, Ľ.P., Richterová, D., Rašplová, V., Trnovec, T., Klánová, J., 2021. Targeted and suspect screening of plasticizers in house dust to assess cumulative human exposure risk. Sci. Total Environ. 00489697. 781, 146667. doi:10.1016/j.scitotenv.2021.146667.

Deng, Z., Hu, J., Qin, D., Li, H., 2010. Simultaneous analysis of hexaconazole, myclobutanil, and tebuconazole residues in apples and soil by SPE clean-up and GC with nitrogen–phosphorus detection. Chromatographia. 0009-5893. 71 (7–8), 679–684. doi:10.1365/s10337-010-1505-1.

Di Lorenzo, R.A., Lobodin, V.V., Cochran, J., Kolic, T., Besevic, S., Sled, J.G., Reiner, E.J., Jobst, K.J., 2019. Fast gas chromatography-atmospheric pressure (photo)ionization mass spectrometry of polybrominated diphenylether flame retardants. Anal. Chim. Acta. 18734324. 1056, 70–78. Available from: http://www.journals.elsevier.com/analytica-chimica-acta/. doi:10.1016/j.aca.2019.01.007.

Díaz, A., Peña-Alvarez, A., 2017. A simple method for the simultaneous determination of pharmaceuticals and personal care products in river sediment by ultrasound-assisted extraction followed by solid-phase microextraction coupled with gas chromatography–mass spectrometry. J. Chromatogr. Sci. 0021-9665. 55 (9), 946–953. doi:10.1093/chromsci/bmx058.

Diera, T., Thomsen, A.H., Tisler, S., Karlby, L.T., Christensen, P., Rosshaug, P.S., Albrechtsen, H.-J., Christensen, J.H., 2023. A non-target screening study of high-density polyethylene pipes revealed rubber compounds as main contaminant in a drinking water distribution system. Water Res. 00431354. 229, 119480. doi:10.1016/j.watres.2022.119480.

Dimzon, I.K., Westerveld, J., Gremmel, C., Frömel, T., Knepper, T.P., de Voogt, P., 2017. Sampling and simultaneous determination of volatile per- and polyfluoroalkyl substances in wastewater treatment plant air and water. Anal. Bioanal. Chem. 1618-2642. 409 (5), 1395–1404. doi:10.1007/s00216-016-0072-1.

Dohmann, J.F., Thiäner, J.B., Achten, C., 2019. Ultrasensitive detection of polycyclic aromatic hydrocarbons in coastal and harbor water using GC-APLI-MS. Mar. Pollut. Bull. 0025326X. 149, 110547. doi:10.1016/j.marpolbul.2019.110547.

Dong, C.-D., Wang, M.-H., Chen, C.-F., Shih, Y.-J., Chang, K.-L., Lee, S.-H., Lin, Y.-L., Wu, C.-H., Chen, C.-W., 2020. Detecting phthalate esters in sludge particulates from wastewater treatment plants. J. Environ. Sci. Health A. 1093-4529. 55 (10), 1233–1240. doi:10.1080/10934529.2020.1780850.

Dubocq, F., Kärrman, A., Gustavsson, J., Wang, T., 2021. Comprehensive chemical characterization of indoor dust by target, suspect screening and nontarget analysis using LC-HRMS and GC-HRMS. Environ. Pollut. 02697491. 276, 116701. doi:10.1016/j.envpol.2021.116701.

Duff, D., Lennard, C., Li, Y., Doyle, C., Edge, K.J., Holland, I., Lothridge, K., Johnstone, P., Beylerian, P., Spikmans, V., 2023. Portable gas chromatography–mass spectrometry method for the in-field screening of organic pollutants in soil and water at pollution incidents. Environ. Sci. Pollut. Res. 16147499. 30 (40), 93088–93102. Available from: https://www.springer.com/journal/11356. doi:10.1007/s11356-023-28648-w.

Ettre, L.S., 1993. Nomenclature for chromatography (IUPAC recommendations 1993). Pure Appl. Chem. 1365-3075. 65 (4), 819–872. doi:10.1351/pac199365040819.

Fang, J., Zhao, H., Zhang, Y., Lu, M., Cai, Z., 2020. Atmospheric pressure chemical ionization in gas chromatography-mass spectrometry for the analysis of persistent organic pollutants. Trends Environ. Anal. Chem. 22141588. 25, e00076. doi:10.1016/j.teac.2019.e00076.

Farajzadeh, M.A., Bahram, M., Jafary, F., Bamorowat, M., 2011. Combination of extraction by silylated vessel-dispersive liquid–liquid microextraction as a high-enrichment factor technique: optimization and application in preconcentration of some triazole pesticides from aqueous samples followed by GC-FID determination. Chromatographia. 0009-5893. 73 (3–4), 393–401. doi:10.1007/s10337-010-1895-0.

Favela, K.A., Hartnett, M.J., Janssen, J.A., Vickers, D.W., Schaub, A.J., Spidle, H.A., Pickens, K.S., 2021. Nontargeted analysis of face masks: comparison of manual curation to automated GCxGC processing tools. J. Am. Soc. Mass Spectrom. 18791123. 32 (4), 860–871. Available from: https://pubs.acs.org/loi/jamsef. doi:10.1021/jasms.0c00318.

García, L., Muro, C., De La Rosa, I., Amador-Muñoz, O., Ponce, M.G., Borja, M., 2020. Comparison of gas chromatography techniques for the analysis of organochlorine pesticides in sediments. Soil Sediment Contam. Int. J. 1532-0383. 29 (3), 257–271. doi:10.1080/15320383.2019.1703099.

Genisoglu, M., Sofuoglu, A., Kurt-Karakus, P.B., Birgul, A., Sofuoglu, S.C., 2019. Brominated flame retardants in a computer technical service: indoor air gas phase, submicron (PM1) and coarse (PM10) particles, associated inhalation exposure, and settled dust. Chemosphere. 18791298. 231, 216–224. Available from: www.elsevier.com/locate/chemosphere. doi:10.1016/j.chemosphere.2019.05.077.

Gonçalves, R., Ribeiro, C., Cravo, S., Cunha, S.C., Pereira, J.A., Fernandes, J.O., Afonso, C., Tiritan, M.E., 2019. Multi-residue method for enantioseparation of psychoactive substances and beta blockers by gas chromatography–mass spectrometry. J. Chromatogr. B. 15700232. 1125, 121731. doi:10.1016/j.jchromb.2019.121731.

González-Mariño, I., Quintana, J.B., Rodríguez, I., Cela, R., 2010. Determination of drugs of abuse in water by solid-phase extraction, derivatisation and gas chromatography-ion trap-tandem mass spectrometry. J. Chromatogr. A 18733778. 1217 (11), 1748–1760. Available from: www.elsevier.com/locate/chroma. doi:10.1016/j.chroma.2010.01.046.

Gripp, L., da Silva Carreira, R., Moreira, D., de Lemos Scofield, A., Massone, C.G., 2022. Method development and application to sediments for multi-residue analysis of organic contaminants using gas chromatography-tandem mass spectrometry. Anal. Bioanal. Chem. 16182650. 414 (19), 5845–5855. Available from: https://link.springer.com/journal/216/volumes-and-issues. doi:10.1007/s00216-022-04148-7.

Große Brinkhaus, S., Thiäner, J.B., Achten, C., 2017. Ultra-high sensitive PAH analysis of certified reference materials and environmental samples by GC-APLI-MS. Anal. Bioanal. Chem. 16182650. 409 (11), 2801–2812. Available from: https://link.springer.com/journal/216/volumes-and-issues. doi:10.1007/s00216-017-0224-y.

Guitart, C., Readman, J.W., 2010. Critical evaluation of the determination of pharmaceuticals, personal care products, phenolic endocrine disrupters and faecal steroids by GC/MS and PTV-GC/MS in environmental waters. Anal. Chim. Acta. 18734324. 658 (1), 32–40. doi:10.1016/j.aca.2009.10.066.

Gustavsson, J., Ahrens, L., Nguyen, M.A., Josefsson, S., Wiberg, K., 2017. Development and comparison of gas chromatography–mass spectrometry techniques for analysis of flame retardants. J. Chromatogr. A. 00219673. 1481, 116–126. doi:10.1016/j.chroma.2016.12.028.

Habib, A., Noriega Landa, E., Holbrook, K.L., Chacon, A.A., Lee, W.-Y., 2024. Green analytical method for perfluorocarboxylic acids (PFCAs) in water of stir bar sorptive extraction coupled with thermal desorption–gas chromatography—mass spectroscopy. Water. 2073-4441. 16 (17), 2543. doi:10.3390/w16172543.

He, P., Aga, D.S., 2019. Comparison of GC-MS/MS and LC-MS/MS for the analysis of hormones and pesticides in surface waters: advantages and pitfalls. Anal. Methods. 17599679. 11 (11), 1436–1448. Available from: http://pubs.rsc.org/en/journals/journal/ay. doi:10.1039/c8ay02774a.

Henderson, W.M., Evich, M.G., Washington, J.W., Ward, T.T., Schumacher, B.A., Zimmerman, J.H., Kim, Y.D., Weber, E.J., Williams, A.C., Smeltz, M.G., Glinski, D.A., 2024. Analysis of legacy and novel neutral per- and polyfluoroalkyl substances in soils from an industrial manufacturing facility. Environ. Sci. Technol. 15205851. 58 (24), 10729–10739. Available from: http://pubs.acs.org/journal/esthag. doi:10.1021/acs.est.3c10268.

Hernández, F., Ibáñez, M., Portolés, T., Cervera, M.I., Sancho, J.V., López, F.J., 2015. Advancing towards universal screening for organic pollutants in waters. J. Hazard. Mater. 18733336. 282, 86–95. Available from: www.elsevier.com/locate/jhazmat. doi:10.1016/j.jhazmat.2014.08.006.

Homem, V., Llompart, M., Vila, M., Ribeiro, A.R.L., Garcia-Jares, C., Ratola, N., Celeiro, M., 2022. Gone with the flow—assessment of personal care products in Portuguese rivers. Chemosphere. 00456535. 293, 133552. doi:10.1016/j.chemosphere.2022.133552.

Huang, Y., You, Y., Wu, M., Han, M., Zhang, J., Gao, W., Xie, D., Chen, H., Ou, H., Song, N., Cheng, C., Zhuang, W., Li, J., Lei, Z., Jin, B., Zhou, Z., Li, M., 2023. Chemical characterization and source attribution of organic pollutants in industrial wastewaters from a Chinese chemical industrial park. Environ. Res. 00139351. 229, 115980. doi:10.1016/j.envres.2023.115980.

Kadhum, H.A., Hammood, M.K., Arif, M.A., 2024. Optimization and validation of a GC-FID/QuEChERS method for quantitative determination of spiromesifen residues in tomato fruits, leaves and soil matrices. Anal. Bioanal. Chem. Res. 2383093X. 11 (4), 447–461. Available from: www.analchemres.org. doi:10.22036/abcr.2024.449512.2077.

Kamiya, Y., Ikemori, F., Ohura, T., 2015. Optimisation of pre-treatment and ionisation for GC/MS analysis for the determination of chlorinated PAHs in atmospheric particulate samples. Int. J. Environ. Anal. Chem. 0306-7319. 95 (12), 1157–1168. doi:10.1080/03067319.2015.1048439.

Lava, R., Aimo, E., Menegus, L., Pojana, G., Marcomini, A., 2013. Determination of benzotrifluoride derivative compounds in groundwater. Anal. Chim. Acta. 00032670. 804, 126–134. doi:10.1016/j.aca.2013.10.006.

Lee, Y., Son, H., Lee, J., Lim, S.H., 2024. Review on micro-gas chromatography system for analysis of multiple low-concentration volatile organic compounds: preconcentration, separation, detection, integration, and challenges. Micro Nano Syst. Lett. 22139621. 12 (1). Available from: https://mnsl-journal.springeropen.com. doi:10.1186/s40486-024-00203-3.

Li, D.X., Gan, L., Bronja, A., Schmitz, O.J., 2015. Gas chromatography coupled to atmospheric pressure ionization mass spectrometry (GC-API-MS): review. Anal. Chim. Acta 18734324. 891, 43–61. Available from: http://www.journals.elsevier.com/analytica-chimica-acta/. doi:10.1016/j.aca.2015.08.002.

Liu, W.L., Hwang, B.H., Li, Z.G., Jen, J.F., Lee, M.R., 2011. Headspace solid phase microextraction in situ supercritical fluid extraction coupled to gas chromatography-tandem mass spectrometry for simultaneous determination of perfluorocarboxylic acids in sediments. J. Chromatogr. A. 18733778. 1218 (43), 7857–7863. doi:10.1016/j.chroma.2011.08.062.

Liu, X., Wu, Y., Zhang, X., Shen, L., Brazeau, A.L., Adams, D.H., Marler, H., Watts, B.D., Chen, D., 2019. Novel dechlorane analogues and possible sources in Peregrine Falcon Eggs and Shark Livers from the Western North Atlantic regions. Environ. Sci. Technol. 0013-936X. 53 (7), 3419–3428. doi:10.1021/acs.est.8b06214.

Löbbecke, S., Pape, A., Montero, L., Uteschil, F., Ayala-Cabrera, J.F., Schmitz, O.J., 2025. Improving the reliability of phthalate esters analysis in water samples by gas chromatography-tube plasma ionization-high-resolution mass spectrometry (GC-TPI-HRMS). Talanta. 00399140. 285, 127388. doi:10.1016/j.talanta.2024.127388.

Loos, R., Tavazzi, S., Mariani, G., Suurkuusk, G., Paracchini, B., Umlauf, G., 2017. Analysis of emerging organic contaminants in water, fish and suspended particulate matter (SPM) in the Joint Danube Survey using solid-phase extraction followed by UHPLC-MS-MS and GC–MS analysis. Sci. Total Environ. 00489697. 607-608, 1201–1212. doi:10.1016/j.scitotenv.2017.07.039.

Łozowicka, B., Jankowska, M., Rutkowska, E., Kaczyński, P., Hrynko, I., 2012. Comparison of extraction techniques by matrix solid phase dispersion and liquid-liquid for screening 150 pesticides from soil, and determination by gas chromatography. Pol. J. Environ. Stud. 12301485. 21 (4), 973–992. Available from: https://www.pjoes.com/Comparison-of-Extraction-Techniques-by-Matrix-r-nSolid-Phase-Dispersion-and-Liquid,88830,0,2.html.

Macneil, A., Li, X., Amiri, R., Muir, D.C.G., Simpson, A., Simpson, M.J., Dorman, F.L., Jobst, K.J., 2022. Gas chromatography-(Cyclic) ion mobility mass spectrometry: a novel platform for the discovery of unknown per-/polyfluoroalkyl substances. Anal. Chem. 15206882. 94 (31), 11096–11103. Available from: http://pubs.acs.org/journal/ancham. doi:10.1021/acs.analchem.2c02325.

Martínez, C., Ramírez, N., Gómez, V., Pocurull, E., Borrull, F., 2013. Simultaneous determination of 76 micropollutants in water samples by headspace solid phase microextraction and gas chromatography–mass spectrometry. Talanta. 00399140. 116, 937–945. doi:10.1016/j.talanta.2013.07.055.

Megson, D., Robson, M., Jobst, K.J., Helm, P.A., Reiner, E.J., 2016. Determination of halogenated flame retardants using gas chromatography with atmospheric pressure chemical ionization (APCI) and a high-resolution quadrupole time-of-flight mass spectrometer (HRqTOFMS). Anal. Chem. 15206882. 88 (23), 11406–11411. Available from: http://pubs.acs.org/journal/ancham. doi:10.1021/acs.analchem.6b01550.

Mercier, F., Gilles, E., Saramito, G., Glorennec, P., Le Bot, B., 2014. A multi-residue method for the simultaneous analysis in indoor dust of several classes of semi-volatile organic compounds by pressurized liquid extraction and gas chromatography/tandem mass spectrometry. J. Chromatogr. A. 00219673. 1336, 101–111. doi:10.1016/j.chroma.2014.02.004.

Migowska, N., Caban, M., Stepnowski, P., Kumirska, J., 2012. Simultaneous analysis of non-steroidal anti-inflammatory drugs and estrogenic hormones in water and wastewater samples using gas chromatography–mass spectrometry and gas chromatography with electron capture detection. Sci. Total Environ. 00489697. 441, 77–88. doi:10.1016/j.scitotenv.2012.09.043.

Minho, L.A.C., Valenzuela, E.F., Cardeal, Z.D.L., Menezes, H.C., 2022. Novel miniaturized passive sampling devices based on liquid phase microextraction equipped with cellulose-grafted membranes for the environmental monitoring of phthalic acid esters in natural waters. Anal. Chim. Acta 18734324. 1231. Available from: http://www.journals.elsevier.com/analytica-chimica-acta/. doi:10.1016/j.aca.2022.340405.

Mok, S., Lee, S., Choi, Y., Jeon, J., Kim, Y.H., Moon, H.-B., 2023. Target and non-target analyses of neutral per- and polyfluoroalkyl substances from fluorochemical industries using GC-MS/MS and GC-TOF: insights on their environmental fate. Environ. Int. 01604120. 182, 108311. doi:10.1016/j.envint.2023.108311.

Mondello, L., Cordero, C., Janssen, H.G., Synovec, R.E., Zoccali, M., Tranchida, P.Q., 2025. Comprehensive two-dimensional gas chromatography–mass spectrometry. Nat. Rev. Methods Primers. 26628449. 5 (1). Available from: https://www.nature.com/nrmp/. doi:10.1038/s43586-024-00379-3.

Moschet, C., Lew, B.M., Hasenbein, S., Anumol, T., Young, T.M., 2017. LC- and GC-QTOF-MS as complementary tools for a comprehensive micropollutant analysis in aquatic systems. Environ. Sci. Technol. 15205851. 51 (3), 1553–1561. Available from: http://pubs.acs.org/journal/esthag. doi:10.1021/acs.est.6b05352.

Nascimento, M.M., Nascimento, M.L., Pereira Dos Anjos, J., Cunha, R.L., da Rocha, G.O., Ferreira Dos Santos, I., Pereira, P.A.D.P., de Andrade, J.B., 2023. A green method for the determination of illicit drugs in wastewater and surface waters-based on a semi-automated liquid-liquid microextraction device. J. Chromatogr. A 18733778. 1710. Available from: https://www.sciencedirect.com/science/journal/00219673. doi:10.1016/j.chroma.2023.464230.

Naudé, Y., Gorst-Allman, P., Rohwer, E., 2015. A cheap and simple passive sampler using silicone rubber for the analysis of surface water by gas chromatography–time of flight mass spectrometry. Water SA. 0378-4738. 41 (2), 182. doi:10.4314/wsa.v41i2.02.

Noche, G.G., Laespada, M.E.F., Pavón, J.L.P., Cordero, B.M., Lorenzo, S.M., 2011. Microextraction by packed sorbent for the analysis of pharmaceutical residues in environmental water samples by in situ derivatization-programmed temperature vaporizer-gas chromatography-mass spectrometry. J. Chromatogr. A. 18733778. 1218 (52), 9390–9396. doi:10.1016/j.chroma.2011.10.094.

Ochiai, N., Ieda, T., Sasamoto, K., Takazawa, Y., Hashimoto, S., Fushimi, A., Tanabe, K., 2011. Stir bar sorptive extraction and comprehensive two-dimensional gas chromatography coupled to high-resolution time-of-flight mass spectrometry for ultra-trace analysis of organochlorine pesticides in river water. J. Chromatogr. A. 00219673. 1218 (39), 6851–6860. doi:10.1016/j.chroma.2011.08.027.

Pintado-Herrera, M.G., González-Mazo, E., Lara-Martín, P.A., 2013. Environmentally friendly analysis of emerging contaminants by pressurized hot water extraction-stir bar sorptive extraction-derivatization and gas chromatography-mass spectrometry. Anal. Bioanal. Chem. 16182650. 405 (1), 401–411. doi:10.1007/s00216-012-6453-1.

Pintado-Herrera, M.G., González-Mazo, E., Lara-Martín, P.A., 2014. Atmospheric pressure gas chromatography-time-of-flight-mass spectrometry (APGC-ToF-MS) for the determination of regulated and emerging contaminants in aqueous samples after stir bar sorptive extraction (SBSE). Anal. Chim. Acta. 18734324. 851 (C), 1–13. Available from: http://www.journals.elsevier.com/analytica-chimica-acta/. doi:10.1016/j.aca.2014.05.030.

Portolés, T., Mol, J.G.J., Sancho, J.V., López, F.J., Hernández, F., 2014. Validation of a qualitative screening method for pesticides in fruits and vegetables by gas chromatography quadrupole-time of flight mass spectrometry with atmospheric pressure chemical ionization. Anal. Chim. Acta. 00032670. 838, 76–85. doi:10.1016/j.aca.2014.06.006.

Price, E.J., Palát, J., Coufaliková, K., Kukučka, P., Codling, G., Vitale, C.M., et al., 2020. RECETOX Exposome HR-[EI+]-MS Library. Zenodo. doi:10.5281/ZENODO.4471217.

Romagnoli, M., Scarparo, A., Catani, M., Giannì, B., Pasti, L., Cavazzini, A., Franchina, F.A., 2023. Development and validation of a GC × GC-ToFMS method for the quantification of pesticides in environmental waters. Anal. Bioanal. Chem. 1618-2642. 415 (18), 4545–4555. doi:10.1007/s00216-023-04686-8.

Rösch, A., Beck, B., Hollender, J., Singer, H., 2019. Picogram per liter quantification of pyrethroid and organophosphate insecticides in surface waters: a result of large enrichment with liquid–liquid extraction and gas chromatography coupled to mass spectrometry using atmospheric pressure chemical ionization. Anal. Bioanal. Chem. 1618-2642. 411 (14), 3151–3164. doi:10.1007/s00216-019-01787-1.

Samaras, V.G., Thomaidis, N.S., Stasinakis, A.S., Lekkas, T.D., 2011. An analytical method for the simultaneous trace determination of acidic pharmaceuticals and phenolic endocrine disrupting chemicals in wastewater and sewage sludge by gas chromatography-mass spectrometry. Anal. Bioanal. Chem. 16182650. 399 (7), 2549–2561. doi:10.1007/s00216-010-4607-6.

Sánchez-Piñero, J., Moreda-Piñeiro, J., Moscoso-Pérez, C., FernándezGonzález, V., Prada-Rodríguez, D., López-Mahía, P., 2021. Development and validation of a multi-pollutant method for the analysis of polycyclic aromatic hydrocarbons, synthetic musk compounds and plasticizers in atmospheric particulate matter (PM2.5). Talanta Open. 26668319. 4, 100057. doi:10.1016/j.talo.2021.100057.

Schreckenbach, S.A., Simmons, D., Ladak, A., Mullin, L., Muir, D.C.G., Simpson, M.J., Jobst, K.J., 2021. Data-independent identification of suspected organic pollutants using gas chromatography-atmospheric pressure chemical ionization-mass spectrometry. Anal. Chem. 15206882. 93 (3), 1498–1506. Available from: http://pubs.acs.org/journal/ancham. doi:10.1021/acs.analchem.0c03733.

Semard, G., Adahchour, M., Focant, J.F., 2009. Chapter 2 basic instrumentation for GC × GC. Compr. Anal. Chem. 0166526X. 55, 15–48. doi:10.1016/S0166-526X(09)05502-0.

Spikmans, V., 2019. The evolution of environmental forensics: from laboratory to field analysis. WIREs Forensic Sci. 2573-9468. 1 (3). doi:10.1002/wfs2.1334.

Steeves, K.L., Bissram, M.J., Kleywegt, S., Stevens, D., Dorman, F.L., Simpson, A.J., Simpson, M.J., Cahill, L.S., Jobst, K.J., 2023. Nontargeted screening reveals fluorotelomer ethoxylates in indoor dust and industrial wastewater. Environ. Int. 18736750. 171. Available from: www.elsevier.com/locate/envint. doi:10.1016/j.envint.2022.107634.

Sühring, R., Barber, J.L., Wolschke, H., Kötke, D., Ebinghaus, R., 2015. Fingerprint analysis of brominated flame retardants and Dechloranes in North Sea sediments. Environ. Res. 10960953. 140, 569–578. Available from: http://www.elsevier.com/inca/publications/store/6/2/2/8/2/1/index.htt. doi:10.1016/j.envres.2015.05.018.

Takeuchi, S., Kojima, H., Saito, I., Jin, K., Kobayashi, S., Tanaka-Kagawa, T., Jinno, H., 2014. Detection of 34 plasticizers and 25 flame retardants in indoor air from houses in Sapporo, Japan. Sci. Total Environ. 00489697. 491-492, 28–33. doi:10.1016/j.scitotenv.2014.04.011.

Tobiszewski, M., Namieśnik, J., 2011. Determination of chlorinated solvents in industrial water and wastewater by DAI–GC–ECD. Anal. Bioanal. Chem. 1618-2642. 399 (10), 3565–3572. doi:10.1007/s00216-011-4676-1.

Tranchida, P.Q., 2018. Comprehensive two-dimensional gas chromatography: a perspective on processes of modulation. J. Chromatogr. A 18733778. 1536, 2–5. Available from: www.elsevier.com/locate/chroma. doi:10.1016/j.chroma.2017.04.039.

Tranchida, P.Q., Aloisi, I., Mondello, L., 2022. Heart-cutting and comprehensive multidimensional gas chromatography: basic principles. Compr. Anal. Chem. 96, 69–92. Available from: http://www.elsevier.com/wps/find/bookseriesdescription.cws_home/BS_CAC/description#description. doi:10.1016/bs.coac.2021.10.003.

Valls-Cantenys, C., Villaverde-de-Sáa, E., Rodil, R., Quintana, J.B., Iglesias, M., Salvadó, V., Cela, R., 2013. Application of polydimethylsiloxane rod extraction to the determination of sixteen halogenated flame retardants in water samples. Anal. Chim. Acta. 18734324. 770, 85–93. doi:10.1016/j.aca.2013.01.059.

Van den Eede, N., Dirtu, A.C., Neels, H., Covaci, A., 2011. Analytical developments and preliminary assessment of human exposure to organophosphate flame retardants from indoor dust. Environ. Int. 01604120. 37 (2), 454–461. doi:10.1016/j.envint.2010.11.010.

Vial, J., Pezous, B., Thiébaut, D., Sassiat, P., Teillet, B., Cahours, X., Rivals, I., 2011. The discriminant pixel approach: a new tool for the rational interpretation of GCxGC-MS chromatograms. Talanta. 00399140. 83 (4), 1295–1301. doi:10.1016/j.talanta.2010.07.059.

Wang, D., Weston, D.P., Ding, Y., Lydy, M.J., 2010. Development of a sample preparation method for the analysis of current-use pesticides in sediment using gas chromatography. Arch. Environ. Contam. Toxicol. 00904341. 58 (2), 255–267. doi:10.1007/s00244-009-9398-4.

Wang, F., Liigand, J., Tian, S., Arndt, D., Greiner, R., Wishart, D.S., 2021a. CFM-ID 4.0: more accurate ESI-MS/MS spectral prediction and compound identification. Anal. Chem. 15206882. 93 (34), 11692–11700. Available from: http://pubs.acs.org/journal/ancham. doi:10.1021/acs.analchem.1c01465.

Wang, K., Dong, F., Pan, X., Wu, X., Xu, J., Zheng, Y., 2024. Quantitative screening of organophosphorus and organochlorine pesticides in water and soil using comprehensive two-dimensional gas chromatography coupled with time-of-flight mass spectrometry. J. Chromatogr. Open. 27723917. 5, 100140. doi:10.1016/j.jcoa.2024.100140.

Wang, L., Cheng, Y., Naidu, R., Chadalavada, S., Bekele, D., Gell, P., Donaghey, M., Bowman, M., 2021b. Application of portable gas chromatography–mass spectrometer for rapid field based determination of TCE in soil vapour and groundwater. Environ. Technol. Innov. 23521864. 21, 101274. doi:10.1016/j.eti.2020.101274.

Wang, Q., Yin, J., Pan, H., Xu, F., Yang, Y., 2014. A novel method based on combining ultrasonic-assisted dispersive solid-phase extraction and low-density solvent dispersive liquid–liquid microextraction (UA-DSPE-LDS-DLLME) for the determination of organophosphorus pesticides in soil. Anal. Methods. 1759-9660. 6 (18), 7335–7342. doi:10.1039/c4ay00521j.

Weinberg, I., Dreyer, A., Ebinghaus, R., 2011. Waste water treatment plants as sources of polyfluorinated compounds, polybrominated diphenyl ethers and musk fragrances to ambient air. Environ. Pollut. 02697491. 159 (1), 125–132. doi:10.1016/j.envpol.2010.09.023.

Wooding, M., Rohwer, E.R., Naudé, Y., 2017. Determination of endocrine disrupting chemicals and antiretroviral compounds in surface water: a disposable sorptive sampler with comprehensive gas chromatography—time-of-flight mass spectrometry and large volume injection with ultra-high performance liquid chromatography–tandem mass spectrometry. J. Chromatogr. A 18733778. 1496, 122–132. Available from: www.elsevier.com/locate/chroma. doi:10.1016/j.chroma.2017.03.057.

Yu, J.T., Bisceglia, K.J., Bouwer, E.J., Roberts, A.L., Coelhan, M., 2012. Determination of pharmaceuticals and antiseptics in water by solid-phase extraction and gas chromatography/mass spectrometry: analysis via pentafluorobenzylation and stable isotope dilution. Anal. Bioanal. Chem. 16182650. 403 (2), 583–591. doi:10.1007/s00216-012-5846-5.

Yu, Y., Liu, X., He, Z., Wang, L., Luo, M., Peng, Y., Zhou, Q., 2016. Development of a multi-residue method for 58 pesticides in soil using QuEChERS and gas chromatography-tandem mass spectrometry. Anal. Methods. 1759-9660. 8 (11), 2463–2470. doi:10.1039/c6ay00337k.

Zhu, L., Bossi, R., Carvalho, P.N., Rigét, F.F., Christensen, J.H., Weihe, P., Bonefeld-Jørgensen, E.C., Vorkamp, K., 2024. Suspect and non-target screening of chemicals of emerging Arctic concern in biota, air and human serum. Environ. Pollut. 18736424. 360. Available from: https://www.sciencedirect.com/science/journal/02697491. doi:10.1016/j.envpol.2024.124605.

Chapter 8

Liquid chromatography-based techniques for environmental analysis

Gabriela Castro[a], Benigno José Sieira[b], Leticia Pérez-Mayán[b], and Rafael Cela[b]

[a]*Department of Analytical Chemistry, Nutrition and Food Chemistry, Aquatic One Health Research Center (ARCUS), Universidade de Santiago de Compostela, Santiago de Compostela, Spain*

[b]*Mestrelab Research Center (CIM), Santiago de Compostela, Spain*

1 Introduction

The growing concern about the presence of contaminants of emerging concern (CECs) in the environment and their potential risks to human health and wildlife, along with the implementation of new regulations restricting these substances, has led to the development of state-of-the-art analytical techniques designed to meet the scientific community's needs for their detection and quantification. Liquid chromatography (LC)-based techniques have emerged as the reference methodology for the determination of CECs with a wide range of polarities, due to their high sensitivity, robustness, and versatility (Aceña et al., 2015; Beccaria and Cabooter, 2020). In the last decades, significant improvements have been achieved, especially in the field of hyphenated LC-mass spectrometry (MS), allowing the identification and quantification of a broad spectrum of CECs in complex samples at increasingly low concentrations, in the range from the microgram to the picogram per liter.

LC improvements, such as ultrahigh-performance liquid chromatography (UHPLC), have greatly enhanced separation efficiency, allowing the resolution of structurally similar compounds in shorter analysis times. Innovations in stationary phases, including core-shell particles and monolithic columns (Khoo and Leow, 2021; Vasconcelos Soares Maciel et al., 2020), have further improved chromatographic performance, making it possible to analyze low-concentration contaminants in complex matrices such as wastewater or soil extracts. In addition, supercritical fluid chromatography (SFC) has gained prominence for its

Contaminants of Emerging Concern in the Environment. https://doi.org/10.1016/B978-0-443-33821-2.10008-8

ability to separate lipophilic and chiral compounds, expanding the scope of LC in environmental analysis (Gazárková et al., 2022; Castro et al., 2024).

Coupling LC with advanced MS instruments, such as tandem mass spectrometry (LC-MS/MS) and high-resolution mass spectrometry (LC-HRMS), has significantly elevated the analytical capabilities for CEC determination (Steiner et al., 2021). HRMS, including Orbitrap and time-of-flight (TOF) mass analyzers, provide accurate mass measurements and the ability to perform suspect and non-target screening, enabling the detection of known, suspected, and unknown contaminants simultaneously (Gago-Ferrero et al., 2020; Hollender et al., 2023; Hollender et al., 2017), and thus providing the complete fingerprint of a sample.

Sample preparation techniques have also evolved alongside LC advancements to handle the complexity of environmental matrices (Kanu, 2021; Martín-Pozo et al., 2019). Preconcentration and extraction methodologies, such as matrix solid-phase dispersion (MSPD), liquid-liquid extraction (LLE), and solid-phase extraction (SPE); their miniaturized versions (i.e., solid-phase microextraction (SPME)) and lately, the hyphenated SPE coupled to LC-MS/MS, have been optimized to align with the sensitivity and selectivity of LC-MS workflows (de la Serna Calleja et al., 2023). Furthermore, matrix-matched calibration and isotope dilution approaches are increasingly used to account for matrix effects, ensuring robust quantification even in challenging environmental samples (Raposo and Barceló, 2021).

Several studies on the fate, occurrence, and behavior of CECs in environmental matrices using LC-MS techniques have been published in the last decades. This chapter dives into the recent innovations in LC-based methods for the analysis of different families of CECs, including pesticides, pharmaceutical and personal care products (PPCPs), per- and polyfluorinated substances (PFAS), and industrial chemicals in environmental samples, emphasizing the interplay between chromatographic advancements and detection technologies. The discussion highlights how these developments have transformed the analysis of diverse environmental matrices, including aqueous and solid complex samples. By exploring the integration of cutting-edge LC techniques into environmental workflows, this chapter sets the stage for understanding how modern analytical tools address the challenges posed by CECs and provide reliable data for environmental monitoring and risk assessment.

2 Fundamentals of LC-MS

LC is the most widely used analytical separation technique due to its versatility, quantitative capabilities, automation potential and, especially, because of its broad field of application. The principle of LC relies on the interaction between the analytes carried in a liquid mobile phase and a stationary phase, making it especially suitable for separating polar, thermally unstable, or low-volatility compounds. Following chromatographic separation, a detector such as a UV or mass spectrometry (MS) detector is required to register the signal. When LC

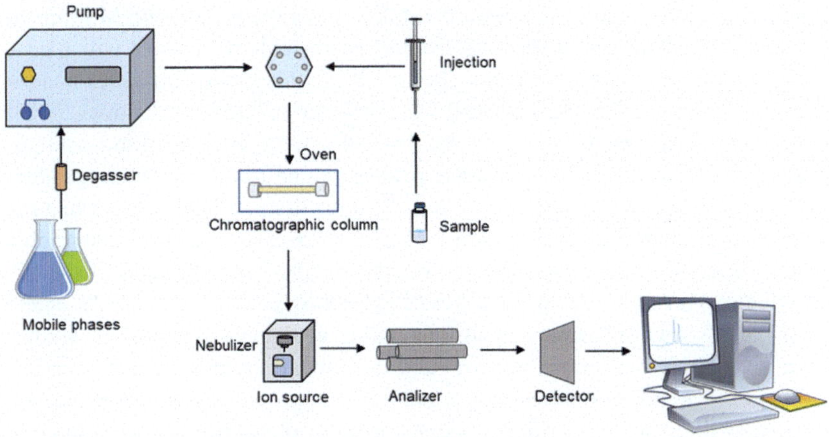

FIG. 8.1 Instrumentation of an LC-MS system.

is coupled to an MS detector, the resulting LC-MS system offers exceptional selectivity, sensitivity, and mass spectral information (Gu, 2019). Fig. 8.1 illustrates the main components of a LC-MS instrument.

In the context of CECs, the huge variety of these substances in the market makes the selection of a single analytical methodology that provides optimal results for all compounds utopian. As a general rule, nonpolar and volatile substances thermically stable are best processed by gas chromatography (GC), while polar and nonvolatile substances are usually analyzed by LC. Whatever the chromatographic approach selected, MS is the preferred detection technique coupled to both GC or LC, because MS (a) allows monitoring many classes of chemicals simultaneously with high sensitivity, (b) may operate in real-life samples without the need of exhaustive clean-up, and (c) can compensate matrix effects satisfactorily.

Given the complexity of most environmental samples in CEC determination, and the large number of samples processed in environmental monitoring campaigns, modern U(H)PLC systems are preferred over conventional HPLC due to their superior resolution and faster elution capabilities. These systems, operating with shorter columns (i.e., 20–30 mm) and smaller particle size (typically from 1.7 to 2 μm) provide very narrow peaks, enhancing sensitivity in detection and resolution, and reducing the total run time, resulting in efficient component separation in a shorter period. In this context, selecting an appropriate stationary phase is crucial to ensure optimal separation of the target analytes. The most common chromatographic modes are normal phase (NP) and reversed-phase chromatography (RP), applied for the determination of polar and nonpolar substances, respectively. While NP chromatography can be useful for separating structural isomers and highly polar compounds such as organophosphorus pesticides, its use in environmental analysis has been reduced due to several limitations. These include low reproducibility, incompatibility with MS ionization sources, and the use of solvents that are not miscible with water, which pose environmental and safety concerns. As for RPLC, this mode dominates quantitative analysis, and it is used in most LC applications (Dong and Boyes, 2018) as it offers great versatility, good compatibility with MS ionization sources, and the ability to use buffered mobile phases, which improves both reproducibility and sensitivity by controlling analyte ionization and pH. Even when the analysis is limited to a specific family of CECs, variations in chromatographic column selection can significantly affect the results. One example of the latter situation can be found in the paper of Huidobro-López et al. (2022), where a total of 40 target pharmaceuticals (including analgesics, antibiotics, antihypertensives, lipid regulators, anticonvulsants, antidepressants, antiarrhythmics, and beta-blockers), hormones, and some lifestyle products were screened in a vegetation filter nonconventional water treatment plant. In this study, two RP stationary phases were evaluated, both with positive and negative polarities in the MS/MS mass analyzer (a 3 × 50 mm Kinetex Biphenyl column, 2.7 μm particle size from Phenomenex, and a 3 × 50 mm Poroshell 120 EC-C18 column with 2.7 μm, from Agilent). Several mobile phase compositions, flow rates, and gradient programs were tested, while temperature and injection volumes were kept identical in all runs. Biphenyl column was able to retain and elute all target analytes detected in positive polarity, except for metformin. This substance presents high water solubility and polarity. While very polar compounds such as metformin show poor retention in RP, in normal phase they may present more adequate retention. However, due to the incompatibility of organic NP solvents with the electrospray ionization (ESI) source commonly used in MS, it is preferred to adapt the conditions in RP, e.g., by using C18 columns with embedded polar groups that improve the retention of hydrophilic analytes. For persistent and mobile organic compounds (PMOCs), such as metformin, some studies (Angeles and Aga, 2020) recommend employing HILIC or aqueous normal phase chromatographic (ANPC) columns (Pesek et al., 2013). However, such columns are not adequate for many other CECs (Egede Frøkjær et al., 2023) and thus, samples need to be processed with different columns in separate runs using also different sample preparation procedures. In general, RP columns allow processing a large number of CECs and thus are the preferred option in most screening studies. Moreover, changes in the mobile phase additives may increase the retention of high-polarity CECs but frequently reduce the sensitivity, so their use requires careful evaluation. In the case of CECs evaluated in negative polarity (Huidobro-López et al., 2022), the Poroshell column gave better results than the biphenyl one. In general, for basic compounds, C18 columns having more hydrophobic interaction power should be recommended.

In recent years, new stationary phases such as those used in mixed-mode chromatography and innovative chromatographic techniques, including supercritical fluid chromatography, have been developed to address the analytical challenges posed by substances for which RPLC is insufficient. These approaches are outlined below.

3 Updated methodologies

3.1 Mixed-mode chromatography

As previously discussed, LC-MS has become one of the most used determination techniques for CEC determination. Its versatility in terms of mobile and stationary phases brings the possibility of combining different analytical conditions, depending on the nature of the analytes. Water is a commonly selected matrix for the determination and monitoring of contaminants. Given its polar nature, nonpolar compounds can be easily extracted and determined using RPLC. However, when it comes to polar substances that fall into the category of PMOCs (Montes et al., 2019), their detection and quantification become an analytical challenge. Traditional RPLC columns are not suitable for highly polar chemicals, so other alternative approaches, such as the hydrophilic interaction liquid chromatography (HILIC) or mixed-mode LC columns, are needed (Montes et al., 2019).

Mixed-mode LC columns (MMLC) combine two different mechanisms of interaction, typically RP and ion-exchange, within a single stationary phase. This combination enables the separation of a wide range of chemicals by leveraging both hydrophobic interactions characteristic of RPLC and electrostatic interactions (charge-based) characteristic of ion exchange. Depending on the analytical application, several MMLC columns can be used. As for the electrostatic interaction, anion (AEX) or cation exchange (CEX), or even a mixture of both (AEX/CEX), the second interaction is commonly based on RP, where HILIC phases are also found in the literature (Bell, 2021). Therefore, by merging both interactions, a comprehensive catalog of MMLC columns may be used (Liu et al., 2014). Additional MMLC columns are described in the literature, where the ion exchange may not be present, as in the case of the combination between HILIC and RP, or even affinity chromatography (AC) with RP (Bell, 2021).

Among MMLC, different types of stationary phases have been described and sorted into four main groups (Fig. 8.2), where the ion exchange remains as the common part (Lesellier et al., 2017; Zhang and Liu, 2016):

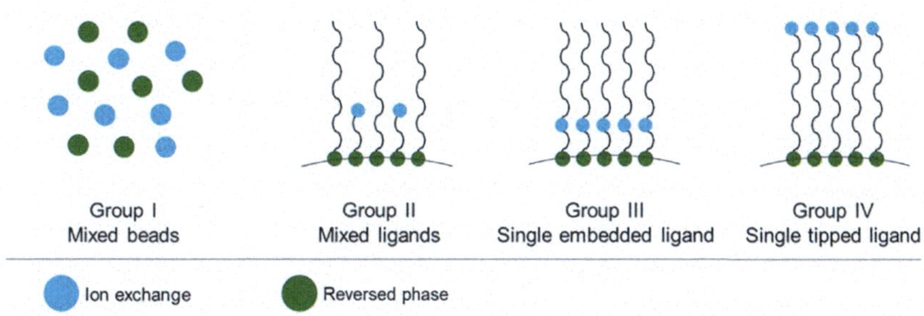

FIG. 8.2 MMLC column types. Mixed-mode LC phases categorized according to their functional groups.

During its use, MMLC setup requires a demanding configuration. As it is based on ion exchange, key parameters, such as pH, are critical during method development, along with other common chromatographic parameters like column temperature, flow rate, or mobile phase composition (Fukushima et al., 2024).

In addition to chromatographic separations, mixed-mode technology is also widely applied in sample preparation techniques such as SPE. Many SPE cartridges are designed with mixed-mode stationary phases, allowing them to exploit both hydrophobic and ionic interactions. This dual-mode capability enables the effective retention and extraction of a broad range of compounds, including nonpolar, polar, and charged species. As a result, mixed-mode approaches are particularly advantageous for processing complex or highly polar sample matrices, enhancing both selectivity and separation efficiency in chromatographic workflows.

Although the application of MMLC has recently gained attention in the literature, this technique was previously proposed with very promising results for the determination of polar compounds as an alternative to traditional chromatographic methods (Zhang and Liu, 2016). Montes et al. proposed a methodology for the determination of 37 polar compounds in water samples, selected by considering their acid-base functionalities. In this study, an MMLC based on a triple phase was used, consisting of AEX, CEX, and RP stationary phases. Mobile phases pH and composition were important parameters in this work, increasing the modifier concentration from 5 to 20 mM along the LC gradient and maintaining a pH of 5.5. Using ion exchange SPE cartridges and MMLC, the extraction and identification of 22 chemicals was described, reporting concentrations ranging from 0.04 to 6.3 ng L^{-1}, and including compounds that have been described for the first time in the literature, as is the case of di-o-tolylguanidine (Montes et al., 2017). Another example of MMLC is described by the same authors, who assessed the presence of 23 PMOCs in water samples, using MM for both sample preparation and analysis. SPE was selected as the main sample pretreatment, using two different MM cartridges: Oasis WCX (based on cation exchange) and Oasis WAX (anion exchange), both combined with HLB stationary phase. Then, samples were analyzed in an LC-MS system equipped with an MMLC column based on RP, CEX, and AEX. Under those conditions, 17 out of 23 compounds were quantified in surface and drinking water samples, reporting concentrations above 120 μg L^{-1} (Montes et al., 2019).

3.2 Online SPE LC-MS

SPE, since first described in the 1950s (Braus et al., 1951), has become the preferred methodology employed for the extraction and concentration of target compounds from environmental liquid samples. The automation of the extraction has enormously evolved, leading to online SPE protocols. Online SPE increases sample throughput and decreases total analysis time and solvent

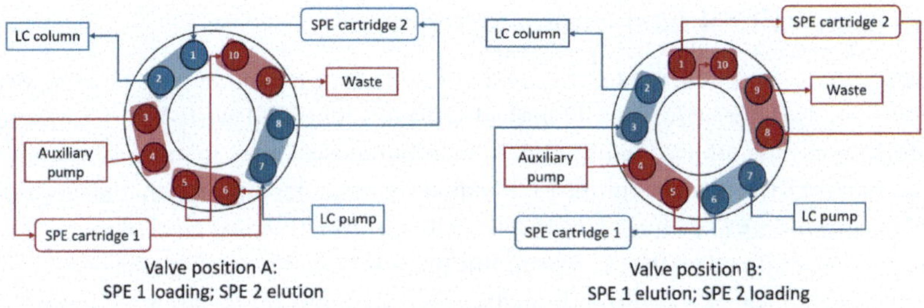

FIG. 8.3 Online SPE valve positions from Agilent instrument. Diagram of the valve port configuration connecting the SPE cartridge, LC chromatographic column, and both the LC and auxiliary pumps.

consumption. In addition, this methodology is adequate when a very limited amount of sample is available. Nevertheless, conditions are less flexible regarding the extraction sorbents and their suitability for the subsequent analysis. Thus, this approach requires an exhaustive method development and trustworthy software to control the process.

The combination of SPE and injection in LC systems can be performed following different configurations:

- An SPE extractor composed of disposable SPE cartridges, connected with the LC system by switching valves (i.e., Symbiosis from Spark Holland (Emmen, The Netherlands)) (Mitchell et al., 2010).
- A dual-column switching setup, which serially connects the extraction or purification column with the separation one by using one or two multichannel switching valves. This configuration is versatile; thus, it can be implemented in every LC system. For instance, the 1290 Infinity III Online SPE System from Agilent (Santa Clara, CA, United States) includes a module with two 10-port (2-position) valves. This module facilitates the use of two reusable SPE cartridges in parallel (Fig. 8.3).

The analysis of complex environmental matrices could introduce undesirable particles into the system and complicate the accurate determination of analytes if a previous filter step is not carried out. Thus, complex matrices should be filtered before the introduction into the system to ensure the correct detection of compounds. Available sorbents vary from the traditional silica-bonded phase, polymeric materials, ion-exchange, to mixed-mode sorbents. The wettability and selectivity of those sorbents were considerably increased by adding different functional groups to the sorbent. For instance, the molecular imprinted polymers (MIPs) were raised as a selective choice for polar compounds.

Recently, several studies have focused on the multianalyte determination of different families of CECs in environmental samples, such as drinking, surface (Mann et al., 2016), and groundwater (Barbieri et al., 2020; Fernández-Fernández et al., 2023; Zhong et al., 2019), and aqueous soil extracts (Enevoldsen and Juhler, 2010) using fully automated online SPE-LC-MS/MS.

3.3 Supercritical fluid chromatography

Supercritical fluid chromatography (SFC) is considered a green analytical technique, an alternative to GC and LC, usually coupled with MS. CO_2 is the preferred mobile phase, combined with other cosolvents to improve analyte separation. Different configurations can be found depending on the application requirements (Si-Hung and Bamba, 2022). In any case, an automatic back pressure regulator (ABPR) is needed to ensure that the mobile phase remains as a single dense phase throughout the chromatographic system (column and connection with the detection technique). When combined with MS through an ionization source, a makeup auxiliary solution is required to improve the efficiency of compound ionization. Methanol is commonly used for this purpose, combined with acid or basic modifiers depending on the chemical characteristics of the analytes (Castro et al., 2024).

The main advantages of this technique are high flow rates, which comprise faster separations due to the higher diffusivity and lower viscosity of supercritical CO_2, and reduced use of organic solvents. SFC also improves separation efficiency and shows shorter equilibration times than those required for chiral or achiral separations carried out with classical LC eluents. Whatever the design of the instrumentation, the hyphenation with MS represents the majority of applications, either for chiral or achiral chromatographic separations (West, 2018). The main chiral stationary phases (CSPs) are polysaccharide derivatives of cellulose and amylose (esters or carbamates coated or immobilized to silica), brush-type (Pirkle-type) CSPs, native and derivatized cyclodextrins and cyclofructans, quinine-based and macrocyclic antibiotics-based CSPs. Regarding achiral analysis, the 2-ethylpyridine bonded silica phase (2-EP) is the most cited, followed by octadecyl-bonded and diol-bonded silica phases. A limited number of studies covering the multiresidue/multiclass determination using SFC are available. The applications are extended to the pharmaceutical (Cobo-Golpe et al., 2022; Wang et al., 2017a), environmental (Pérez-Mayán et al., 2021; Svan et al., 2015), and food commodity (Cutillas et al., 2019; Di et al., 2023) fields.

4 Application to the determination of CECs

4.1 Pesticides

Pesticides are extensively used in agriculture, industry, and urban applications. Their persistence, degradation, and mobility mainly depend on their physicochemical properties and the ecosystem where they are applied. Consequently, research has focused on the analysis of pesticides and their transformation products (TPs) in environmental and food commodities to monitor contamination and assess ecological risks. Notably, some studies have proven the harmful effects of azole compounds in aquatic organisms (Bhagat et al., 2021).

Analytical methodologies for the determination of pesticides have evolved to cover a wide range of compounds in complex matrices at trace concentrations. This is critical to comply with the current regulations on drinking water quality (Directive [EU] 2020/2184) and the Watch List for water (Decision [EU] 2022/1307). Although regulations for solid environmental samples are still lacking, those substances included in the Watch list are currently applied to the land compartment.

As for sample preparation, the complexity of the environmental matrix often requires a pretreatment step to ensure effective extraction and accurate detection. For liquid samples, SPE is the standard methodology. In the last decade, online SPE has enabled highly automated workflows, reducing solvent use and sample manipulation by the operator, while improving precision (Section 3.2). Alternative techniques include dispersive liquid-liquid microextraction (DLLME) and SPME. As for solid samples, pressurized liquid extraction (PLE) or QuEChERS (quick, easy, cheap, effective, rugged, and safe) are common, often preceded by drying or lyophilization, and sieving. Among them, PLE is the preferred method as it enhances automation, efficiency, and reduces solvent consumption.

Chromatographic separation and detection are typically achieved using LC-MS or LC-MS/MS. Most applications employ RP columns, covering the determination of low to medium polarity pesticides. In the case of highly polar compounds, as described in previous sections, MMLC and HILIC columns are the preferred ones. SFC is a green alternative to chiral and achiral separations, obtaining faster separations and reducing organic solvent use. Detection is usually performed with an ESI source and instruments such as QqQ, QTOF, or Q-Orbitrap, depending on the goal (targeted vs. non-targeted analysis).

Several authors have addressed the presence of pesticides in environmental samples. López-Vázquez et al. proposed an automated methodology for the analysis of glyphosate (GLY) and five additional anionic pesticides in water (López-Vázquez et al., 2023). GLY is a nonselective herbicide extensively used in agricultural and urban areas, with aminomethylphosphonic acid (AMPA) as its main TP. This methodology also included the fungicide Fosetyl-Al, the herbicide glufosinate (GLU), and its TPs N-acetyl glufosinate (NAG) and 3-(methylphosphinic) propionic acid (MPPA). Due to their polarity and metal-binding properties, these compounds are difficult to extract. Derivatization with FMOC-Cl is commonly performed; however, this approach involves slow kinetics and tedious sample preparation and purification. Moreover, Fosetyl and MPPA cannot be extracted by this method, since they do not react with FMOC-Cl. To address this, a workflow involving filtration and direct injection in an online SPE instrument tailored by UPLC-MS/MS was selected. Sample filtration was carried out with 0.22 μm PTFE filters and directly injected into the instrument, which was previously cleaned with citric acid to prevent interaction between analytes and metal cations. A PEEK-lined column (Metrosep A Supp 6, strong anionic exchange column

(150 x 2 mm, 5 μm)) minimized metal interaction. Chromatographic separation used acetonitrile and ammonium bicarbonate, with ESI in positive mode, except for AMPA and Fosetyl. The proposed methodology was validated in soft and hard water. As expected, hard water interfered with detection and thus, GLY was undetectable, while AMPA and MPPA signals were reduced by ~90%. Still, the method enabled up to 300 extraction cycles per cartridge and achieved LOQs between 5 and 20 ng L^{-1}. Different water samples, including ultrapure water, surface water (streams and rivers), mineral water (bottled water), and tap water were analyzed with the latter methodology, revealing the widespread presence of AMPA (detection frequency 88%), followed by GLY (37%) and Fosetyl. GLY was detected in concentrations up to 3000 ng L^{-1}. The obtained results were lower than those reported in the USA and South America but higher than those in German residential areas. The prevalence of AMPA underscores its stability compared to its parent compound. High concentrations near sewage treatment plant discharges suggest additional AMPA sources, such as phosphates in detergents. A related study by Fernández-Fernández et al. applied a similar online SPE-LC-MS/MS methodology to monitor 73 CECs, including pesticides and pharmaceuticals (Section 4.2), in surface and groundwater (Fernández-Fernández et al., 2023).

As for solid samples, particularly soils from vineyards, present different analytical challenges. Vineyards are permanent crops receiving constant application of commercial formulations to reduce the effects of different undesirable pests. In this regard, research is focused on the presence of those residues in wine, but not in the soil. Thus, the potential pesticide accumulation in soil poses a risk to microbial diversity, plant health, and resistance development in fungi. Pérez-Mayán et al. presented a comprehensive methodology for the determination of 41 pesticides in vineyard soil to assess their presence, persistence, and time-course trends (Pérez-Mayán et al., 2020). The proposed methodology was based on PLE and UHPLC-QqQ as sample preparation and determination techniques, respectively. Samples were collected from nine vineyards across three Galician wine regions between 2017 and 2019. Sampling was carried out before vine growth cycles to monitor compound dissipation. Results revealed total concentrations ranging from 60 to 2800 ng g^{-1} (Fig. 8.4). Pesticides such as dimethomorph (DIM), carbendazim (CAR), myclobutanil (MYC), pyraclostrobin (PYRA), fluopicolide (FLUO), and imidacloprid (IMI) demonstrated the highest concentrations.

As for dissipation, from October to March, total pesticide concentrations decreased by ~40%. Nevertheless, individual dissipation factors varied from 12% (DIM) to 75% for benalaxyl (BEN), with seven pesticides showing values below 50%. Thus, half of the quantified pesticides showed a potential risk to be accumulated if applied in consecutive campaigns. Cymoxanil was notable for its rapid degradation, detected only at the end of summer. Seasonal inputs from foliage and rainfall may also contribute to pesticide presence during winter.

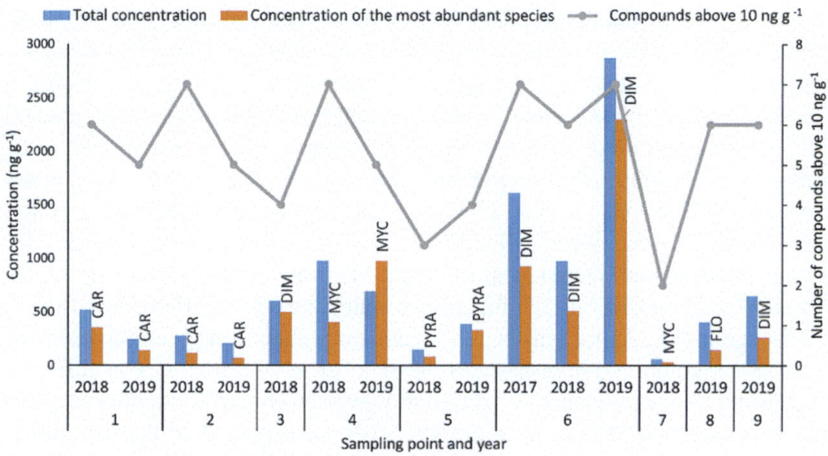

FIG. 8.4 Total pesticide residue concentration (ng g^{-1}) and number of pesticides above 10 ng g^{-1} in vineyard soils before treatment. *(Reproduced with permission from Pérez-Mayán, L., Ramil, M., Cela, R., Rodríguez, I., 2020. Multiresidue procedure to assess the occurrence and dissipation of fungicides and insecticides in vineyard soils from Northwest Spain. Chemosphere. 0045-6535. 261, 127696. doi:10.1016/j.chemosphere.2020.127696.)*

4.2 Pharmaceuticals

The pervasive presence of pharmaceuticals in the environment has emerged as a significant concern, primarily due to their continuous introduction through various anthropogenic activities. These compounds, often referred to as CECs, originate from sources such as domestic sewage, industrial discharges, agricultural runoff, and hospital effluents. Their persistence and potential bioactivity necessitate sensitive and accurate analytical methods for detection and quantification in complex environmental matrices. To that, LC-MS/MS has become the cornerstone technique for such analyses.

In recent years, there have been significant advances in LC-MS methodologies tailored to the environmental analysis of pharmaceuticals. Particularly, the development of UHPLC coupled to MS/MS has improved separation efficiency, sensitivity, and reduced analysis time. Castro et al. presented new methodologies for the quantification of cardiovascular drugs in environmental water samples (Castro et al., 2019) and sewage sludge (Castro et al., 2018a) based on UHPLC-QqQ-MS analysis. In the case of aquatic samples, the authors proposed a methodology based on SPE, using a mixed-mode cartridge for the extraction of 7 sartan compounds commonly used in hypertension treatments. As for sewage sludge, the sample preparation selected was the matrix-solid phase dispersion (MSPD) using C_{18} as dispersant for the extraction of 9 cardiovascular drugs. In both works, the proposed methodologies were validated in terms of extraction efficiency, matrix effects, and accuracy. LOQs in both cases remained below 5 μg L^{-1}.

Antibiotics are another critical class of pharmaceuticals to include in monitoring studies, as their accumulation in environmental compartments can contribute to the development and spread of antimicrobial resistance, posing a serious threat to human, animal, and ecosystem health. Fernandes et al. studied the occurrence of antibiotics and antidepressants in surface waters and sediments in the North of Portugal (Fernandes et al., 2020). To that, the authors proposed SPE using Strata-X (200 mg) cartridges and QuEChERS for the extraction of 27 pharmaceuticals from aquatic and solid samples, respectively. As for determination techniques, LC-QqQ-MS was selected, achieving LOQs between 0.225 and 40.4 ng L^{-1} for surface water and between 0.007 and 18.1 ng g^{-1} for sediments. As for wastewater treatment plants, Rodríguez-Mozaz et al. presented a comprehensive monitoring of a range of antibiotics in different wastewater treatment plants in 7 different European countries. Sample preparation was based on SPE with an OASIS HLB cartridge, while the determination of 53 antibiotics was carried out by UPLC-Qtrap-MS using an Acquity HSS T3 column (50 × 2.1 mm, 1.7 μm) (Rodriguez-Mozaz et al., 2020). iLOQs ranged from 17.7 to 46.9 ng L^{-1}. Regarding extraction methodology, recoveries achieved for all antibiotics stayed between 60% and 120%. Concentrations of antibiotics ranged between 3.2 and 597 ng L^{-1} for the different classes and countries. Azithromycin and ciprofloxacin presented the highest concentrations among all countries. Aydin et al. assessed the occurrence, ecotoxicological risks, and temporal variations of 38 pharmaceuticals in other environmentally relevant and complex sample: sewage sludge (Aydın et al., 2022). Among the pharmaceutical classes targeted, antibiotics, analgesics, anti-inflammatories, and antidepressants were selected. In this case, ultrasound-assisted extraction (UAE) with methanol was chosen as sample preparation technique, and SPE as cleanup step. Due to the complexity of the sample, two different SPE cartridges were needed. OASIS MCX was used for the extraction of acetaminophen and acetylsalicylic acid, while OASIS HLB was applied for the extraction of the other substances. As for determination, LC-QqQ-MS using a Poroshell 120 EC-C18 column (100 × 3 mm, 2.7 μm) (Aydın et al., 2022). Obtained iLOQs ranged from 0.006 to 64.1 ng mL^{-1}. Recoveries were evaluated at two different concentrations and stayed between 68% and 96% (RSD < 8%). According to the results presented by the authors, once again, the antibiotics azithromycin and ciprofloxacin presented the highest median concentrations, together with the analgesic, acetaminophen. Mohapatra et al. recently published a comprehensive review on the occurrence of pharmaceuticals in drinking water, surface water, and groundwater (Mohapatra et al., 2025). The main conclusion from their review is that SPE using an OASIS HLB cartridge, combined with RPLC-ESI-QqQ (either in LC or UPLC format) as determination technique represents the preferred analytical approach for detecting pharmaceuticals in aquatic environments. Among the most frequently detected compounds are antibiotics such as azithromycin and sulfamethoxazole, the analgesic acetaminophen, caffeine, and the anti-inflammatory drug diclofenac. Notably, these substances presented consistently high median concentrations in surface water. Even more surprisingly, their median concentrations in drinking water often exceeded those detected in groundwater, underscoring the urgent need for comprehensive risk assessment (Mohapatra et al., 2025).

Effective sample preparation is crucial for accurate LC-MS analysis. In general, SPE remains the predominant technique for aquatic samples due to its efficiency in concentrating analytes and removing interferences. Innovations in SPE, such as the development of novel sorbent materials and the miniaturization of extraction devices, have enhanced the selectivity and sensitivity of the methods (Cortés-Bautista et al., 2024). A notable development is the combination of online SPE with LC-MS/MS systems. As previously described in Section 3.2, this automation streamlines the sample preparation process, reducing solvent consumption and enhancing throughput. Several authors have developed methodologies for the analysis of pharmaceuticals in aquatic samples. For example, Fernández-Fernández et al. presented a fully automated online SPE-LC-MS/MS methodology capable of detecting 73 organic pollutants, including pharmaceuticals and pesticides, in surface and groundwater samples. Two different SPE cartridges were tested: reversed-phase PLRP-S (12.5 × 2.1 mm) cartridges, containing polystyrene-divinyl benzene (PS-DVD) as sorbent (particle size 15–20 μm), and mixed-mode, reversed-phase and weak-cation exchange, Strata-X-CW (20 × 2 mm) cartridges, containing a polymeric weak cationic exchanger (25 μm). The latter one led to wider peaks for basic compounds; thus, reversed-phase PLRP-S was selected as sorbent. Under optimal conditions, water samples were ready for analysis after filtration through 0.22 μm PTFE hydrophilic filters and acidification with 0.1% formic acid. The analytical procedure achieved limits of quantification between 1 and 10 ng L^{-1}, with a throughput of 1 sample per 30 min, providing accurate recoveries for >90% of the 73 selected compounds, and thus, significantly reducing the sample preparation time and the cost of analysis per sample (Fernández-Fernández et al., 2023). López-García et al. focused on the study of 37 psychoactive substances in wastewater using online SPE tailored to LC-MS/MS. In this case, 5 mL of 10-fold diluted wastewater sample was loaded into a wettable polymeric sorbent PLRPs (cross-linked styrene- divinylbenzene polymer, 10 × 2 mm, 15–25 μm particle size) and analyzed in an LC combined with a hybrid triple quadrupole-linear ion trap (QqLIT) mass spectrometer. mLOQs of the optimized methodology ranged from 0.6 to 228 ng L^{-1} (López-García et al., 2018).

Despite all the advancements, several challenges persist in the LC-MS analysis of pharmaceuticals in environmental samples. Matrix effects, resulting from coeluting substances in complex environmental matrices, can suppress or enhance the ionization of target analytes, significantly affecting the accuracy, sensitivity, and reliability of the obtained results (Williams et al., 2023). To mitigate these effects, the use of isotope-labeled internal standards and matrix-matched calibration curves is recommended (Raposo and Barceló, 2021). Another consideration is the identification and quantification of transformation products resulting from the degradation of parent pharmaceuticals (Montone et al., 2024). These metabolites can exhibit different physicochemical properties and biological activities, necessitating comprehensive analytical strategies. HRMS techniques, such as QTOF and Orbitrap analyzers, have been employed to elucidate the structures of unknown compounds and monitor their occurrence in the environment instead of monitoring the parent compound (Hernández et al., 2014).

4.3 Per- and polyfluoroalkyl substances

Per- and polyfluoroalkyl substances (PFAS) are a group of synthetic organic chemicals extensively used in industrial applications and consumer products due to their unique physicochemical properties (Glüge et al., 2020). Structurally, PFAS are characterized by a carbon chain of varying length in which one or all hydrogen atoms are replaced by fluorine atoms, forming highly stable carbon–fluorine bonds (Buck et al., 2011). These compounds often include functional groups at the end of the fluorocarbon chain, such as carboxylic or sulfonic acids, which confer strong acidity, thermal stability, and surfactant properties (Wang et al., 2017b). Commonly found in items like firefighting foams, textiles, cookware, and cosmetics, PFAS are environmentally persistent, resistant to degradation, bioaccumulative, and highly mobile in ecosystems, and thus they are often referred to as *forever chemicals* (Joudan and Lundgren, 2022). Their widespread presence in the environment and association with adverse health problems, such as cardiovascular and kidney diseases (Jian et al., 2017), have raised concerns. Regulatory actions, including those under the Stockholm Convention, have targeted PFAS as PFOS, PFOA, PFHxS, and long-chain perfluorocarboxylic acids (PFCAs, $C > 9$) (UNEP, 2021), prompting a shift toward the development and use of alternative substances, such as short-chain PFAS ($C < 6$), their sulfonic and carboxylic derivatives (i.e., perfluorobutanesulfonic acid, PFBS and perfluorobutanoic acid, PFBA), and the main precursors of those substances, the fluorotelomer sulfonic acids (FTS) (Zhong et al., 2021).

Over the past few decades, PFAS have been widely detected in drinking water (Coggan et al., 2019; Domingo and Nadal, 2019), representing a primary source of human exposure (Cordner et al., 2019). PFAS have also been detected in other environmental compartments, including air, rain, snow, rivers, lakes, surface water, and sediments, wastewater, and sludge. The monitoring and detection of PFAS, including newly identified compounds, pose increasing analytical challenges due to their physicochemical properties and the extremely low concentrations required to comply with evolving regulatory limits. EU Drinking Water Directive 2020/2184 established a concentration of 0.1 μg/L for the sum of PFAS (DIRECTIVE (EU) 2020/2184, 2020).

In the context of environmental monitoring, LC-MS/MS methodologies have been optimized for the simultaneous quantification of multiple PFAS in diverse matrices. RPLC, commonly using C18 columns, remains the standard for PFAS analysis. Typical mobile phases consist of ultrapure water (aqueous phase, A) and either methanol or acetonitrile (organic phase, B), both containing different percentages of modifier, such as ammonium acetate (5–20 mM), formic acid, or acetic acid (0.1%–0.2%). Chromatographic gradient may be optimized to enhance selectivity, peak shape, and resolution. Since PFAS are commonly used in PTFE tubing and packaging, laboratory contamination can occur both during sample preparation and from analytical instruments. To minimize background contamination from HPLC components, particularly the tubing, it is recommended to install a delay column between the LC pump and the injection valve to discriminate the peak coming from the sample from the contribution of the mobile phase, which is shifted to slightly longer retention times. Teymoorian et al. provided a comprehensive review on current analytical techniques for PFAS detection and quantification in drinking water (Teymoorian et al., 2023), including comparisons with EPA method 533 (EPA US, 2018). Huerta et al. developed and validated a method for the quantification of 15 PFAS in aquatic environmental samples, employing UHPLC-QqQ-MS with a C18 column for the analysis of PFAS (C4–C10) (Huerta et al., 2022). Following the same strategy, other authors have targeted a higher number of compounds. For instance, Coggan et al. presented a single analytical methodology for the determination of 53 legacy and emerging PFAS ($C > 4$) in wastewater, surface, and drinking water, also using a C18 column (2.1×50 mm, 1.8 μm) (Coggan et al., 2019). In these complex samples and under those conditions, the quantification of PFBA is challenging due to this substance is highly polar and does not interact strongly with the nonpolar C18 stationary phase. This results in very short retention times, often coeluting with the solvent front or other matrix components, reducing sensitivity and selectivity.

While RP columns perform effectively for long-chain PFAS, they present limitations in retaining short and ultrashort-chain PFAS (C1–C4) (Table 8.1). To address these limitations, alternative chromatographic approaches have been introduced, including HILIC and ion-exchange columns, which have improved the separation of these compounds. For instance, Sørmo et al. investigated the decomposition and emission of 56 PFAS from organic waste subjected to dry pyrolysis using three analytical approaches. Method A involved the analysis of 41 PFAS by UHPLC-QqQ-MS using a Kinetex C18 column (30×2.1 mm, 1.3 μm), while Methods B and C employed RPLC and HILIC for the quantification of short and ultrashort PFAS (Sørmo et al., 2023).

As for drinking water, several authors have developed different methodologies for the determination of PFAS to meet the requirements of Directive 2020/2184/EU. Dasu et al. investigated the occurrence of 14 PFAS in drinking water samples using a novel methodology based on offline SPE for initial sample clean-up, followed by online preconcentration using a WAX column. PFAS analysis was performed by using column-switching HPLC–MS/MS (Dasu et al., 2017). According to the results, PFOA and PFOS presented the highest median concentrations, 10 and 7.6 ng L^{-1}, respectively. Building on a

TABLE 8.1 Summary of selected LC-MS/MS applications to the determination of PFAS in environmental samples.

#	PFAS	Matrix	Sample preparation	Detection	Stationary phase	LOQ	Reference
1	41 PFAS (C > 4) 11 precursors of PFAS 4 ultrashort PFAS (C < 4)	Organic Waste	UAE	UHPLC-QqQ-MS	Kinetex C18 (2.1 × 30 mm, 1.3 μm) XSelect HSS T3 (2.1 50 mm, 3.5 μm) Acquity BEH Amide (2.1 × 100 mm, 1.7 μm)	0.01–5.00 ng mL^{-1}	(Sørmo et al., 2023)
2	40 PFAS	Soil	UAE	UHPLC-QqQ-MS	Kinetex C18 (2.1 × 30 mm, 1.3 μm)	0.01–5.00 ng mL^{-1}	(Hubert et al., 2023)
3	53 PFAS	Aquatic samples	SPE (WAX 150 mg)	LC-MS/MS	C18 RRHD (4.6 × 50 mm, 3.5 μm)	0.35–26.0 ng L^{-1}	(Coggan et al., 2019)
4	28 PFAS (20 PFAS, five fluorotelomers, and three emerging PFAS)	Industrial wastewater and river water	SPE (WAX 200 mg)	HPLC-QqQ-MS	ZORBAX XDB C18 (4.6 × 150 mm, 3.5 μm)	0.54–6.81 ng L^{-1}	(Kim et al., 2021)
5	15 PFAS	Aquatic samples	SPE (WAX 500 mg and HLB 500 mg)	UHPLC-QqQ-MS	Zorbax Eclipse Plus C18 (3.0 × 50 mm, 1.8 μm)	1.52–58.9 ng L^{-1}	(Huerta et al., 2022)
6	7 ultra-short PFAS	Aquatic samples	Direct injection	UHPLC-QqQ-MS	Raptor Polar X (2.1 × 50 mm, 2.7 μm)	1.30–17.0 ng L^{-1}	(Liang et al., 2023)

7	24 PFAS	Plant tissues	UAE + SPE (WAX 150 mg)	UHPLC-QqQ-MS	Phenomenex Kinetex C18 (2.1 × 100 mm, 1.7 μm)	0.18–8.10 ng g^{-1}	(Nassazzi et al., 2022)
8	31 PFAS	River sediment	UAE + SPE (Envicarb)	UHPLC-QqQ-MS	ACQUITY UPLC BEH C18 (100 x 2.1 mm, 1.7 μm)	0.006–0.024 ng g^{-1}	(Zhong et al., 2021)
9	32 PFAS	Feathers	UAE	UHPLC-QqQ-MS	Kinetex C18 (2.1 × 30 mm, 1.3 μm)	0.20–18.0 ng g^{-1}	(Zhang et al., 2023)
10	14 PFAS	Drinking water	SPE (HLB 225 mg)	LC-QqQ-MS	Inertsustain C18 (2.1 × 50 mm, 3 μm)	0.59–3.40 ng L^{-1}	(Dasu et al., 2017)
11	20 PFAS	Drinking water	SPE (WAX 150 mg)	UHPLC-QqQ-MS	Zorbax Extend -C18 RRHT (50 × 2.1 mm × 1.8 μm)	0.05–0.30 ng L^{-1}	(López-Vázquez et al., 2024)

similar strategy, López-Vázquez et al. developed and compared three different methods for analyzing 20 PFAS in both tap and bottled water samples, based on online and offline SPE, and direct injection (López-Vázquez et al., 2024). Both offline and online SPE protocols employed WAX cartridges for PFAS extraction, while quantification was performed by UHPLC-MS/MS. Among the three approaches, online SPE demonstrated the worst performance due to persistent blank contamination, even after installing a delay column. Between the remaining two, offline SPE was the best approach, as it was the only methodology capable of achieving the mLOQs required by the Directive. In this study, PFBA and PFHxA showed the highest median concentrations, 3.1 and 1.60 ng L^{-1}, respectively. As for ultrashort-chain PFAS, Liang et al. developed a rapid analytical methodology for the quantification of ultrashort PFAS ($C < 4$) in potable and nonpotable waters, including trifluoroacetic acid (TFA), using a hybrid HILIC/ion exchange column (Liang et al., 2023) (Table 8.1). When working with ultrashort-chain PFAS, matrix interferences critically affect the selectivity and the sensitivity, and therefore, potentially lead to false positives. In the present study, signal suppression between 14% and 30% was observed. To mitigate these effects, stable isotopes were employed as internal standards. The proposed methodology proved effective, achieving the separation of the 7 target PFAS in <4 min. The obtained results indicated that TFA was ubiquitous in tap waters in concentrations from 120 to 500 ng L^{-1}.

4.4 Industrial chemicals

In addition to previously discussed CECs, such as pesticides, pharmaceuticals, and PFAS, industrial activities also constitute a significant source of CECs in the environment. Many manufacturing processes involve the use of chemicals which can be subsequently released into the environment, particularly through wastewater discharges, and thus posing potential risks to aquatic life, and ultimately, to human health. Even though water samples are the main matrix where contaminants may be detected, solid samples are also an endpoint for CECs coming from industrial processes (EPA, 2025).

Regarding determination, and following the same trends as the abovementioned CECs, LC-MS has become the main determination technique, due to its versatility and robustness in the identification and monitoring of several compounds. Reemstma et al. presented a methodology for the analysis of benzotriazoles, widely used in industry as anticorrosive agents, but also discharged through household wastewater. Highlighting their moderate-high polarity, benzotriazoles are released into water samples, becoming a challenge for the aquatic ecosystem. In addition, due to their low degradation rate, these substances are prone to reach drinking water after the disinfection steps in treatment plants. LC-MS was used in analysis after an SPE sample pretreatment (groundwater samples), or even by direct injection (surface and wastewater samples), for the determination of Tolyltriazole (TTri) isomers (Reemtsma et al., 2010). In this work, a diphenyl LC column (2 × 150 mm, 3 μm) was used for a suitable separation of TTri chemicals, achieving LOQs ranging from 0.05 to 0.25 μg L^{-1} in all sample types. According to the presented results, 1 H-Benzotriazole demonstrated the highest concentrations in the range of 7–18 μg L^{-1}, highlighting the presence of 4-TTri in all surface waters from rivers with >600 km length (Reemtsma et al., 2010).

More contaminants derived from industrial activities are the well-known 1,3-di-o-tolylguanidine (DTG) and 1,3-diphenylguanidine (DPG). Used as rubber accelerator in vulcanization processes, both DTG and DPG are widely used, registering a production in the range of 100–10,000 tons per year for each compound. However, limited information is available regarding their presence in environmental samples, being DTG detected in water for the first time by Montes et al. (2017). As previously mentioned in Section 3.1, this methodology was focused on the analysis of PMOCs and used MM-SPE and MMLC for the analysis of these substances. A study conducted by Sieira et al. investigates both DTG and DPG to identify potential TPs formed during disinfection treatments at WWTPs, using LC-MS/MS). The authors identified >30 TPs (Sieira et al., 2020) with a relative formation yield of ≈30%. (Eco)toxicity assessment of all TPs was evaluated in silico to assess their toxicity in comparison to the parent compounds. To this end, quantitative structure–activity relationship (QSAR) models and bioluminescence tests with *Vibrio Fisheri* were employed. The study confirmed an increased toxicity for most of the Cl-TPs compared to DTG and DPG, particularly evident in the aquatic-related tests (*Daphnia Magna LC50*). Another well-known family of CECs commonly used in the plastic industry is the phthalates, which are also described as an example of an environmental threat. During manufacturing processes, phthalates are discharged into water, leading to potential issues for the contamination of water and soil, and even reaching drinking water. A study from Bach et al. described the potential migration of CECs from polyethylene (PET) plastic bottles to drinking water stored in them using LC-MS methodologies (Bach et al., 2012). This study highlights the presence of Butylated Hydroxytoluene, known as BHT, a phenolic antioxidant used in the plastic industry as an additive in plastic packaging, or even used as a food additive. This chemical was detected in the bottled water at concentrations ranging from 22 to 38 μg L^{-1}, which is higher than those concentrations reported in water contained in glass bottles. In the same study, Bach also investigated the concentration of another relevant CEC, the plasticizer Bisphenol-A (BPA). In this case, the authors reported BPA concentrations ranged from 18 to 324 ng L^{-1} in the analyzed drinking water (Bach et al., 2012).

5 Current and future trends

The evolution of LC-MS techniques continues to redefine environmental analysis, especially in the detection and quantification of CECs. Recent advances have been focused on improving analytical throughput, sensitivity, and data processing capabilities. As environmental challenges evolve, so does the application of LC-MS, with several future trends defining its trajectory. These advancements include (i) the expansion of suspect and non-target screening approaches, (ii) development of multiclass and multiresidue methodologies, (iii) emphasis on harmonization and standardization, (iv) miniaturization and automation of LC-MS systems, (v) integration of machine learning and artificial intelligence in data analysis and interpretation, and (vi) implementation of environmental risk assessment. All these advances are crucial for addressing the complex challenges posed by the presence of CECs in the environment.

5.1 Expansion of suspect and non-target screening approaches

Traditionally, analytical methods used in monitoring studies have relied on targeted approaches that focus solely on the detection of known substances, typically based on a predefined list of compounds. While these methods offer high sensitivity, they are completely blind to any substances not included on that list. However, given the dynamic and evolving nature of environmental contaminants, a more comprehensive and holistic approach is essential.

To address this need, suspect and non-targeted screening analysis are becoming increasingly important (Chapter 9), as they enable the detection of unknown compounds. HRMS, either based on the use of TOF or Orbitrap mass analyzers, supports these advanced methodologies by recording full-spectrum data, allowing for retrospective analysis as new contaminants emerge. The most common operational modes of HRMS systems for suspect and non-target screening studies are data-dependent (DDA) and data-independent (DIA) acquisition modes (Castro et al., 2021; Martínez-Bueno et al., 2019). DDA performs a first scan and selects those species with the highest intensity, isolates the precursor ions in the first MS analyzer, and then performs their fragmentation to obtain a product ion spectrum (MS/MS spectra). Thus, DDA combines accurate MS scan spectra with product ion spectra for a preselected list of precursor ions. As for DIA, this acquisition mode does not provide real MS/MS spectra. Instead, all the ions formed during ionization and entering the mass analyzer are fragmented in the collision cell by applying different collision energies. Thus, pseudo-MS/MS spectra are recorded for precursor and fragment ions.

Regardless of the method used, the reliability of compound identification, especially in complex matrices, depends on the spectral quality. High mass accuracy and clean isotopic profiles for both precursor and fragment ions are essential for correctly matching compounds to those compiled in the database. Fig. 8.5 reproduced from Wang et al. (2023) shows the workflow in the screening of CECs using DIA and DDA.

In steps I and II, raw data files are processed to extract non-target features and align chromatograms along the samples. In step III, features with mass intensity higher than a defined threshold, with significant differences against

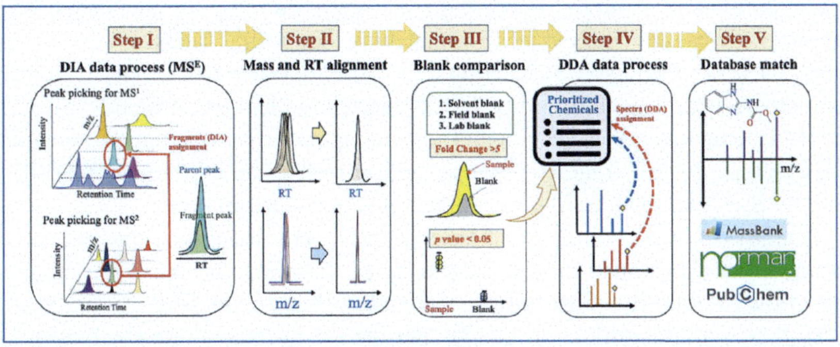

FIG. 8.5 Workflow of non-target screening in environmental samples based on DDA and DIA data. *(Reproduced with permission from Wang, R., Yan, Y., Liu, H., Li, Y., Jin, M., Li, Y., Tao, R., Chen, Q., Wang, X., Zhao, B., Xie, D., 2023. Integrating data-dependent and data-independent non-target screening methods for monitoring emerging contaminants in the Pearl River of Guangdong Province, China. Sci. Total Environ. 0048-9697. 891, 164445. doi:10.1016/j.scitotenv.2023.164445.)*

blanks and present in samples above several times the control set, are retained. These retained features are now assigned with DDA data and compared to databases (e.g., Massbank, or the NORMAM SusDat databases).

Although the databases available for step V in Fig. 8.5 are extensive, containing reference data for thousands of compounds, and continuously improving to enable rapid sample screening and identification of potentially matching CECs, this final step in the workflow usually requires a reference retention time for accurate comparison. Acquiring reference standards for suspect compounds is often expensive, and some of them are not even commercially available (e.g., metabolites and/or transformation products). In addition, the retention data provided by databases may not always be directly comparable to the separation data produced in the screening campaigns due to differences in chromatographic procedures performed. In such cases, retention time prediction may help in the confirmation of suspects. These predictive approaches range from the simple log-P based model to more elaborate quantitative structure-retention relationships (QSRR) or learning machine approaches (Boelrijk et al., 2023; Richardson et al., 2021).

Castro et al. have developed a non-targeted analytical methodology to investigate the presence of CECs in sewage sludge. To achieve this, DDA and DIA acquisition strategies using a UHPLC-QTOF-MS system were investigated (Castro et al., 2021). According to the obtained results, DDA performed slightly better than DIA when working in iterative mode. Most of the compounds identified in the study were pharmaceuticals or their human excretion metabolites. Another example of such a workflow is presented by García-Vara et al. (2023), who developed a prioritization procedure to identify the most relevant CECs in an agricultural area. Samples collected in summer and winter provided a total of 2773 and 2257 candidates after the first LC-HRMS data processing, respectively.

When compounds naturally occurring (i.e., not anthropogenic chemicals) and features with less than three fragments matching the spectral database were eliminated, these numbers were reduced to 723 and 580, respectively. After filtering compounds' identities considering CECs reported in literature, a final list of 129 and 145 compounds was obtained. These compounds were assigned by exact mass, isotopic fit, and fragmentation patterns against spectral libraries using MS^2 data, with an additional estimation of the plausible retention time.

Quantification in non-targeted analysis (qNTA) is also evolving to enable semiquantitative estimation of those substances detected through NTA, regardless of the availability of analytical standards. When commercial standards are accessible, semiquantification can be performed by matrix-matched calibration, which involves spiking the obtained extracts with increasing concentrations of the selected species (Castro et al., 2021). In cases where standards are not available, alternative approaches rely on correlating ionization efficiencies with molecular descriptors. This approach allows to estimate concentrations of unknown compounds, bridging the gap between qualitative and quantitative analysis (Tisler et al., 2024).

5.2 Development of multiclass and multiresidue methodologies

Sample preparation remains a critical step in LC-MS analysis, especially when dealing with complex environmental matrices, which often contain a wide variety of contaminants from multiple chemical classes. Innovations in sample preparation techniques, such as online SPE and related methodologies, have significantly improved the recovery and concentration of trace-level contaminants, thereby enhancing the sensitivity and accuracy of detection. To meet current directives for CECs monitoring in drinking water and wastewater from urban WWTPs, there is a growing shift toward the development of multiclass and multiresidue LC-MS methods that can simultaneously detect a broad spectrum of compounds. Such comprehensive approaches enhance efficiency and provide a more holistic understanding of environmental contamination (Castro et al., 2018b; Paíga et al., 2021; Petrie et al., 2016). Recently, Gutiérrez et al. have presented a new analytical methodology for the analysis of 75 CECs belonging to 7 contaminant classes in two different residues, sewage sludge and waste from electrical and electronic equipment (WEEE) (Gutiérrez-Martín et al., 2025). The proposed methodology combined UAE and SPE as a cleanup technique, tailored to UPLC-QqQ-MS. This approach constitutes a powerful tool for managing complex waste streams within a circular economy framework and contributes to improved chemical risk assessment in waste treatment and recycling processes.

5.3 Emphasis on harmonization and standardization

As LC-MS techniques become indispensable in environmental analysis, the need for reliable and comparable data is increasingly critical. Harmonizing protocols ensures data comparability across studies and laboratories, supporting regulatory acceptance and enabling collaborative research to better characterize human and environmental exposure to chemicals (Lennon et al., 2024). However, monitoring methodologies for CECs are often poorly validated, lacking harmonization or standardization, and thus are only applicable under specific conditions. This undermines data comparability among the different studies conducted worldwide (Schwesig et al., 2011; Van Winckel et al., 2021). Existing validation guidance is often limited in scope, addressing only specific methods or matrices, as is the case with the SANTE guidelines for pesticide residues in food and feed (SANTE/EU, 2015). To improve consistency, QC procedures should be systematically integrated within a general validation framework applicable to diverse methodologies. In response, several EU and US organizations have reunited efforts to develop comprehensive method validation guidelines for both monitoring and biomonitoring of CECs across various matrices. Their efforts extend beyond monitoring harmonized methodologies to include QA/QC protocols for suspect and non-target screening. These initiatives aim to improve the evaluation of feature detection sensitivity, reproducibility, accuracy, precision, and the consistency of data preprocessing in human biomonitoring, food safety, and environmental analysis (Lennon et al., 2024).

5.4 Miniaturization and automation of LC-MS systems

Technological advances are leading to the development of miniaturized and automated LC-MS systems. The main idea of LC miniaturization is the reduction of the internal diameter (i.d.) of chromatographic columns. While conventional HPLC columns typically have diameters of 4.6–3.2 mm, miniaturized formats include micro-LC (1.5–0.5 mm i.d.), nano-LC (150–10 μm i.d.), and even open-tubular LC systems with capillary diameters as small as 50–5 μm (Cortés-Bautista et al., 2024). Operating at lower flow rates not only reduces solvent consumption and waste generation, aligning with green chemistry principles, but also enables more sensitive analysis due to decreased radial diffusion and increased analyte concentration, scaling inversely with the square of the column radius ($1/r^2$). However, this gain in sensitivity can be offset by the reduced injectable volumes, which are limited by the smaller amount of stationary phase in miniaturized columns (Vasconcelos Soares Maciel et al., 2020; Cortés-Bautista et al., 2024). Miniaturization also brings the advantage of requiring lower sample volumes, a critical factor when dealing with limited or precious samples, as in bioanalysis or trace environmental studies. Automation plays a complementary role by streamlining workflows, reducing manual intervention, and minimizing human error (Cortés-Bautista et al., 2024). These improvements not only enhance analytical reliability but also increase analytical throughput, making LC-MS more practical for high-frequency or routine applications. Together, miniaturization and automation are expanding the accessibility of LC-MS to laboratories and field settings previously constrained by the size, complexity, or cost of traditional systems.

All these innovations led to the development of compact, portable instruments capable of performing high-sensitivity analysis at the point of need (Rahimi et al., 2020). Such systems are particularly advantageous in fields like environmental monitoring, where real-time data acquisition and reduced sample degradation are critical for accurate assessment. As these technologies continue to mature, they promise to broaden the applicability of LC-MS, bridging the gap between laboratory precision and field-based practicality.

5.5 Integration of machine learning and artificial intelligence in data analysis and interpretation

Despite the many advantages offered by suspect and NTA analysis, significant limitations related to the data processing remain. The vast volume of datasets generated by LC-MS can complicate interpretation and hinder the identification of unknown compounds. Machine learning (ML) and artificial intelligence (AI) are emerging as powerful tools for processing, extracting, and interpreting complex spectral data produced in these analyses. These technologies, based on tools previously developed for metabolomics, lipidomics, narcotics, and other forensic studies, can improve pattern recognition, enable the prediction of compound behavior, and streamline the identification process (Canchola et al., 2025; Jin et al., 2025; Kavianpour et al., 2024; Russo et al., 2024), thereby accelerating environmental monitoring efforts (Jin et al., 2025). Building comprehensive spectral databases is a time- and resources-consuming process, and often lags the evolving needs of analytical screening. As an alternative, compound databases such as PubChem offer valuable information, even though they are focused on compound properties rather than mass spectra data. To extract meaningful information from these databases, molecular fingerprints must be linked to mass spectra. Recent advances in both supervised and unsupervised approaches have been developed to generate molecular fingerprints for ESI-MS and EI-MS using natural language processing (NLP), treating mass spectra as text documents, where peaks and neutral losses are interpreted as words. Neural networks (NN) are trained to learn and predict the context of these words and compute spectral similarity scores in the embedding space (Wang et al., 2016).

Another promising approach is the computational prediction of mass spectra through in silico fragmentation methods. These methods, which are particularly useful for compounds not included in the existing spectral databases, have recently included graph neural network (GNN), showing promising results (Zhang et al., 2022). Unlike conventional ab initio methods, in silico fragmentation tools based on ML models can predict spectral intensities more rapidly and accurately, enabling the expansion of existing MS libraries as an alternative to fingerprint approaches. For those substances not included in any public databases, neither in silico nor fingerprint-based approaches can be applied, so the de novo structure generation has emerged as a promising solution in the field of NTA, which is especially relevant in the field of novel drug development.

Effective preprocessing of MS spectra also enhances the quality of analytical results. Recent studies using convolutional neural networks (CNNs) have demonstrated excellent results in improving spectral data interpretation (Mottershead and Miller, 2023). However, improving any ML approach in advancing NTA in environmental science hinges on the access to large and quality datasets for model training. Not only must these repositories be publicly available, but also quality-controlled and accompanied by detailed metadata describing experimental conditions, analytical methods, and data processing steps. Some initiatives have already been started to provide such repositories (Kavianpour et al., 2024; Song et al., 2025).

5.6 Implementation of environmental risk assessment (ERA)

Beyond detection, there is a growing emphasis on assessing the environmental and health risks associated with CECs. The main objective of ERA is to identify the hazards and the exposure to CECs in order to evaluate the effects and consequences of a nonaccidental release. In addition, it provides a quantitative estimation of the event probability and compares it with agreed criteria. LC-MS data contribute to understanding the fate, transport, and toxicity of contaminants, informing risk assessments and guiding regulatory decisions. Integrating LC-MS findings with ecological and toxicological data enhances the overall evaluation of environmental risks. Recently, Tarazona et al. have published a landscape-based ERA for pesticides (Tarazona et al., 2024) including in the assessment of different considerations such as the individual and combined effects of all pesticides active in substances used in a specific area, together with the potential interactions with other environmental stress factors and population recovery after suffering harmful effects and their adaptation to pesticide stress. To this end, concentrations obtained in monitoring studies of CECs in the environment are key to supporting this spatial and temporal exposure.

Acknowledgments

G. Castro acknowledges the economic support received from project ED431C2021/06, funded by Xunta de Galicia, and CALIMAR from the network REDEMAR, Consellería do Mar, Xunta de Galicia, through the European Maritime, Fisheries and Aquaculture Fund (EMFAF). B. J. Sieira, L. Pérez-Mayán, and R. Cela thank the creation of the Mestrelab Research Center (CIM), subsidized by the Axencia Galega de Innovación. Operation funded by the Xunta de Galicia, through the business aid program for the creation and integration of new business research centers 001_IN853D_2022.

References

DIRECTIVE (EU) 2020/2184, 2020. Available from: https://eur-lex.europa.eu/legal-content/EN/TXT/PDF/?uri=CELEX:32020L2184.

Aceña, J., Stampachiacchiere, S., Pérez, S., Barceló, D., 2015. Advances in liquid chromatography–high-resolution mass spectrometry for quantitative and qualitative environmental analysis. Anal. Bioanal. Chem. 1618-2642. 407 (21), 6289–6299. doi:10.1007/s00216-015-8852-6.

Angeles, L.F., Aga, D.S., 2020. Catching the elusive persistent and mobile organic compounds: novel sample preparation and advanced analytical techniques. Trends Environ. Anal. Chem. 22141588. 25. Available from: http://www.journals.elsevier.com/trends-in-environmental-analytical-chemistry/. doi:10.1016/j.teac.2019.e00078.

Aydın, S., Ulvi, A., Bedük, F., Aydın, M.E., 2022. Pharmaceutical residues in digested sewage sludge: occurrence, seasonal variation and risk assessment for soil. Sci. Total Environ. 00489697. 817, 152864. doi:10.1016/j.scitotenv.2021.152864.

Bach, C., Dauchy, X., Chagnon, M.C., Etienne, S., 2012. Chemical compounds and toxicological assessments of drinking water stored in polyethylene terephthalate (PET) bottles: a source of controversy reviewed. Water Res. 18792448. 46 (3), 571–583. Available from: www.elsevier.com/locate/watres. doi:10.1016/j.watres.2011.11.062.

Barbieri, M.V., Monllor-Alcaraz, L.S., Postigo, C., López de Alda, M., 2020. Improved fully automated method for the determination of medium to highly polar pesticides in surface and groundwater and application in two distinct agriculture-impacted areas. Sci. Total Environ. 18791026. 745. Available from: www.elsevier.com/locate/scitotenv. doi:10.1016/j.scitotenv.2020.140650.

Beccaria, M., Cabooter, D., 2020. Current developments in LC-MS for pharmaceutical analysis. Analyst. 0003-2654. 145 (4), 1129–1157. doi:10.1039/c9an02145k.

Bell, D.S., 2021. Modern trends in mixed-mode liquid chromatography (LC) columns. LC-GC N. Am. 15275949. 39 (2), 56–60. Available from: www.chromatographyonline.com. doi:10.56530/lcgc.na.qo9881n6.

Bhagat, J., Singh, N., Nishimura, N., Shimada, Y., 2021. A comprehensive review on environmental toxicity of azole compounds to fish. Chemosphere. 00456535. 262, 128335. doi:10.1016/j.chemosphere.2020.128335.

Boelrijk, J., van Herwerden, D., Ensing, B., Forré, P., Samanipour, S., 2023. Predicting RP-LC retention indices of structurally unknown chemicals from mass spectrometry data. J. Cheminform. 1758-2946. 15 (1). doi:10.1186/s13321-023-00699-8.

Braus, H., Middleton, F., Walton, G., 1951. Orangic chemical compounds in raw and filtered surface waters. Anal. Chem. 0003-2700. 23 (8), 1160–1164. doi:10.1021/ac60056a031.

Buck, R.C., Franklin, J., Berger, U., Conder, J.M., Cousins, I.T., Voogt, P.D., Jensen, A.A., Kannan, K., Mabury, S.A., van Leeuwen, S.P.J., 2011. Perfluoroalkyl and polyfluoroalkyl substances in the environment: terminology, classification, and origins. Integr. Environ. Assess. Manag. 15513793. 7 (4), 513–541. doi:10.1002/ieam.258.

Canchola, A., Tran, L.N., Woo, W., Tian, L., Lin, Y.H., Chou, W.C., 2025. Advancing non-target analysis of emerging environmental contaminants with machine learning: current status and future implications. Environ. Int. 18736750. 198. Available from: https://www.sciencedirect.com/science/journal/01604120. doi:10.1016/j.envint.2025.109404.

Castro, G., Carpinteiro, I., Rodríguez, I., Cela, R., 2018a. Determination of cardiovascular drugs in sewage sludge by matrix solid-phase dispersion and ultra-performance liquid chromatography tandem mass spectrometry. Anal. Bioanal. Chem. 1618-2642. 410 (26), 6807–6817. doi:10.1007/s00216-018-1268-3.

Castro, G., Pérez-Mayán, L., Rodríguez-cabo, T., Rodríguez, I., Ramil, M., Cela, R., 2018b. Multianalyte, high-throughput liquid chromatography tandem mass spectrometry method for the sensitive determination of fungicides and insecticides in wine. Anal. Bioanal. Chem. 1618-2642. 410 (3), 1139–1150. doi:10.1007/s00216-017-0724-9.

Castro, G., Ramil, M., Cela, R., Rodríguez, I., 2021. Identification and determination of emerging pollutants in sewage sludge driven by UPLC-QTOF-MS data mining. Sci. Total Environ. 00489697. 778, 146256. doi:10.1016/j.scitotenv.2021.146256.

Castro, G., Ramil, M., Rodríguez, I., 2024. Supercritical fluid chromatography with mass spectrometry for the determination of chiral pollutants. Compr. Anal. Chem. 407, 6289–6299. Available from: https://www.sciencedirect.com/science/handbooks/0166526X. doi:10.1016/bs.coac.2024.08.010.

Castro, G., Rodríguez, I., Ramil, M., Cela, R., 2019. Selective determination of sartan drugs in environmental water samples by mixed-mode solid-phase extraction and liquid chromatography tandem mass spectrometry. Chemosphere. 00456535. 224, 562–571. doi:10.1016/j.chemosphere.2019.02.137.

Cobo-Golpe, M., Ramil, M., Cela, R., Rodríguez, I., 2022. Supercritical fluid chromatography time-of-flight mass spectrometry enantiomeric determination of basic drugs in sewage samples. J. Chromatogr. A. 00219673. 1673, 463088. doi:10.1016/j.chroma.2022.463088.

Coggan, T.L., Anumol, T., Pyke, J., Shimeta, J., Clarke, B.O., 2019. A single analytical method for the determination of 53 legacy and emerging per- and polyfluoroalkyl substances (PFAS) in aqueous matrices. Anal. Bioanal. Chem. 16182650. 411 (16), 3507–3520. Available from: http://link.springer.de/link/service/journals/00216/index.htm. doi:10.1007/s00216-019-01829-8.

Cordner, A., De La Rosa, V.Y., Schaider, L.A., Rudel, R.A., Richter, L., Brown, P., 2019. Guideline levels for PFOA and PFOS in drinking water: the role of scientific uncertainty, risk assessment decisions, and social factors. J. Expo. Sci. Environ. Epidemiol. 1559064X. 29 (2), 157–171. Available from: http://www.nature.com/jes/index.html. doi:10.1038/s41370-018-0099-9.

Cortés-Bautista, S., Molins-Legua, C., Campíns-Falcó, P., 2024. Miniaturized liquid chromatography in environmental analysis. A review. J. Chromatogr. A. 00219673. 1730, 465101. doi:10.1016/j.chroma.2024.465101.

Cutillas, V., Murcia-Morales, M., Gómez-Ramos, M.D.M., Taha, S.M., Fernández-Alba, A.R., 2019. Supercritical fluid chromatography coupled to tandem mass spectrometry for the analysis of pesticide residues in dried spices. Benefits and drawbacks. Anal. Chim. Acta 18734324. 1059, 124–135. Available from: http://www.journals.elsevier.com/analytica-chimica-acta/. doi:10.1016/j.aca.2019.01.010.

Dasu, K., Nakayama, S.F., Yoshikane, M., Mills, M.A., Wright, J.M., Ehrlich, S., 2017. An ultra-sensitive method for the analysis of perfluorinated alkyl acids in drinking water using a column switching high-performance liquid chromatography tandem mass spectrometry. J. Chromatogr. A 18733778. 1494, 46–54. Available from: www.elsevier.com/locate/chroma. doi:10.1016/j.chroma.2017.03.006.

de la Serna Calleja, M.Á., Bolado, S., Jiménez, J.J., López-Serna, R., 2023. Performance critical comparison of offline SPE, online SPE, and direct injection for the determination of CECs in complex liquid environmental matrices. Microchem. J. 0026265X. 187. Available from: www.elsevier.com/inca/publications/store/6/2/0/3/9/1. doi:10.1016/j.microc.2023.108395.

Di, S., Zhao, H., Liu, Z., Lu, Y., Xu, H., Qi, P., Wang, Z., Wang, X., 2023. Stereoselective analysis of chiral succinate dehydrogenase inhibitors (SDHIs) in foods of plant origin and animal origin by supercritical fluid chromatography-tandem mass spectrometry (SFC-MS/MS). Food Chem. 03088146. 411, 135452. doi:10.1016/j.foodchem.2023.135452.

Domingo, J.L., Nadal, M., 2019. Human exposure to per- and polyfluoroalkyl substances (PFAS) through drinking water: a review of the recent scientific literature. Environ. Res. 00139351. 177, 108648. doi:10.1016/j.envres.2019.108648.

Dong, M.W., Boyes, B.E., 2018. Modern trends and best practices in mobile-phase selection in reversed-phase chromatography. LC-GC Eur. 14716577. 31 (10), 572–583. Available from: http://www.chromatographyonline.com/lcgc/issue/issueList.jsp?id=2555.

Egede Frøkjær, E., Rüsz Hansen, H., Hansen, M., 2023. Non-targeted and suspect screening analysis using ion exchange chromatography-Orbitrap tandem mass spectrometry reveals polar and very mobile xenobiotics in Danish drinking water. Chemosphere. 339, 139745. doi:10.1016/j.chemosphere.2023.139745

Enevoldsen, R., Juhler, R.K., 2010. Perfluorinated compounds (PFCs) in groundwater and aqueous soil extracts: using inline SPE-LC-MS/MS for screening and sorption characterisation of perfluorooctane sulphonate and related compounds. Anal. Bioanal. Chem. 16182650. 398 (3), 1161–1172. doi:10.1007/s00216-010-4066-0.

EPA US, 2018. Method 533: determination of per-and polyfluoroalkyl substances in drinking water by isotope dilution anion exchange solid phase extraction and liquid chromatography/tandem mass spectrometry. Available from: https://www.epa.gov/dwanalyticalmethods/method-533-determination-and-polyfluoroalkyl-substances-drinking-water-isotope.

EPA, 2025. WWW document contaminants of emerging concern including pharmaceuticals and personal care products. Available from: https://www.epa.gov/wqc/contaminants-emerging-concern-including-pharmaceuticals-and-personal-care-products.

Fernandes, M.J., Paíga, P., Silva, A., Llaguno, C.P., Carvalho, M., Vázquez, F.M., Delerue-Matos, C., 2020. Antibiotics and antidepressants occurrence in surface waters and sediments collected in the north of Portugal. Chemosphere. 18791298. 239. Available from: www.elsevier.com/locate/chemosphere. doi:10.1016/j.chemosphere.2019.124729.

Fernández-Fernández, V., Ramil, M., Cela, R., Rodríguez, I., 2023. Occurrence and risk assessment of pesticides and pharmaceuticals in viticulture impacted watersheds from Northwest Spain. Chemosphere. 00456535. 341, 140098. doi:10.1016/j.chemosphere.2023.140098.

Fukushima, T., Koishi, M., Sakamoto, T., Onozato, M., 2024. Use of commercial mixed-mode stationary phases and sorbents in the high-performance liquid chromatography analysis and solid-phase extraction of ionized and hydrophilic bioactive compounds. Molecules. 1420-3049. 29 (10), 2341. doi:10.3390/molecules29102341.

Gago-Ferrero, P., Bletsou, A.A., Damalas, D.E., Aalizadeh, R., Alygizakis, N.A., Singer, H.P., Hollender, J., Thomaidis, N.S., 2020. Wide-scope target screening of >2000 emerging contaminants in wastewater samples with UPLC-Q-ToF-HRMS/MS and smart evaluation of its performance through the validation of 195 selected representative analytes. J. Hazard. Mater. 18733336. 387. Available from: www.elsevier.com/locate/jhazmat. doi:10.1016/j.jhazmat.2019.121712.

García-Vara, M., Orlando-Véliz, D., Bonansea, R.I., Postigo, C., López de Alda, M., 2023. Prioritization of organic contaminants in a reclaimed water irrigation system using wide-scope LC-HRMS screening. J. Hazard. Mater. 18733336. 459. Available from: www.elsevier.com/locate/jhazmat. doi:10.1016/j.jhazmat.2023.132119.

Gazárková, T., Plachká, K., Svec, F., Nováková, L., 2022. Current state of supercritical fluid chromatography-mass spectrometry. TrAC Trends Anal. Chem. 01659936. 149, 116544. doi:10.1016/j.trac.2022.116544.

Glüge, J., Scheringer, M., Cousins, I.T., Dewitt, J.C., Goldenman, G., Herzke, D., Lohmann, R., Ng, C.A., Trier, X., Wang, Z., 2020. An overview of the uses of per- and polyfluoroalkyl substances

(PFAS). Environ. Sci. Process. Impacts. 20507895. 22 (12), 2345–2373. Available from: http://pubs.rsc.org/en/journals/journal/em. doi:10.1039/d0em00291g.

Gu, C., 2019. LC/MS: fundamentals, perspectives, and applications. In: Dong, M.W., (Ed.), HPLC and UHPLC for Practicing Scientists. Wiley, pp. 147–175. doi:10.1002/9781119313786.ch6.

Gutiérrez-Martín, D., Castro, G., González, S.V., Arp, H.P.H., López-Serna, R., Asimakopoulos, A.G., 2025. A comprehensive analytical approach to monitoring selected emerging contaminant classes in sewage sludge and e-waste from Norway. Environ. Chem. Ecotoxicol. 25901826. 7, 901–909. Available from: https://www.sciencedirect.com/science/journal/25901826. doi:10.1016/j.enceco.2025.05.004.

Hernández, F., Ibáñez, M., Bade, R., Bijlsma, L., Sancho, J.V., 2014. Investigation of pharmaceuticals and illicit drugs in waters by liquid chromatography-high-resolution mass spectrometry. TrAC Trends Anal. Chem. 01659936. 63, 140–157. doi:10.1016/j.trac.2014.08.003.

Hollender, J., Schymanski, E.L., Ahrens, L., Alygizakis, N., Béen, F., Bijlsma, L., Brunner, A.M., Celma, A., Fildier, A., Fu, Q., Gago-Ferrero, P., Gil-Solsona, R., Haglund, P., Hansen, M., Kaserzon, S., Kruve, A., Lamoree, M., Margoum, C., Meijer, J., Merel, S., Rauert, C., Rostkowski, P., Samanipour, S., Schulze, B., Schulze, T., Singh, R.R., Slobodnik, J., Steininger-Mairinger, T., Thomaidis, N.S., Togola, A., Vorkamp, K., Vulliet, E., Zhu, L., Krauss, M., 2023. Norman guidance on suspect and non-target screening in environmental monitoring. Environ. Sci. Eur. 21904715. 35 (1). Available from: https://enveurope.springeropen.com. doi:10.1186/s12302-023-00779-4.

Hollender, J., Schymanski, E.L., Singer, H.P., Ferguson, P.L., 2017. Nontarget screening with high resolution mass spectrometry in the environment: ready to go? Environ. Sci. Technol. 15205851. 51 (20), 11505–11512. Available from: http://pubs.acs.org/journal/esthag. doi:10.1021/acs.est.7b02184.

Hubert, M., Arp, H.P.H., Hansen, M.C., Castro, G., Meyn, T., Asimakopoulos, A.G., Hale, S.E., 2023. Influence of grain size, organic carbon and organic matter residue content on the sorption of per- and polyfluoroalkyl substances in aqueous film forming foam contaminated soils—implications for remediation using soil washing. Sci. Total Environ. 18791026. 875. Available from: www.elsevier.com/locate/scitotenv. doi:10.1016/j.scitotenv.2023.162668.

Huerta, B., McHugh, B., Regan, F., 2022. Development and application of an LC-MS method to the determination of poly- and perfluoroalkyl substances (PFASs) in drinking, sea and surface water samples. Anal. Methods. 1759-9660. 14 (21), 2090–2099. doi:10.1039/d2ay00300g.

Huidobro-López, B., López-Heras, I., Alonso-Alonso, C., Martínez-Hernández, V., Nozal, L., de Bustamante, I., 2022. Analytical method to monitor contaminants of emerging concern in water and soil samples from a non-conventional wastewater treatment system. J. Chromatogr. A. 00219673. 1671, 463006. doi:10.1016/j.chroma.2022.463006.

Jian, J.M., Guo, Y., Zeng, L., Liang-Ying, L., Lu, X., Wang, F., Zeng, E.Y., 2017. Global distribution of perfluorochemicals (PFCs) in potential human exposure source–a review. Environ. Int. 18736750. 108, 51–62. Available from: www.elsevier.com/locate/envint. doi:10.1016/j.envint.2017.07.024.

Jin, Z.L., Chen, L., Wang, Y., Shi, C.T., Zhou, Y., Xia, B., 2025. Application of machine learning in LC-MS-based non-targeted analysis. TrAC Trends Anal. Chem. 18793142. 189. Available from: https://www.sciencedirect.com/science/journal/01659936. doi:10.1016/j.trac.2025.118243.

Joudan, S., Lundgren, R.J., 2022. Taking the "F" out of forever chemicals. Science. 10959203. 377 (6608), 816–817. Available from: https://www.science.org/doi/10.1126/science.add1813. doi:10.1126/science.add1813.

Kanu, A.B., 2021. Recent developments in sample preparation techniques combined with high-performance liquid chromatography: a critical review. J. Chromatogr. A 18733778. 1654. Available from: www.elsevier.com/locate/chroma. doi:10.1016/j.chroma.2021.462444.

Kavianpour, B., Piadeh, F., Gheibi, M., Ardakanian, A., Behzadian, K., Campos, L.C., 2024. Applications of artificial intelligence for chemical analysis and monitoring of pharmaceutical and personal care products in water and wastewater: a review. Chemosphere. 00456535. 368, 143692. doi:10.1016/j.chemosphere.2024.143692.

Khoo, H.T., Leow, C.H., 2021. Advancements in the preparation and application of monolithic silica columns for efficient separation in liquid chromatography. Talanta. 00399140. 224. Available from: https://www.journals.elsevier.com/talanta. doi:10.1016/j.talanta.2020.121777.

Kim, K.Y., Ndabambi, M., Choi, S., Oh, J.E., 2021. Legacy and novel perfluoroalkyl and polyfluoroalkyl substances in industrial wastewater and the receiving river water: temporal changes in relative abundances of regulated compounds and alternatives. Water Res. 18792448. 191. Available from: www.elsevier.com/locate/watres. doi:10.1016/j.watres.2021.116830.

Lennon, S., Chaker, J., Price, E.J., Hollender, J., Huber, C., Schulze, T., Ahrens, L., Béen, F., Creusot, N., Debrauwer, L., Dervilly, G., Gabriel, C., Guérin, T., Habchi, B., Jamin, E.L., Klánová, J., Kosjek, T., Bizec, B.L., Meijer, J., Mol, H., Nijssen, R., Oberacher, H., Papaioannou, N., Parinet, J., Sarigiannis, D., Stravs, M.A., Tkalec, Ž., Schymanski, E.L., Lamoree, M., Antignac, J.P., David, A., 2024. Harmonized quality assurance/quality control provisions to assess completeness and robustness of MS1 data preprocessing for LC-HRMS-based suspect screening and non-targeted analysis. TrAC Trends Anal. Chem. 18793142. 174. Available from: https://www.sciencedirect.com/science/journal/01659936. doi:10.1016/j.trac.2024.117674.

Lesellier, E., West, C., Lemasson, E., Hennig, P., Bertin, S., 2017. WWW document. Special issue LCGC iinternational mixed-mode chromatography—a review. Available from: http://https://www.chromatographyonline.com/view/mixed-mode-chromatography-review.

Liang, S.H., Steimling, J.A., Chang, M., 2023. Analysis of ultrashort-chain and short-chain (C1 to C4) per- and polyfluorinated substances in potable and non-potable waters. J. Chromatogr. Open. 27723917. 4. Available from: https://www.journals.elsevier.com/journal-of-chromatography-open. doi:10.1016/j.jcoa.2023.100098.

Liu, X., Tracy, M., Aich, U., Pohl, C., 2014. Exploring mixed-mode chromatography: column chemistry, properties, and applications. Available from: https://documents.thermofisher.com/TFS-Assets/CMD/posters/PN-21137-LC-Mixed-Mode-Chromatography-Pittcon2015-PN21137-EN.pdf.

López-García, E., Mastroianni, N., Postigo, C., Barceló, D., López de Alda, M., 2018. A fully automated approach for the analysis of 37 psychoactive substances in raw wastewater based on on-line solid phase extraction-liquid chromatography-tandem mass spectrometry. J. Chromatogr. A. 00219673. 1576, 80–89. doi:10.1016/j.chroma.2018.09.038.

López-Vázquez, J., Montes, R., Rodil, R., Cela, R., Martínez-Pontevedra, J.Á., Pena, M.T., Quintana, J.B., 2024. Determination of regulated perfluoroalkyl substances (PFAS) in drinking water according to directive 2020/2184/EU. Environ. Sci. Pollut. Res. 16147499. Available from: https://www.springer.com/journal/11356. doi:10.1007/s11356-024-34852-z.

López-Vázquez, J., Pérez-Mayán, L., Fernández-Fernández, V., Cela, R., Rodríguez, I., 2023. Direct, automated and sensitive determination of glyphosate and related anionic pesticides in environmental water samples using solid-phase extraction on-line combined with liquid chromatography tandem mass spectrometry. J. Chromatogr. A. 00219673. 1687, 463697. doi:10.1016/j.chroma.2022.463697.

Mann, O., Pock, E., Wruss, K., Wruss, W., Krska, R., 2016. Development and validation of a fully automated online-SPE–ESI–LC–MS/MS multi-residue method for the determination of different classes of pesticides in drinking, ground and surface water. Int. J. Environ. Anal. Chem. 0306-7319. 96 (4), 353–372. doi:10.1080/03067319.2016.1160381.

Martín-Pozo, L., de Alarcón-gómez, B., Rodríguez-Gómez, R., García-Córcoles, M.T., Çipa, M., Zafra-Gómez, A., 2019. Analytical methods for the determination of emerging contaminants in sewage sludge samples. A review. Talanta. 00399140. 192, 508–533. Available from: https://www.journals.elsevier.com/talanta. doi:10.1016/j.talanta.2018.09.056.

Martínez-Bueno, M.J., Gómez Ramos, M.J., Bauer, A., Fernández-Alba, A.R., 2019. An overview of non-targeted screening strategies based on high resolution accurate mass spectrometry for the identification of migrants coming from plastic food packaging materials. TrAC Trends Anal. Chem. 01659936. 110, 191–203. doi:10.1016/j.trac.2018.10.035.

Mitchell, R.J., Christian, R., Hughes, H., Miah, A., Walker, D.K., 2010. The application of fully automated on-line solid phase extraction in bioanalysis. J. Pharm. Biomed. Anal. 07317085. 52 (1), 86–92. doi:10.1016/j.jpba.2009.11.002.

Mohapatra, S., Tong, X., Mukherjee, S., Dubey, M., Subhash, S., Luhua, Y., van der Hoek, J.P., Gin, K.Y.H., 2025. Comprehensive insights on the detection, occurrence and modelling of pharmaceuticals in surface water, groundwater, and drinking water treatment plants. J. Hazard. Mater. Adv. 27724166. 18. Available from: https://www.sciencedirect.com/science/journal/27724166. doi:10.1016/j.hazadv.2025.100707.

Montes, R., Aguirre, J., Vidal, X., Rodil, R., Cela, R., Quintana, J.B., 2017. Screening for polar chemicals in water by trifunctional mixed-mode liquid chromatography-high resolution mass spectrometry. Environ. Sci. Technol. 15205851. 51 (11), 6250–6259. Available from: http://pubs.acs.org/journal/esthag. doi:10.1021/acs.est.6b05135.

Montes, R., Rodil, R., Cela, R., Quintana, J.B., 2019. Determination of persistent and mobile organic contaminants (PMOCs) in water by mixed-mode liquid chromatography-tandem mass spectrometry. Anal. Chem. 15206882. 91 (8), 5176–5183. Available from: http://pubs.acs.org/journal/ancham. doi:10.1021/acs.analchem.8b05792.

Montone, C.M., Giannelli Moneta, B., Laganà, A., Piovesana, S., Taglioni, E., Cavaliere, C., 2024. Transformation products of antibacterial drugs in environmental water: identification approaches based on liquid chromatography-high resolution mass spectrometry. J. Pharm. Biomed. Anal. 07317085. 238, 115818. doi:10.1016/j.jpba.2023.115818.

Mottershead, K., Miller, T.H., 2023. Application of deep learning to support peak picking during non-target high resolution mass spectrometry workflows in environmental research. Environ. Sci. Adv. 27547000. 2 (6), 877–885. Available from: www.rsc.org/journals-books-databases/about-journals/environmental-science-advances. doi:10.1039/d3va00005b.

Nassazzi, W., Lai, F.Y., Ahrens, L., 2022. A novel method for extraction, clean-up and analysis of per- and polyfluoroalkyl substances (PFAS) in different plant matrices using LC-MS/MS. J. Chromatogr. B 1873376X. 1212. Available from: www.elsevier.com/inca/publications/store/5/0/2/6/8/9. doi:10.1016/j.jchromb.2022.123514.

Paíga, P., Sousa, S., Vera, J., Bitencourt, L., Vieira, J., Jorge, S., Silva, J.G., Correia, M., Domingues, V.F., Delerue-Matos, C., 2021. Multi-residue analysis of fifty pesticides in river waters and in wastewaters. Environ. Sci. Pollut. Res. 16147499. 28 (47), 66787–66803. Available from: https://link.springer.com/journal/11356. doi:10.1007/s11356-021-15134-4.

Pérez-Mayán, L., Ramil, M., Cela, R., Rodríguez, I., 2020. Multiresidue procedure to assess the occurrence and dissipation of fungicides and insecticides in vineyard soils from Northwest Spain. Chemosphere. 00456535. 261, 127696. doi:10.1016/j.chemosphere.2020.127696.

Pérez-Mayán, L., Ramil, M., Cela, R., Rodríguez, I., 2021. Supercritical fluid chromatography-mass spectrometric determination of chiral fungicides in viticulture-related samples. J. Chromatogr. A. 00219673. 1644, 462124. doi:10.1016/j.chroma.2021.462124.

Pesek, J.J., Matyska, M.T., Boysen, R.I., Yang, Y., Hearn, M.T.W., 2013. Aqueous normal-phase chromatography using silica-hydride-based stationary phases. TrAC Trends Anal. Chem. 18793142. 42, 64–73. Available from: www.elsevier.com/locate/trac. doi:10.1016/j.trac.2012.09.016.

Petrie, B., Youdan, J., Barden, R., Kasprzyk-Hordern, B., 2016. Multi-residue analysis of 90 emerging contaminants in liquid and solid environmental matrices by ultra-high-performance liquid chromatography tandem mass spectrometry. J. Chromatogr. A. 00219673. 1431, 64–78. doi:10.1016/j.chroma.2015.12.036.

Rahimi, F., Chatzimichail, S., Saifuddin, A., Surman, A.J., Taylor-Robinson, S.D., Salehi-Reyhani, A., 2020. A review of portable high-performance liquid chromatography: the future of the field? Chromatographia. 16121112. 83 (10), 1165–1195. Available from: http://www.chromatographia.de/. doi:10.1007/s10337-020-03944-6.

Raposo, F., Barceló, D., 2021. Challenges and strategies of matrix effects using chromatography-mass spectrometry: an overview from research versus regulatory viewpoints. TrAC Trends Anal. Chem. 01659936. 134, 116068. doi:10.1016/j.trac.2020.116068.

Reemtsma, T., Miehe, U., Duennbier, U., Jekel, M., 2010. Polar pollutants in municipal wastewater and the water cycle: occurrence and removal of benzotriazoles. Water Res. 00431354. 44 (2), 596–604. doi:10.1016/j.watres.2009.07.016.

Richardson, A.K., Chadha, M., Rapp-Wright, H., Mills, G.A., Fones, G.R., Gravell, A., Stürzenbaum, S., Cowan, D.A., Neep, D.J., Barron, L.P., 2021. Rapid direct analysis of river water and machine learning assisted suspect screening of emerging contaminants in passive sampler extracts. Anal. Methods. 17599679. 13 (5), 595–606. Available from: http://pubs.rsc.org/en/journals/journal/ay. doi:10.1039/d0ay02013c.

Rodriguez-Mozaz, S., Vaz-Moreira, I., Varela Della Giustina, S., Llorca, M., Barceló, D., Schubert, S., Berendonk, T.U., Michael-Kordatou, I., Fatta-Kassinos, D., Martinez, J.L., Elpers, C., Henriques, I., Jaeger, T., Schwartz, T., Paulshus, E., O'Sullivan, K., Pärnänen, K.M.M., Virta, M., Do, T.T., Walsh, F., Manaia, C.M., 2020. Antibiotic residues in final effluents of European wastewater treatment plants and their impact on the aquatic environment. Environ. Int. 18736750. 140. Available from: www.elsevier.com/locate/envint. doi:10.1016/j.envint.2020.105733.

Russo, F.F., Nowatzky, Y., Jaeger, C., Parr, M.K., Benner, P., Muth, T., Lisec, J., 2024. Machine learning methods for compound annotation in non-targeted mass spectrometry—a brief overview of fingerprinting, in silico fragmentation and de novo methods. Rapid Commun. Mass Spectrom. 10970231. 38 (20). Available from: http://onlinelibrary.wiley.com/journal/10.1002/(ISSN)1097-0231. doi:10.1002/rcm.9876.

SANTE/EU, 2015. Legal deposit 1 42 guidance document on analytical quality control and method validation procedures for pesticides residues analysis in food and feed. Available from: https://www.eurl-pesticides.eu/docs/public/tmplt_article.asp?CntID=727.

Schwesig, D., Borchers, U., Chancerelle, L., Dulio, V., Eriksson, U., Farré, M., Goksoyr, A., Lamoree, M., Leonards, P., Wegener, J.W., Lepom, P., Leverett, D., O'Neill, A., Robinson, R., Silharova, K., Tolgyessy, P., Slobodnik, J., Tutundjian, R., Westwood, D., 2011. A harmonized European framework for method validation to support research on emerging pollutants. TrAC Trends Anal. Chem. 18793142. 30 (8), 1233–1242. doi:10.1016/j.trac.2011.03.015.

Si-Hung, L., Bamba, T., 2022. Current state and future perspectives of supercritical fluid chromatography. TrAC Trends Anal. Chem. 01659936. 149, 116550. doi:10.1016/j.trac.2022.116550.

Sieira, B.J., Montes, R., Touffet, A., Rodil, R., Cela, R., Gallard, H., Quintana, J.B., 2020. Chlorination and bromination of 1,3-diphenylguanidine and 1,3-di-o-tolylguanidine: kinetics, transformation products and toxicity assessment. J. Hazard. Mater. 18733336. 385. Available from: www.elsevier.com/locate/jhazmat. doi:10.1016/j.jhazmat.2019.121590.

Song, W.-Y., Seo, S., Kim, T.-Y., 2025. Taming the unknown: advancing non-targeted analysis through collaborative data sharing and standardized data repositories. Environ. Sci. Technol. Lett. 12 (1), 5–7. doi:10.1021/acs.estlett.4c01031.

Sørmo, E., Castro, G., Hubert, M., Licul-Kucera, V., Quintanilla, M., Asimakopoulos, A.G., Cornelissen, G., Arp, H.P.H., 2023. The decomposition and emission factors of a wide range of PFAS in diverse, contaminated organic waste fractions undergoing dry pyrolysis. J. Hazard. Mater. 18733336. 454. Available from: www.elsevier.com/locate/jhazmat. doi:10.1016/j.jhazmat.2023.131447.

Steiner, D., Malachová, A., Sulyok, M., Krska, R., 2021. Challenges and future directions in LC-MS-based multiclass method development for the quantification of food contaminants. Anal. Bioanal. Chem. 1618-2642. 413 (1), 25–34. doi:10.1007/s00216-020-03015-7.

Svan, A., Hedeland, M., Arvidsson, T., Jasper, J.T., Sedlak, D.L., Pettersson, C.E., 2015. Rapid chiral separation of atenolol, metoprolol, propranolol and the zwitterionic metoprolol acid using supercritical fluid chromatography-tandem mass spectrometry—application to wetland microcosms. J. Chromatogr. A 18733778. 1409, 251–258. Available from: www.elsevier.com/locate/chroma. doi:10.1016/j.chroma.2015.07.075.

Tarazona, J.V., de Alba-gonzalez, M., Bedos, C., Benoit, P., Bertrand, C., Crouzet, O., Dagès, C., Dorne, J.L.C., Fernandez-Agudo, A., Focks, A., Gonzalez-Caballero, M.D.C., Kroll, A., Liess, M., Loureiro, S., Ortiz-Santaliestra, M.E., Rasmussen, J.J., Royauté, R., Rundlöf, M., Schäfer, R.B., Short, S., Siddique, A., Sousa, J.P., Spurgeon, D., Staub, P.F., Topping, C.J., Voltz, M., Axelman, J., Aldrich, A., Duquesne, S., Mazerolles, V., Devos, Y., 2024. A conceptual framework for landscape-based environmental risk assessment (ERA) of pesticides. Environ. Int. 18736750. 191. Available from: https://www.sciencedirect.com/science/journal/01604120. doi:10.1016/j.envint.2024.108999.

Teymoorian, T., Munoz, G., Duy, S.V., Liu, J., Sauvé, S., 2023. Tracking PFAS in drinking water: a review of analytical methods and worldwide occurrence trends in tap water and bottled water. ACS ES&T Water. 26900637. 3 (2), 246–261. Available from: https://pubs.acs.org/page/aewcaa/about.html. doi:10.1021/acsestwater.2c00387.

Tisler, S., Kilpinen, K., Pattison, D.I., Tomasi, G., Christensen, J.H., 2024. Quantitative nontarget analysis of CECs in environmental samples can be improved by considering all mass adducts. Anal. Chem. 15206882. 96 (1), 229–237. Available from: http://pubs.acs.org/journal/ancham. doi:10.1021/acs.analchem.3c03791.

UNEP, 2021. Proposal to List Long-Chain Perfluorocarboxylic Acids, Their Salts and Related Compounds in Annexes A, B And/or C to the Stockholm Convention on Persistent Organic Pollutants, vol. 2. UNEP, pp. 1–13.

Van Winckel, T., Cools, J., Vlaeminck, S.E., Joos, P., Van Meenen, E., Borregán-Ochando, E., Van Den Steen, K., Geerts, R., Vandermoere, F., Blust, R., 2021. Towards harmonization of water quality management: a comparison of chemical drinking water and surface water quality standards around the globe. J. Environ. Manage. 03014797. 298, 113447. doi:10.1016/j.jenvman.2021.113447.

Vasconcelos Soares Maciel, E., de Toffoli, A.L., Sobieski, E., Domingues Nazário, C.E., Lanças, F.M., 2020. Miniaturized liquid chromatography focusing on analytical columns and mass spectrometry: a review. Anal. Chim. Acta. 18734324. 1103, 11–31. Available from: http://www.journals.elsevier.com/analytica-chimica-acta/. doi:10.1016/j.aca.2019.12.064.

Wang, M., Carver, J.J., Phelan, V.V., Sanchez, L.M., Garg, N., Peng, Y., Nguyen, D.D., Watrous, J., Kapono, C.A., Luzzatto-Knaan, T., Porto, C., Bouslimani, A., Melnik, A.V., Meehan, M.J., Liu, W.T., Crüsemann, M., Boudreau, P.D., Esquenazi, E., Sandoval-Calderón, M., Kersten, R.D., Pace, L.A., Quinn, R.A., Duncan, K.R., Hsu, C.C., Floros, D.J., Gavilan, R.G., Kleigrewe, K., Northen, T., Dutton, R.J., Parrot, D., Carlson, E.E., Aigle, B., Michelsen, C.F., Jelsbak,

L., Sohlenkamp, C., Pevzner, P., Edlund, A., McLean, J., Piel, J., Murphy, B.T., Gerwick, L., Liaw, C.C., Yang, Y.L., Humpf, H.U., Maansson, M., Keyzers, R.A., Sims, A.C., Johnson, A.R., Sidebottom, A.M., Sedio, B.E., Klitgaard, A., Larson, C.B., Boya, C.A.P., Torres-Mendoza, D., Gonzalez, D.J., Silva, D.B., Marques, L.M., Demarque, D.P., Pociute, E., O'Neill, E.C., Briand, E., Helfrich, E.J.N., Granatosky, E.A., Glukhov, E., Ryffel, F., Houson, H., Mohimani, H., Kharbush, J.J., Zeng, Y., Vorholt, J.A., Kurita, K.L., Charusanti, P., McPhail, K.L., Nielsen, K.F., Vuong, L., Elfeki, M., Traxler, M.F., Engene, N., Koyama, N., Vining, O.B., Baric, R., Silva, R.R., Mascuch, S.J., Tomasi, S., Jenkins, S., Macherla, V., Hoffman, T., Agarwal, V., Williams, P.G., Dai, J., Neupane, R., Gurr, J., Rodríguez, A.M.C., Lamsa, A., Zhang, C., Dorrestein, K., Duggan, B.M., Almaliti, J., Allard, P.M., Phapale, P., Nothias, L.F., Alexandrov, T., Litaudon, M., Wolfender, J.L., Kyle, J.E., Metz, T.O., Peryea, T., Nguyen, D.T., VanLeer, D., Shinn, P., Jadhav, A., Müller, R., Waters, K.M., Shi, W., Liu, X., Zhang, L., Knight, R., Jensen, P.R., Palsson, B., Pogliano, K., Linington, R.G., Gutiérrez, M., Lopes, N.P., Gerwick, W.H., Moore, B.S., Dorrestein, P.C., Bandeira, N., 2016. Sharing and community curation of mass spectrometry data with global natural products social molecular networking. Nat. Biotechnol. 15461696. 34 (8), 828–837. Available from: http://www.nature.com/nbt/index.html. doi:10.1038/nbt.3597.

Wang, M., Wang, Y.H., Avula, B., Radwan, M.M., Wanas, A.S., Mehmedic, Z., van Antwerp, J., ElSohly, M.A., Khan, I.A., 2017a. Quantitative determination of cannabinoids in cannabis and cannabis products using ultra-high-performance supercritical fluid chromatography and diode array/mass spectrometric detection. J. Forensic Sci. 15564029. 62 (3), 602–611. doi:10.1111/1556-4029.13341.

Wang, R., Yan, Y., Liu, H., Li, Y., Jin, M., Li, Y., Tao, R., Chen, Q., Wang, X., Zhao, B., Xie, D., 2023. Integrating data dependent and data independent non-target screening methods for monitoring emerging contaminants in the Pearl River of Guangdong Province, China. Sci. Total Environ. 00489697. 891, 164445. doi:10.1016/j.scitotenv.2023.164445.

Wang, Z., Dewitt, J.C., Higgins, C.P., Cousins, I.T., 2017b. A never-ending story of per- and polyfluoroalkyl substances (PFASs)? Environ. Sci. Technol. 51 (5), 2508–2518. doi:10.1021/acs.est.6b04806.

West, C., 2018. Current trends in supercritical fluid chromatography. Anal. Bioanal. Chem. 1618-2642. 410 (25), 6441–6457. doi:10.1007/s00216-018-1267-4.

Williams, M.L., Olomukoro, A.A., Emmons, R.V., Godage, N.H., Gionfriddo, E., 2023. Matrix effects demystified: strategies for resolving challenges in analytical separations of complex samples. J. Sep. Sci. 16159314. 46 (23). Available from: http://onlinelibrary.wiley.com/journal/10.1002/(ISSN)1615-9314. doi:10.1002/jssc.202300571.

Zhang, B., Zhang, J., Xia, Y., Chen, P., Wang, B., 2022. Prediction of electron ionization mass spectra based on graph convolutional networks. Int. J. Mass Spectrom. 13873806. 475, 116817. doi:10.1016/j.ijms.2022.116817.

Zhang, J., Jaspers, V.L.B., Røe, J., Castro, G., Kroglund, I.B., Gonzalez, S.V., Østnes, J.E., Asimakopoulos, A.G., 2023. Per- and poly-fluoroalkyl substances in Tawny Owl (Strix aluco) feathers from Trøndelag, Norway. Sci. Total Environ. 18791026. 903. Available from: www.elsevier.com/locate/scitotenv. doi:10.1016/j.scitotenv.2023.166213.

Zhang, K., Liu, X., 2016. Mixed-mode chromatography in pharmaceutical and biopharmaceutical applications. J. Pharm. Biomed. Anal. 07317085. 128, 73–88. doi:10.1016/j.jpba.2016.05.007.

Zhong, H., Zheng, M., Liang, Y., Wang, Y., Gao, W., Wang, Y., Jiang, G., 2021. Legacy and emerging per- and polyfluoroalkyl substances (PFAS) in sediments from the East China Sea and the Yellow Sea: occurrence, source apportionment and environmental risk assessment. Chemosphere. 00456535. 282, 131042. doi:10.1016/j.chemosphere.2021.131042.

Zhong, M., Wang, T., Qi, C., Peng, G., Lu, M., Huang, J., Blaney, L., Yu, G., 2019. Automated online solid-phase extraction liquid chromatography tandem mass spectrometry investigation for simultaneous quantification of per- and polyfluoroalkyl substances, pharmaceuticals and personal care products, and organophosphorus flame retardants in environmental waters. J. Chromatogr. A. 00219673. 1602, 350–358. doi:10.1016/j.chroma.2019.06.012.

Chapter 9

Suspect and nontarget screening in environmental analysis: Principles, current directions and real-world applications

Alberto Celma[a,e], Lidia Belova[b], Rubén Gil-Solsona[c], and Adrià Sunyer-Caldú[d]

[a]*Department of Aquatic Sciences and Assessment, Swedish University of Agricultural Sciences, Uppsala, Sweden*

[b]*Toxicological Centre, University of Antwerp, Antwerp, Belgium*

[c]*Institute of Environmental Assessment and Water Research (IDAEA-CSIC), Barcelona, Spain*

[d]*Science for Life Laboratory, Department of Environmental Science (ACES, Exposure & Effects), Stockholm University, Stockholm, Sweden*

[e]*Environmental and Public Health Analytical Chemistry (Q-AMS), Research Institute fos Pesticides and Water (IUPA), University Jaume I, Castellón de la Plana, Castellón, Spain*

1 Introduction

Over the last 15 years, the technological progress achieved in high-resolution mass spectrometry (HRMS) instrumentation and data treatment streamlined outstanding advances in the field of suspect (SS) and nontarget screening (NTS) of contaminants of emerging concern (CECs) in environmental samples (Hernández et al., 2019; Hollender et al., 2017; Schymanski et al., 2015). The different instrument configurations and versatile working modes currently available have given researchers substantial flexibility for innovative developments and applications fostering the larger implementation of SS and NTS. A search on the peer-reviewed scientific literature indexing platform Scopus with the keywords "Target/Suspect/Nontarget + screening + contaminants + environment" shows an increasing trend for published SS and NTS studies, especially over the last 5 years (Fig. 9.1). Even to the point where SS and NTS studies balanced out target screening publications for the first time in 2024.

Target analysis remains a highly valuable strategy when focusing on limited number of substances with available reference standards. However, the versatility of SS and NTS approaches by HRMS instrumentation offers powerful

Contaminants of Emerging Concern in the Environment. https://doi.org/10.1016/B978-0-443-33821-2.00009-8

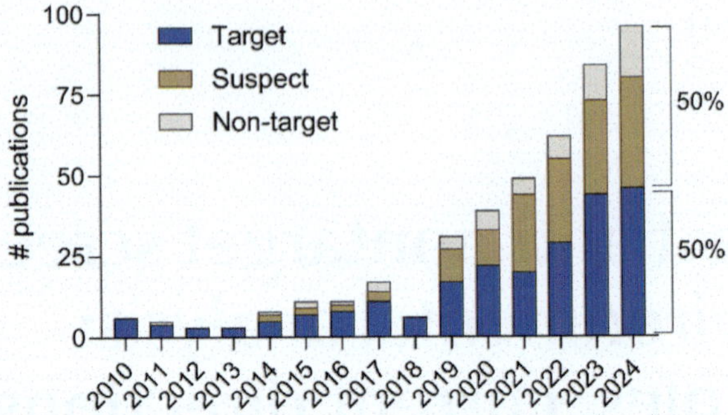

FIG. 9.1 Temporal trend of scientific literature making use of target (blue), suspect (dark brown) and nontarget (pale brown) screening approaches for environmental analyses. *(Data obtained from a document search in Scopus using the keywords "screening + contaminants + environment + Target/Suspect/Nontarget" present in the publication title, abstract or keywords sections.)*

means to investigate a much larger number of substances or chemical classes, for which no standards are available a priori. This strength is of utmost value and has driven the rapid expansion of numerous SS and NTS applications (Hernández et al., 2019; Hollender et al., 2023; Krauss et al., 2010; Menger et al., 2020).

In this chapter, the fundamentals of SS and NTS as well as the different instrumental configurations available to ensure a robust screening performance will be discussed in-depth. The most recent and innovative developments in the field will also be highlighted, including the hyphenation of ion mobility separation with HRMS (IMS-HRMS), the use of machine learning and/or in silico tools, and the application of effect-directed analysis for feature prioritization. In addition, special attention will be given to the emerging fields of semiquantitation and the evaluation of human exposure to environmental contaminants.

2 Fundamentals and instrumentation

2.1 General concept of HRMS data acquisition

Mass spectrometry has become the cornerstone for contamination profiling across environmental, food, and clinical analyses. While tandem low-resolution instruments (e.g., triple quadrupole, QqQ) have historically offered high sensitivity and robustness, they require an a priori selection of analytes, functioning essentially as mass filters. Such preselection limits their application in wide-scope screening analyses. In contrast, HRMS records all detectable ions, enabling SS and NTS across complex matrices without prior knowledge of the compounds (Lee, 2012).

Recent advances in HRMS have significantly improved sensitivity, making it competitive with classical low-resolution platforms for targeted analysis. HRMS instruments measure the mass-to-charge ratio (*m/z*) of all ionizable molecules above the detection threshold, offering an accurate and comprehensive chemical snapshot of the sample. By providing exact mass measurements and enabling tandem fragmentation, HRMS facilitates structural elucidation without the need for reference standards. Combined with extensive databases, HRMS has become indispensable for several research fields, for example, environmental exposure assessment (or exposomics), emerging contaminant studies, and retrospective analysis.

HRMS acquisition involves recording high-resolution, high-accuracy spectral data, allowing for the virtual detection of all ionizable CECs present in the injected sample. And thus, it allows for retrospective searches, prediction of molecular formulas, and structure elucidation via fragmentation patterns and database comparisons (Dührkop et al., 2019; Helmus et al., 2021).

2.2 Chemical coverage and domain

The concept of NTS chemical coverage refers to the breadth of molecules that can be detected under given experimental conditions. This coverage is primarily determined by the ionization technique used, the chromatographic separation, the polarity mode (positive or negative), sample preparation and the nature of the sample matrix (Hollender et al., 2023). In general, mass spectrometry platforms can detect a wide variety of chemical classes, including pharmaceuticals, pesticides, industrial chemicals, metabolites, and natural products.

However, there is no single method providing complete coverage to the whole chemical universe that NTS workflows aim to analyze and, therefore, a combination of techniques is necessary to maximize coverage. Typically, semipolar and polar compounds are well-covered under standard electrospray ionization (ESI) conditions, while nonpolar compounds require different ionization strategies (e.g., APCI or EI) with gas chromatographic (GC) separation. However, it is important to notice that nonionizable chemicals are still outside of the HRMS umbrella, unless other sample treatment techniques (i.e., derivatization) are applied (Hollender et al., 2023).

2.3 Sample collection and treatment for SS and NTS

Notwithstanding the importance of ionization technique in expanding the chemical coverage, a proper sample collection and sample treatment remain crucial steps in HRMS-based SS and NTS workflows, since they are critical points where potential analytes of interest may be lost. Samples should be collected in a way that preserves chemical integrity, avoids contamination, and is compatible with subsequent analytical methods.

For environmental matrices (e.g., water, air, soil), the use of clean (ideally baked if these are glass containers) or prerinsed containers is essential. Plastic containers should be avoided for lipophilic compounds. Ensuring that samples are kept cold or frozen when appropriate helps minimize degradation. In human biomonitoring, urine, blood, and saliva are commonly used, with anticoagulants or preservatives added when necessary. Sample volume and replicates should also be considered based on the sensitivity of the HRMS system and the goals of the study. Quality control (QC) samples, field blanks, and transport blanks should accompany every sampling campaign to account for potential contamination or losses during transport. When a collection of a field blank is not possible (i.e. sampling of tissues in an operating room), a rinse of the material used for sampling with ultrapure water can be performed instead (Caballero-Casero et al., 2021).

Once samples are collected, analytes need to be transferred to an extract compatible with the chosen technique, aiming to remove interferences while preserving the broadest range of chemicals covered. For this reason, generic extraction protocols are recommended in SS and NTS to maintain chemical diversity. Common sample preparation strategies include solid-phase extraction (SPE), protein precipitation, dilution, and liquid-liquid extraction (LLE). SPE is particularly useful for aqueous samples, as it concentrates analytes and removes matrix components, though the use of more than a single extraction phase enhances the chemical coverage (Black et al., 2023). For biological samples, protein precipitation using organic solvents (e.g., acetonitrile or methanol) is often applied, while sonication in solvent is also commonly used when extracting solid matrices (i.e. tissues, dust or air filters). Sample treatment must avoid introducing contaminants that may interfere with HRMS analysis. All reagents, solvents, and labware must be of high purity, and procedural blanks should be included to identify background contamination.

2.4 Chromatography and mass instruments

The instruments used for SS and NTS analyses are typically a hyphenation of a chromatographic system with a high-resolution mass spectrometer (normally a tandem Q-TOF or Q-Orbitrap), coupled via an interphase (ESI) or ionization source (APCI, EI). Chromatographic separation remains essential in HRMS workflows to reduce ion suppression by separating analytes of interest from matrix components and therefore increasing both identification ability and confidence (Lee, 2012).

Since HRMS requires gas-phase ions, ESI is typically used for the hyphenation of LC and HRMS. This interphase converts liquid-phase ions into gas phase by drying the solvent in microdrops generated in situ (Smyth, 2005) while atmospheric pressure photoionization (APPI) is useful for less polar analytes (Hanold et al., 2004), especially in environmental and food samples. On the contrary, GC provides gas-phase compounds in neutral phase, requiring ionization through techniques such as CI/APCI or EI (directly in EI, which implies high fragmentation, or indirectly by ionizing other molecules in the medium, as CI/APCI). EI is more widely used, as the availability of extensive spectral libraries (i.e. NIST) facilitates easier compound identification compared to CI/APCI (Lee, 2012).

Liquid chromatography (LC) and gas chromatography (GC) are the most commonly used separation techniques, depending on compound volatility and polarity. LC is generally applied to polar and semipolar compounds and is often used with electrospray ionization. Reverse-phase LC (e.g., C18 columns) is the standard, though hydrophilic interaction liquid chromatography (HILIC) may be used for very polar compounds. GC is preferred for volatile, nonpolar compounds and typically paired with electron ionization (EI), but also the possibility of ambient pressure sources (APCI) has been explored in the last years.

Gradient elution is employed in LC to resolve a wide range of compounds within a single run. For NTS, method robustness and reproducibility are prioritized over speed, especially for retrospective analysis, so generic methodologies are usually applied, featuring linear gradients and run times typically exceeding 15–20 min.

The most used HRMS instruments include Time-of-Flight (TOF) and Orbitrap, in tandem with low-resolution MS instruments such as quadrupoles. Each HRMS platform has distinct characteristics in terms of resolution, accuracy, dynamic range, and scan speed, which may also lead to different results depending on the chemicals of interest. QTOF mass analyzers offer high-resolution full scan and MS/MS capabilities with fast acquisition rates, making them suitable for both qualitative and quantitative analyses (Bhavyasri et al., 2023). They provide accurate mass measurements and are well-suited for data-dependent and data-independent acquisition. Orbitrap analyzers deliver very high resolution and mass accuracy, which is critical for distinguishing isobaric compounds. Orbitrap-based systems are ideal for metabolomics, exposomics, and complex mixture analyses, where mass accuracy and resolving power are essential.

The analysis of small molecules (<1000 Da) is normally better with Orbitrap instruments, while for the analysis of proteins TOFs are normally preferred. In addition, other analyzers are also available, such as Fourier transform ion cyclotron resonance (FT-ICR), which offer even higher resolution, though they are less commonly used due to their complexity and enormous operational costs.

2.5 HRMS acquisition modes

HRMS supports multiple data acquisition modes that influence how spectral information is collected and interpreted. The main strategies for MS^2 acquisition are data-dependent acquisition (DDA) and data-independent acquisition (DIA) (Hollender et al., 2023) (Fig. 9.2). Data-dependent acquisition (DDA) selects precursor ions (usually based on intensity) for MS/MS fragmentation during the run. While this allows rich structural information while minimizing interference from unrelated fragment ions, it can bias results toward more abundant compounds, potentially missing low-abundance features. Data-independent acquisition (DIA) overcomes this by fragmenting all ions within a predefined *m/z* range, generating

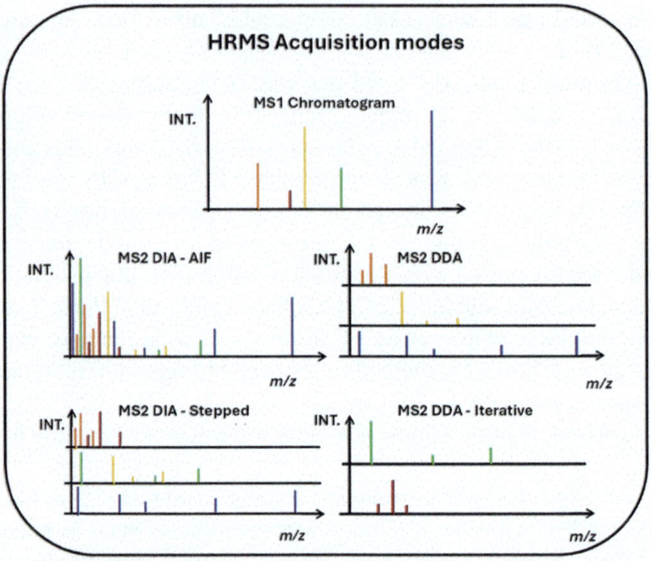

FIG. 9.2 Different acquisition modes in HRMS. From the same MS^1, DIA creates a high collision energy function with all the information together, while stepped DIA divides it in different acquisitions (despite there is still overlap), while DDA isolates only few ions from MS^1, and iterative DDA isolates more with extra injections.

comprehensive MS/MS data. However, additional processing is needed for associating fragment ions with parent ions, which may introduce errors.

In the last years, new strategies have appeared to make these acquisition modes more powerful. A summary of them is listed here:

1. *Stepped-DIA:* it is a modification of classical DIA, where instead of acquiring only one window at high collision energy for the whole mass range (e.g. *m/z* 50–1000), the first quadrupole isolates narrower *m/z* windows (e.g. *m/z* 50–200, 200–400, 400–750, 750–1000). This means that fragment ions associated to a parent with *m/z* 300 will only appear in the second window, therefore reducing the possibility of false assignment of fragments.
2. *Iterative DDA*: the main limitation of DDA is that only a fraction of the ions is selected in a single run, so MS^2 data for the nonselected ions is not available. However, the reinjection of the same sample avoiding the isolation of the parent ions with already acquired MS^2 helps enhancing the amount of information obtained. However, several sample reinjections are required, so this strategy has significant limitations when the sample set to analyze is large, due to time issues. This challenge can be addressed by the preparation of pooled samples which are then analyzed in iterative DDA.
3. *Postacquisition inclusion lists for DDA:* in order to help performing wide-scope SS analysis, a promising open-source strategy has been introduced recently (Giné et al., 2021). HERMES requires an MS^1 acquisition of

the sample of interest. From the MS^1 data, a list of molecular formulas of interest (e.g., from a priority-substance list) can be generated. HERMES uses this list to calculate the exact mass for each formula and search extracted ion chromatograms (XICs) for those *m/z* values across the entire chromatogram. Whenever HERMES detects a peak matching one of the target masses, it logs its retention time and produces one (or usually more) inclusion lists ready to be imported into your instrument—allowing to perform MS^2 experiments directly for those lists without any further MS^1 scans, obtaining information for the chemicals of interest. However, reinjection of the sample is still needed.

Recently, ASTRAL instruments commercialized by Thermo Scientific have hyphenated Orbitrap mass analyzers with a multireflecting TOF. Such configuration allows to get in-parallel information of low collision energy spectra (with the Orbitrap) and small-window (i.e., 2 Da, with frequencies of around 200 scans per second) high collision energy spectra (with the TOF), generating virtually clean MS/MS spectra for virtually all the ions in the whole mass range every few seconds (2–3 s), which may be categorized as a hybrid between DDA and DIA.

3 Suspect and nontarget workflows

HRMS generates highly complex raw datasets, often comprising hundreds of thousands of chemical features per sample (Lai et al., 2024). Manually reviewing such an immense volume of signals is impractical, making the use of automated or semiautomated data processing workflows essential. These workflows can vary significantly depending on multiple factors—including the research objective, sample matrix complexity, instrument configuration, and data acquisition strategy. Nevertheless, most SS and NTS approaches follow a broadly similar sequence of steps, aimed at maximizing chemical coverage while minimizing analytical bias. This section outlines the general structure of such workflows (Fig. 9.3), from data acquisition through processing and feature annotation, with an emphasis on maintaining an untargeted perspective at each stage to ensure detection of as many relevant analytes as possible.

3.1 Untargeted acquisition vs. untargeted processing

In NTS, the distinction between untargeted acquisition and untargeted processing is essential but often misunderstood. Both aim to maximize chemical coverage, yet they apply to different stages of the workflow. *Untargeted acquisition* refers to the instrumental strategy during data collection, where the mass spectrometer captures data without predefined precursor selection. As discussed in Section 2.5, this is typically done via full-scan MS^1 followed by MS^2 acquisition using DIA or DDA. While DDA can provide cleaner spectra, it is not fully untargeted—low-abundance features may be missed. Tools such as iterative inclusion/exclusion lists can improve coverage, but combining DDA and DIA

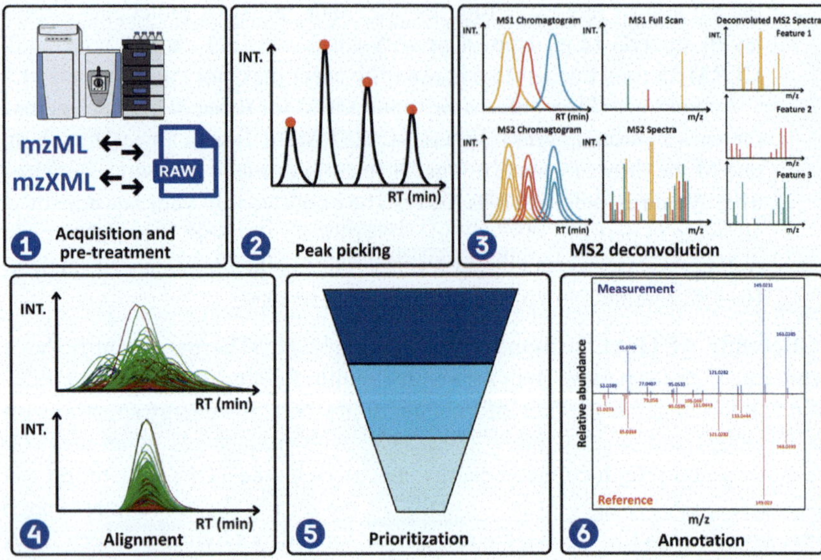

FIG. 9.3 Generic nontarget workflow. Common steps of a generic nontarget workflow, from acquisition through processing and feature annotation.

remains the best approach due to their complementary strengths. These methods ensure that a wide range of ions—irrespective of their expected presence or abundance—are captured, enabling retrospective analysis and annotation of unexpected compounds. The goal is to record the most chemically comprehensive dataset possible, often trading selectivity and sensitivity for inclusivity. In contrast, *untargeted processing* refers to how the raw data is evaluated postacquisition. Even if the acquisition itself is broad, processing will always introduce bias—for example, by filtering out low-abundance signals, using limited spectral libraries, or applying thresholds based on expected compound behavior. True untargeted processing seeks to avoid these constraints by using generic workflows for peak detection, alignment, and feature deconvolution, followed by compound annotation using open databases, in silico predictions, or cheminformatics tools. This ensures that data interpretation remains as inclusive and discovery-based as the acquisition. Ultimately, the strength of an NTS workflow lies in how well it combines both untargeted acquisition and untargeted processing. A targeted or semitargeted strategy at either step can narrow chemical space and limit the discovery of unknown or unexpected contaminants, transformation products, or novel biomarkers.

3.2 Generic workflow description

3.2.1 File conversion

Before any data processing can begin, HRMS raw files must often be converted into formats compatible with downstream software tools. While vendor-specific software can handle proprietary formats, open-source workflows typically require conversion to open formats such as .mzML or .mzXML. Tools such as ProteoWizard and MSConvert are commonly used for this step (Chambers et al., 2012; Holman et al., 2014). Though not part of data analysis itself, this step is essential to ensure accessibility, interoperability, and reproducibility in nontarget workflows.

3.2.2 Peak picking

Peak picking is the process of detecting signal features—representing potential chemical compounds—from raw spectral data. It transforms continuous mass spectrometry signals into discrete features based on *m/z* and retention time (RT), assigning intensity or area values to each. This step often involves thresholds for intensity and signal-to-noise ratio to distinguish true signals from background noise. Accurate peak detection is critical, as it forms the foundation for all downstream analyses. Algorithms may differ depending on whether MS^1 or MS^2 data is processed, and whether the data is in profile (continuous signal) or centroid (discrete peaks) format—the latter being a simplified representation often used to reduce data size and complexity.

3.2.3 MS^2 deconvolution

In datasets that include MS^2 spectra—particularly those generated from DIA or complex samples—fragments from multiple precursor ions often appear within the same scan. MS^2 deconvolution addresses this issue by separating the mixed spectra and correctly assigning fragments to their corresponding precursor features. This process refines and enhances the MS^2 spectra, leading to more reliable and higher-quality spectral matching. MS^2 deconvolution is often skipped when processing DDA data, as the MS^2 spectra are typically clean and derived directly from the precursor of interest. However, it becomes a crucial step in the analysis of DIA data, where it significantly enhances both the confidence and accuracy of compound annotation. To achieve this, algorithms typically leverage RT alignment, coelution behavior, and intensity correlation patterns.

3.2.4 Alignment

Slight variations in RT across multiple runs can cause features representing the same compound to misalign, making direct comparisons difficult. Alignment corrects for these shifts by matching corresponding features across samples, enabling meaningful and accurate comparisons. This process ensures consistent tracking of the same chemical across different samples or time points—an essential step in studies involving longitudinal data or large sample sets. Alignment is typically performed at the feature level using *m/z* and RT, and can be further refined by incorporating isotopic patterns and adduct groupings when available. A representative sample—often a QC injection repeated throughout the run—can serve as reference for alignment. After alignment, it is common practice to filter out features that provide redundant information, such as adducts or in-source fragments. These can often be readily identified based on retention time, mass differences, and intensity correlations, such as done by CAMERA algorithm (Kuhl et al., 2012). Removing them reduces data complexity and avoids unnecessary processing in subsequent analytical steps.

3.3 Processing software

These workflows require sophisticated tools to implement all the described steps, typically in the form of specialized software. The software landscape for NTS is broadly divided into vendor-specific solutions—developed by instrument manufacturers—and open-source platforms created by the scientific community. Both categories offer distinct advantages and limitations, and the choice between them depends on factors such as study design, available resources, and user expertise. However, open-source tools should be prioritized whenever possible, as they foster transparency, reproducibility, and collaborative development. They also offer greater flexibility and interoperability across analytical platforms—an essential feature in the rapidly evolving field of NTS. The most widely used open-source software tools for NTS workflows include patRoon (Helmus et al., 2021), Mzmine (Schmid et al., 2023), XCMS (Tautenhahn et al., 2012), openMS (Röst et al., 2016), and MS-DIAL (Tsugawa et al., 2015). XCMS, OpenMS, and patRoon are integrated into R or Python environments, enabling flexible scripting and advanced statistical analysis. In contrast, MZmine and MS-DIAL are user-friendly applications with graphical interfaces. Among them, MS-DIAL stands out for its versatility, as it uniquely supports both LC-MS (DDA and DIA modes) and GC-MS (EI and CI modes), making it one of the most comprehensive tools currently available.

3.4 Feature prioritization

Once the feature list is obtained, a wide range of strategies can be applied to filter and prioritize features, depending on the study's objectives. The most common approach is using online databases or suspect lists of certain families of compounds or contaminants suspected to occur in the samples (further discussed later in the following section). Structural characteristics can also guide prioritization. Isotopic pattern filtering helps identify halogenated compounds such as chlorinated or brominated chemicals, which often have environmental or toxicological relevance. In addition, mass defect filtering is useful for detecting features with specific elemental compositions, particularly when hydrogen atoms are replaced with heteroatoms. Normalized mass defect plots are commonly used to spot homologous series differing by repeating units (e.g., $-CH_2-$ or $-CF_2-$), such as in PFAS or hydrocarbon chains (Liu et al., 2019; Tang et al., 2023). Feature filtering can also be based on characteristic fragments or neutral losses, aiding the identification of compound classes with known substructures or functional groups. For example, specific ions or neutral losses can help detect polychlorinated biphenyls (PCBs), flame retardants, or other chemical families with characteristic structures (Guo et al., 2013). In both environmental and biological studies, statistical comparisons between sample groups—such as locations, time points, or treatment conditions—can help isolate features with differential presence or abundance. This supports the identification of trends, sources of contamination, or potential markers for exposure, transformation, or bioaccumulation.

3.5 Compound annotation strategies

The most common and straightforward strategy is matching feature lists to database entries to obtain tentative annotations. Open-access databases offer extensive and continuously expanding compound collections. Commonly used open-source databases include MassBank Europe (Neumann et al., 2026), MassBank North America, KEGG (Subramaniam, 2002), or GNPS (Wang et al., 2016). Comprehensive commercial options, such as mzCloud (Thermo Scientific, 2025) or Wiley/NIST (NIST mass spectral library, 2025) are also available and widely used. Although spectral libraries are continually expanding to keep pace with the rapid advancements in SS and NTS research and the identification of CECs classes, their limited compound coverage and suboptimal spectral quality often hamper the reliability of compound annotation. Implementation of quality control measures for spectral quality can address these issues (Oberacher et al., 2020).

In addition, specialized databases exist for specific research areas—for example, toxicology (EPA Comptox Chemical Dashboard (Williams et al., 2017), Toxic Exposome Database (Wishart et al., 2015), exposomics (PubChemLite (Bolton et al., 2020), The European Human Biomonitoring Dashboard (Gilles et al., 2021), or environmental science (CECscreen (Meijer et al., 2021), NORMAN Suspect List Exchange (SLE) (Taha et al., 2022)), among others. In addition to commercially available standards, researchers often develop in-house suspect lists tailored to specific chemical groups, such as transformation products or compound classes of interest. These lists are central to SS workflows, where expected compounds are searched for in the dataset without confirmed reference standards. A notable contributor in this area is the NORMAN Network (Dulio et al., 2018), which supports open-access tools and resources for suspect screening. Their Suspect List Exchange (NORMAN-SLE) (Taha et al., 2022) compiles over 100 curated lists with >120,000 substances, while the Digital Sample Freezing Platform (DSFP) (Alygizakis et al., 2019) enables retrospective analysis of stored HRMS data. However, while resources for LC-HRMS are well developed, databases compatible with GC-HRMS—apart from NIST—remain scarce (Casey et al., 2023).

Despite database comparisons, most detected features remain unannotated. When MS^2 data is available but compound identity is still unknown, in silico fragmentation tools offer a powerful solution by generating candidate structures. Popular tools such as SIRIUS + CSI:FingerID (Dührkop et al., 2019), CFM-ID (Allen et al., 2014), MetFrag (Ruttkies et al., 2016), and MS-FINDER (Lai et al., 2018) simulate how molecules fragment and match predicted spectra to experimental data to suggest possible structures or formulas. Among them, SIRIUS v5 stands out for integrating AI and deep learning—its CSI:FingerID module uses neural networks trained on large spectral libraries to predict molecular fingerprints and improve structure ranking accuracy.

Additional powerful tools for annotation, which are openly available within the GNPS platform, include Feature-Based Molecular Networking (FBMN) (Nothias et al., 2020), classical Molecular Networking (MN), and Network Annotation Propagation (NAP) (da Silva et al., 2018). These approaches leverage the similarity among MS/MS spectra to cluster structurally related compounds and facilitate annotation propagation across unknown features. FBMN builds networks that connect and cluster together features with similar fragmentation patterns, aiding to identify the structure of unannotated features clustering with others with known structure. NAP enhances this by integrating in silico fragmentation predictions to propose candidate structures for unannotated features based on their network neighbors, effectively improving coverage and confidence in compound identification. Together, these tools enable semiautomated, scalable annotation of complex datasets by exploring relationships within the chemical space of the sample.

When compound annotation is not the primary goal, generic workflows focus on comparing sample groups and exploring the chemical landscape. These approaches aim to identify trends in feature abundance, such as distinguishing exposure groups or tracking temporal changes. Statistical methods further analyze feature intensities and distributions across groups, offering insights into environmental exposures, disease markers, or chemical transformations. This approach is particularly useful in exploratory studies focused on linking chemical features to conditions without requiring full identification.

3.6 Quality assurance and quality control (QA/QC) measures in SS and NTS

Given the high complexity of untargeted acquisition and data processing approaches, these workflows call for comprehensive quality control measures to assess the reliability of the reported compound annotations. Minimum standards for QA/QC and method validation are well defined and established in the field of targeted analyses. However, such implementations face major difficulties in the SS and NTS field as analytes are not a priori defined and their reference standards are lacking. Initiatives within the NORMAN and Human biomonitoring for Europe (HBM4EU) networks have proposed guidelines for QA/QC measures in SS and NTS studies aiming to increase harmonization of these measures in future applications (Hollender et al., 2023).

QA/QC measures should cover the whole analytical workflow, starting from sample collection through sample preparation, instrumental analysis and, finally, data processing and interpretation. Sampling strategies should aim at maximizing the representation of environmental chemical patterns while minimizing both compound losses and contamination. The latter should be avoided by using high-quality solvents and sampling devices, preferably made of glass, stainless steel, or plasticizer-free polymers.

Sample collection should be accompanied by the inclusion of field blank samples allowing to assess possible contamination introduced by the sampling containers or handling process. Prior to sample extraction, samples are recommended to be fortified with isotopically labeled internal standards (IS). Similar to quantitative targeted studies, IS are utilized to assess matrix effects and account for potential analyte losses or degradation during sample preparation and storage. Their use in SS and NTS studies can be further expanded to support the assessment of RT and signal stability (Ng et al., 2020), mass accuracy and facilitate sample alignment during data processing (see Section 3.2.3).

QA/QC procedures for sample preparation can be further expanded by the inclusion of QC samples, composed of blank matrix or, if the latter is not available, synthetic matrix, spiked with a set of QC compounds. This will allow the assessment of analyte recoveries, inter- and intraday variability, method specificity and true positive annotation rates. The spiked QC compounds can be composed of analytes representative for compound classes of interest and/or expected to be present in the analyzed matrix. Alternatively, QC mixtures proposed specifically for SS and NTS applications can be used (Knolhoff et al., 2021). If widely implemented across different laboratories such mixtures can facilitate method harmonization as well as comparison and interlaboratory reproducibility of screening results.

Spiked QC samples implemented during sample preparation can be used to assess the coverage of the instrumental method allowing to estimate which compound classes potentially remain undetected due to poor separation or ionization efficiency. Further QC measures recommended within instrumental analysis include the randomization of the sample injection order, inclusion of solvent blanks for carry-over assessment, periodical reinjection of QC samples to ensure stable sensitivity.

Harmonized QA/QC procedures for all steps of data processing in SS and NTS studies are lacking, given the high variety of available software tools (Hohrenk et al., 2020). Guidelines for the assessment of quality and robustness of MS^1 data have been introduced recently and await wide implementation in the field (Lennon et al., 2024). Applied processing workflows should be kept consistent throughout all analysis batches and be reported in a detailed and transparent manner. At the end of an SS or NTS workflow, a list of annotated compounds is composed. Since the confidence of annotation is dependent on the quality of the data obtained for each compound and can vary between the latter, a harmonized and transparent system for the reporting of annotation confidence is needed. Available schemes are discussed in the following section.

3.7 Data reporting

The first scheme for the reporting of annotation confidence in SS and NTS studies was introduced by Schymanski et al. (2014). Since then, it has been widely adopted in the field. Following this scheme, annotated structures are reported with a confidence level (CL) ranging between one and five (Fig. 9.4). CL1 is achieved if a compound is unequivocally confirmed through comparison with a reference standard. If no reference standard of the annotated compound is available, CL2 to CL5 (with decreasing annotation confidence) is assigned based on the available data.

Nevertheless, the assignment of confidence levels requires expert judgment and is consequently subject to a degree of subjectivity. This might lead to discrepancies in CL assignments between studies. This issue was addressed through the introduction of clear guidelines for CL assignments following a scoring system (Alygizakis et al., 2023). In addition, separate annotation schemes tailored to specific compound classes have been introduced with the scheme for PFAS annotations introduced by Charbonnet et al. being a widely adopted example (Charbonnet et al., 2022). For GC-HRMS–based SS and NTS studies, the scoring framework of Koelmel et al. provides guidelines for the communication of compound annotations (Koelmel et al., 2022).

Available QA/QC guidelines unanimously advise on transparent and detailed reporting of the chosen communication scheme of annotation confidence and the considerations and applied cut-off values leading to particular confidence levels. To further harmonize the reporting of SSA/NTS results and provide tools for reviewers' assessments of research quality, a Nontargeted Analysis Study

		Data requirements	
CL1	**Confirmed structure** by reference standard	MS, MS^2, RT, Reference standard	CCS ≤ 2%
CL2	**Probable structure** a) library spectrum match b) diagnostic evidence	a) MS, MS^2, Library MS^2 b) MS, MS^2, characteristic MS^2	$CCS_{library}$ ≤ 2% $CCS_{predicted}$
CL3	**Tentative structure** Substructure, isomers, class	MS, MS^2, experimental MS^2	$CCS_{predicted}$ CCS-*m/z* trendlines
CL4	**Molecular formula**	MS, isotope/adduct	CCS
CL5	**Exact mass**	MS	CCS

FIG. 9.4 Annotation confidence level scales (CL) for LC-HRMS and IMS-LC-HRMS workflows. Confidence levels (CL) used for the communication of identification confidence for compounds identified through target, suspect or nontarget screening based on the schemes proposed by Schymanski et al. (2014) and Celma et al. (2020).

Reporting Tool has been introduced and widely adopted in the scientific community (Peter et al., 2021).

3.8 FAIR data sharing—repositories

Open-source data-sharing repositories for SS- and NTS-derived raw data and results from studies on environmental contaminants play a critical role in the implementation of Findability, Accessibility, Interoperability, and Reuse (FAIR) principles. The use of open-source data processing tools and/or R or Python-based approaches (Section 3.3) facilitates open sharing of underlying processing scripts on repositories such as Zenodo or GitHub.

The provision of raw HRMS data, allowing the assessment of reproducibility of data processing approaches, alongside the corresponding publication is a recommended practice even though it is not yet widely implemented in the field.

Extensive open-source data repositories have been established within the NORMAN network including the NORMAN suspect list exchange and the digital sample freezing platform (DSFP) latter of which contains LC- and GC-HRMS–derived data provided by partner laboratories (Alygizakis et al., 2019).

4 Latest technical developments and innovations in SS and NTS analyses

The field of SS and NTS is rapidly evolving with outstanding technical innovations as well as the hyphenation and/or combination of orthogonal techniques. The inclusion of ion mobility separation as an additional separation dimension, the utilization of machine learning strategies or the combination of biological and chemical analyses for toxicants identification are examples which will be further discussed in the following sections.

4.1 Ion mobility separation hyphenated to high resolution mass spectrometry

Ion-mobility separation (IMS) measures the drift of ions through an (inert) buffer gas under the influence of an electric field and was first described in 1898 (Zeleny, 1898). Already in the early development stages, IMS was coupled to a mass spectrometer, providing an additional separation dimension (Cohen and Karasek, 1970). Initially, applications of IMS-HRMS were limited to in-house developed instrumentation in an academic setting (Wu et al., 1998). In 2006, the first commercially available IMS-HRMS instrument, the Synapt IMS time-of-flight MS, was introduced by Waters (Pringle et al., 2007). Since then, several instruments from different vendors were introduced incorporating different principles of ions' mobility measurement. Most commonly used instrumental set-ups include drift-tube IMS (DTIMS), travelling wave IMS (TWIMS), structures for lossless ion manipulations (SLIM), and trapped IMS (TIMS) each of which is implemented in a commercially available instrument.

Generally, IMS measures an ion's mobility through a buffer gas under the influence of a weak electric field. Under ideal measurement conditions, an analyte's mobility (K) is defined as the ratio between the ion's apparent drift velocity ($v_d = L/t_d$, where L is the length of the drift tube and t_d is the drift time) and the applied electric field (E, [$V \cdot m^{-1}$]), as displayed in Eq. (9.1).

$$K = \frac{v_d}{E} \tag{9.1}$$

From the reduced mobility K_0 (corresponding to mobility K, normalized to standardized pressure (101.325 Pa) and temperature (273.15 K) (Gabelica et al., 2019), a collision cross section (Ω) value describing the rotationally averaged surface of an ion can be calculated. This is based on a modified version of the fundamental zero field limit equation, commonly referred to as the Mason-Schamp equation (Mason and Schamp, 1958; Revercomb and Mason, 1975) where k_b is the Boltzmann constant, z is the charge state, m_i is the mass of the ion, m_b is the mass of the drift gas, and N is the number of gas molecules per unit volume:

$$\Omega = \frac{1}{K_0} \cdot \frac{(18\pi)^{1/2}}{16} \cdot \frac{ze}{(k_b T)^{1/2}} \cdot \left[\frac{1}{m_i} + \frac{1}{m_b}\right]^{1/2} \cdot \frac{1}{N} \tag{9.2}$$

DTIMS is the only instrumental set-up allowing a primary measurement of an ion's mobility from which the CCS value can be directly derived based on Eq. (9.2). The calculation of CCS values on other established instrumental platforms (TIMS, TWIMS, single-field DTIMS) requires the use of external calibrants with available reference CCS values.

As IMS measurements operate on a millisecond time scale, they can easily be implemented between a chromatographic separation (seconds to milliseconds) and mass analysis (in microseconds). This hyphenation provides several benefits for SS and NTS studies including the improvement of spectral quality in DIA fragmentation through drift-time alignment of parent and fragmentation ions and increased sensitivity through separation of interfering matrix components in the drift time dimension (Celma et al., 2021; Menger et al., 2022). The main focus of IMS for SS and NTS studies in the environmental field was laid on the utilization of CCS values as an additional identification parameter. This development was facilitated by early implementation of CCS values in the well-established scheme used for the reporting of identification confidence introduced by Schymanski et al. (Fig. 9.4) (Schymanski et al., 2014; Celma et al., 2020). Such applications commonly rely on the comparison of experimental CCS values with reference CCS databases allowing feature filtering based on established ΔCCS thresholds. Within the exposomics field, for example, numerous CCS databases for various classes of xenobiotics and their metabolites have been reported (Belova et al., 2021; Joseph et al., 2025; Mullin et al., 2020; Song et al., 2022). Cumulated databases including, for example, AllCCS (Zhang et al., 2023), the Unified CCS Compendium (Picache et al., 2019), or recent implementations of IMS data in PubChem (Elapavalore et al., 2025) aim at a consolidated collection of CCS supporting the parallel use of several databases. This often results in database transfer between different IMS instrumental set-ups raising the question of interinstrumental reproducibility of mobility measurements and CCS calculations. These questions have been addressed for several combinations of IMS instrumentations (Belova et al., 2022; Dodds et al., 2025; Feuerstein et al., 2022; Hinnenkamp et al., 2018). Despite the ever-growing number of experimental reference CCS databases, their coverage remains limited, especially for newly discovered environmental contaminants or their metabolites. To fill this gap, several machine learning-based prediction models have been introduced aiming to provide CCS values for which no reference standards and/or reference CCS value is available (Zhang et al., 2023; Bijlsma et al., 2017; Celma et al., 2022).

4.2 Use of machine learning and in silico tools

Machine learning has revolutionized the SS and NTS applications over the recent years. With one of the main setbacks of untargeted studies being the lack of reference standards, the need for additional evidence to support analytical findings becomes of paramount importance. In this regard, the use of in silico tools based on artificial intelligence can foster, smooth, and even permit the structural elucidation of a candidate, enhance the identification confidence, or provide an estimation of its concentration in a sample. Following, we will discuss some major developments.

4.2.1 Retention time prediction

Retention time (RT) reflects how a compound interacts with both mobile and stationary phases under specific chromatographic conditions, making it a critical parameter in NTS. In discovery-driven NTS workflows, tentatively identified CECs or transformation products (TPs) often lack known RT data. However, advances in traditional quantitative structure–retention relationship (QSRR) models and modern machine learning (ML) now enable in silico RT prediction, offering a powerful tool for narrowing down candidate structures. Current prediction models link molecular descriptors (e.g., logP, polar surface area, hydrogen bonding) and fingerprints with chromatographic retention using regression or ML algorithms. The use of retention time indices (RTIs) instead of raw RT enhances comparability across systems, enabling annotation verification even under differing chromatographic conditions.

RT prediction tools were initially developed for metabolomics applications in LC, such as Retip (Bonini et al., 2020) (ML-based) and QSRR Automator (Naylor et al., 2020) (QSRR-based), and have since expanded to more general setups. Notably, Aalizadeh et al. (2021b) introduced a method combining calibrant injection with ML-based prediction to validate annotations, after building a calibration curve, implemented in the NORMAN Digital Sample Freezing Platform (Alygizakis et al., 2019). Additional tools such as MultiConditionRT (Souihi et al., 2022) and RT-predict (Zakir et al., 2025) have since emerged. Recently, Boelrijk et al. (Boelrijk et al., 2023) proposed RT prediction using MS^1 fragmentation data. In GC workflows, while models often use boiling point, vapor pressure, or Kovats retention indices (RI), the availability of RI in databases reduces the need for predictive modeling compared to LC.

RT prediction significantly reduces the number of candidate structures, especially when spectral libraries are incomplete. It offers orthogonal evidence alongside accurate mass and MS^1 data and can be tailored to specific chromatographic setups. However, limitations remain. RT is highly method-dependent—affected by column, solvent, gradient, temperature, and ionization—making predictions context-specific. Limited, diverse training data hampers generalizability, and incorrect chemical representations (e.g., stereoisomers, tautomers) can skew predictions. In addition, models often underperform for polar or ionizable compounds due to insufficient descriptor coverage or training datasets population.

Incorporating RT prediction into NTS workflows adds valuable selectivity and strengthens structural assignment by providing an independent criterion to support or refute candidate identities. As open-source tools and curated datasets grow, model accuracy continues to improve, particularly with deep learning integration. Broader adoption, however, requires standardization and transparent performance reporting. Collaborative efforts, such as those led by the NORMAN Network (Dulio et al., 2018) or EU projects, are essential for building cross-platform RT databases that underpin robust, generalizable models.

4.2.2 *In silico-based semiquantification of features in SS and NTS*

One of the main limitations of SS and NTS compared to targeted analysis lies in their qualitative nature. While targeted analysis enables the identification and quantification of known compounds using authentic reference standards, SS and NTS approaches typically lack this certainty. Nonetheless, converting qualitative detections into quantitative information is essential for risk assessment, as concentration levels are crucial for evaluating potential toxicity in environmental or human health contexts (Kruve, 2020).

The ideal strategy involves confirming the presence of a compound using an authentic standard, followed by its quantification. However, this is not always feasible due to the presence of unknown transformation products, metabolites, or compounds that are prohibitively expensive to purchase or synthesize, or even unavailable. To overcome these challenges, researchers have adopted semiquantification, which aims to estimate concentrations using surrogate compounds—chemicals with similar instrument response—rather than exact standards. This approach relies on the concept of ionization efficiency, that is, the percentage of neutral analyte molecules that are converted into gas-phase ions during the ionization step.

Despite its utility, semiquantification can introduce considerable uncertainty. As discussed in Malm et al. (2024), several strategies have been evaluated:

1. *Use of parent compounds to estimate concentrations of TPs*: While conceptually straightforward, this approach showed a median error of 3.4× (mean: 13×), indicating frequent over- or underestimation. Moreover, parent compounds might not always be available.
2. *Use of structurally similar compounds*: This strategy yielded similar errors (mean: 17×, median: 3.4×) but is more versatile, as suitable analogs can often be sourced for analysis.
3. *Use of close-eluting compounds*: This approach was the least reliable (mean error: 150×), showing that proximity in retention time does not imply similar ionization behavior. However, it has been the most commonly applied so far for its simplicity.
4. *Ionization efficiency prediction models:* Machine learning approaches such as random forest or multiple linear regression have shown the greatest promise. Studies such as Liigand et al. (2020) and Aalizadeh et al. (2021a, 2022) reported mean errors around 5.5× and 11×, respectively—generally within one order of magnitude. These methods currently represent the state-of-the-art for semiquantitative estimation in SS and NTS.

Still, important limitations persist. Matrix effects are not yet integrated into most predictive models (which are typically trained in pure solvent), recovery rates are often neglected, and in-source fragmentation can result in underestimation. These issues must be addressed to increase the robustness and transferability of semiquantification in routine applications.

4.2.3 Toxicity prediction

Beyond concentration levels, the ultimate goal of environmental and exposomics analysis is to evaluate the potential toxicity of detected compounds. Recent studies have made substantial advances in predicting toxicity directly from HRMS data, without requiring chemical quantification or even full structural elucidation.

A notable contribution is MS2Tox, developed by Peets et al. (2022), which uses machine learning to predict acute toxicity (LC_{50}) for fish from raw MS^1 spectra. The approach relies on structural fingerprints generated by SIRIUS + CSI:FingerID (Dührkop et al., 2015; Hoffmann et al., 2022) and was trained on curated toxicity datasets from CompTox. The models achieved root mean square errors (RMSE) of around 1 log-mM for five different endpoints, with best performance observed for fish static LC_{50}. Validations using spiked samples and MassBank spectra confirmed the method's effectiveness, enabling fast hazard prioritization of unknown compounds without the need for authentic standards or additional bioassays.

In a follow-up study, the same authors extended this approach to predict structural alerts—substructural units associated with known toxic effects—directly from MS^1 spectra, bypassing the fingerprinting step. These models were applied to real environmental water samples and successfully flagged potentially hazardous features, even across different instrument platforms. This methodology offers a scalable solution for prioritizing unknowns in NTS workflows, especially in fields such as exposomics, environmental surveillance, and effect-directed analysis (see Section 4.3).

Despite the promise of these tools, several limitations must be considered:

1. *MS^1 spectral quality is critical*—poor or noisy spectra can compromise predictions.
2. *Annotation software plays a central role*—since these models are built upon annotations, any inaccuracy can propagate.
3. *Blanks and background contamination must be carefully reviewed*—signals arising from procedural blanks may lead to false positives if not properly managed.

In conclusion, while tools such as MS2Tox and structural alert classifiers have transformed the field by enabling risk-based prioritization without full identification, careful implementation and validation are essential to harness their full potential.

4.3 Effect-directed analysis

Feature prioritization is key for a successful and meaningful application of NTS analysis of complex matrices. Ideally, NTS should be able to decipher the complete chemical profile of complex matrices; however, in reality, the annotation of every single feature is unfeasible and extremely time consuming. Thus, incorporating tools that help discriminating relevant features to focus the annotation and identification efforts is of great importance (Hernández et al., 2019; Hollender et al., 2023; Dulio et al., 2018). In this sense, the combination of NTS with biological analysis can assist in speeding up the process by discarding non-biologically relevant samples.

Effect-directed analysis (EDA) has recently emerged as a promising iterative approach that combines biological and chemical analyses to identify toxic chemicals in environmental samples. When combined with sample fractionation, EDA allows narrowing down the sample fraction in which toxic activity can be observed and thus focus the identification efforts on features eluting in a narrowed RT window (Dopp et al., 2019; Liu et al., 2024; Radke et al., 2025).

In brief, the EDA workflow often starts with an overall toxicity assessment per sample (Fig. 9.5) which permits that samples which have not shown any overall toxic effect can be unranked for the untargeted toxicants search. Later, samples for which a certain toxicity threshold has been exceeded are

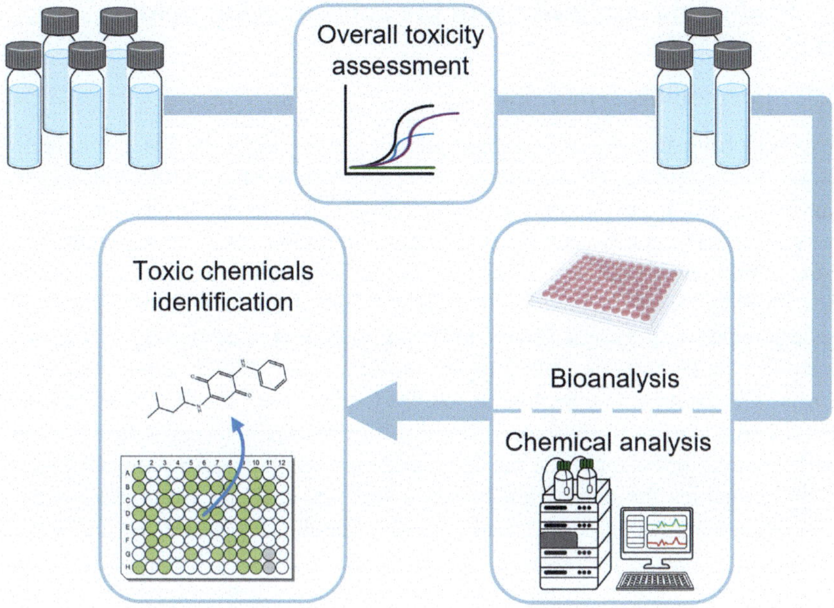

FIG. 9.5 Schematic workflow for effect-directed analysis approaches. Starting with overall toxicity assessment, samples can be prioritized for fractionations. Feature prioritization is based on active fractions.

fractionated by means of liquid chromatography and both biological and chemical analyses are performed in the fractions, allowing the detection of which sample fractions are active to the selected bioassay, and thus detecting in which RT window the toxicant is being eluted. If needed, the cycle fractionation—bioanalysis—chemical analysis could be iteratively repeated with different chromatographic chemistries to separate isobaric interferents and facilitate the toxic chemical identification. Moreover, the application of a panel of different bioassays will also allow identifying toxicants for different endpoints and expand the chemical and toxicological coverage of the approach (Liu et al., 2024; Luo et al., 2024).

Recently, an alternative approach within EDA incorporates multivariate statistics to combine identified patterns in bioanalysis with mass spectrometric features (either temporal, spatial, or other kinds of trends) to prioritize hitherto unknown features (Alvarez-Mora et al., 2025b), making EDA even more powerful. However, the so-called virtual-EDA requires a very comprehensive sample set with sufficient variability in both bioactivity and feature intensities (Alvarez-Mora et al., 2025b).

Overall, EDA has shown a strong potential for unraveling toxic chemical compositions in complex matrices as well as identifying previously unreported chemicals by combining state-of-the-art feature identification strategies with bioanalysis feature prioritization.

5 Real-world applications of the latest developments in SS and NTS

5.1 Ion mobility separation

Profiting from the above-described advantages provided by IMS, this technique has significantly facilitated the annotation of environmental contaminants in various sample types, especially for the analysis of per- and polyfluoroalkyl substances (PFAS). This compound class shows characteristic trendlines for the relationship between CCS values and *m/z* as a result of a high degree of fluorination leading to increased mass without significant enlargement of the gaseous shape of the ion (i.e., increased density). CCS-*m/z* trendlines have been described for several classes of PFAS in the context of large CCS databases (Joseph et al., 2025; Dodds et al., 2020). These characteristic trends have furthermore been used for feature filtering within nontargeted metabolomic analyses: the described clustering of PFAS in the CCS-*m/z* domain enables the distinction between features deriving from endogenous (nonhalogenated) compounds and potential xenobiotics (Foster et al., 2022). Finally, IMS-based analyses of PFAS provide examples for the separation of isomeric structures based on differences in their gaseous mobilities. For example, a study analyzing influent and effluent samples from >140 wastewater treatment plants reported the annotation of 38 branched PFAS isomers revealing high structural variability which might be overlooked if only chromatographic separation is used (Mu et al., 2023). Given the numerous advantages provided by IMS for PFAS analysis, the widely used PFAS-specific data processing and annotation software tool FluoroMatch was expanded allowing the implementation of CCS data into the identification workflow (Smolinski et al., 2025).

Similar to the annotation of PFAS based on characteristic IMS-derived data, CCS-*m/z* trendlines can also be used as an additional annotation parameter for other classes of environmental contaminants. A suspect screening study on quaternary ammonium compounds (QACs), representing a class of surface-active compounds commonly used in cleaning and disinfecting products, compared CCS data obtained for suspect QACs with CCS-*m/z* trendlines established from legacy QACs, for which reference standards were available. Suspect QACs matched established trendlines showing that IMS facilitates the annotation of contaminant classes covering wide homologue series.

Following the rapid developments of in silico CCS prediction tools, their implementation in SS and NTS studies is increasing. Several applications have described the use of predicted CCS data for feature prioritization, increased coverage of compounds for which experimental reference CCS data is lacking (e.g., transformation products, metabolites etc.) or the reduction of candidate structures (Bijlsma et al., 2019; Song et al., 2023).

5.2 Semiquantification and risk assessment

As previously mentioned, the growing adoption of SS and NTS has significantly expanded the role of cheminformatics in environmental and human exposure studies. This approach enables researchers to extract valuable semiquantitative information from a vast array of detected chemicals. Importantly, such information can also be used to infer toxicity data, which is highly relevant for assessing the potential hazard of these substances. Several recent studies have demonstrated the potential of combining semiquantification and toxicity prediction tools within NTS workflows to improve risk assessment efforts.

1. One such example is presented by Castaño-Ortiz et al. (2024), who explored the use of HRMS-based SS and NTS as complementary or alternative strategies to classical targeted analysis for detecting pharmaceuticals and their TPs in the environment. Their study covered multiple environmental matrices, including water, sediment, biofilm, and fish tissue. While traditional targeted analysis detected 41 pharmaceutical compounds, SS and NTS identified 38 additional chemicals, of which 14 were unique to the NTS approach. Notably, TPs were often found at higher concentrations than their parent compounds. Semiquantification was employed using predicted ionization efficiencies to estimate chemical concentrations by both strategies (Fig. 9.6A). Although this approach introduces uncertainty—potentially exceeding tenfold errors due to variable ionization efficiency, matrix effects, and the absence of calibration curves (Malm et al., 2024)—it still provides valuable data for early-stage screening and prioritization. The study also highlighted the need for matrix-specific models to mitigate the strong influence of matrix effects on quantification accuracy.
2. Another example is found in the study by Gutiérrez-Martín et al. (2023), which applied semiquantitative tools trained specifically on human urine samples. These models enabled the detection and quantification of various chemicals, including both phase I and phase II human metabolites (Fig. 9.6B), without requiring prior deconjugation steps—thanks to the enhanced acquisition capabilities of HRMS. The authors reported the successful semiquantification of 35 chemicals, with estimated concentrations ranging from 49% to 175% of the actual measured values. This result underscores the potential of semiquantitative strategies in complex biological matrices, provided that models are specifically trained for those environments.
3. A further advancement in the field was described by Sepman et al. (2023), who developed MS2Quant, a machine learning-based quantification model that bypasses the need for full structural identification (Fig. 9.6C). Using MS^1 fragmentation spectra, the *xgbTree* regression algorithm was trained on ionization efficiency data for 1191 unique

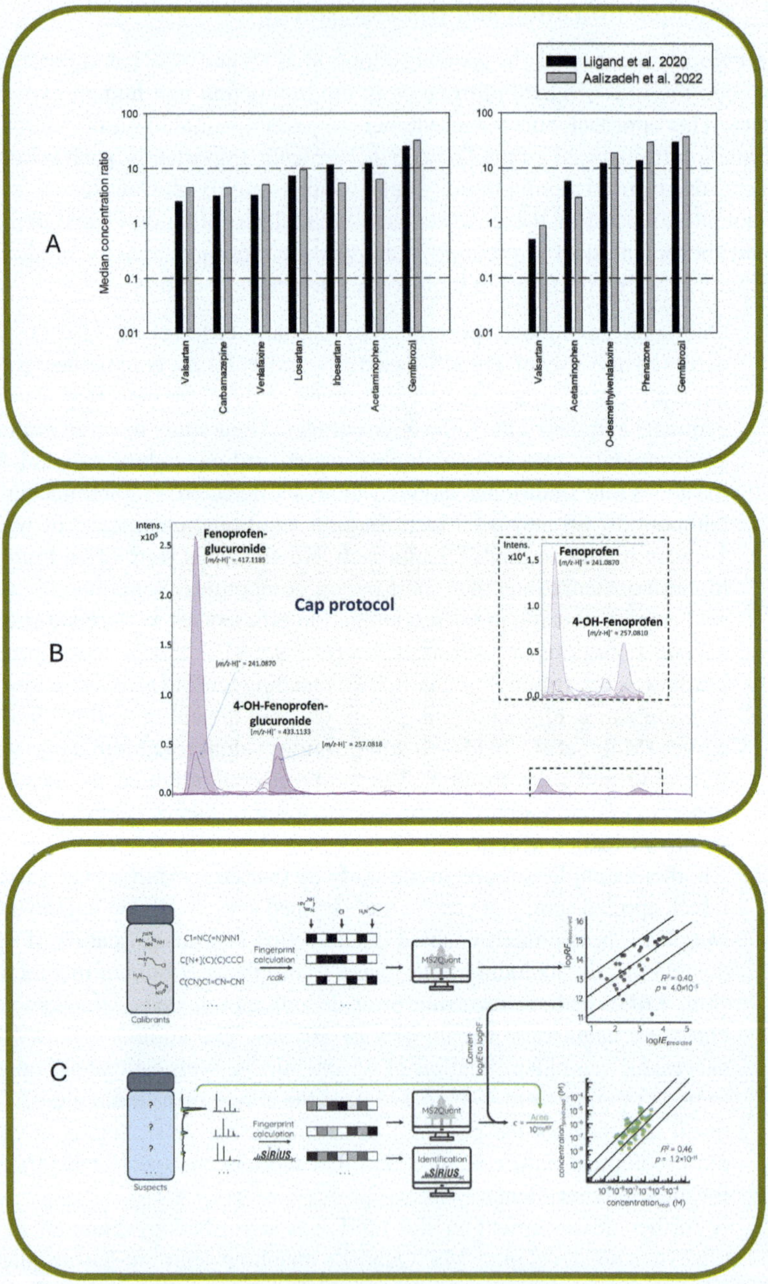

FIG. 9.6 Semiquantification case studies. Case studies demonstrate the applicability of semi-quantification in (A) environmental samples, (B) human fluids and (C) toxicity prediction without identifying chemicals.

chemicals spanning eight orders of magnitude. The model achieved prediction errors between 4× and 7×—comparable to models requiring complete structural information. With the help of tools such as SIRIUS+CSI:FingerID, MS2Quant can predict structural fingerprints from MS^1 data alone, allowing for exposure estimation of unidentified, yet detected, chemicals.

Collectively, these studies highlight the powerful role of semiquantification and in silico structure prediction in enhancing the utility of NTS workflows. Despite some limitations, these tools represent a promising foundation for early warning systems in both environmental monitoring and human biomonitoring applications, enabling more proactive and informed responses to emerging chemical risks.

5.3 Effect-directed analysis

While EDA strategy is very promising, its application is also very challenging (Liu et al., 2024; Alvarez-Mora et al., 2025b) and, as a consequence, there is a limited number of authentic EDA studies due to the high level of instrumental sophistication needed. However, over the recent years, some successful applications have been published in which the full power of both bioanalytical and chemical tools is applied. Following, three main contributions to the field are highlighted:

1. Recently, Baygildiev et al. conducted a study in surface and well water from the Upper Rhine in order to identify bioactive compounds that could be classified as persistent, mobile and toxic (PMT) (Baygildiev et al., 2025). To this end, a panel of multiple bioassays was implemented. However, authors facilitated the identification of bioactive chemicals by using bioassay-specific suspect lists instead of noncurated broad lists which might yield several false positive annotations. Such bioassay-specific suspect lists were created based on in-house databases of active chemicals for the bioassays under study. This suspect list curation prior to structural elucidation by means of NTS workflows, permitted the identification of several chemicals in the samples.
2. Similarly, another study was able to significantly reduce the number of features initially detected in an EDA-NTS study of hospital wastewater by predicting the toxicity of the spotted features (Alvarez-Mora et al., 2025a). While NTS initially revealed over 5000 features of potential interest, toxicity prediction based on molecular fingerprints derived from MS^1 spectra, without spectral annotation, by means of MLinvitroTOX (Arturi and Hollender, 2023) narrowed down the list of features to further investigate to approximately only 250. In addition, Alvarez-Mora et al. estimated semiquantitatively the concentration of the identified chemicals to explain the observed activity, resulting in one of the more

complete and comprehensive studies for EDA and NTS making use of several complementary tools.

3. Lately, another EDA study became a milestone in the environmental analysis since it identified a major toxicant being released from vehicle tires as the reason for an enhanced mortality of coho salmon in small creeks (Tian et al., 2021). Apart from the high environmental relevance of this study, it also showcased how the combination of iterative EDA applications can hugely influence the identification capabilities. Therein, Tian et al. utilized 6 consecutive chromatographic fractionation—bioanalysis iterations until and unequivocal isolation and structural elucidation of the toxic chemical was possible, reducing the number of features detected by NTS in the bioactive fraction from over 2000 to only 4 after consecutive iterations with different chromatographic column chemistries.

These cases herein described are examples of EDA applications that represent a step-forward into a more widely spread implementation of EDA to focus, prioritize and narrow down analytical efforts in NTS for the identification of toxic and relevant contaminants.

5.4 Chemical exposome

The use of NTS has become a powerful strategy for the discovery of unknown CECs across diverse matrices. One growing application is the study of the human chemical exposome, which seeks to characterize all environmental exposures throughout an individual's lifespan—including many CECs. NTS has been applied to prioritize and identify unknown chemicals in matrices linked to human exposure, such as drinking water, air, sewage sludge, and human biospecimens such as serum or placenta. This section highlights four case studies that illustrate the potential of NTS when rigorously applied in exposomics research.

1. Water is a direct exposure route, yet the diversity of CECs present is often poorly characterized. To address this, Bonnefille et al. (2023) applied complementary ionization techniques, electrospray (ESI) and atmospheric pressure chemical ionization (APCI), to urban and rural surface waters in Bangladesh. This dual approach captured 19,427 and 7365 unique features from ESI and APCI, respectively. Using OPLS-DA for feature prioritization, they tentatively identified and confirmed over 100 CECs differentiating rural and urban sites (Fig. 9.7A). Their work highlights the value of combining analytical techniques with intelligent prioritization to transform large NTS datasets into actionable exposure insights.
2. Gil-Solsona et al. (2021) hypothesized that the presence of xenobiotics in urban sludge could be indicative of their bioaccumulation in humans. To assess that, they used HRMS-based approaches to characterize maternal blood and placenta samples of a birth cohort of people residing in the same area serving the WWTP where they sampled sewage sludge. They

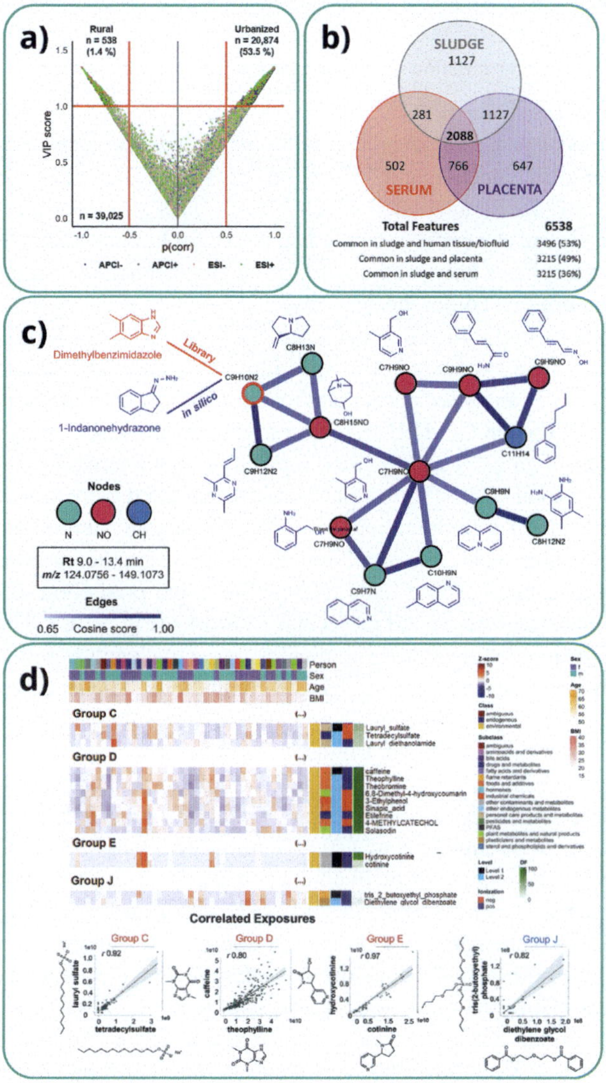

FIG. 9.7 Case studies showcasing the application of advanced NTS approaches within exposomics to prioritize, annotate, and interpret features of interest. (A) Scatterplot illustrating APCI- and ESI-derived features from surface water samples, differentiating rural and urban areas (Bonnefille et al., 2023). (B) Venn diagram showing overlapping features between placenta, serum, and sewage sludge, supporting sludge as a proxy matrix for human exposure assessment (Gil-Solsona et al., 2021). (C) Example of a molecular cluster from PM samples analyzed by LC-HRMS, following in silico structural prediction using FBMN and NAP tools (Papazian et al., 2022). (D) Zoomed sections of a hierarchical cluster analysis heatmap, highlighting correlated exposure profiles across 46 individuals (Sdougkou et al., 2024). *(From Bonnefille, B., Karlsson, O., Rian, M.B., Raqib, R., Parvez, F., Papazian, S., Islam, M.S., Martin, J.W., 2023. Nontarget analysis of polluted surface waters in Bangladesh using open science workflows. Environ. Sci. Technol. 0013-936X. 57 (17), 6808–6824. doi:10.1021/acs.est.2c08200, licensed under CC-BY 4.0. From Gil-Solsona, R., Nika, M.-C., Bustamante, M., Villanueva, C.M., Foraster, M., Cosin-Tomás, M., Alygizakis, N., Gómez-Roig, M.D., Llurba-Olive, E., Sunyer, J., Thomaidis, N.S., Dadvand, P., Gago-Ferrero, P., 2021. The potential of sewage sludge to predict and evaluate the human chemical exposome. Environ. Sci. Technol. Lett. 2328-8930. 8 (12), 1077–1084. doi:10.1021/acs.estlett.1c00848, licensed under CC-BY 4.0. From Sdougkou, K., Papazian, S., Bonnefille, B., Xie, H., Edfors, F., Fagerberg, L., Uhlén, M., Bergström, G., Martin, L.J., Martin, J.W., 2024. Longitudinal exposomics in a multiomic wellness cohort reveals distinctive and dynamic environmental chemical mixtures in blood. Environ. Sci. Technol. 0013-936X. 58 (37). doi:10.1021/acs.est.4c05235, licensed under CC-BY 4.0.)*

observed substantial overlaps: 67% of sludge-associated features also appeared in placenta, and 55% in serum (Fig. 9.7B). After applying SS, they confidently annotated 42 chemicals. This study demonstrates that urban sludge can serve as a practical proxy for human chemical exposure and exemplifies how NTS can trace the environmental-human exposure continuum in a comprehensive way.

3. Despite its major health impacts, air remains one of the least chemically characterized exposure pathways (Sunyer-Caldú et al., 2026). Papazian et al. (2022) analyzed particulate matter (PM) from the Indian subcontinent by LC- and GC-HRMS to investigate the presence of organic structures through NTS and in silico molecular characterization. While only 0.5% of total 60,030 molecular features extracted were annotated, after application of a computational workflow based on cheminformatics and computation tools, structural annotation could be increased up to 17% of total features. These tools included in silico prediction of molecular formulas, (feature-based) molecular networks (Fig. 9.7C). This molecular profiling distinguished chemical patterns in clean vs. polluted air and uncovered persistent anthropogenic compounds. The study exemplifies how combining NTS with computational exposomics tools can unlock insights from complex environmental mixtures.
4. Recent advances in analytical and computational tools have enabled unprecedented chemical profiling of human biospecimens. In a landmark study, Sdougkou et al. (2024) conducted a longitudinal NTS analysis of blood samples from 46 adults collected over six time points across two years. By combining SS and NTS with rigorous data processing, numerous substances not previously reported in humans were discovered and semiquantified, as similarly demonstrated in other exposomics studies (Guerrini et al., 2024; Plassmann et al., 2018; Sunyer-Caldú et al., 2023; Xie et al., 2024), highlighting the exposome's chemical complexity. Handling such vast datasets requires sophisticated statistical and visualization strategies. After confidently annotating 519 compounds, the authors applied hierarchical cluster analysis (HCA) based on Pearson correlations to uncover coexposure patterns (Fig. 9.7D). For instance, caffeine clustered with its metabolites theobromine and theophylline (Group D), while cotinine correlated closely with its derivative 3-hydroxy-cotinine (Group E). This approach not only validated known metabolic relationships but also revealed unexpected chemical coexposures, demonstrating the transformative potential of combining NTS with advanced longitudinal designs in human exposomics.

Together, these studies highlight how NTS, when applied within exposomics frameworks, moves beyond chemical detection toward meaningful exposure interpretation. While a majority of features still remain unidentified, the path toward a comprehensive exposome characterization is clearer than ever, enhanced by continuous analytical and computational advances.

References

NIST mass spectral library, 2025.

Aalizadeh, R., Alygizakis, N.A., Schymanski, E.L., Krauss, M., Schulze, T., Ibáñez, M., McEachran, A.D., Chao, A., Williams, A.J., Gago-Ferrero, P., Covaci, A., Moschet, C., Young, T.M., Hollender, J., Slobodnik, J., Thomaidis, N.S., 2021b. Development and application of liquid chromatographic retention time indices in HRMS-based suspect and nontarget screening. Anal. Chem. 0003-2700. 93 (33), 11601–11611. doi:10.1021/acs.analchem.1c02348.

Aalizadeh, R., Nikolopoulou, V., Alygizakis, N., Slobodnik, J., Thomaidis, N.S., 2022. A novel workflow for semi-quantification of emerging contaminants in environmental samples analyzed by LC-HRMS. Anal. Bioanal. Chem. 1618-2642. 414 (25), 7435–7450. doi:10.1007/s00216-022-04084-6.

Aalizadeh, R., Panara, A., Thomaidis, N.S., 2021a. Development and application of a novel semi-quantification approach in LC-QToF-MS analysis of natural products. J. Am. Soc. Mass Spectrom. 18791123. 32 (6), 1412–1423. Available from: https://pubs.acs.org/loi/jamsef. doi:10.1021/jasms.1c00032.

Allen, F., Pon, A., Wilson, M., Greiner, R., Wishart, D., 2014. CFM-ID: a web server for annotation, spectrum prediction and metabolite identification from tandem mass spectra. Nucleic Acids Res. 1362-4962. 42 (W1), W94. doi:10.1093/nar/gku436.

Alvarez-Mora, I., Arturi, K., Béen, F., Buchinger, S., El Mais, A.E.R., Gallampois, C., Hahn, M., Hollender, J., Houtman, C., Johann, S., Krauss, M., Lamoree, M., Margalef, M., Massei, R., Brack, W., Muz, M., 2025b. Progress, applications, and challenges in high-throughput effect-directed analysis for toxicity driver identification—is it time for HT-EDA? Anal. Bioanal. Chem. 1618-2642. 417 (3), 451–472. doi:10.1007/s00216-024-05424-4.

Alvarez-Mora, I., Muratuly, A., Johann, S., Arturi, K., Jünger, F., Huber, C., Hollert, H., Krauss, M., Brack, W., Muz, M., 2025a. High-throughput effect-directed analysis of androgenic compounds in hospital wastewater: identifying effect drivers through non-target screening supported by toxicity prediction. Environ. Sci. Technol. 0013-936X. doi:10.1021/acs.est.4c09942.

Alygizakis, N.A., Oswald, P., Thomaidis, N.S., Schymanski, E.L., Aalizadeh, R., Schulze, T., Oswaldova, M., Slobodnik, J., 2019. Norman digital sample freezing platform: a European virtual platform to exchange liquid chromatography high resolution-mass spectrometry data and screen suspects in "digitally frozen" environmental samples. TrAC Trends Anal. Chem. 18793142. 115, 129–137. Available from: www.elsevier.com/locate/trac. doi:10.1016/j.trac.2019.04.008.

Alygizakis, N., Lestremau, F., Gago-Ferrero, P., Gil-Solsona, R., Arturi, K., Hollender, J., Schymanski, E.L., Dulio, V., Slobodnik, J., Thomaidis, N.S., 2023. Towards a harmonized identification scoring system in LC-HRMS/MS based non-target screening (NTS) of emerging contaminants. TrAC Trends Anal. Chem. 01659936. 159, 116944. doi:10.1016/j.trac.2023.116944.

Arturi, K., Hollender, J., 2023. Machine learning-based hazard-driven prioritization of features in nontarget screening of environmental high-resolution mass spectrometry data. Environ. Sci. Technol. 0013-936X. 57 (46), 18067–18079. doi:10.1021/acs.est.3c00304.

Baygildiev, T., Meijer, J., Cenijn, P., Riegel, M., Arp, H.P.H., Lamoree, M., Hamers, T., 2025. Identification of polar bioactive substances in the upper rhine using effect-directed analysis. Water Res. 00431354. 268 (Pt A), 122607. doi:10.1016/j.watres.2024.122607.

Belova, L., Caballero-Casero, N., Van Nuijs, A.L.N., Covaci, A., 2021. Ion mobility-high-resolution mass spectrometry (IM-HRMS) for the analysis of contaminants of emerging concern

(CECs): database compilation and application to urine samples. Anal. Chem. 15206882. 93 (16), 6428–6436. Available from: http://pubs.acs.org/journal/ancham. doi:10.1021/acs.analchem.1c00142.

Belova, L., Celma, A., Van Haesendonck, G., Lemière, F., Sancho, J.V., Covaci, A., van Nuijs, A.L.N., Bijlsma, L., 2022. Revealing the differences in collision cross section values of small organic molecules acquired by different instrumental designs and prediction models. Anal. Chim. Acta. 00032670. 1229, 340361. doi:10.1016/j.aca.2022.340361.

Bhavyasri, K., Saileela, S., Sumakanth, M., 2023. High-resolution mass spectroscopy (HRMS)-a review. Int. J. Curr. Pharm. Res. 0975-7066. 9–13. doi:10.22159/ijcpr.2023v15i3.3008.

Bijlsma, L., Bade, R., Celma, A., Mullin, L., Cleland, G., Stead, S., Hernandez, F., Sancho, J.V., 2017. Prediction of collision cross-section values for small molecules: application to pesticide residue analysis. Anal. Chem. 0003-2700. 89 (12), 6583–6589. doi:10.1021/acs.analchem.7b00741.

Bijlsma, L., Berntssen, M.H.G., Merel, S., 2019. A refined nontarget workflow for the investigation of metabolites through the prioritization by in silico prediction tools. Anal. Chem. 15206882. 91 (9), 6321–6328. Available from: http://pubs.acs.org/journal/ancham. doi:10.1021/acs.analchem.9b01218.

Black, G., Lowe, C., Anumol, T., Bade, J., Favela, K., Feng, Y.-L., Knolhoff, A., Mceachran, A., Nuñez, J., Fisher, C., Peter, K., Quinete, N.S., Sobus, J., Sussman, E., Watson, W., Wickramasekara, S., Williams, A., Young, T., 2023. Exploring chemical space in non-targeted analysis: a proposed ChemSpace tool. Anal. Bioanal. Chem. 1618-2642. 415 (1), 35–44. doi:10.1007/s00216-022-04434-4.

Boelrijk, J., van Herwerden, D., Ensing, B., Forré, P., Samanipour, S., 2023. Predicting RP-LC retention indices of structurally unknown chemicals from mass spectrometry data. J. Cheminform. 1758-2946. 15 (1). doi:10.1186/s13321-023-00699-8.

Bolton, E., Schymanski, E., Kondic, T., Thiessen, P., Zhang, J., 2020. Zenodo PubChemLite for Exposomics (Pubchemlite.0.3.0). Zenodo. doi:10.5281/zenodo.4183801.

Bonini, P., Kind, T., Tsugawa, H., Barupal, D.K., Fiehn, O., 2020. Retip: retention time prediction for compound annotation in untargeted metabolomics. Anal. Chem. 15206882. 92 (11), 7515–7522. Available from: http://pubs.acs.org/journal/ancham. doi:10.1021/acs.analchem.9b05765.

Bonnefille, B., Karlsson, O., Rian, M.B., Raqib, R., Parvez, F., Papazian, S., Islam, M.S., Martin, J.W., 2023. Nontarget analysis of polluted surface waters in Bangladesh using open science workflows. Environ. Sci. Technol. 0013-936X. 57 (17), 6808–6824. doi:10.1021/acs.est.2c08200.

Caballero-Casero, N., Belova, L., Vervliet, P., Antignac, J.-P., Castaño, A., Debrauwer, L., López, M.E., Huber, C., Klanova, J., Krauss, M., Lommen, A., Mol, H.G.J., Oberacher, H., Pardo, O., Price, E.J., Reinstadler, V., Vitale, C.M., van Nuijs, A.L.N., Covaci, A., 2021. Towards harmonised criteria in quality assurance and quality control of suspect and non-target LC-HRMS analytical workflows for screening of emerging contaminants in human biomonitoring. TrAC Trends Anal. Chem. 01659936. 136, 116201. doi:10.1016/j.trac.2021.116201.

Casey, J.S., Jackson, S.R., Ryan, J., Newton, S.R., 2023. The use of gas chromatography—high resolution mass spectrometry for suspect screening and non-targeted analysis of per- and polyfluoroalkyl substances. J. Chromatogr. A 1693. doi:10.1016/j.chroma.2023.463884

Castaño-Ortiz, J.M., Gago-Ferrero, P., Barceló, D., Rodríguez-Mozaz, S., Gil-Solsona, R., 2024. HRMS-based suspect screening of pharmaceuticals and their transformation products in multiple environmental compartments: an alternative to target analysis? J. Hazard. Mater. 03043894. 465, 132974. doi:10.1016/j.jhazmat.2023.132974.

Celma, A., Ahrens, L., Gago-Ferrero, P., Hernández, F., López, F., Lundqvist, J., Pitarch, E., Sancho, J.V., Wiberg, K., Bijlsma, L., 2021. The relevant role of ion mobility separation in LC-HRMS based screening strategies for contaminants of emerging concern in the aquatic environment. Chemosphere. 00456535. 280, 130799. doi:10.1016/j.chemosphere.2021.130799.

Celma, A., Bade, R., Sancho, J.V., Hernandez, F., Humphries, M., Bijlsma, L., 2022. Prediction of retention time and collision cross section (CCS H+, CCS H–, and CCS Na+) of emerging contaminants using multiple adaptive regression splines. J. Chem. Inf. Model. 1549-9596. 62 (22), 5425–5434. doi:10.1021/acs.jcim.2c00847.

Celma, A., Sancho, J.V., Schymanski, E.L., Fabregat-Safont, D., Ibáñez, M., Goshawk, J., Barknowitz, G., Hernández, F., Bijlsma, L., 2020. Improving target and suspect screening high-resolution mass spectrometry workflows in environmental analysis by ion mobility separation. Environ. Sci. Technol. 0013-936X. 54 (23), 15120–15131. doi:10.1021/acs.est.0c05713.

Chambers, M.C., MacLean, B., Burke, R., Amodei, D., Ruderman, D.L., Neumann, S., Gatto, L., Fischer, B., Pratt, B., Egertson, J., Hoff, K., Kessner, D., Tasman, N., Shulman, N., Frewen, B., Baker, T.A., Brusniak, M.Y., Paulse, C., Creasy, D., Flashner, L., Kani, K., Moulding, C., Seymour, S.L., Nuwaysir, L.M., Lefebvre, B., Kuhlmann, F., Roark, J., Rainer, P., Detlev, S., Hemenway, T., Huhmer, A., Langridge, J., Connolly, B., Chadick, T., Holly, K., Eckels, J., Deutsch, E.W., Moritz, R.L., Katz, J.E., Agus, D.B., MacCoss, M., Tabb, D.L., Mallick, P., 2012. A cross-platform toolkit for mass spectrometry and proteomics. Nat. Biotechnol. 15461696. 30 (10), 918–920. doi:10.1038/nbt.2377.

Charbonnet, J.A., McDonough, C.A., Xiao, F., Schwichtenberg, T., Cao, D., Kaserzon, S., Thomas, K.V., Dewapriya, P., Place, B.J., Schymanski, E.L., Field, J.A., Helbling, D.E., Higgins, C.P., 2022. Communicating confidence of per- and polyfluoroalkyl substance identification via high-resolution mass spectrometry. Environ. Sci. Technol. Lett. 23288930. 9 (6), 473–481. Available from: http://pubs.acs.org/page/estlcu/about.html. doi:10.1021/acs.estlett.2c00206.

Cohen, M.J., Karasek, F.W., 1970. Plasma chromatography—a new dimension for gas chromatography and mass spectrometry. J. Chromatogr. Sci. 0021-9665. 8 (6), 330–337. doi:10.1093/chromsci/8.6.330.

da Silva, R.R., Wang, M., Nothias, L.-F., van der Hooft, J.J.J., Caraballo-Rodríguez, A.M., Fox, E., Balunas, M.J., Klassen, J.L., Lopes, N.P., Dorrestein, P.C., 2018. Propagating annotations of molecular networks using in silico fragmentation. PLoS Comput. Biol. 14 (4). doi:10.1371/journal.pcbi.1006089.

Dodds, J.N., Ford, L.C., Ryan, J.P., Solosky, A.M., Rusyn, I., Baker, E.S., 2025. Evaluating ion mobility data acquisition, calibration, and processing for small molecules: a cross-platform assessment of drift tube and traveling wave methodologies. J. Am. Soc. Mass Spectrom. 18791123. 36 (7), 1456–1466. Available from: https://pubs.acs.org/loi/jamsef. doi:10.1021/jasms.5c00056.

Dodds, J.N., Hopkins, Z.R., Knappe, D.R.U., Baker, E.S., 2020. Rapid characterization of per- and polyfluoroalkyl substances (PFAS) by ion mobility spectrometry-mass spectrometry (IMS-MS). Anal. Chem. 15206882. 92 (6), 4427–4435. Available from: http://pubs.acs.org/journal/ancham. doi:10.1021/acs.analchem.9b05364.

Dopp, E., Pannekens, H., Itzel, F., Tuerk, J., 2019. Effect-based methods in combination with state-of-the-art chemical analysis for assessment of water quality as integrated approach. Int. J. Hyg. Environ. Health. 14384639. 222 (4), 607–614. doi:10.1016/j.ijheh.2019.03.001.

Dührkop, K., Fleischauer, M., Ludwig, M., Aksenov, A.A., Melnik, A.V., Meusel, M., Dorrestein, P.C., Rousu, J., Böcker, S., 2019. SIRIUS 4: a rapid tool for turning tandem mass spectra into metabolite structure information. Nat. Methods. 1548-7091. 16 (4), 299–302. doi:10.1038/s41592-019-0344-8.

Dührkop, K., Shen, H., Meusel, M., Rousu, J., Böcker, S., 2015. Searching molecular structure databases with tandem mass spectra using CSI:FingerID. Proc. Natl. Acad. Sci. U. S. A. 0027-8424. 112 (41), 12580–12585. doi:10.1073/pnas.1509788112.

Dulio, V., van Bavel, B., Brorström-Lundén, E., Harmsen, J., Hollender, J., Schlabach, M., Slobodnik, J., Thomas, K., Koschorreck, J., 2018. Emerging pollutants in the EU: 10 years of Norman in support of environmental policies and regulations. Environ. Sci. Eur. 2190-4707. 30 (1). doi:10.1186/s12302-018-0135-3.

Elapavalore, A., Ross, D.H., Grouès, V., Aurich, D., Krinsky, A.M., Kim, S., Thiessen, P.A., Zhang, J., Dodds, J.N., Baker, E.S., Bolton, E.E., Xu, L., Schymanski, E.L., 2025. PubChemLite plus collision cross section (CCS) values for enhanced interpretation of nontarget environmental data. Environ. Sci. Technol. Lett. 2328-8930. 12 (2), 166–174. doi:10.1021/acs.estlett.4c01003.

Feuerstein, M.L., Hernández-Mesa, M., Kiehne, A., Bizec, B.L., Hann, S., Dervilly, G., Causon, T., 2022. Comparability of steroid collision cross sections using three different IM-HRMS technologies: an interplatform study. J. Am. Soc. Mass Spectrom. 18791123. 33 (10), 1951–1959. Available from: https://pubs.acs.org/loi/jamsef. doi:10.1021/jasms.2c00196.

Foster, M., Rainey, M., Watson, C., Dodds, J.N., Kirkwood, K.I., Fernández, F.M., Baker, E.S., 2022. Uncovering PFAS and other xenobiotics in the dark metabolome using ion mobility spectrometry, mass defect analysis, and machine learning. Environ. Sci. Technol. 0013-936X. 56 (12), 9133–9143. doi:10.1021/acs.est.2c00201.

Gabelica, V., Shvartsburg, A.A., Afonso, C., Barran, P., Benesch, J.L.P., Bleiholder, C., Bowers, M.T., Bilbao, A., Bush, M.F., Campbell, J.L., Campuzano, I.D.G., Causon, T., Clowers, B.H., Creaser, C.S., De Pauw, E., Far, J., Fernandez-Lima, F., Fjeldsted, J.C., Giles, K., Groessl, M., Hogan, C.J., Hann, S., Kim, H.I., Kurulugama, R.T., May, J.C., McLean, J.A., Pagel, K., Richardson, K., Ridgeway, M.E., Rosu, F., Sobott, F., Thalassinos, K., Valentine, S.J., Wyttenbach, T., 2019. Recommendations for reporting ion mobility mass spectrometry measurements. Mass Spectrom. Rev. 0277-7037. 38 (3), 291–320. doi:10.1002/mas.21585.

Gil-Solsona, R., Nika, M.-C., Bustamante, M., Villanueva, C.M., Foraster, M., Cosin-Tomás, M., Alygizakis, N., Gómez-Roig, M.D., Llurba-Olive, E., Sunyer, J., Thomaidis, N.S., Dadvand, P., Gago-Ferrero, P., 2021. The potential of sewage sludge to predict and evaluate the human chemical exposome. Environ. Sci. Technol. Lett. 2328-8930. 8 (12), 1077–1084. doi:10.1021/acs.estlett.1c00848.

Gilles, L., Govarts, E., Rambaud, L., Vogel, N., Castano, A., Lopez, M.E., Martin, L.R., Koppen, G., Remy, S., Vrijheid, M., Montazeri, P., Birks, L., Sepai, O., Stewart, L., Fiddicke, U., Loots, I., Knudsen, L.E., Kolossa-Gehring, M., Schoeters, G., 2021. HBM4EU combines and harmonises human biomonitoring data across the EU, building on existing capacity-The HBM4EU survey. Int. J. Hyg. Environ. Health, 237, 113809. doi:10.1016/j.ijheh.2021.113809.

Giné, R., Capellades, J., Badia, J.M., Vughs, D., Schwaiger-Haber, M., Alexandrov, T., Vinaixa, M., Brunner, A.M., Patti, G.J., Yanes, O., 2021. HERMES: a molecular-formula-oriented method to target the metabolome. Nat. Methods. 1548-7091. 18 (11), 1370–1376. doi:10.1038/s41592-021-01307-z.

Guerrini, C., Sunyer-Caldú, A., Gil-Solsona, R., Escribano, J., Vinaixa, M., Gago-Ferrero, P., Ramírez, N., 2024. Early-life chemical exposome: comprehensive strategies for wide-scope screening of organic compounds. TrAC Trends Anal. Chem. 01659936. 180, 117903. doi:10.1016/j.trac.2024.117903.

Guo, T., LaBelle, B., Petreas, M., Park, J.-S., 2013. Mass spectrometric characterization of halogenated flame retardants. Rapid Commun. Mass Spectrom. 09514198. 27 (13), 1437–1449. doi:10.1002/rcm.6588.

Gutiérrez-Martín, D., Restrepo-Montes, E., Golovko, O., López-Serna, R., Aalizadeh, R., Thomaidis, N.S., Marquès, M., Gago-Ferrero, P., Gil-Solsona, R., 2023. Comprehensive profiling and semi-quantification of exogenous chemicals in human urine using HRMS-based strategies. Anal. Bioanal. Chem. 1618-2642. 415 (29-30), 7297–7313. doi:10.1007/s00216-023-04998-9.

Hanold, K.A., Fischer, S.M., Cormia, P.H., Miller, C.E., Syage, J.A., 2004. Atmospheric pressure photoionization. 1. General properties for LC/MS. Anal. Chem. 76 (10), 2842–2851. doi:10.1021/ac035442i.

Helmus, R., ter Laak, T.L., van Wezel, A.P., de Voogt, P., Schymanski, E.L., 2021. patRoon: open source software platform for environmental mass spectrometry based non-target screening. J. Cheminform. 17582946. 13 (1). Available from: https://jcheminf.biomedcentral.com/. doi:10.1186/s13321-020-00477-w.

Hernández, F., Bakker, J., Bijlsma, L., de Boer, J., Botero-Coy, A.M., Bruinen de Bruin, Y., Fischer, S., Hollender, J., Kasprzyk-Hordern, B., Lamoree, M., López, F.J., Laak, T.L.T., van Leerdam, J.A., Sancho, J.V., Schymanski, E.L., de Voogt, P., Hogendoorn, E.A., 2019. The role of analytical chemistry in exposure science: focus on the aquatic environment. Chemosphere. 00456535. 222, 564–583. doi:10.1016/j.chemosphere.2019.01.118.

Hinnenkamp, V., Klein, J., Meckelmann, S.W., Balsaa, P., Schmidt, T.C., Schmitz, O.J., 2018. Comparison of CCS values determined by traveling wave ion mobility mass spectrometry and drift tube ion mobility mass spectrometry. Anal. Chem. 0003-2700. 90 (20), 12042–12050. doi:10.1021/acs.analchem.8b02711.

Hoffmann, M.A., Nothias, L.F., Ludwig, M., Fleischauer, M., Gentry, E.C., Witting, M., Dorrestein, P.C., Dührkop, K., Böcker, S., 2022. High-confidence structural annotation of metabolites absent from spectral libraries. Nat. Biotechnol. 15461696. 40 (3), 411–421. Available from: http://www.nature.com/nbt/index.html. doi:10.1038/s41587-021-01045-9.

Hohrenk, L.L., Itzel, F., Baetz, N., Tuerk, J., Vosough, M., Schmidt, T.C., 2020. Comparison of software tools for liquid chromatography-high-resolution mass spectrometry data processing in nontarget screening of environmental samples. Anal. Chem. 15206882. 92 (2), 1898–1907. Available from: http://pubs.acs.org/journal/ancham. doi:10.1021/acs.analchem.9b04095.

Hollender, J., Schymanski, E.L., Ahrens, L., Alygizakis, N., Béen, F., Bijlsma, L., Brunner, A.M., Celma, A., Fildier, A., Fu, Q., Gago-Ferrero, P., Gil-Solsona, R., Haglund, P., Hansen, M., Kaserzon, S., Kruve, A., Lamoree, M., Margoum, C., Meijer, J., Merel, S., Rauert, C., Rostkowski, P., Samanipour, S., Schulze, B., Schulze, T., Singh, R.R., Slobodnik, J., Steininger-Mairinger, T., Thomaidis, N.S., Togola, A., Vorkamp, K., Vulliet, E., Zhu, L., Krauss, M., 2023. Norman guidance on suspect and non-target screening in environmental monitoring. Environ. Sci. Eur. 2190-4715. 35 (1). doi:10.1186/s12302-023-00779-4.

Hollender, J., Schymanski, E.L., Singer, H.P., Ferguson, P.L., 2017. Nontarget screening with high resolution mass spectrometry in the environment: ready to go? Environ. Sci. Technol. 15205851. 51 (20), 11505–11512. Available from: http://pubs.acs.org/journal/esthag. doi:10.1021/acs.est.7b02184.

Holman, J.D., Tabb, D.L., Mallick, P., 2014. Employing ProteoWizard to convert raw mass spectrometry data. Curr. Protoc. Bioinform. 1934340X. 46. doi:10.1002/0471250953.bi1324s46.

Joseph, K.M., Boatman, A.K., Dodds, J.N., Kirkwood-Donelson, K.I., Ryan, J.P., Zhang, J., Thiessen, P.A., Bolton, E.E., Valdiviezo, A., Sapozhnikova, Y., Rusyn, I., Schymanski, E.L., Baker, E.S., 2025. Multidimensional library for the improved identification of per- and polyfluoroalkyl substances (PFAS). Sci. Data. 20524463. 12 (1). Available from: https://www.nature.com/sdata/. doi:10.1038/s41597-024-04363-0.

Knolhoff, A.M., Premo, J.H., Fisher, C.M., 2021. A proposed quality control standard mixture and its uses for evaluating nontargeted and suspect screening LC/HR-MS method performance.

Anal. Chem. 15206882. 93 (3), 1596–1603. Available from: http://pubs.acs.org/journal/ancham. doi:10.1021/acs.analchem.0c04036.

Koelmel, J.P., Xie, H., Price, E.J., Lin, E.Z., Manz, K.E., Stelben, P., Paige, M., Papazian, S., Okeme, J., Jones, D.P., Barupal, D., Bowden, J., Rostkowski, P., Pennell, K.D., Nikiforov, V., Wang, T., Hu, X., Lai, Y., Miller, G.W., Walker, D.I., Martin, J.W., Godri Pollitt, K.J., 2022. An actionable annotation scoring framework for gas chromatography—high resolution mass spectrometry (GC-HRMS). ChemRxiv. 25732293. Available from: https://chemrxiv.org. doi:10.26434/chemrxiv-2022-dr5tc.

Krauss, M., Singer, H., Hollender, J., 2010. LC–high resolution MS in environmental analysis: from target screening to the identification of unknowns. Anal. Bioanal. Chem. 1618-2642. 397 (3), 943–951. doi:10.1007/s00216-010-3608-9.

Kruve, A., 2020. Strategies for drawing quantitative conclusions from nontargeted liquid chromatography–high-resolution mass spectrometry analysis. Anal. Chem. 0003-2700. 92 (7), 4691–4699. doi:10.1021/acs.analchem.9b03481.

Kuhl, C., Tautenhahn, R., Böttcher, C., Larson, T.R., Neumann, S., 2012. CAMERA: an integrated strategy for compound spectra extraction and annotation of liquid chromatography/mass spectrometry data sets. Anal. Chem. 00032700. 84 (1), 283–289. doi:10.1021/ac202450g.

Lai, Y., Koelmel, J.P., Walker, D.I., Price, E.J., Papazian, S., Manz, K.E., Castilla-Fernández, D., Bowden, J.A., Nikiforov, V., David, A., Bessonneau, V., Amer, B., Seethapathy, S., Hu, X., Lin, E.Z., Jbebli, A., McNeil, B.R., Barupal, D., Cerasa, M., Xie, H., Kalia, V., Nandakumar, R., Singh, R., Tian, Z., Gao, P., Zhao, Y., Froment, J., Rostkowski, P., Dubey, S., Coufalíková, K., Seličová, H., Hecht, H., Liu, S., Udhani, H.H., Restituito, S., Tchou-Wong, K.-M., Lu, K., Martin, J.W., Warth, B., Godri Pollitt, K.J., Klánová, J., Fiehn, O., Metz, T.O., Pennell, K.D., Jones, D.P., Miller, G.W., 2024. High-resolution mass spectrometry for human exposomics: expanding chemical space coverage. Environ. Sci. Technol. 0013-936X. 58 (29), 12784–12822. doi:10.1021/acs.est.4c01156.

Lai, Z., Tsugawa, H., Wohlgemuth, G., Mehta, S., Mueller, M., Zheng, Y., Ogiwara, A., Meissen, J., Showalter, M., Takeuchi, K., Kind, T., Beal, P., Arita, M., Fiehn, O., 2018. Identifying metabolites by integrating metabolome databases with mass spectrometry cheminformatics. Nat. Methods. 1548-7091. 15 (1), 53–56. doi:10.1038/nmeth.4512.

Lee, M.S., 2012. Mass Spectrometry Handbook. Wiley. doi:10.1002/9781118180730.

Lennon, S., Chaker, J., Price, E.J., Hollender, J., Huber, C., Schulze, T., Ahrens, L., Béen, F., Creusot, N., Debrauwer, L., Dervilly, G., Gabriel, C., Guérin, T., Habchi, B., Jamin, E.L., Klánová, J., Kosjek, T., Le Bizec, B., Meijer, J., Mol, H., Nijssen, R., Oberacher, H., Papaioannou, N., Parinet, J., Sarigiannis, D., Stravs, M.A., Tkalec, Ž., Schymanski, E.L., Lamoree, M., Antignac, J.-P., David, A., 2024. Harmonized quality assurance/quality control provisions to assess completeness and robustness of MS1 data preprocessing for LC-HRMS-based suspect screening and non-targeted analysis. TrAC Trends Anal. Chem. 01659936. 174, 117674. doi:10.1016/j.trac.2024.117674.

Liigand, J., Wang, T., Kellogg, J., Smedsgaard, J., Cech, N., Kruve, A., 2020. Quantification for non-targeted LC/MS screening without standard substances. Sci. Rep. 2045-2322. 10 (1). doi:10.1038/s41598-020-62573-z.

Liu, J., Xiang, T., Song, X.-C., Zhang, S., Wu, Q., Gao, J., Lv, M., Shi, C., Yang, X., Liu, Y., Fu, J., Shi, W., Fang, M., Qu, G., Yu, H., Jiang, G., 2024. High-efficiency effect-directed analysis leveraging five high level advancements: a critical review. Environ. Sci. Technol. 0013-936X. 58 (23), 9925–9944. doi:10.1021/acs.est.3c10996.

Liu, Y., D'Agostino, L.A., Qu, G., Jiang, G., Martin, J.W., 2019. High-resolution mass spectrometry (HRMS) methods for nontarget discovery and characterization of poly- and per-fluoroalkyl substances (PFASs) in environmental and human samples. TrAC Trends Anal. Chem. 18793142. 121. Available from: www.elsevier.com/locate/trac. doi:10.1016/j.trac.2019.02.021.

Luo, W., Chou, L., Cui, Q., Wei, S., Zhang, X., Guo, J., 2024. High-efficiency effect-directed analysis (EDA) advancing toxicant identification in aquatic environments: latest progress and application status. Environ. Int. 01604120. 190, 108855. doi:10.1016/j.envint.2024.108855.

Malm, L., Liigand, J., Aalizadeh, R., Alygizakis, N., Ng, K., Frøkjær, E.E., Nanusha, M.Y., Hansen, M., Plassmann, M., Bieber, S., Letzel, T., Balest, L., Abis, P.P., Mazzetti, M., Kasprzyk-Hordern, B., Ceolotto, N., Kumari, S., Hann, S., Kochmann, S., Steininger-Mairinger, T., Soulier, C., Mascolo, G., Murgolo, S., Garcia-Vara, M., de Alda, M.L., Hollender, J., Arturi, K., Coppola, G., Peruzzo, M., Joerss, H., van der Neut-marchand, C., Pieke, E.N., Gago-Ferrero, P., Gil-Solsona, R., Licul-Kucera, V., Roscioli, C., Valsecchi, S., Luckute, A., Christensen, J.H., Tisler, S., Vughs, D., Meekel, N., Andújar, B.T., Aurich, D., Schymanski, E.L., Frigerio, G., Macherius, A., Kunkel, U., Bader, T., Rostkowski, P., Gundersen, H., Valdecanas, B., Davis, W.C., Schulze, B., Kaserzon, S., Pijnappels, M., Esperanza, M., Fildier, A., Vulliet, E., Wiest, L., Covaci, A., Schönleben, A.M., Belova, L., Celma, A., Bijlsma, L., Caupos, E., Mebold, E., Le Roux, J., Troia, E., de Rijke, E., Helmus, R., Leroy, G., Haelewyck, N., Chrastina, D., Verwoert, M., Thomaidis, N.S., Kruve, A., 2024. Quantification approaches in non-target LC/ESI/HRMS analysis: an interlaboratory comparison. Anal. Chem. 0003-2700. 96 (41), 16215–16226. doi:10.1021/acs.analchem.4c02902.

Mason, E.A., Schamp, H.W., 1958. Mobility of gaseous lons in weak electric fields. Ann. Phys. 1096035X. 4 (3), 233–270. doi:10.1016/0003-4916(58)90049-6.

Meijer, J., Lamoree, M., Hamers, T., Antignac, J.-P., Hutinet, S., Debrauwer, L., Covaci, A., Huber, C., Krauss, M., Walker, D.I., Schymanski, E.L., Vermeulen, R., Vlaanderen, J., 2021. An annotation database for chemicals of emerging concern in exposome research. Environ. Int. 01604120. 152, 106511. doi:10.1016/j.envint.2021.106511.

Menger, F., Celma, A., Schymanski, E.L., Lai, F.Y., Bijlsma, L., Wiberg, K., Hernández, F., Sancho, J.V., Ahrens, L., 2022. Enhancing spectral quality in complex environmental matrices: supporting suspect and non-target screening in zebra mussels with ion mobility. Environ. Int. 01604120. 170, 107585. doi:10.1016/j.envint.2022.107585.

Menger, F., Gago-Ferrero, P., Wiberg, K., Ahrens, L., 2020. Wide-scope screening of polar contaminants of concern in water: a critical review of liquid chromatography-high resolution mass spectrometry-based strategies. Trends Environ. Anal. Chem. 22141588. 28, e00102. doi:10.1016/j.teac.2020.e00102.

Mu, H., Wang, J., Chen, L., Hu, H., Wang, J., Gu, C., Ren, H., Wu, B., 2023. Identification and characterization of diverse isomers of per- and polyfluoroalkyl substances in Chinese municipal wastewater. Water Res. 00431354. 230, 119580. doi:10.1016/j.watres.2023.119580.

Mullin, L., Jobst, K., DiLorenzo, R.A., Plumb, R., Reiner, E.J., Yeung, L.W.Y., Jogsten, I.E., 2020. Liquid chromatography-ion mobility-high resolution mass spectrometry for analysis of pollutants in indoor dust: identification and predictive capabilities. Anal. Chim. Acta. 00032670. 1125, 29–40. doi:10.1016/j.aca.2020.05.052.

Naylor, B.C., Catrow, J.L., Maschek, J.A., Cox, J.E., 2020. QSRR automator: a tool for automating retention time prediction in lipidomics and metabolomics. Metabolites. 10 (6). doi:10.3390/metabo10060237.

Neumann, S., Meier, R., Wenk, M., Elapavalore, A., Nishioka, T., Schulze, T., Stravs, M., Tsugawa, H., Matsuda, F., Schymanski, E.L., 2026. MassBank: An Open and FAIR Mass Spectral Data Resource. Nucleic. Acids Res. 54(D1), D601–D606. doi:10.1093/nar/gkaf1193.

Ng, B., Quinete, N., Gardinali, P.R., 2020. Assessing accuracy, precision and selectivity using quality controls for non-targeted analysis. Sci. Total Environ. 00489697. 713, 136568. doi:10.1016/j.scitotenv.2020.136568.

Nothias, L.F., Petras, D., Schmid, R., Dührkop, K., Rainer, J., Sarvepalli, A., Protsyuk, I., Ernst, M., Tsugawa, H., Fleischauer, M., Aicheler, F., Aksenov, A.A., Alka, O., Allard, P.M., Barsch, A., Cachet, X., Caraballo-Rodriguez, A.M., Da Silva, R.R., Dang, T., Garg, N., Gauglitz, J.M., Gurevich, A., Isaac, G., Jarmusch, A.K., Kameník, Z., Kang, K.B., Kessler, N., Koester, I., Korf, A., Le Gouellec, A., Ludwig, M., Martin H, C., McCall, L.I., McSayles, J., Meyer, S.W., Mohimani, H., Morsy, M., Moyne, O., Neumann, S., Neuweger, H., Nguyen, N.H., Nothias-Esposito, M., Paolini, J., Phelan, V.V., Pluskal, T., Quinn, R.A., Rogers, S., Shrestha, B., Tripathi, A., van der Hooft, J.J.J., Vargas, F., Weldon, K.C., Witting, M., Yang, H., Zhang, Z., Zubeil, F., Kohlbacher, O., Böcker, S., Alexandrov, T., Bandeira, N., Wang, M., Dorrestein, P.C., 2020. Feature-based molecular networking in the GNPS analysis environment. Nat. Methods. 15487105. 17 (9), 905–908. Available from: http://www.nature.com/nmeth/. doi:10.1038/s41592-020-0933-6.

Oberacher, H., Sasse, M., Antignac, J.-P., Guitton, Y., Debrauwer, L., Jamin, E.L., Schulze, T., Krauss, M., Covaci, A., Caballero-Casero, N., Rousseau, K., Damont, A., Fenaille, F., Lamoree, M., Schymanski, E.L., 2020. A European proposal for quality control and quality assurance of tandem mass spectral libraries. Environ. Sci. Eur. 2190-4707. 32 (1). doi:10.1186/s12302-020-00314-9.

Papazian, S., D'Agostino, L.A., Sadiktsis, I., Froment, J., Bonnefille, B., Sdougkou, K., Xie, H., Athanassiadis, I., Budhavant, K., Dasari, S., Andersson, A., Gustafsson, Ö., Martin, J.W., 2022. Nontarget mass spectrometry and in silico molecular characterization of air pollution from the Indian subcontinent. Commun. Earth Environ. 2662-4435. 3 (1). doi:10.1038/s43247-022-00365-1.

Peets, P., Wang, W.-C., MacLeod, M., Breitholtz, M., Martin, J.W., Kruve, A., 2022. MS2Tox machine learning tool for predicting the ecotoxicity of unidentified chemicals in water by nontarget LC-HRMS. Environ. Sci. Technol. 0013-936X. 56 (22), 15508–15517. doi:10.1021/acs.est.2c02536.

Peter, K.T., Phillips, A.L., Knolhoff, A.M., Gardinali, P.R., Manzano, C.A., Miller, K.E., Pristner, M., Sabourin, L., Sumarah, M.W., Warth, B., Sobus, J.R., 2021. Nontargeted analysis study reporting tool: a framework to improve research transparency and reproducibility. Anal. Chem. 15206882. 93 (41), 13870–13879. Available from: http://pubs.acs.org/journal/ancham. doi:10.1021/acs.analchem.1c02621.

Picache, J.A., Rose, B.S., Balinski, A., Leaptrot, K.L., Sherrod, S.D., May, J.C., McLean, J.A., 2019. Collision cross section compendium to annotate and predict multi-omic compound identities. Chem. Sci. 20416539. 10 (4), 983–993. Available from: http://pubs.rsc.org/en/journals/journal/sc. doi:10.1039/c8sc04396e.

Plassmann, M.M., Fischer, S., Benskin, J.P., 2018. Nontarget time trend screening in human blood. Environ. Sci. Technol. Lett. 23288930. 5 (6), 335–340. Available from: http://pubs.acs.org/page/estlcu/about.html. doi:10.1021/acs.estlett.8b00196.

Pringle, S.D., Giles, K., Wildgoose, J.L., Williams, J.P., Slade, S.E., Thalassinos, K., Bateman, R.H., Bowers, M.T., Scrivens, J.H., 2007. An investigation of the mobility separation of some peptide and protein ions using a new hybrid quadrupole/travelling wave IMS/oa-ToF instrument. Int. J. Mass Spectrom. 13873806. 261 (1), 1–12. doi:10.1016/j.ijms.2006.07.021.

Radke, M.J., Cresswell, S.L., Leusch, F.D.L., 2025. Combining non-targeted high resolution mass spectrometry with effect-directed analysis to identify contaminants of emerging concern in the field of ecotoxicology: a systematic quantitative literature review. Sci. Total Environ.

18791026. 972. Available from: https://www.sciencedirect.com/science/journal/00489697. doi:10.1016/j.scitotenv.2025.179122.

Revercomb, H.E., Mason, E.A., 1975. Theory of plasma chromatography/gaseous electrophoresis. Review. Anal. Chem. 0003-2700. 47 (7), 970–983. doi:10.1021/ac60357a043.

Röst, H.L., Sachsenberg, T., Aiche, S., Bielow, C., Weisser, H., Aicheler, F., Andreotti, S., Ehrlich, H.C., Gutenbrunner, P., Kenar, E., Liang, X., Nahnsen, S., Nilse, L., Pfeuffer, J., Rosenberger, G., Rurik, M., Schmitt, U., Veit, J., Walzer, M., Wojnar, D., Wolski, W.E., Schilling, O., Choudhary, J.S., Malmström, L., Aebersold, R., Reinert, K., Kohlbacher, O., 2016. OpenMS: a flexible open-source software platform for mass spectrometry data analysis. Nat. Methods. 15487105. 13 (9), 741–748. Available from: http://www.nature.com/nmeth/. doi:10.1038/nmeth.3959.

Ruttkies, C., Schymanski, E.L., Wolf, S., Hollender, J., Neumann, S., 2016. MetFrag relaunched: incorporating strategies beyond in silico fragmentation. J. Cheminform. 17582946. 8 (1). Available from: http://www.jcheminf.com/. doi:10.1186/s13321-016-0115-9.

Schmid, R., Heuckeroth, S., Korf, A., Smirnov, A., Myers, O., Dyrlund, T.S., Bushuiev, R., Murray, K.J., Hoffmann, N., Lu, M., Sarvepalli, A., Zhang, Z., Fleischauer, M., Dührkop, K., Wesner, M., Hoogstra, S.J., Rudt, E., Mokshyna, O., Brungs, C., Ponomarov, K., Mutabdžija, L., Damiani, T., Pudney, C.J., Earll, M., Helmer, P.O., Fallon, T.R., Schulze, T., Rivas-Ubach, A., Bilbao, A., Richter, H., Nothias, L.-F., Wang, M., Orešič, M., Weng, J.-K., Böcker, S., Jeibmann, A., Hayen, H., Karst, U., Dorrestein, P.C., Petras, D., Du, X., Pluskal, T., 2023. Integrative analysis of multimodal mass spectrometry data in MZmine 3. Nat. Biotechnol. 1087-0156. 41 (4), 447–449. doi:10.1038/s41587-023-01690-2.

Schymanski, E.L., Jeon, J., Gulde, R., Fenner, K., Ruff, M., Singer, H.P., Hollender, J., 2014. Identifying small molecules via high resolution mass spectrometry: communicating confidence. Environ. Sci. Technol. 15205851. 48 (4), 2097–2098. doi:10.1021/es5002105.

Schymanski, E.L., Singer, H.P., Slobodnik, J., Ipolyi, I.M., Oswald, P., Krauss, M., Schulze, T., Haglund, P., Letzel, T., Grosse, S., Thomaidis, N.S., Bletsou, A., Zwiener, C., Ibáñez, M., Portolés, T., De Boer, R., Reid, M.J., Onghena, M., Kunkel, U., Schulz, W., Guillon, A., Noyon, N., Leroy, G., Bados, P., Bogialli, S., Stipaničev, D., Rostkowski, P., Hollender, J., 2015. Non-target screening with high-resolution mass spectrometry: critical review using a collaborative trial on water analysis. Anal. Bioanal. Chem. 16182650. 407 (21), 6237–6255. Available from: http://link.springer.de/link/service/journals/00216/index.htm. doi:10.1007/s00216-015-8681-7.

Sdougkou, K., Papazian, S., Bonnefille, B., Xie, H., Edfors, F., Fagerberg, L., Uhlén, M., Bergström, G., Martin, L.J., Martin, J.W., 2024. Longitudinal exposomics in a multiomic wellness cohort reveals distinctive and dynamic environmental chemical mixtures in blood. Environ. Sci. Technol. 0013-936X. 58 (37). doi:10.1021/acs.est.4c05235.

Sepman, H., Malm, L., Peets, P., MacLeod, M., Martin, J., Breitholtz, M., Kruve, A., 2023. Bypassing the identification: MS2Quant for concentration estimations of chemicals detected with nontarget LC-HRMS from MS 2 data. Anal. Chem. 0003-2700. 95 (33), 12329–12338. doi:10.1021/acs.analchem.3c01744.

Smolinski, R., Koelmel, J.P., Stelben, P., Weil, D., Godri, D., Schiessel, D., Kummer, M., Stow, S.M., Mohsin, S., Royer, L., McKenzie-Coe, A., Lubinsky, T., DeBord, D., Chevallier, O., Rennie, E.E., Godri Pollitt, K.J., McDonough, C., 2025. FluoroMatch IM: an interactive software for PFAS analysis by ion mobility spectrometry. Environ. Sci. Technol. 0013-936X. 59 (13), 6636–6648. doi:10.1021/acs.est.4c13846.

Smyth, W., 2005. Recent studies on the electrospray ionisation mass spectrometric behaviour of selected nitrogen-containing drug molecules and its application to drug analysis using liquid chromatography–electrospray ionisation mass spectrometry. J. Chromatogr. B. 15700232. 824 (1–2), 1–20. doi:10.1016/j.jchromb.2005.07.018.

Song, X.C., Canellas, E., Dreolin, N., Goshawk, J., Lv, M., Qu, G., Nerin, C., Jiang, G., 2023. Application of ion mobility spectrometry and the derived collision cross section in the analysis of environmental organic micropollutants. Environ. Sci. Technol. 15205851. 57 (51), 21485–21502. Available from: http://pubs.acs.org/journal/esthag. doi:10.1021/acs.est.3c03686.

Song, X.C., Canellas, E., Dreolin, N., Goshawk, J., Nerin, C., 2022. A collision cross section database for extractables and leachables from food contact materials. J. Agric. Food Chem. 15205118. 70 (14), 4457–4466. Available from: http://pubs.acs.org/journal/jafcau. doi:10.1021/acs.jafc.2c00724.

Souihi, A., Mohai, M.P., Palm, E., Malm, L., Kruve, A., 2022. MultiConditionRT: predicting liquid chromatography retention time for emerging contaminants for a wide range of eluent compositions and stationary phases. J. Chromatogr. A 1666. doi:10.1016/j.chroma.2022.462867

Subramaniam, K.M., 2002. The KEGG database. In: Novartis Foundation Symposium, vol. 247. Japan, pp. 91–103.

Sunyer-Caldú, A., Peiró, A., Díaz, M., Ibáñez, L., Gil-Solsona, R., Gago-Ferrero, P., Silvia Diaz-cruz, M., 2023. Target analysis and suspect screening of UV filters, parabens and other chemicals used in personal care products in human cord blood: prenatal exposure by mother-fetus transfer. Environ. Int. 01604120. 173, 107834. doi:10.1016/j.envint.2023.107834.

Sunyer-Caldú, A., et al. 2026. Silicone-foam passive air samplers for combined target and nontarget chemical profiling and toxicity assessment of airborne exposomes

. Environ. Sci. Technol. 60(7), 5628–5644. doi:10.1021/acs.est.5c1661310.1021/acs.est.5c16613.s00110.1021/acs.est.5c16613.s002.

Taha, H.M., Aalizadeh, R., Alygizakis, N., Antignac, J.P., Arp, H.P.H., Bade, R., Baker, N., Belova, L., Bijlsma, L., Bolton, E.E., Brack, W., Celma, A., Chen, W.L., Cheng, T., Chirsir, P., Čirka, Ľ., D'Agostino, L., Feunang, Y.D., Dulio, V., Fischer, S., Gago-Ferrero, P., Galani, A., Geueke, B., Głowacka, N., Glüge, J., Groh, K., Grosse, S., Haglund, P., Hakkinen, P.J., Hale, S.E., Hernández, F., Janssen, E.M.L., Jonkers, T., Kiefer, K., Kirchner, M., Koschorreck, J., Krauss, M., Krier, J., Lamoree, M.H., Letzel, M., Letzel, T., Li, Q., Little, J., Liu, Y., Lunderberg, D.M., Martin, J.W., McEachran, A.D., McLean, J.A., Meier, C., Meijer, J., Menger, F., Merino, C., Muncke, J., Muschket, M., Neumann, M., Neveu, V., Ng, K., Oberacher, H., O'Brien, J., Oswald, P., Oswaldova, M., Picache, J.A., Postigo, C., Ramírez, N., Reemtsma, T., Renaud, J., Rostkowski, P., Rüdel, H., Salek, R., Samanipour, S., Scheringer, M., Schliebner, I., Schulz, W., Schulze, T., Sengl, M., Shoemaker, B.A., Sims, K., Singer, H., Singh, R.R., Sumarah, M., Thiessen, P.A., Thomas, K.V., Torres, S., Trier, X., van Wezel, A.P., Vermeulen, R.C.H., Vlaanderen, J.J., von der Ohe, P.C., Wang, Z., Williams, A.J., Willighagen, E.L., Wishart, D.S., Zhang, J., Thomaidis, N.S., Hollender, J., Slobodnik, J., Schymanski, E.L., 2022. The norman suspect list exchange (NORMAN-SLE): facilitating European and worldwide collaboration on suspect screening in high resolution mass spectrometry. Research Square Available from: https://www.researchsquare.com/browse. doi:10.21203/rs.3.rs-1902466/v1.

Tang, C., Liang, Y., Wang, K., Liao, J., Zeng, Y., Luo, X., Peng, X., Mai, B., Huang, Q., Lin, H., 2023. Comprehensive characterization of per- and polyfluoroalkyl substances in wastewater by liquid chromatography-mass spectrometry and screening algorithms. Npj Clean Water. 2059-7037. 6 (1). doi:10.1038/s41545-023-00220-6.

Tautenhahn, R., Patti, G.J., Rinehart, D., Siuzdak, G., 2012. XCMS online: a web-based platform to process untargeted metabolomic data. Anal. Chem. 00032700. 84 (11), 5035–5039. doi:10.1021/ac300698c.

Thermo Scientific, 2025.

Tian, Z., Zhao, H., Peter, K.T., Gonzalez, M., Wetzel, J., Wu, C., Hu, X., Prat, J., Mudrock, E., Hettinger, R., Cortina, A.E., Biswas, R.G., Kock, F.V.C., Soong, R., Jenne, A., Du, B., Hou, F., He, H., Lundeen, R., Gilbreath, A., Sutton, R., Scholz, N.L., Davis, J.W., Dodd, M.C., Simpson, A., McIntyre, J.K., Kolodziej, E.P., 2021. A ubiquitous tire rubber–derived chemical induces acute mortality in coho salmon. Science. 0036-8075. 371 (6525), 185–189. doi:10.1126/science.abd6951.

Tsugawa, H., Cajka, T., Kind, T., Ma, Y., Higgins, B., Ikeda, K., Kanazawa, M., VanderGheynst, J., Fiehn, O., Arita, M., 2015. MS-DIAL: data-independent MS/MS deconvolution for comprehensive metabolome analysis. Nat. Methods. 1548-7091. 12 (6), 523–526. doi:10.1038/nmeth.3393.

Wang, M., Carver, J.J., Phelan, V.V., Sanchez, L.M., Garg, N., Peng, Y., Nguyen, D.D., Watrous, J., Kapono, C.A., Luzzatto-Knaan, T., Porto, C., Bouslimani, A., Melnik, A.V., Meehan, M.J., Liu, W.-T., Crüsemann, M., Boudreau, P.D., Esquenazi, E., Sandoval-Calderón, M., Kersten, R.D., Pace, L.A., Quinn, R.A., Duncan, K.R., Hsu, C.-C., Floros, D.J., Gavilan, R.G., Kleigrewe, K., Northen, T., Dutton, R.J., Parrot, D., Carlson, E.E., Aigle, B., Michelsen, C.F., Jelsbak, L., Sohlenkamp, C., Pevzner, P., Edlund, A., McLean, J., Piel, J., Murphy, B.T., Gerwick, L., Liaw, C.-C., Yang, Y.-L., Humpf, H.-U., Maansson, M., Keyzers, R.A., Sims, A.C., Johnson, A.R., Sidebottom, A.M., Sedio, B.E., Klitgaard, A., Larson, C.B., Boya P, C.A., Torres-Mendoza, D., Gonzalez, D.J., Silva, D.B., Marques, L.M., Demarque, D.P., Pociute, E., O'Neill, E.C., Briand, E., Helfrich, E.J.N., Granatosky, E.A., Glukhov, E., Ryffel, F., Houson, H., Mohimani, H., Kharbush, J.J., Zeng, Y., Vorholt, J.A., Kurita, K.L., Charusanti, P., McPhail, K.L., Nielsen, K.F., Vuong, L., Elfeki, M., Traxler, M.F., Engene, N., Koyama, N., Vining, O.B., Baric, R., Silva, R.R., Mascuch, S.J., Tomasi, S., Jenkins, S., Macherla, V., Hoffman, T., Agarwal, V., Williams, P.G., Dai, J., Neupane, R., Gurr, J., Rodríguez, A.M.C., Lamsa, A., Zhang, C., Dorrestein, K., Duggan, B.M., Almaliti, J., Allard, P.-M., Phapale, P., Nothias, L.-F., Alexandrov, T., Litaudon, M., Wolfender, J.-L., Kyle, J.E., Metz, T.O., Peryea, T., Nguyen, D.-T., VanLeer, D., Shinn, P., Jadhav, A., Müller, R., Waters, K.M., Shi, W., Liu, X., Zhang, L., Knight, R., Jensen, P.R., Palsson, B.Ø., Pogliano, K., Linington, R.G., Gutiérrez, M., Lopes, N.P., Gerwick, W.H., Moore, B.S., Dorrestein, P.C., Bandeira, N., 2016. Sharing and community curation of mass spectrometry data with global natural products social molecular networking. Nat. Biotechnol. 1087-0156. 34 (8), 828–837. doi:10.1038/nbt.3597.

Williams, A.J., Grulke, C.M., Edwards, J., McEachran, A.D., Mansouri, K., Baker, N.C., Patlewicz, G., Shah, I., Wambaugh, J.F., Judson, R.S., Richard, A.M., 2017. The CompTox chemistry dashboard: a community data resource for environmental chemistry. J. Cheminform. 17582946. 9 (1). Available from: http://www.jcheminf.com/. doi:10.1186/s13321-017-0247-6.

Wishart, D., Arndt, D., Pon, A., Sajed, T., Guo, A.C., Djoumbou, Y., Knox, C., Wilson, M., Liang, Y., Grant, J., Liu, Y., Ali Goldansaz, S., Rappaport, S.M., 2015. T3DB: the toxic exposome database. Nucleic Acids Res. 1362-4962. 43 (D1), D928. doi:10.1093/nar/gku1004.

Wu, C., Siems, W.F., Asbury, G.R., Hill, H.H., 1998. Electrospray ionization high-resolution ion mobility spectrometry-mass spectrometry. Anal. Chem. 70 (23), 4929–4938. doi:10.1021/ac980414z.

Xie, H., Sdougkou, K., Bonnefille, B., Papazian, S., Bergdahl, I.A., Rantakokko, P., Martin, J.W., 2024. Chemical exposomics in human plasma by lipid removal and large-volume injection gas chromatography–high-resolution mass spectrometry. Environ. Sci. Technol. 0013-936X. 58 (40), 17592–17605. doi:10.1021/acs.est.4c05942.

Zakir, M., LeVatte, M.A., Wishart, D.S., 2025. RT-Pred: a web server for accurate, customized liquid chromatography retention time prediction of chemicals. J. Chromatogr. A 18733778. 1747. Available from: https://www.sciencedirect.com/science/journal/00219673. doi:10.1016/j.chroma.2025.465816.

Zeleny, J., 1898. VI on the ratio of the velocities of the two ions produced in gases by Röntgen radiation; and on some related phenomena. Dublin Philos. Mag. J. Sci 46, 120–154.

Zhang, H., Luo, M., Wang, H., Ren, F., Yin, Y., Zhu, Z.-J., 2023. AllCCS2: curation of ion mobility collision cross-section atlas for small molecules using comprehensive molecular representations. Anal. Chem. 0003-2700. 95 (37), 13913–13921. doi:10.1021/acs.analchem.3c02267.

Chapter 10

Portable techniques and sensors for environmental analysis

Saidy Cristina Ayala-Durán, Matheus Schiavon Kronka, Hernán Dario Rojas-Mantilla, Maria Valnice Boldrin Zanoni, and Paulo Clairmont Feitosa de Lima Gomes
Department of Analytical, Physical Chemistry and Inorganic Chemistry, Institute of Chemistry, National Institute of Alternative Technologies for Detection, Toxicological Assessment and Removal of Micropollutants and Radioactive—INCT-DATREM, São Paulo State University (UNESP), São Paulo, Brazil

1 Introduction

The recent progress of Industry 4.0 technologies has triggered great potential in the field of portable analytical devices capable of performing fast, reliable, and low-cost analyses, offering a very promising alternative to traditional laboratory-based analytical systems. These applications have benefited from improving the Internet, commercial electronics, and mobile technologies, thus creating compact platforms capable of performing and replicating complex analyses safely and precisely. By combining hardware miniaturization with software integration, analytical tools are becoming more accessible, democratizing scientific knowledge and enabling diverse applications in the areas of health, food safety, and environmental monitoring (Barton et al., 2022).

To better understand the behavior of the scientific landscape regarding portable analytical devices, a co-occurrence network was generated between the main concepts in the field using data collection of the last two decades extracted from the Scopus database. The search strategy applied combined the keywords "portable-devices" AND "smartphone-based" AND "industry 4.0" OR "Internet-of-things OR "Bluetooth" OR "wireless-networks" OR "point-of-care" yielding >1526 indexed documents, including research articles, reviews, and book chapters dated from 2004 to 2025. A bibliometric analysis using **VOSviewer** was instructed only to include terms that appeared at least ten times and had significant interconnections. Fig. 10.1 presents the bibliometric results obtained in the period.

The co-occurrence analysis highlights a strong interconnection between portable devices and emerging technologies, including the Internet of Things (IoT), wireless communication, and, more recently, artificial intelligence (AI) and

Contaminants of Emerging Concern in the Environment. https://doi.org/10.1016/B978-0-443-33821-2.00001-3

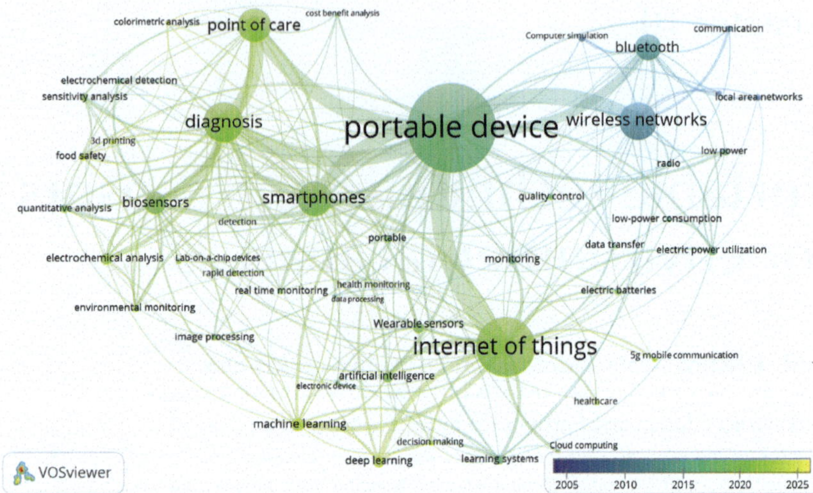

FIG. 10.1 Co-occurrence network of keywords obtained from Scopus using the terms"portable-devices" AND "smartphone-based" AND "industry 4.0" OR "internet-of-things" OR "Bluetooth" OR "wireless-networks" OR "point-of-care" on Scopus between 2004 and 2025. The network map was generated using VOSviewer software, using terms from the SCOPUS database with at least 10 co-occurrences. The resulting clusters highlight the central role of portable devices concerning wireless communication technologies, smart systems, and analytical applications.

machine learning (ML). The central role of the "portable device" reflects its importance for applied fields such as biosensors, electrochemical analysis, environmental diagnostics, and real-time monitoring. Some keywords include "smartphones", "Bluetooth", "data transfer", and "low energy consumption", emphasizing the importance of connectivity and energy efficiency to enable the autonomous and mobile operation of these devices. In summary, we can see how the convergence of these technologies drives the development of more intelligent, autonomous, connected sensors capable of collecting, processing, and transmitting data in real-time precisely and efficiently. Additionally, integrating AI and ML means these devices can learn from data over time, optimizing their performance and expanding their applications (Barton et al., 2022; Javaid et al., 2022; Bersani et al., 2022).

Smartphones are powerful, portable, and continuously connected miniaturized computers capable of collecting private information. They presented technological advances that involve the constant updating of their operating systems, increased storage capacity, integration of high definition (HD) cameras, easy internet connectivity, artificial intelligence (AI) assistance, global positioning system (GPS), fast data transmission via Wi–Fi or Bluetooth, and near field communication (NFC), as well as their portability and widespread use in daily life (Geballa-Koukoula et al., 2023; Zahra et al., 2022).

Their application as detectors or interfaces in biosensors (SbSs), whether optical or electrochemical, enables their use in various fields, including healthcare, food safety, and environmental monitoring. Currently, these devices have the potential to automate data processing and handling without the need for expensive equipment or highly trained personnel. Smartphones simplify laboratory protocols and popularize their use among individuals with little training, marking a significant advancement in the democratization of knowledge and facilitating access to analytical techniques (Ross et al., 2023).

For instance, electrochemical sensors that utilize a portable potentiostat as a power source and are connected to a computer may be susceptible to voltage, current, and power output variations, resulting in increased noise in measurements, particularly when connected via Wi–Fi or Bluetooth. In the same scenario, Arduino technology has become one of the most popular and widely used open-source hardware platforms designed to create DIY (Do It Yourself) projects. Based on the programmable ATmega328P microcontroller with various analog and digital ports, it allows the prototyping of multipurpose devices, facilitates process automation, and effectively controls parameters. Additionally, it enables the development of sensors, data collection, and management while supporting connections with other modules, sensors, and commercial electronic components. It is even compatible with more sophisticated equipment, such as computers, tablets, and smartphones, enabling fast communication and real-time data transmission (Cadeado et al., 2022; Kondaveeti et al., 2021).

Arduino programming is conducted in C/C++ using the "Arduino IDE" (Integrated Development Environment) software, which is regarded as an user-friendly environment that is easy to learn and available for free online (Kondaveeti et al., 2021). It is also compatible with other programming systems like Python, providing advantages such as simple learning, faster prototyping, ready-to-use libraries, and easy integration with different technologies.

Photodiodes and photoresistors are light-sensitive electronic components that operate differently. Photodiodes are silicon-based p-n junction semiconductors that create a depletion region, eliminating mobile charge carriers and enabling light conversion into electrical current. When exposed to light, a photon is absorbed, generating an electron-hole pair. Under reverse bias, an electric field forms across the depletion region, causing the charge carriers to move toward the electrons (e^-) or the holes (h^+). This movement results in photon absorption, generating an electric current flow proportional to the incident light intensity (Tavernier and Steyaert, 2011; Bielecki et al., 2022). PIN photodiodes are also characterized by enhanced collection efficiency due to a lightly doped region in the p-n junction (Okhonin et al., 2015). This photocurrent increases in proportion to the rate of absorbed photons.

Photoresistors, also known as photocells or LDRs (Light-Dependent Resistors), are sensors that change their electrical resistance as light strikes them. They exhibit specific characteristics such as surface defects, crystallographic orientation, and grain size (Adams-Mcnichol et al., 2019). Among the advantages are their good stability (they exhibit memory effect); however, they cannot support large electrical currents and have issues with limited photocurrent speed. Due to their low cost, robustness, and miniaturization, these electronic

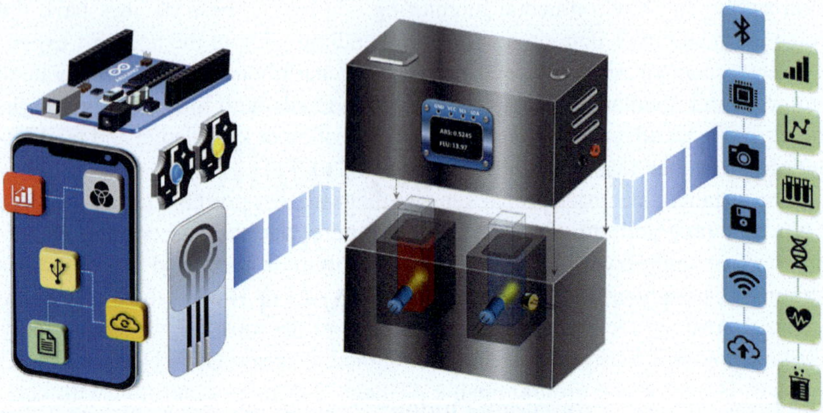

FIG. 10.2 Use of commercial electronic components in creating portable devices, showing the advantages of connectivity, storage, and reproducibility for their application in environmental analysis.

components become highly attractive for analytical applications, especially in the creation of miniaturized devices such as photometers and fluorometers (Bamdad and Bamdad, 2025; Di Nonno and Ulber, 2022; Song et al., 2017), sensors (Butt et al., 2024; Khudaverdyan et al., 2021), biosensors (Thakur et al., 2024; Sohrabi et al., 2021), and bioluminescent devices (Zappi et al., 2021; Mehta et al., 2023).

The growing concern over environmental pollution caused by contaminants of emerging concern (CECs) and their detrimental effects on ecosystems and human health has driven the development of advanced methods for environmental monitoring. Portable techniques and sensors have emerged as valuable tools in this context, offering real-time, on-site analytical capabilities with high sensitivity and precision. In contrast to conventional laboratory-based methods, portable sensors provide the distinct advantage of being deployable directly in the field, enabling continuous and immediate monitoring of environmental matrices such as air, water, and soil (Javaid et al., 2022). Electrochemical portable devices have gained prominence in environmental analysis due to their ability to provide rapid, sensitive, and selective detection of a wide range of contaminants. Their advantages lie in the low-cost instrumentation, minimal sample preparation, and potential for on-site analysis, allowing real-time monitoring in diverse and challenging environments (Geballa-Koukoula et al., 2023). Fig. 10.2 illustrates the main electronic resources for developing portable devices for detecting CECs in environmental analyses.

2 Procedure

The development and prototyping of devices for techniques such as fluorescence spectroscopy, bioluminescence, and colorimetric tests used in the detection and monitoring of CECs require verifying parameters, including wavelengths (λ), bandwidth, type, and intensity of the radiation source, as well as sensor sensitivity.

For example, Sethu et al. (2023) utilized a PDMS lab-on-chip mold designed using the free software Tinkercad and 3D printing. The PDMS chip contained a reaction well interconnected by channels that allowed fluid to move through the component. SFH756 LEDs and SFH350V photodiodes connected to an Arduino microcontroller were used as an optical detection system. The photodetector captures the intensity of the light transmitted through the reaction well, and a quad IC LM324 amplifies the voltage. Finally, the voltages acquired by the Arduino microcontroller are correlated with the concentration of gentamicin.

Several parameters are essential for the methodology used in analyzing digital images. They must be considered, such as the model and brand of the camera used, shutter speed, focus, balance, angle, and aperture of the camera lens (Lima et al., 2017; Soares et al., 2023). When using devices such as smartphones, it is essential to specify the model, brand, and settings to be applied, including the type of lighting, distance, stability, focus, capture area, and flash status (on or off). These parameters must be reported to ensure adequate sensitivity and reproducibility, as well as the accuracy and linearity of the results (Soares et al., 2023; Oliveira et al., 2023). Additionally, free software applications such as Open Camera and Color Grab, among others, are used to adjust brightness, regulate color intensity, and determine the measurement range of the channels, thereby optimizing sensitivity and data processing. Other parameters, such as, for example, brightness, intensity, or chromatic scale, as well as optimization in the optical system through the use of converging lenses, or elimination of external light inside the device, allow an improvement in image acquisition and quality, ensuring analyses with greater precision and reproducibility.

Bergua et al. (2022) developed a smartphone-portable method for biological and chemical analyses. Each image was captured using the smartphone at an ISO of 1000 and various integration times, then analyzed using ImageJ (version 1.53 h). In quantifying the toxicity of pentachlorophenol (PCP) on *A. fischeri*, the intensity of the bacteria's bioluminescence was measured. A 12 × 8 matrix of circles was created and printed on poly(ethylene terephthalate) (PET) with dimensions of 12.5 × 8.2 cm using a wax printer (Xerox ColorQube 8570DN) in black, which provides high contrast. Finally, the Arduino board is an efficient microcontroller, enabling easy access and control of the prototype's components, including LEDs, stepper motors, and temperature and humidity sensors, while facilitating data management .

Jiang et al. (2022) obtained by mixing QDs $CdTe_5/Tb^{3+}$ with NFX in a quartz cuvette for 10 s and at room temperature from the fluorescence ratio I_{545}/I_{645}, with NFX concentration (0–1200 nmol L^{-1}), and digital images were obtained by photographs taken using a Canon EOS 700D digital camera. The solution is deposited in a multiwell plate, followed by incubation for 30 min at 30°C in the dark, with a color change from red to yellow with the concentration of NFX. The smartphone camera (iPhone X, USA) captures fluorescent images under irradiation from a portable UV lamp (AGL-9406, 365 nm), which are converted into RGB values by the color recognition APP based on the G/R value of the smartphone.

Electrochemical sensors convert chemical energy into an electrical signal through signal transduction. This conversion is mediated by a transducer, usually an electrode, which interacts directly with the chemical environment at the electrode-electrolyte interface. Consequently, the geometry of the electrochemical cell is crucial to these studies (Karthik et al., 2022; Stradiotto et al., 2003). Conventionally, electrochemical cells are composed of three electrodes: the working electrode (where transduction takes place), the reference electrode (which provides a pattern potential), and the counter electrode (which allows the current flow) all submerged in a conductive electrolyte solution. Bearing in mind the growing demand for portable devices and decentralized analyses, the miniaturization of the electrochemical cell has become a strategic field in this development. Many studies have focused on using sensors based on printed electrodes (SPEs), which integrate three electrodes into a single disposable substrate, usually plastic or paper. These devices operate with tiny sample droplets, facilitating rapid and in situ analysis (Lee et al., 2018; Nery and Kubota, 2013). In parallel, the progress made in microfabrication and microfluidic technologies has enabled the development of the so-called Lab-on-a-Chip (LoC). These miniaturized systems integrate microfluidic channels, electrochemical sensors, reagent compartments, and, in some cases, even data processing and wireless communication modules. LoC devices represent a fully portable lab platform capable of complex analysis in tiny volumes with high automation and efficiency (Javaid et al., 2022; Mohammed et al., 2015).

Using a potentiostat is also crucial in electrochemical studies because it allows precise control of the experimental parameters that control redox processes, such as the applied potential, scan speed, pulse frequency, and current application time. These parameters are essential for optimizing the electrochemical response, adjusting the sensitivity and selectivity of the sensor, studying reaction mechanisms, and obtaining reliable analytical curves. Compact, low-cost electronic equipment has also been developed to bring portable devices onto the market by integrating operational amplifiers, A/D conversion chips, D/A conversion chips, impedance analysis chips, and other circuit modules (Ainla et al., 2018; Benachio et al., 2021; Vidotto et al., 2023; Cruz et al., 2014). Sensit Smart is a commercial example of a miniaturized potentiostat that connects directly to Android smartphones via USB. Despite its small size, Sensit Smart can perform various electroanalytical techniques such as cyclic voltammetry (CV), square wave voltammetry (SWV), differential pulse voltammetry (DPV), and chronoamperometry, being ideal for field applications in areas such as environmental monitoring. Similarly, systems developed with microcontrollers such as Arduino and ESP32 have made it possible to produce educational and functional DIY potentiostats with Bluetooth connectivity and integration with smartphone or notebook interfaces.

Table 10.1 highlights portable devices applied in CECs determination.

TABLE 10.1 CECs determined using different types of portable devices.

Device	Microcontroller	Sample volume/ mL or μL	Sample type	Methods and techniques	CECs	Price /$	Detection limit (LOD)	Ref.
Smartphone sensing platform	*	*	Lake water Tap Water Honey	**Fluorescence** (G/R) channel	Norfloxacin	*	13.7 nmol L^{-1}	(Jiang et al., 2022)
Smartphone-integrated with Arduino	Arduino Nano	100 μL 50 μL 100 μL 50 μL	Milli-Q Water	**Colorimetric** ELISA Tests. **Bioluminescence** Assays. **Turbidimetric** analysis **Fluorescence** Tests.	**Antibodies** SARS-CoV-2 Human IgG Pentachlorophenol Penicillin Ampicillin Amoxicillin Kanamycin **Dyes.** (Q21341MP) (Q21321MP)	$300	31.4 ± 0.3 ng mL^{-1} 5.1 ± 0.5 ng mL^{-1} 5.61 ± 3.20 mg L^{-1} 30.04 ± 21.9 μg mL^{-1} 0.23 ± 0.06 μg mL^{-1} 1.54 ± 0.44 μg mL^{-1} 4.55 ± 1.70 μg mL^{-1} 0.1 to 3 nmol L^{-1}	(Bergua et al., 2022)
Portable optical system	Arduino UNO	100 μL	Tap water	**Colorimetric** analysis	Gentamicin	*	2.6 μmol L^{-1}	(Sethu et al., 2023)

Continued

TABLE 10.1 CECs determined using different types of portable devices—cont'd

Device	Microcontroller	Sample volume/ mL or μL	Sample type	Methods and techniques	CECs	Price /$	Detection limit (LOD)	Ref.
Wearable glove sensors coupled to a miniaturized potentiostat	PalmSens3 potentiostat coupled to laptop And Sensit Smart potentiostat coupled to smartphone	100 μL	Real Effluents (Tap water, lake water, whiskey and orange juice)	**Amperometric** square wave voltammetry	Picric acid 4-nitrophenol Diazepam	**$0.23** (each sensor)	0.24 $\mu mol\ L^{-1}$ 0.35 $\mu mol\ L^{-1}$ 0.06 $\mu mol\ L^{-1}$	(Corsato et al., 2024)
Screen-printed electrodes coupled to a miniaturized potentiostat	smartphone-controlled potentiostat	40 mL	PBS buffer 50 mmol L^{-1}	**Amperometric** cyclic voltammetry	Isocarbophos Chlorpyrifos Trichlorfon	*	0.1 $\mu g\ L^{-1}$	(Shi et al., 2018)

* This information has not been reported.

3 Fundamentals and instrumentation

3.1 Smartphones

The optical sensor of a smartphone works by emitting a beam of light, either visible or infrared, onto an object. It measures the amount of light reflected, converting this data into electronic signals that are then interpreted by the device (Soares et al., 2023; Chen et al., 2021). Optical analytical techniques, such as reflectance, fluorescence, absorbance, bioluminescence, and colorimetric analysis, can be easily assessed using smartphones, which function as smart detectors.

This chemical imaging analysis technique is based on the RGB and other color spaces such as CMYK, HSB/HSL, CIE XYZ, L*a*b* e YUV. The RGB is the most widely used color space in image-capturing devices like smartphones. Its concept is based on the combination of the main colors: red, green and blue, enabling the generation of pixels with variations between 0 and 255 per channel, providing a response of up to 16 million combinations that can be associated with chemical characteristics of the analyzed substance, such as, for example, the concentration. Image processing steps from programs such as Photoshop, ImageJ or MatLab are necessary, allowing adequate image processing to be analyzed and facilitating tasks such as contour extraction, segmentation of the work area, and improvements in lighting, ensuring adequate extraction of the information present in each pixel. Finally, the pixels obtained undergo calibration models from statistical methods such as Multivariate Regression or Principal Component Analysis (PCA), which allow an effective relationship between the color evaluation and the sample concentration (Soares et al., 2023; Oliveira et al., 2023). Finally, model validation is performed through mean square error calculations or cross-validation (Kim et al., 2017; Fan et al., 2021).

Other smartphone applications include serving as interfaces for real-time control of various measurements on devices, such as potentiostats, and facilitating data transmission via Bluetooth or NFC. This allows for the efficient capture and management of data obtained during an experiment (Sun and Hall, 2019).

Currently, the development of smartphone applications (apps) that transform these devices into even more efficient detection systems for chemical analysis is an innovative approach. In this context, Rosa et al. (2022) developed the free REDGIM app, available for the Android system, which can be used with various spectrometric and colorimetric methods. The app can be downloaded for free from Google Play, enabling multivariate image analysis applied to chemometric methods based on the light intensity of photos in the RGB format. Operated on the smartphone, REDGIM also allows adjustments to the position, geometric shape, and corrections of the average intensities of regions of interest (ROI) in previously captured images, simplifying subsequent processing. It shows great promise as an alternative for chemical image analysis via smartphone in multivariate systems, offering quick response, easy updates, and potential for application in other colorimetric matrices.

Smartphone applications are designed to serve as interfaces for data collection and acquisition, which can be controlled, semi-controlled, or uncontrolled. These applications may be developed as integrated solutions by third parties or customized open-source or proprietary software. However, the main issue lies in privacy risks, software licensing restrictions, or patent and intellectual property constraints, which could compromise data security by facilitating access for analysis. For this reason, open-source applications such as those licensed under MIT, Apache, and GNU offer the most reliable, transparent, and traceable solutions. These licenses establish clear agreements between developers and users while maintaining a simple and accessible framework. Among the advantages of third-party applications, open-source options like Open Camera and proof-of-concept applications stand out. They allow modifications to camera functions such as shutter speed, focus, and white balance, enabling manual adjustments to enhance SbS sensitivity. Additionally, they support raw data export and third-party processing while remaining open-source and free to use (Geballa-Koukoula et al., 2023; Ross et al., 2023; Upadhyay et al., 2024).

After smartphone data acquisition, the image must be processed, transformed, and normalized online or offline. The absence of a well-defined framework to outline the steps of this analysis, combined with a scarcity of bibliographic references, makes it challenging to achieve consistent results. Normalization involves removing noise to minimize interferences that could compromise an image's or video's quality. This process helps correct uneven lighting, calibrate the camera, and mitigate effects such as dust on the lens. Additionally, it allows for background subtraction and eliminates unnecessary data for analysis. Researchers highlight the importance and need to use a standardized database and methodologies, such as SPECTACLE, which enables smartphone camera calibration based on linearity, polarization, ISO speed variation, and RGB spectral response parameters (Ross et al., 2023; Upadhyay et al., 2024).

The emergence of Artificial Intelligence (AI) is considered another valuable resource that, when combined with devices such as smartphones, optimizes image processing. In this regard, Bhuiyan et al. (2022) used a system that, through the integration of technologies such as AI, smartphones, and microfluidic platforms, facilitated the processing and analysis of images obtained in enzyme assays by ELISA. The programmed and automated use of AI avoided frequent problems in this type of process, such as the presence of air bubbles or inadequate filling of the tank where the reaction takes place. The aforementioned validates the technological advantages of devices such as smartphones when used as smart sensors, denoting characteristics such as their portability, ease of use, and accessibility in fields such as environmental analysis.

3.2 Platform arduino

Arduino prototyping boards feature an ATmega328 microcontroller (8 bits), considered the system's brain, which allows for the management and control of ports (input/output) and communication. They operate from low supply voltages, between 3.0 and 12 V. Data transmission is carried out via serial communication, facilitating communication and sending of information between devices. Digital ports allow connection to different types of commercial electronic components or sensors, such as LEDs, photoresistors, and servo motors, among many others (Cadeado et al., 2022).

Being an open-source system, a large community creates and freely shares numerous modifications and source code programming, and countless libraries and support systems are available on the Internet. Together with newer versions like the Arduino Uno Wi–Fi, Arduino Mega 2560, Arduino Due, Arduino Mega ADK, Arduino Pro Mini, and Arduino Nano, these boards are considered advanced prototyping platforms. They offer features such as Wi–Fi and Bluetooth connectivity, simplifying the development of various sensors and monitoring systems. Additionally, they support the integration of sensors, including cameras and proximity sensors, enabling image processing and providing accurate measurements over greater distances. AI resources like Machine Learning (ML) allow a system to learn and improve autonomously (Kondaveeti et al., 2021). Integration with communication protocols such as MQTT or HTTP, along with using cloud platforms for real-time data analysis, is also possible. Furthermore, hardware expansion is achievable through the implementation of new shields like GPS, touch display, and sound, or the addition of modules for LDR photoresistors, pH sensors, integrated humidity and temperature modules, IR LED modules with built-in IR receivers, and Hall effect modules, among many others. These capabilities make it easier to generate devices and prototypes, which is particularly interesting in the field of analytical instrumentation, mainly significantly when associated with spectrophotometric techniques, where the use of optical systems, sensors, and highly refined technological devices is required, offering improved reproducibility and sensitivity while maintaining production costs at an accessible level (Gubsky, 2023; Hayden et al., 2022).

These enhanced capabilities and features have not gone unnoticed in the substance detection and monitoring of CECs, where their use is increasingly common. In the specialized literature, the association of Arduino technology with the development of various devices is reported, such as fluorimeters, spectrophotometers, colorimetric electronic sensors, portable biosensors, and potentiostat/galvanostatic (Granica and Tymecki, 2019; Hooda et al., 2021; Sheta et al., 2022; Omar et al., 2024; Elias, 2024).

In general, Arduino technology for all the mentioned works provided rapid and precise data acquisition and real-time data collection, as well as the adaptation of sensors and components such as photodiodes and LEDs, allowing the adjustment of various measurement parameters such as intensity, current variation, and potential with precision and automation. In addition, rapid and effective signal amplification, communication, and connectivity were also facilitated when integrated with smartphones, enabling remote data reception or cloud storage for later processing, resulting in portable, robust, and low-cost devices.

3.3 Electronic components

Electronic components such as photoresistors, photodiodes, and LEDs are essential for the creation of photometric devices, which are based on the Lambert-Beer Law, Eq. (10.1),

$$A = \varepsilon bC \tag{10.1}$$

Which establishes the relationship between absorbance (*A*), the molar absorption coefficient (ε), the molar concentration of the analyte (*C*), and the optical path length (*b*).

Photometers consist of a light-emitting diode (LED) as the source of electromagnetic radiation, a cuvette for the sample, a photodetector to register the signal, and a monitor or interface device (Bastan et al., 2023; Seetasang and Kaneta, 2021). The operating principle is based on the emission of light by the LED, which passes through the microcuvette and is focused on the photodiode to generate a voltage output. However, in some cases, the system operates based on the Paired-Emitter-Detector-Diode (PEDD) principle, involving two LEDs. One is used conventionally as a light source, while the second acts as a partially selective light detector when operating in reverse mode, exploiting the internal photoelectric effect.

The PDDE (potentiometric), it allows the establishment of a relationship between current and voltage of diodes in forward bias, reverse bias, or no bias, involving the ideal Shockley diode equation, Eq. (10.2),

$$I = \text{IS}\left[\exp\left(-\frac{E}{ET}\right) - 1\right] \tag{10.2}$$

Which relates to the diode current (*I*), the saturation current (Is), and the voltage across the diode (*ET*) (Tymecki et al., 2008). The energy emitted by the LED is directly proportional to the electrical energy flowing through the p-n junction, relating the light intensity incident on the LED detector (IL) with the current (IC) supplied by the LED emitter. This equation can be simplified into a Shockley exponential equation, Eq. (10.3),

$$E = a + b\log(\text{IC}) = a + b\log(\text{IL}) \tag{10.3}$$

Where *a* and *b* as constants that can be simplified from an equation in logarithmic form, taking into account that $\text{IL} = I_{\text{max}}$ with absorbance equal to 0. Eq. (10.4),

$$E_{\text{max}} = a + b\log(I_{\text{max}}) \tag{10.4}$$

Determining the voltage in the sample measured by the LED detector is defined by Eq. (10.5),

$$\Delta E_{\text{sample}} = E_{\text{max}} - E_{\text{sample}} = b\log\left(\frac{I_{\text{max}}}{I_{\text{sample}}}\right) \tag{10.5}$$

According to the Lambert-Beer law, this is a definition of the absorbance of the sample (A_{sample}) (Tymecki et al., 2008). Eq. (10.6), so:

$$\Delta E_{sample} = bA_{sample} = b\varepsilon lc \tag{10.6}$$

The theoretical foundation for the determination of absorbance when developing miniaturized devices such as photometers is based on the experimental determination of absorbance, which is given by the logarithm of the transmitted radiation (I), relative to the incident radiation (I_0), Eq. (10.7),

$$A = -\log(I / I_0) \tag{10.7}$$

When an LDR photodetector is used, it is inversely proportional to the intensity (I) of the light, Eq. (10.8),

$$I = \frac{k}{R} \tag{10.8}$$

This relationship allows the correlation of absorbance with resistance, Eq. (10.9),

$$A = -\log\left(\frac{R_0}{R}\right) \tag{10.9}$$

R_0 and R are the electrical resistance of the LDR photodetector in the absence and presence of light absorption, respectively.

3.4 Electrochemical sensors and biosensors

While signals in other conventional sensors are based on absorption intensity, mass changes, or thermal effects, electrochemical sensors' signals rely on the electron transfer between the analyte and the electrode material at the electrode-electrolyte interface. This electron transfer occurs through redox reactions, generating quantifiable electrical signals that are directly related to the concentration of the target analyte. A classic example of this mechanism is the ferricyanide/ferrocyanide ($Fe(CN)_6^{3-}/Fe(CN)_6^{4-}$) redox couple, often used as a model system to study electron transfer kinetics and electrode behavior. This redox pair is frequently employed in electrochemical characterization techniques due to its well-defined and reversible electrochemical properties (Hanrahan et al., 2004).

Two key equations—the Nernst equation (Eq. 10.10) and the Butler-Volmer equation (Eq. 10.11)—govern the theoretical foundation of electrochemical processes (Bard and Rubinstein, 2003; Bard and Faulkner, 2001).

$$E = E^0 - \frac{RT}{zF}\ln Q \tag{10.10}$$

E is the reduction potential (V), $E°$ is the standard potential of reaction (V), R is the universal gas constant, T is the temperature (K), z is the ionic charge (mols of e^-), F is the Faraday constant (C mol^{-1}) and Q the reaction quotient.

$$j = j_0\left[\exp\left(\frac{\alpha_A nF}{RT}\eta\right) - \exp\left(-\frac{\alpha_c nF}{RT}\eta\right)\right] \quad (10.11)$$

j is the current density (A m^{-2}), j_0 is the exchange current density (A m^{-2}), E is the electrode potential (V), and T is the temperature (K), n is the number of electrons transferred, F is the Faraday constant, R is the gas constant, and α_A and α_C are the anodic and cathodic transfer coefficients (dimensionless) and η: activation overpotential.

According to the Nernst equation, the equilibrium potential of a redox couple can be described as a function of the oxidized and reduced species concentrations. By using this equation, thermodynamic aspects of electrochemical systems can be predicted and analyzed, such as redox potentials, chemical equilibrium, and the impact of changes in the concentration of the analyte. On the other hand, the Butler-Volmer equation establishes a kinetic model for electron transfer at the electrode interface, which relates the current density transferred at the applied overpotential and the speed constants of the redox reaction. This equation is essential for studying reaction mechanisms, electron transfer rates, and charge transfer resistance, which are fundamental in ensuring optimal sensor performance (Karthik et al., 2022; Stradiotto et al., 2003; Hitchman and Berlouis, 1995; Janata, 2009; Miyake, 2003).

The electrochemical sensors can be classified by the technique used and the intended results. These include amperometric, potentiometric, and conductometric sensors. Amperometric and potentiometric sensors have gained particular prominence thanks to their rapid analysis capacity, low cost, and portability. Amperometric sensors feature prominently in various fields, including medical diagnostics for the detection of biological markers, process control in the chemical and food industries, and environmental monitoring of heavy metals, pesticides, and CECs in wastewater (Karthik et al., 2022; Arya and Bhansali, 2011; Wang et al., 2019). For this type of sensor, voltammetric techniques have been widely used thanks to their sensitivity in providing detailed information on the redox processes at the electrode/solution interface. In cyclic voltammetry (CV), conducted under diffusion-controlled conditions, the Randles-Ševčík equation relates the peak current (i_p) to the square root of the scan speed (v) and the analyte concentration (C), as described in Eq. (10.12):

$$i_p = (2.69x10^5)n^{\frac{3}{2}}AD^{\frac{1}{2}}Cv^{\frac{1}{2}} \quad (10.12)$$

N represents the number of electrons transferred in the redox reaction, A is the electrode area (cm^2), F represents the Faraday constant (C mol^{-1}), D is the diffusion coefficient (cm^2 s^{-1}), R *is* the Gas constant (J K^{-1} mol^{-1}), and T is the temperature (K).

The versatility of voltammetric techniques employed in electrochemical sensors is noteworthy considering that the signal and sensitivity to different targets can be optimized by controlling the scan rate. On the other hand, once the matrix becomes complex or the detection potentials overlap, requiring greater refinement and sensitivity of the analysis, more advanced techniques such as square wave voltammetry (SWV) and differential pulse voltammetry (DPV) are preferred. By minimizing the capacitive current normally found in these techniques, the analysis presents improved accuracy and sensitivity (Wong et al., 2018; Dos Santos et al., 2024).

Even though numerous studies have highlighted the development of dynamic and miniaturized electrochemical sensors, significant challenges in remain designing portable devices for environmental applications. Key issues include ensuring the long-term stability of the sensor in adverse environmental conditions, high selectivity detection on complex sample matrices, and sensing analytes at trace levels. Further barriers, such as sensor contamination and calibration demands, hinder their practical implementation. In this context, new materials such as nanostructured materials, metals, and bioelectrodes have been developed to address these challenges, offering higher surface areas and improved electron transfer kinetics (Thakur et al., 2024; Wong et al., 2018; Dos Santos et al., 2024; Wachholz Junior et al., 2025; Mello and Kubota, 2002).

One of the key challenges in terms of improving the electrochemical signal in complex matrices is the accurate detection of target compounds that are susceptible to interference in the sample. To meet these challenges, the development of biosensors has been identified as a promising area for greater specificity. Electrochemical biosensors are analytical devices that combine a biological component on the electrode surface. The principle of transduction is also fundamental for bioelectrochemical sensors, but through the incorporation of biological recognition elements—such as enzymes, antibodies, DNA chains, or cells—into the electrode surface. These devices' specificity derives from the natural selectivity of biological components, which can recognize and interact with the target analyte precisely. The integration with electrochemical devices makes it possible to perform rapid and direct measurements of biological recognition reactions, such as antigen-antibody binding or enzyme-substrate interaction, by varying the current, potential, or charge (Agarwal and Csóka, 2019; Thakur et al., 2024; Decarli et al., 2022; Pauliukaite et al., 2010; Jadhao et al., 2022).

4 Case of studies—examples of the application to CECs

4.1 Case of study 1

Bergua et al. (2022) developed a portable device used to perform fluorescence, colorimetry, bioluminescence, and turbidity analyses applied to the detection of toxins, evaluation of pharmaceutical compounds (penicillin, ampicillin, kanamycin, amoxicillin), and commercial fluorophore dyes (red quantum dots (QDs) Q21341MP, green QDs Q21321MP, and blue graphene QDs). This device uses a smartphone that functions as an optical reader, enabling image acquisition. Inside the portable device, a 96-well microplate is placed, integrated with a stepper motor that enables movement and agitation up to 160 rpm, a Bluetooth module, an Arduino microcontroller, and temperature and humidity sensors. The device's optical system was designed using four white LEDs for colorimetric tests, a UV SMD LED (365 nm) for fluorescence analysis, excitation, and emission filters, and a concave lens that facilitates fine focal distance adjustment between the samples and the smartphone.

Fluorescence tests using three dyes as a study model demonstrated that the device allows selective detection of the different RGB secondary wavelengths associated with the emission spectra of each dye, as well as it was possible identify two fluorophore compounds, under the conditions, simultaneously. Limit of detection (LOD) of 0.1 to 3.0 mol L^{-1} and linear range between 0.1 to 300 mol L^{-1} were achieved. Bioluminescence analysis using the bacteria *A. fischeri* applied to detect the pesticide pentachlorophenol (PCP) at a concentration of 9.0 ± 2.0 mg L^{-1}, promoted 50% inhibition of the bioluminescence signal. These results are close to those obtained with commercial devices (4.0 ± 0.3 mg L^{-1}). Bacterial resistance in the presence of antibiotics was evaluated, relating the growth of the bacterial population to the variation in turbidity. For this purpose, four different bacteria (*A. fischeri, E. coli ATCC11303, E. coli ATCC25922, and Salmonella*) were used in the presence of different concentrations (0 to 64 μg mL^{-1}) of penicillin, ampicillin, amoxicillin, and kanamycin. On the other hand, colorimetric analysis allowed the detection of the biomarker (N of SARS-CoV-2). The method achieved a LOD of 31.4 ng mL^{-1} evaluated in the linear range between 30 and 1000 ng L^{-1}, a result comparable to those obtained when using commercial plates under similar experimental conditions (LOD of 8.74 ng mL^{-1}, and linear range of 10 to 1000 ng mL^{-1}).

The developed device presented considerable advantages over commercial systems, highlighting its portability and ease of production. It was designed with dimensions of 27 × 18 × 7 cm and cost less than $300. The device uses economical and easily acquired components, which ensures greater accessibility, representing a significant advantage over bulkier and more expensive commercial systems. Furthermore, its design facilitates maintenance and upgrades when necessary, as well as component replacement (Jadhao et al., 2022). The device was developed to perform different types of analyses with precision and reliability, and it is suitable for multiplex analyses with various types of contaminants of emerging interest. However, its main disadvantage is limited sensitivity, associated with using of expensive components and high-tech industrial manufacturing.

Sethu et al. (2023) developed a portable optical system for the real-time detection of aminoglycoside antibiotics. The system is based on Lab-On-Chip (LoC) technology, which involves miniaturizing and integrating an analysis laboratory's functions into a single polydimethylsiloxane (PDMS) chip. This chip, combined with optical fibers, signal processing systems, and colorimetric assays based on Fe^{2+}-mediated lipid peroxidation, enables the detection of gentamicin. The system showed a working range between 8 and 50 μmol L^{-1} of gentamicin, with a linear coefficient of 0.997 and a LOD of 2.6 μmol L^{-1}.

Gentamicin was spiked in tap water, and a recovery of over 90% was achieved, validating its detection in actual samples. The system demonstrated great potential in detecting antibiotics because the colored product is generated after 5 min of reaction, enabling rapid analysis. All these features demonstrate its potential for traditional analytical equipment.

4.2 Case of study 2

Jiang et al. (2022) introduced a multiplex chroma probe as a ratiometric sensor using a single fluorophore composed of CdTe quantum dots doped with Tb^{3+} (CdTe/Tb^{3+} QDs) for fluorescent detection of norfloxacin (NFX), an antibiotic widely used in the treatment of infections. The photoluminescence principle of the dual-emission optical sensor relies on the emission peak of CdTe/Tb^{3+} QDs at 645 nm (red), which, when chelated with NFX (a conjugated ring and planar structure), forms a Schiff base. This interaction leads to energy transfer from the intramolecular π^*-π luminescent transition of NFX to Tb^{3+}, resulting in emission at 545 nm (yellow). This enables detection through a noticeable color variation, shifting from red to yellow as the NFX concentration increases, making the sensor more sensitive to visual detection. The quantitative and visual detection results for NFX were compared using a smartphone as a color recognition instrument (RGB) in G/R lines, which captures and converts fluorescence colors in real time.

The results demonstrated that the QDs CdTe/Tb^{3+}-NFX sensor was strongly correlated with the ratiometric intensity ratio I_{545}/I_{645} (6:1) and could be used within a working range of 0 to 1200 nmol L^{-1} of NFX, with R^2 0.996 and a LOD of 6.03 nmol L^{-1} (1925 ng L^{-1}). The sensor's stability was evaluated in terms of pH (4.0–9.0) and temperature (4°C–80°C), with pH 7.0 and room temperature being the optimal conditions without stability loss. The selectivity of the probe was assessed in the presence of interferents with overlapping fluorescence spectra, including analytes such as pesticides (thiram, chlorpyrifos, and dipterex), heavy metals (Pb^{2+}, Cr^{3+}, Hg^{2+}), and antibiotics (tetracycline, gentamicin, chloramphenicol, and amoxicillin), with no interference observed. Compared to smartphone-based results, an increase in LOD was noted (13.7 nmol L^{-1}), with R^2 0.995 and a working range of 1 to 12 μmol L^{-1}. The probe was applied to tap water, lake water (Chaohu, China), and honey samples, achieving recoveries between 94% and 104%, with relative deviations ranging from 1.11% to 3.01%.

Using fluorescent chromatic probes to detect the antibiotic demonstrated significant advantages, such as rapid visual analysis, stability (pH and temperature), high selectivity, and no intensity fluctuations. Furthermore, it allows the study of the analysis levels of analytes at the nmol L^{-1}, and the smartphone application permits the portability of this method. This sensor is suitable for real sample analysis, making it a fast, sensitive, and efficient method. However, some disadvantages include the potential for operational errors in individuals with color blindness, as it relies on visual detection. Additionally, sensor reuse is not mentioned, although it could be used with other analytes that exhibit fluorescence in the same region.

4.3 Case of study 3

Shi et al. (2018) showed promising results by using an electrochemical biosensor of Prussian blue-modified dual-channel screen-printed electrode (DSPE) towards the fast detection of organophosphorus pesticide residues in water. This innovative biosensor was performed using an enzymatic inhibition mechanism, in which the pesticide acetylcholinesterase (AChE) was inhibited, thus driving measurable changes in the current response. Integrated into a handheld potentiostat monitored via a smartphone, the device performs real-time analysis in the field, without requiring complex laboratory equipment. The sensor has demonstrated high sensitivity, offering values of limits of quantification (LOQ) to 10^{-7} g mL^{-1} for pesticides such as isocarbophos, chlorpyrifos and trichlorfon. The author also showed that the design sensor was successfully tested on real agricultural samples with recovery rates corresponding to regulatory standards. This work highlights the viability of integrating a sensitive and low-cost device for on-site pesticide detection in public health applications.

A significant contribution to the field of portable sensors was recently reported by Corsato et al. (2024), who developed a wearable electrochemical sensor showing a practical application of portable devices for detecting CECs in real-world samples. The glove-based sensors were produced using a stencil printing technique on nitrile gloves, chosen for their durability, flexibility, and chemical resistance. To improve the electrochemical performance of the working electrodes (WE), they used multi-walled carbon nanotubes (MWCNT) and sulfonated polystyrene poly(3,4-ethylenedioxythiophene) (PEDOT). Samples were taken directly by touching or sliding the surfaces using the fingertip sensors or by submerging the sensors in liquid samples. The wearable sensor demonstrated remarkable versatility, effectively analyzing dust and liquid samples, showing its adaptability to different environmental conditions and sample types. Furthermore, the data acquisition was facilitated by coupling the wearable sensor to a miniaturized Sensit Smart potentiostat controlled by a smartphone. The wearable sensor exhibited excellent sensitivity for the detection of picric acid (PA), 4-nitrophenol (4-NP), and diazepam (DZ), reaching LOD of 0.24, 0.35, and 0.06 μmol L^{-1}, respectively. Most importantly, the wearable device was successfully applied to real-world samples such as lake water, tap water, orange juice, and whisky, proving its practical viability beyond controlled laboratory conditions. The detection of PA and 4-NP in water samples and of DZ in drinks generated recovery rates of between 90.7% and 114.0%, demonstrating the practical benefits of the sensors for environmental analysis. These portable sensors are ideal for in-field applications, providing rapid, real-time results with high accuracy and reproducibility. The low cost (approximately $0.23 per glove containing four sensors) and ease of use make them a promising tool for environmental monitoring and public safety control, especially in situations requiring immediate responses, such as industrial spills.

Despite significant advancements recently attached, developing portable electrochemical sensors still faces challenges, including issues with long-term stability, sensor miniaturization without compromising sensitivity, and more robust materials that can withstand diverse environmental conditions and maintain accuracy in complex real-world matrices. These challenges highlight that there is still an avenue for advancement in developing portable electrochemical sensing devices, paving the way for further research and innovation.

AI disclosure

During the preparation of this work the author(s) used Grammarly in order to Grammarly was used to revise the English language adopted in this chapter. After using this tool/service, the author(s) reviewed and edited the content as needed and take(s) full responsibility for the content of the publication.

References

Adams-Mcnichol, A.L., Shiell, R.C., Ellis, D.A., 2019. Accurate, photoresistor-based, student-built photometer and its application to the forensic analysis of dyes. J. Chem. Educ. 19381328. 96 (6), 1143–1151. Available from: http://pubs.acs.org/loi/jceda8. doi:10.1021/acs.jchemed.8b00862.

Agarwal, C., Csóka, L., 2019. Recent advances in paper-based analytical devices: a pivotal step forward in building next-generation sensor technology. In: Sustainable Polymer Composites and Nanocomposites. Springer International Publishing, Hungary, pp. 479–517. Available from: https://link.springer.com/chapter/10.1007/978-3-030-05399-4_18. doi:10.1007/978-3-030-05399-4_18.

Ainla, A., Mousavi, M.P.S., Tsaloglou, M.N., Redston, J., Bell, J.G., Fernández-Abedul, M.T., Whitesides, G.M., 2018. Open-source potentiostat for wireless electrochemical detection with smartphones. Anal. Chem. 15206882. 90 (10), 6240–6246. Available from: http://pubs.acs.org/journal/ancham. doi:10.1021/acs.analchem.8b00850.

Arya, S.K., Bhansali, S., 2011. Lung cancer and its early detection using biomarker-based biosensors. Chem. Rev. 15206890. 111 (11), 6783–6809. doi:10.1021/cr100420s.

Bamdad, F., Bamdad, A.M., 2025. Converting a power transistor to an optical transducer as an alternative to a commercial photo-detector for the construction of a simple and low-cost photometer. Phys. Educ. 13616552. 60 (1). Available from: https://iopscience.iop.org/journal/0031-9120. doi:10.1088/1361-6552/ad9808.

Bard, A.J., Faulkner, L.R., 2001. Electrochemical Methods: Fundamentals and Applications, John Wiley & Sons, Inc.

Bard, A.J., Rubinstein, I., 2003. Electroanalytical chemistry: a series of advances, Vol. 22, pp. 309. New York: Marcel Dekker, Inc., ISBN 0-203-91315-9.

Barton, M., Budjac, R., Tanuska, P., Gaspar, G., Schreiber, P., 2022. Identification overview of industry 4.0 essential attributes and resource-limited embedded artificial-intelligence-of-things devices for small and medium-sized enterprises. Appl. Sci. 2076-3417. 12 (11), 5672. doi:10.3390/app12115672.

Bastan, N., Ahmadi, M., Madrakian, T., Afkhami, A., Khalili, S., Majidi, M., Moradi, M., 2023. A paired emitter–detector diode-based photometer for the determination of sodium hypochlorite adulteration in milk. Sci. Rep. 13, 6217.

Benachio, I., Lobato, A., Gonçalves, L.M., 2021. Employing molecularly imprinted polymers in the development of electroanalytical methodologies for antibiotic determination. J. Mol. Recognit. 10991352. 34 (3). Available from: http://onlinelibrary.wiley.com/journal/10.1002/(ISSN)1099-1352. doi:10.1002/jmr.2878.

Bergua, J.F., Álvarez-Diduk, R., Idili, A., Parolo, C., Maymó, M., Hu, L., Merkoçi, A., 2022. Low-cost, user-friendly, all-integrated smartphone-based microplate reader for optical-based biological and chemical analyses. Anal. Chem. 15206882. 94 (2), 1271–1285. Available from: http://pubs.acs.org/journal/ancham. doi:10.1021/acs.analchem.1c04491.

Bersani, C., Ruggiero, C., Sacile, R., Soussi, A., Zero, E., 2022. Internet of things approaches for monitoring and control of smart greenhouses in industry 4.0. Energies. 1996-1073. 15 (10), 3834. doi:10.3390/en15103834.

Bhuiyan, N.H., Hong, J.H., Uddin, M.J., Shim, J.S., 2022. Artificial intelligence-controlled microfluidic device for fluid automation and bubble removal of immunoassay operated by a smartphone. Anal. Chem. 15206882. 94 (9), 3872–3880. Available from: http://pubs.acs.org/journal/ancham. doi:10.1021/acs.analchem.1c04827.

Bielecki, Z., Achtenberg, K., Kopytko, M., Mikołajczyk, J., Wojtas, J., Rogalski, A., 2022. Review of photodetectors characterization methods. Bull. Pol. Acad. Sci. Tech. Sci. 2300-1917. 70 (2), 140534. doi:10.24425/bpasts.2022.140534.

Butt, M.A., Mateos, X., Piramidowicz, R., 2024. Photonics sensors: a perspective on current advancements, emerging challenges, and potential solutions (Invited). Phys. Lett. A. 03759601. 516, 129633. doi:10.1016/j.physleta.2024.129633.

Cadeado, A.N.S., MacHado, C.C.S., Oliveira, G.C., Silva, D.A.S.E., Muñoz, R.A.A., Silv, S.G., 2022. Internet of things as a tool for sustainable analytical chemistry: a review. J. Braz. Chem. Soc. 16784790. 33 (7), 681–692. Available from: http://static.sites.sbq.org.br/jbcs.sbq.org.br/pdf/2021-0453RV_ChemSustainDev.pdf. doi:10.21577/0103-5053.20220048.

Chen, W., Yao, Y., Chen, T., Shen, W., Tang, S., Lee, H.K., 2021. Application of smartphone-based spectroscopy to biosample analysis: a review. Biosens. Bioelectron. 18734235. 172. Available from: www.elsevier.com/locate/bios. doi:10.1016/j.bios.2020.112788.

Corsato, P.C.R., de Lima, L.F., Paschoarelli, M.V., de Araujo, W.R., 2024. Electrochemical sensing at the fingertips: wearable glove-based sensors for detection of 4-nitrophenol, picric acid and diazepam. Chemosphere. 18791298. 363. Available from: https://www.sciencedirect.com/science/journal/00456535. doi:10.1016/j.chemosphere.2024.142771.

Cruz, A.F.D., Norena, N., Kaushik, A., Bhansali, S., 2014. A low-cost miniaturized potentiostat for point-of-care diagnosis. Biosens. Bioelectron. 18734235. 62, 249–254. Available from: www.elsevier.com/locate/bios. doi:10.1016/j.bios.2014.06.053.

Decarli, N.O., Zapp, E., de Souza, B.S., Santana, E.R., Winiarski, J.P., Vieira, I.C., 2022. Biosensor based on laccase-halloysite nanotube and imidazolium zwitterionic surfactant for dopamine determination. Biochem. Eng. J. 1873295X. 186. Available from: www.elsevier.com/locate/bej. doi:10.1016/j.bej.2022.108565.

Di Nonno, S., Ulber, R., 2022. Portuino—a novel portable low-cost arduino-based photo- and fluorimeter. Sensors. 22, 7916.

Dos Santos, A.M., Feitosa, M.H.A., Wong, A., Fatibello-Filho, O., Sotomayor, M.D.P.T., Moraes, F.C., 2024. Functionalized graphene, quantum dots, and PEDOT:PSS based screen-printed electrode for the endocrine disruptor bisphenol A determination. Sens. Actuators B Chem. 09254005. 399, 134745. Available from: https://www.journals.elsevier.com/sensors-and-actuators-b-chemical. doi:10.1016/j.snb.2023.134745.

Elias, J.S., 2024. An improved potentiostat/galvanostat for undergraduate-designed electrochemical laboratories. J. Chem. Educ. 19381328. 101 (4), 1703–1710. Available from: http://pubs.acs.org/loi/jceda8. doi:10.1021/acs.jchemed.3c01044.

Fan, Y., Li, J., Guo, Y., Xie, L., Zhang, G., 2021. Digital image colorimetry on smartphone for chemical analysis: a review. Measurement. 02632241. 171, 108829. doi:10.1016/j.measurement.2020.108829.

Geballa-Koukoula, A., Ross, G.M.S., Bosman, A.J., Zhao, Y., Zhou, H., Nielen, M.W.F., Rafferty, K., Elliott, C.T., Salentijn, G.I.J., 2023. Best practices and current implementation of emerging smartphone-based (bio)sensors—part 2: development, validation, and social impact. TrAC Trends Anal. Chem. 01659936. 161, 116986. doi:10.1016/j.trac.2023.116986.

Granica, M., Tymecki, Ł., 2019. Analytical aspects of smart (phone) fluorometric measurements. Talanta. 00399140. 197, 319–325. doi:10.1016/j.talanta.2019.01.032.

Gubsky, S., 2023. Development of low-cost arduino-based equipment for analytical and educational applications. Eng. Proc. 26734591. 48 (1). Available from: https://www.mdpi.com/2673-4591/48/1/8. doi:10.3390/CSAC2023-14893.

Hanrahan, G., Patil, D.G., Wang, J., 2004. Electrochemical sensors for environmental monitoring: design, development and applications. J. Environ. Monit. 14640325. 6 (8), 657–664. doi:10.1039/b403975k.

Hayden, D., Anacleto, S., Archonta, D.E., Khalil, N., Pennella, A., Qureshi, S., Séguin, A., Tabatabaei, N., 2022. Arduino-based sensing platform for rapid, low-cost, and high-sensitivity detection and quantification of analytes in fluidic samples. Eng. Proc. 26734591. 27 (1). Available from: www.mdpi.com/journal/engproc. doi:10.3390/ecsa-9-13277.

Hitchman, M.L., Berlouis, L.E.A., 1995. Electrochemical methods. In: Process Analytical Chemistry. Springer Netherlands, Dordrecht, pp. 217–258. doi:10.1007/978-94-011-0591-0_7.

Hooda, P., Sami, M.A., Hassan, U., 2021. Point-of-care 3-D printed spectrophotometer for therapeutic drug monitoring in tuberculosis patients. IEEE Sens. Lett. 24751472. 5 (9). Available from: http://ieeexplore.ieee.org/xpl/mostRecentIssue.jsp?punumber=7782634. doi:10.1109/LSENS.2021.3106828.

Jadhao, P.R., Ahmad, E., Pant, K.K., Nigam, K.D.P., 2022. Advancements in the field of electronic waste recycling: critical assessment of chemical route for generation of energy and valuable products coupled with metal recovery. Sep. Purif. Technol. 13835866. 289, 120773. doi:10.1016/j.seppur.2022.120773.

Janata, J., 2009. Principles of Chemical Sensors, second ed. Springer.

Javaid, M., Haleem, A., Singh, R.P., Suman, R., Gonzalez, E.S., 2022. Understanding the adoption of industry 4.0 technologies in improving environmental sustainability. Sustain. Oper. Comput. 26664127. 3, 203–217. Available from: https://www.sciencedirect.com/science/journal/26664127. doi:10.1016/j.susoc.2022.01.008.

Jiang, R., Lin, D., Zhang, Q., Li, L., Yang, L., 2022. Multiplex chroma-response based fluorescent smartphone sensing platform for rapid and visual quantitative determination of antibiotic residues. Sens. Actuators B Chem. 09254005. 350, 130902. doi:10.1016/j.snb.2021.130902.

Karthik, V., Selvakumar, P., Kumar, P.S., Satheeskumar, V., Vijaysunder, M.G., Hariharan, S., Antony, K., 2022. Recent advances in electrochemical sensor developments for detecting emerging pollutant in water environment. Chemosphere. 00456535. 304, 135331. doi:10.1016/j.chemosphere.2022.135331.

Khudaverdyan, S., Vaseashta, A., Ayvazyan, G., Khachatryan, M., Atvars, A., Lapkis, M., Rudenko, S., 2021. On the semiconductor spectroscopy for identification of emergent contaminants in transparent mediums. In: Water Safety, Security and Sustainability. Advanced Sciences and Technologies for Security Applications. Springer, Cham, pp. 663–689. Available from: https://link.springer.com/bookseries/5540. doi:10.1007/978-3-030-76008-3_29.

Kim, S.D., Koo, Y., Yun, Y., 2017. A smartphone-based automatic measurement method for colorimetric pH detection using a color adaptation algorithm. Sensors. 1424-8220. 17 (7), 1604. doi:10.3390/s17071604.

Kondaveeti, H.K., Kumaravelu, N.K., Vanambathina, S.D., Mathe, S.E., Vappangi, S., 2021. A systematic literature review on prototyping with arduino: applications, challenges, advantages, and limitations. Comput. Sci. Rev. 15740137. 40. Available from: http://www.journals.elsevier.com/computer-science-review. doi:10.1016/j.cosrev.2021.100364.

Lee, V.B.C., Mohd-Naim, N.F., Tamiya, E., Ahmed, M.U., 2018. Trends in paper-based electrochemical biosensors: from design to application. Anal. Sci. 13482246. 34 (1), 7–18. Available from: https://www.jstage.jst.go.jp/article/analsci/34/1/34_7/_pdf/-char/en. doi:10.2116/analsci.34.7.

Lima, M.J.A., Nascimento, C.F., Rocha, F.R.P., 2017. Feasible photometric measurements in liquid-liquid extraction by exploiting smartphone-based digital images. Anal. Methods. 17599679. 9 (14), 2220–2225. Available from: http://www.rsc.org/Publishing/Journals/AY/About.asp. doi:10.1039/c7ay00388a.

Mehta, H., Patel, P., Mukherjee, A., Munshi, N.S., 2023. Biotechnological advances in detection of contaminants from wastewater. Clean Soil Air Water. 18630669. 51 (3). Available from: http://onlinelibrary.wiley.com/journal/10.1002/(ISSN)1863-0669. doi:10.1002/clen.202100439.

Mello, L.D., Kubota, L.T., 2002. Review of the use of biosensors as analytical tools in the food and drink industries. Food Chem. 03088146. 77 (2), 237–256. doi:10.1016/S0308-8146(02)00104-8.

Miyake, M., 2003. Electrochemical functions. In: Carbon Alloys: Novel Concepts to Develop Carbon Science and Technology. Elsevier Science, Oxford, pp. 435–445. Available from: http://www.sciencedirect.com/science/book/9780080441634. doi:10.1016/B978-008044163-4/50026-7.

Mohammed, M.I., Haswell, S., Gibson, I., 2015. Lab-on-a-chip or chip-in-a-lab: challenges of commercialization lost in translation. Procedia Technol. 22120173. 20, 54–59. doi:10.1016/j.protcy.2015.07.010.

Nery, E.W., Kubota, L.T., 2013. Sensing approaches on paper-based devices: a review. Anal. Bioanal. Chem. 16182650. 405 (24), 7573–7595. doi:10.1007/s00216-013-6911-4.

Okhonin, S., Gureev, M., Sallin, D., Appel, J., Koukab, A., Kvasov, A., Pastre, M., Polzik, E.S., Tagantsev, A.K., Uddegard, F., Kayal, M., 2015. A dynamic operation of a PIN photodiode. Appl. Phys. Lett. 0003-6951. 106 (3). doi:10.1063/1.4906488.

Oliveira, Franco, M., Suarez, W.T., Pereira, G.R.P., Vilanculo, C.B., Vieira, M.C.R., Santos, V.B., Almeida, J.P., 2023. Using colorimetric spot test and digital imaging-based technique for volatile acidity determination in cachaça with the aid of a smartphone. Microchem. J. 187, 108416.

Omar, R., Yuan, M., Wang, J., Sublaban, M., Saliba, W., Zheng, Y., Haick, H., 2024. Self-powered freestanding multifunctional microneedle-based extended gate device for personalized health monitoring. Sens. Actuators B Chem. 09254005. 398, 134788. doi:10.1016/j.snb.2023.134788.

Pauliukaite, R., Ghica, M.E., Fatibello-Filho, O., Brett, C.M.A., 2010. Electrochemical impedance studies of chitosan-modified electrodes for application in electrochemical sensors and biosensors. Electrochim. Acta. 00134686. 55 (21), 6239–6247. doi:10.1016/j.electacta.2009.09.055.

Rosa, T., Siqueira, B., Costa, R., Reis, J., Pacheco, W., Moura, P., Romão, W., Filgueiras, P., 2022. Redgim como aplicativo de smartphone para aplicações quimiométricas por meio de análise de imagens: um uso em pls. Quím. Nova. 16787064. 45 (5). doi:10.21577/0100-4042.20170861.

Ross, G.M.S., Zhao, Y., Bosman, A.J., Geballa-Koukoula, A., Zhou, H., Elliott, C.T., Nielen, M.W.F., Rafferty, K., Salentijn, G.I.J., 2023. Best practices and current implementation of emerging smartphone-based (bio)sensors—part 1: data handling and ethics. TrAC Trends Anal. Chem. 01659936. 158, 116863. doi:10.1016/j.trac.2022.116863.

Seetasang, S., Kaneta, T., 2021. Portable two-color photometer based on paired light emitter detector diodes and its application to the determination of paraquat and diquat. Microchem. J. 0026265X. 171, 106777. doi:10.1016/j.microc.2021.106777.

Sethu, N., Krishnakumar, S., Mitra, V., Tagad, C., Vyas, R., 2023. Design and development of a novel colorimetric assay and a portable optical system for the detection of aminoglycoside antibiotics. Sens. Actuators Rep. 26660539. 5, 100151. doi:10.1016/j.snr.2023.100151.

Sheta, S.M., Abdelelmoaty, A.S., Abu Hashish, H.M., kamel, A.M., Abd-Elzaher, M.M., El-Sheikh, S.M., 2022. Smart prototype for an electronic color sensor device for visual simultaneous detection of macrofuran based on a coated paper strip. Anal. Bioanal. Chem. 16182650. 414 (29-30), 8379–8388. Available from: https://www.springer.com/journal/216. doi:10.1007/s00216-022-04374-z.

Shi, Q., Teng, Y., Zhang, Y., Liu, W., 2018. Rapid detection of organophosphorus pesticide residue on Prussian blue modified dual-channel screen-printed electrodes combing with portable potentiostat. Chin. Chem. Lett. 10018417. 29 (9), 1379–1382. doi:10.1016/j.cclet.2017.11.023.

Soares, S., Fernandes, G.M., Rocha, F.R.P., 2023. Smartphone-based digital images in analytical chemistry: why, when, and how to use. TrAC Trends Anal. Chem. 18793142. 168. Available from: www.elsevier.com/locate/trac. doi:10.1016/j.trac.2023.117284.

Sohrabi, H., Hemmati, A., Majidi, M.R., Eyvazi, S., Jahanban-Esfahlan, A., Baradaran, B., Adlpour-Azar, R., Mokhtarzadeh, A., de la Guardia, M., 2021. Recent advances on portable sensing and biosensing assays applied for detection of main chemical and biological pollutant agents in water samples: a critical review. TrAC Trends Anal. Chem. 01659936. 143, 116344. doi:10.1016/j.trac.2021.116344.

Song, E.Y., Fitzpatrick, G.J., Lee, K.B., 2017. Smart sensors and standard-based interoperability in smart grids. IEEE Sens. J. 1530437X. 17 (23), 7723–7730. Available from: http://ieeexplore.ieee.org/xpl/RecentIssue.jsp?punumber=7361. doi:10.1109/JSEN.2017.2729893.

Stradiotto, N.R., Yamanaka, H., Zanoni, M.V.B., 2003. Electrochemical sensors: a powerful tool in analytical chemistry. J. Braz. Chem. Soc. 01035053. 14 (2), 159–173. Available from: http://jbcs.sbq.org.br/. doi:10.1590/S0103-50532003000200003.

Sun, A.C., Hall, D.A., 2019. Point-of-care smartphone-based electrochemical biosensing. Electroanalysis. 15214109. 31 (1), 2–16. Available from: http://onlinelibrary.wiley.com/journal/10.1002/(ISSN)1521-4109. doi:10.1002/elan.201800474.

Tavernier, F., Steyaert, M., 2011. High-speed optical receivers with integrated photodiode in nanoscale CMOS. Analog Circuits Signal Process. 5. 10.1007/978-1-4419-9925-2

Thakur, A., Malik, V., Gupta, 2024. Paper-based biosensors—from fabrication to applications: a review. Chem. Pap. 79 (1), 43–73. doi:10.1007/S11696-024-03804-9.

Tymecki, Ł., Pokrzywnicka, M., Koncki, R., 2008. Paired emitter detector diode (PEDD)-based photometry—an alternative approach. Analyst. 00032654. 133 (11), 1501–1504. Available from: http://pubs.rsc.org/en/journals/journal/an. doi:10.1039/b807127f.

Upadhyay, S., Kumar, A., Srivastava, M., Srivastava, A., Dwivedi, A., Singh, R.K., Srivastava, S.K., 2024. Recent advancements of smartphone-based sensing technology for diagnosis, food safety analysis, and environmental monitoring. Talanta. 00399140. 275, 126080. doi:10.1016/j.talanta.2024.126080.

Vidotto, L.H.B., Wachholz Junior, D., Kubota, L.T., 2023. A simple and low-cost portable potentiostat with real-time data sharing for wireless electrochemical analyses. J. Electroanal. Chem. 15726657. 937. Available from: https://www.journals.elsevier.com/journal-of-electroanalytical-chemistry. doi:10.1016/j.jelechem.2023.117414.

Wachholz Junior, D., Deroco, P.B., Hryniewicz, B.M., Kubota, L.T., 2025. Strategies for electrochemical point-of-care biosensors. Annu. Rev. Anal. Chem. 1936-1327. doi:10.1146/annurev-anchem-071124-103739.

Wang, M., Mohanty, S.K., Mahendra, S., 2019. Nanomaterial-supported enzymes for water purification and monitoring in point-of-use water supply systems. Acc. Chem. Res. 15204898.

52 (4), 876–885. Available from: http://pubs.acs.org/journal/achre4. doi:10.1021/acs.accounts.8b00613.

Wong, A., Santos, A.M., Fatibello-Filho, O., 2018. Simultaneous determination of paracetamol and levofloxacin using a glassy carbon electrode modified with carbon black, silver nanoparticles and PEDOT:PSS film. Sens. Actuators, B: Chem. 09254005. 255, 2264–2273. doi:10.1016/j.snb.2017.09.020.

Zahra, Q.U.A., Mohsan, S.A.H., Shahzad, F., Qamar, M., Qiu, B., Luo, Z., Zaidi, S.A., 2022. Progress in smartphone-enabled aptasensors. Biosens. Bioelectron. 18734235. 215. Available from: www.elsevier.com/locate/bios. doi:10.1016/j.bios.2022.114509.

Zappi, D., Coronado, E., Soljan, V., Basile, G., Varani, G., Turemis, M., Giardi, M.T., 2021. A microbial sensor platform based on bacterial bioluminescence (luxAB) and green fluorescent protein (gfp) reporters for in situ monitoring of toxicity of wastewater nitrification process dynamics. Talanta. 00399140. 221, 121438. doi:10.1016/j.talanta.2020.121438.

Part 4

Quality assurance, quality control, and validation in environmental analysis

Chapter 11

Quality assurance, quality control and validation in environmental analysis

Tim Boogaerts[a,b,*], Fatima den Ouden[a,b,*], Lidia Belova[a], Maarten Roggeman[a,b], Celine Gys[a,b], Adrian Covaci[a,b], and Alexander van Nuijs[a,b]

[a]*Toxicological Center, University of Antwerp, Wilrijk, Belgium*

[b]*Exposome Center of Excellence, University of Antwerp, Wilrijk, Belgium*

1 Introduction

Since 1950, >140,000 new chemicals have been synthesized, many of which have become widely dispersed in the built and natural environment and are responsible for nearly universal exposure (Landrigan et al., 2018). To understand and monitor the chemical pollution space and drive changes in pollution policy, comprehensive information on contaminants affecting human populations is highly required. This is hindered by the limited availability of monitoring data, making it challenging for policymakers to assess the full extent of chemical pollution, the impact on public health and safety, and the effectiveness of governmental mandates (Sauvé and Desrosiers, 2014). Current monitoring strategies are often limited to a narrow range of chemicals for which surveillance is mandatory (Bopp et al., 2018; Comero et al., 2020). This is in contrast with the major knowledge gaps remaining for the identification and quantification of the broad spectrum of chemicals causing human exposure and impact on the environment (Comero et al., 2020). Therefore chemicals flagged due to their potential ecological or (eco) toxicological risks may need to be measured to create comprehensive information that can be considered for future regulations to support effective environmental risk assessment and optimal regulatory decision-making.

Human external exposures (i.e., the dose of a chemical to which humans are exposed through a specific pathway) and internal exposure (i.e., the fraction of that chemical dose that enters the human body) can be assessed with different methodologies, including analysis of environmental samples (e.g.,

---

*Joint first author

Contaminants of Emerging Concern in the Environment. https://doi.org/10.1016/B978-0-443-33821-2.00005-0

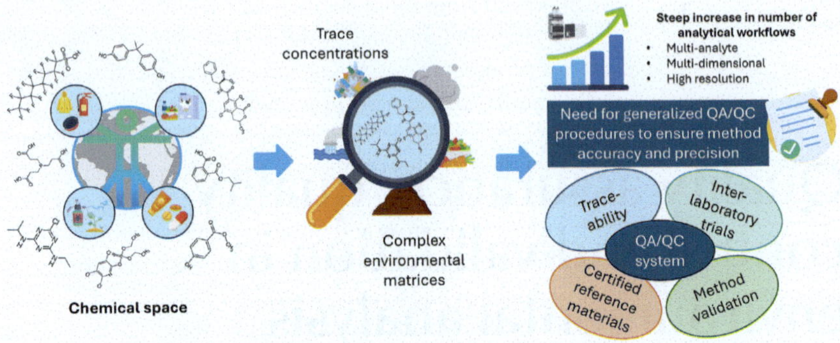

FIG. 11.1 The increasing chemical space and measurement methods warrants the need for appropriate quality control and assurance systems.

dust, food, air) and human biomonitoring (HBM) (e.g., serum, tissues, urine), respectively (Sabbioni et al., 2022; Vorkamp et al., 2023). These approaches heavily rely on the availability of sensitive and accurate analytical workflows for the measurement of trace concentrations of these chemicals and their metabolites in different environmental and biological matrices in the presence of a wide range of interferences (Sabbioni et al., 2022). Due to these analytical challenges, numerous analytical sample preparation, detection and quantification methods have been employed for the measurement of a broad variability of analytes in complex environmental samples (Fig. 11.1). Prior to sample analysis, multiple sampling steps (e.g., collection, field treatment, storage) are involved in environmental monitoring, which could amplify the risk for analytical errors. This warrants the need for generic quality assurance/quality control (QA/QC) procedures applicable for multiple purposes since the limited implementation of these measures results in lower quality and reliability of environmental data and, in turn, less effective risk-assessment and decision-making (Konieczka, 2018; Quevauviller and Maier, 2007). QA involves all systematic and planned initiatives (i.e., *before* the analysis) carried out to test the adequate performance of the analytical workflow and guarantee the reliability of the analysis and its ability to meet quality requirements (Quevauviller and Maier, 2007). To ensure these high quality standards, QC steps are implemented, including the use of standard operating procedures, whole process blanks, calibrations, etc. (i.e., *during and after* the analysis) (Broadhurst et al., 2018; Barwick, 2018).

Even though attention to QA/QC has considerably grown in the last decade (as evidenced by the emergence of extensive guidelines, accreditation systems, and interlaboratory testing schemes) (van Nuijs et al., 2018; Nübler et al., 2022), it has sometimes been overlooked due to the increasing workload that analytical laboratories face (Lescord and Lepage, 2025). This highlights the need for generic routine control programs applicable to different analytical tests. These

QA/QC systems should be capable of tracing and estimating the accuracy and precision of the analytical measurements and involve different key elements such as the use of certified reference materials (CRMs), analytical method validation, traceability of the measurements, participation in interlaboratory trials, and assessment of the overall uncertainty (Konieczka, 2018). The need for QA/QC systems is further amplified by globalization and the requirement to compare results in different international laboratories to reduce discrepancies between different analytical procedures (Konieczka, 2018). This also means that a holistic approach to quality management is necessary to ensure traceability and comparability on an international scale, which requires finetuning of the program designs. At this stage, there are no harmonized QA/QC guidelines for environmental monitoring, and there is no consensus in the system suitability and QC approaches for evaluating system performance and data reporting quality (Esteban López et al., 2021). Furthermore, the vast majority of data from chemical monitoring activities is not systematically collected and coherently managed (Comero et al., 2020). This is amplified by the wide range of analytical procedures, matrices and the absence of "tailor-made" systems to adapt with the wide range of environmental monitoring parameters (Comero et al., 2020).

This chapter aims to evaluate the different key aspects of QA/QC systems for analytical workflows for different environmental matrices. It will not provide an in-depth description of all available international standards for QA/QC (e.g., IUPAC, ISO, EURACHEM) since this information is explained in depth in the documentation of the corresponding organizations. Instead, we will highlight some main aspects and difficulties for analytical workflows for environmental monitoring and list the most relevant guidelines for adopting appropriate QA/QC in environmental chemistry laboratories. In this sense, we will formulate general recommendations for QA/QC in environmental monitoring (Section 2). Since chromatography and mass spectrometry have been widely adopted as detection and quantification methods in environmental chemistry, we will provide an in-depth focus on targeted and nontargeted approaches using these methodologies (Section 3). Finally, we will compile evidence and insights from different case studies (e.g., interlaboratory trails) and identify key priorities for the adaptation of more appropriate QA/QC measures in future research (Section 4). This chapter does not consist of a systematic literature review of all QA/QC measures, but it focuses on relevant real-life examples.

2 Fundamentals, procedures, and instrumentation

2.1 General concepts and considerations for QA/QC systems

QA/QC systems are pivotal for the success of any analytical investigation and critical for acquiring high-quality analytical data for environmental monitoring purposes (Broadhurst et al., 2018). When QA/QC measures are not adequately reported, environmental data will be less useful for other analytical laboratories,

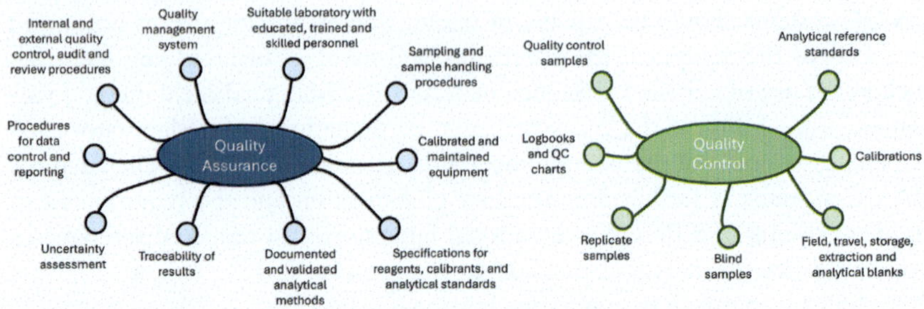

FIG. 11.2 Key elements of quality assurance and quality control systems.

conflicting with one of the main purposes of publishing research findings (Lescord and Lepage, 2025). Fig. 11.2 gives a general overview of all activities that are an integral part of QA/QC systems. Comprehensive QA schemes should include the validation of analytical methods, internal and external quality control measures, standard operating procedures for sample collection, handling and analysis, training of laboratory personnel, traceability of analytical results and uncertainty assessment (Konieczka, 2018; Quevauviller and Maier, 2007; Broadhurst et al., 2018). To achieve this, a wide range of international quality standards (i.e., GLP, IUPAC, ISO) are available for analytical laboratories. Table 11.1 gives a short summary of the written standards that are highly relevant in the field of environmental and analytical chemistry (Valcárcel and Lucena, 2013). A detailed description of these guidelines can be found elsewhere (Barwick, 2018; International Organization for Standardization, 2015, 2016, 2017; Organisation for Economic Co-operation and Development, OECD, 1998; Thompson and Wood, 1995).

It should be noted that not all QA/QC measures are necessary for every type of analysis. For example, the inclusion of blanks at different stages of the analytical chain (e.g., field, travel, reagent) becomes very important for environmental chemicals that are widely used in everyday products and processes (e.g., PFAS, PFRs, phthalates) to assess background contamination (Lescord and Lepage, 2025). However, this issue becomes less problematic for the measurements of other chemical classes (e.g., pharmaceuticals and personal care products) in specific environmental matrices. In addition, the measurement of specific human metabolic excretion products is less likely to be affected by contamination compared to the parent compounds. Therefore, QA/QC measures should be finetuned based on the method limitations and end points of the analysis. Once an analytical method is validated according to suitable guidelines (e.g., ICH, EPA, FDA), appropriate QA/QC measurements should be performed as part of ongoing QC efforts to assess the performance criteria (e.g., calibration, accuracy, precision, contamination) (Lescord and Lepage, 2025; European Medicines Agency, 2022; Environmental Protection Agency, 2025; US Food and Drug Administration, 2021).

TABLE 11.1 Key examples of relevant international quality standards.

Guide	Description	Key aspects	Citation
ISO 9001:2015	Quality management systems	Specifies the criteria for a quality management system for organizations that need to prove their ability to consistently provide products and services that meet customer and applicable statutory and regulatory needs	(International Organization for Standardization, 2015)
ISO/IEC 17025:2017	General requirements for the competence of testing and calibration laboratories	Outlines the basic criteria for laboratories to ensure competent, impartial, and consistent operations	(International Organization for Standardization, 2017)
ISO 17034:2016	General requirements for the competence of reference material producers	Defines the general needs for the competence and consistent operations of reference materials producers	(International Organization for Standardization, 2016)
ISO/IEC Guide 99:2007	International vocabulary of metrology—Basic and general concepts and associated terms (VIM)	Terminology dictionary containing designations and definitions relevant for the field of metrology	(International Organization for Standardization, 2007)
ISO/IEC 17043:2023	Conformity assessment—General requirements for the competence of proficiency testing providers	This guidance document outlines the basic criteria for the competence and impartiality of proficiency testing providers and consistent operations for proficiency testing schemes	(International Organization for Standardization, 2023)
ISO 13528:2022	Statistical methods for use in proficiency testing by interlaboratory comparison	This report gives a detailed description of appropriate statistical methods for proficiency testing providers to design proficiency testing strategies and analyze the corresponding data from participating laboratories.	(International Organization for Standardization, 2022)

Continued

TABLE 11.1 Key examples of relevant international quality standards–cont'd

Guide	Description	Key aspects	Citation
OECD ENV/MC/CHEM(98)17	Principles of Good Laboratory Practice and compliance monitoring	Promote the quality and validity of test data used in assessing the safety of chemicals and chemical products	(Organisation for Economic Co-operation and Development, OECD, 1998)
IUPAC	Harmonized Guidelines on internal quality control in analytical chemistry laboratories	Provides guidance on the implementation of internal quality control in analytical laboratories and harmonizes QC procedures for various fields (e.g., clinical biochemistry, environmental studies etc.)	(Thompson and Wood, 1995)
IUPAC Technical report 2006	The International Harmonized protocol for the proficiency testing of analytical chemistry laboratories	Provides an agreed protocol on the design, execution, and interpretation of collaborative studies by the International Standardizing Organization (ISO) and IUPAC	(Thompson et al., 2006)
EURACHEM/CITAC	Guide to quality in analytical chemistry	Provides analytical laboratories with guidance on the best practice for analytical operations and covers both qualitative and quantitative analysis on a nonroutine and routine basis	(Barwick, 2018)

Appropriate acceptance criteria should be defined prior to the sample analysis. These acceptance criteria should be met to assure adequate accuracy and precision of the analytical methods (e.g., a ±20% relative standard deviation (RSD) threshold between replicates) (Lescord and Lepage, 2025). In the analytical protocol, the upper and lower method detection and quantification limits should be reported, as well as how to handle measurements that fall outside of the calibration range. QA/QC results should be transparently discussed with special attention to how poor QA/QC results were handled. Potential sources of uncertainty (e.g., systematic and random errors) at different stages during the analytical chain should be properly identified and, where possible, minimized (Quevauviller and Maier, 2007). Therefore QA/QC processes should encompass all different analytical steps, including chemical monitoring planning, sampling, sample transportation and storage, sample extraction and analysis, and data analysis and reporting (Comero et al., 2020).

In the following subsections, we will discuss the key components of QA/QC systems, including traceability of results, use of CRMs, method validation and inter-laboratory studies. Special attention is given to the implementation of these elements within the context of environmental chemistry, where unique challenges and limitations (e.g., absence of matrix blanks, unavailability of matrix-specific CRMs) can limit the applicability of traditional QA/QC procedures. We are aware that some sections only provide a focused summary of specific QA/QC features and that there are a multitude of detailed investigations available that take a deeper dive into these aspects. However, the main aim of this chapter is to highlight some key aspects and drivers for the implementation of QA/QC measures in analytical workflows applied for environmental monitoring studies.

2.2 Traceability

According to the International Vocabulary of Metrology (VIM), metrological traceability can be defined as "the property of a measurement result whereby the result can be related to a reference through a documented unbroken chain of calibrations, each contributing to the measurement uncertainty" (International Organization for Standardization, 2007). In essence, it is a continuous process that ensures no steps are omitted in an analytical workflow, which could compromise the effectiveness of the whole procedure (Konieczka, 2018). The intra- and inter-lab comparability of analytical measurements depends on, among other factors, traceability to a consistent and universally agreed set of measurement units and scales (Ellison and Williams, 2019). In QA/QC systems, traceability should be known for each step in the whole analytical workflow including sample extraction steps and it should be guaranteed through calibration (both instrumental and analytical) with appropriate reference standards. In this sense, analytical method validation and uncertainty analysis are critical steps in assuring the overall quality of the chemical measurements. Ensuring that the whole procedure is traceable is a major challenge for analytical methods for environmental monitoring (Konieczka, 2018). For the analysis of trace concentrations of chemicals in environmental matrices, the chain of connection with standards is broken by the introduction of sample processing steps that chemically and physically modify the sample properties. Traceability is further complicated by the occurrence of cointerferences and the difficulty to obtain homogeneous samples because of variations in the matrix composition across different sampling locations and periods. Another key feature influencing the traceability of chemical measurements is the stability of the analytes of interest in the respective environmental matrices under different (storage) conditions. Traceability in environmental analysis can be achieved by comparing the analytical result with reference measurements and by comparing with (certified) reference materials (CRMs). Reference materials (RMs) are essential for traceability assurance, enabling the estimation of the accuracy of new analytical platforms, comparison of different analytical methods and interlaboratory testing. In the following sections, we will discuss different aspects of CRMs, analytical method validation and interlaboratory testing in relationship to metrological traceability.

2.3 Sampling, sample storage, and preparation

Since **Part 2 and 3** delve into greater detail in sampling strategies and sample extraction for environmental monitoring, this section will only briefly focus on QA/QC aspects related to sample collection, handling, and analysis. Fig. 11.3 gives an overview of the different QC measures throughout the different stages of the preanalytical and analytical chain. For sampling, reproducible strategies

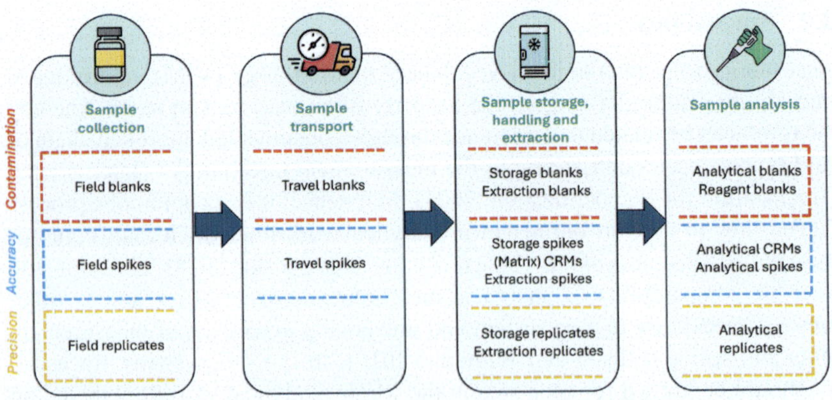

FIG. 11.3 Overview of the QA/QC measures throughout the preanalytical and analytical chain. *(Adapted from Lescord, G.L., Lepage, A.T., 2025. Making waves: report your quality assurance and control (QA/QC) information when publishing analytical data. Water Res. 1879-2448. 278. Available from: https://www.sciencedirect.com/science/journal/00431354. doi:10.1016/j.watres.2025.123167, with permission.)*

should be used to minimize sample contamination and loss of analytes resulting from the sampling method (Schulze et al., 2020). An important consideration in sampling is to choose a sample with a composition that represents the characteristics of the area that they have been taken from. Appropriate sampling containers should be used, which depend on the physicochemical properties of both the matrix and the target compounds. This includes the material of the container (e.g., transparency and chemical composition), the cleaning procedure, addition of specific preservatives and the type of sealing needed (e.g., no air above the sample, gas-tight) (NORMAN Network, 2009). The containers should also be labeled clearly and unambiguously to safeguard proper traceability.

In addition, documentation should include an extensive list of all planned field and laboratory QC samples for each analytical measurement and for each sample medium/matrix. Blanks and spiked matrix can be included as QC measures during sample collection and sample transport. Field blanks usually represent analyte-free matrix which is taken to the sample site and put in the container in the field after which it is preserved and transported in the same way as the samples (Environmental Protection Agency, 2017). Field spikes consist of matrix spiked with a known concentration, which are taken to the sampling site, transferred to the sampling container and preserved and transported as the samples to assess analytical loss during sample collection and transportation (Sandstrom and Lewis, 2009). A travel blank consists of a blank matrix sample that is taken to the sampling site and then transported back without undergoing any sampling procedures. This can be used to assess contamination during transportation (Environmental Protection Agency, 2017; Queensland Government, 2009). A travel spike can be used to determine analyte loss during transportation. Blank matrix is spiked before sampling and taken to the sampling site and back, unopened. Traveling blanks and spikes are mostly implemented for volatile compounds (Queensland Government, 2009).

In addition, appropriate sample storage and handling procedures should be implemented to avoid compound transformation and loss due to adsorption to the container materials. To ensure stability and integrity, the appropriate selection and pretreatment of sampling containers is crucial. In addition, it is preferred to have a short time interval between sample collection and analysis. Preservation methods such as pH-control, freezing, chemical addition, and filtration can be applied but can have limited effects for some target chemicals (NORMAN Network, 2009). For a wide range of target chemicals, in-sample stability data is lacking for specific matrix types. Especially for retrospective analysis of environmental matrices, the long-term stability of the analytes should be evaluated under specific sample preservation conditions.

Depending on the composition of the matrix and the target compounds, sample pretreatment might be needed in the form of filtration, homogenization, or preconcentration (NORMAN Network, 2009). Before sampling, step-by-step instructions should be available for the homogenization, filtration, preservation and/or composition of field samples. QA/QC measures during sample extraction are important to check if all target analytes were successfully extracted from the sample matrix, to evaluate cross-contamination during extractions and to assess if the target chemicals were fully converted into detectable forms (e.g., during sample derivatization) (Lescord and Lepage, 2025). For example, spiking blank matrix with known concentrations of target analytes pre- and postextraction allows evaluation of extraction efficiency. A method blank should be included to assess the contamination during sample extraction. This blank ideally consists of analyte-free matrix which is extracted in parallel with the field samples (Environmental Protection Agency, 2017). Each sample extraction method is associated with extraction bias originating from its selectivity (Schulze et al., 2020).

2.4 Certified reference materials

According to Lippa et al. (2022), RMs are artifact-based measurement standards characterized by known compositions of specific physical, chemical, or biological properties. These standards are typically classified by their intended use (e.g., calibration, quality control, method validation) and cover a range of designs, from matrix-based materials derived from natural sources to matrix-free standards such as pure substances, standard solutions, or mixtures. The key differences between CRMs and uncertified RMs lie in the level of characterization of the chemical properties, content, and associated supporting records (Konieczka, 2018). CRMs are accompanied by official documentation from an authoritative body (e.g., National Institute of Standards and Technology), providing one or more assigned reference values with associated uncertainties and traceability, whereas RMs may lack such certification and metrological details. CRMs and RMs play a pivotal role in different aspects of QA/QC systems to evaluate the reliability and reproducibility of measurements, such as (i) the validation of intra- and interday accuracy and precision in analytical procedures, (ii) interlaboratory comparisons, (iii) estimation of the uncertainty associated with measurement values, and (iv) documenting the traceability of measurement results.

QA/QC systems for environmental analysis should leverage these RMs to assess and reduce variability in analytical measurements associated with sample extraction, instrumental analysis, system suitability, compound identification, and data processing (Lippa et al., 2022). For this reason, their widescale and routine application could aid in a better understanding of sources of uncertainty and the control of metrological bias between published results. In this sense, the use of RMs enhances the comparability of results and conclusions between laboratories and individual studies, even though a certain degree of variability is unavoidable as a result of interlaboratory differences in procedures and instrumentation.

A key consideration for CRMs and uncertified RMs is that they should be representative for their intended function, meaning that they should be similar to unknown/study samples with regard to matrix composition, analyte concentration, type and concentration of matrix interferences, and physical state of the material (Konieczka, 2018; Lippa et al., 2022). In addition, these standard materials should be homogenous and possess long-term stability under defined storage conditions. Their characterization should encompass the chemical identity, mass concentration, and metrological traceability of specific components, including information on relevant impurities, physicochemical properties, and safety regulations (Lippa et al., 2022).

Depending on the analytical application, different selection criteria for CRMs have to be considered including their availability, uncertainty of the assigned reference value, the concentration range, the traceability of the assigned reference value, the cost and the matrix composition. CRMs provide a well-defined target value assigned by the producer to benchmark against to evaluate analytical method performance. The homogeneity and stability of CRMs enables an interlaboratory comparison of results to enhance the reproducibility within the scientific community and externally validate published results. Matrix CRMs are unique batches of specifically characterized material that may be unavailable from other sources and could play a crucial role in the evaluation and validation of new sample extraction methods (Wise and Phillips, 2019).

As indicated by Göen et al. (2012), the use of CRMs to assess the reliability of quantitative measurement is more restricted in environmental monitoring and human biomonitoring because of their limited availability. A few suppliers of proficiency tests for environmental biomarkers are capable of offering relevant biomarker/matrix combinations with representative environmental exposure levels. Therefore the participation in interlaboratory trials for environmental monitoring is imperative to obtain representative information on the comparability and accuracy of study results (Wise and Phillips, 2019; Göen et al., 2012). For many compound classes in specific matrices, there are no CRMs available (Tadić et al., 2025; Olivares et al., 2018). However, RMs without certification might be available. In these cases, it might be necessary to use RMs for validation even though this will decrease the metrological traceability. For matrix CRMs, characterization is more difficult because of the inherent complexity of the mixtures and broad range of matrix constituents. In addition, even when specific matrix CRM exist for the determination of trace organic chemicals in environmental samples, numerous studies do not employ them for method validation or when reporting quantitative data in environmental samples. In this sense, expansion of the number of available matrix CRMs is crucial to assess the reliability and comparability of analytical data and for management of the environment, including decision-making for remediation policies. To address this community need, the development of these RMs is being orchestrated by commercial companies and government agencies to facilitate a wider distribution among international research communities to enable coordinated and quality-controlled data acquisition (Lippa et al., 2022).

2.5 Analytical method validation

The main purpose of analytical method validation is to establish its traceability to a recognized and quality-controlled reference (pure standard or matrix CRM) (Quevauviller and Maier, 2007). An analytical method validation strategy evaluates different aspects of the procedure's performance to assure the quality of the measured results, ensuring that the analytical method is fit for its intended purpose. In other words, a method validation study is crucial to obtain sufficient evidence that the analytical method fulfills a described set of performance criteria, including sensitivity, specificity, accuracy, precision, matrix effects, calibration range, stability, carry-over, and dilution integrity (Quevauviller and Maier, 2007). QA/QC systems support the assessment of the accuracy of the calibration, the frequency needed for calibration and the level of contamination (Lescord and Lepage, 2025).

For Pharmaceutical Quality Systems, the US Food and Drug Administration (2021) and European Medicines Agency (2022) established guidance documents based on the International Council for Harmonization (ICH) recommendations. However, dedicated method validation guidelines for specific environmental matrices are often lacking. As a result, new (bio)analytical methods for these compounds and matrices are typically validated using more general recommendations and validation criteria which might differ between laboratories. Even though the existing ICH guidelines serve as a reliable starting point, there is a clear need for tailored and standardized validation strategies to address the challenges of specific environmental matrices (e.g., contamination of control blanks, matrix matching for calibration, absence of blank matrix). For this reason, the NORMAN association developed specific method validation schemes for environmental monitoring (see Section 4.2).

Traditionally, for the preparation of calibration curves and QC samples, blank biological matrix is spiked with reference standards containing the analytes of interest (European Medicines Agency, 2022). The use of matrix-matched calibration curves and internal standard (IS) calibration is commonly employed to effectively overcome matrix effects (Cortese et al., 2020). For bioanalytical method validation using chromatography coupled to mass spectrometry, Matuszewski et al. (2003) provide a framework to assess the degree of matrix suppression or enhancement based on the prerequisite that blank matrix is available through a postextraction standard addition. However, for environmental analysis, blank matrix might not always be available because of the widespread occurrence of organic pollutants in environmental samples. In addition, environmental samples commonly harbor a wide range of matrix interferences that could potentially be coextracted with the analytes of interest, potentially affecting the sensitivity, reproducibility and accuracy of analytical methods. Environmental samples could also show high variability in the presence of matrix interferences depending on the extraction method, contaminant type and sample origin (Thakare et al., 2016). For these reasons, alternative validation strategies should be employed to robustly compensate for matrix effects (Cortese et al., 2020; Thakare et al., 2016). These approaches have also been employed for the analysis of endogenous compounds in biological samples (Thakare et al., 2016):

- Standard addition

The analyte of interest is spiked in extracts from each sample at different concentrations to construct a separate calibration curve for each individual sample. Each studied sample is divided into identical aliquots which are spiked with increasing known amount of the target analyte. One nonspiked sample is included which can be regarded as the blank sample. By applying standard addition, the background presence or variance in matrix effect between different samples is accounted for. This method of compensating for matrix effects is very effective since the specific matrix for each sample is used rather than a pooled matrix. Therefore this method allows quantification of target compounds in a sample without applying background subtraction. However, it also requires large sample volumes and since each sample will have its own calibration curve, depending on the amount of samples and the analytical run time this method might be extremely time- and labor-intensive. Therefore this approach is not considered suitable for routine analysis (Cortese et al., 2020; Thakare et al., 2016; Visconti et al., 2023).

- Background subtraction

In background subtraction, the concentrations of a target analyte in a representative matrix or pooled matrix are subtracted from the concentrations of samples spiked with known standard concentrations. Using this method, the same calibration curve can be used for all samples. However, when applying background subtraction, the increase in peak area between the "blank" samples and the lowest concentrations of standard solutions should be significantly high (i.e., at least 15%–20% of the background peak area). This means that the lower limit of quantification (LLOQ) of the method might be limited by the background concentrations of the target compound. To overcome this, matrix may be diluted by adding a certain volume of solvent or buffer solution with similar characteristics as the matrix to the blank matrix, before spiking with the standard. However, this will mean that the matrix composition differs between the studied samples and the calibration. Therefore, before applying this method, extraction recoveries between matrix and diluted matrix should be determined and yield comparable results. Another limitation of this method is that the background levels in the matrix that is used as blank matrix might vary due to intra- and intersample variation which results in highly variable results and low reproducibility when using pooled matrix (Cortese et al., 2020; Thakare et al., 2016; Visconti et al., 2023).

- Surrogate matrix

This approach uses surrogate matrices such as neat solvent, stripped matrices or artificial matrices to represent a blank matrix. Neat solvents can be composed of water or mobile phase solvents. Stripped matrix is obtained by treatment of the real matrix (e.g., using activated charcoal as an absorbent of target analytes) to generate analyte-free surrogate matrices that can be used as blank for construction of calibration curves. This method does not work for all analytes and additionally, the matrix composition and characteristics might change due to the applied treatment. Artificial matrix is prepared from scratch to reproduce the real matrix in terms of composition, analyte solubility, and salt content but without any target analytes present. Artificial matrices can mainly be applied for matrices primarily composed of water such as saliva, urine, or cerebrospinal fluid. Application of surrogate matrix is widely used and will result in direct and sensitive quantification of analytes. However, if the surrogate matrix is not similar enough to the "real" matrix, the recoveries and the signal of the analyte might differ between the surrogate matrix and the study samples. Before applying this method it should therefore be demonstrated that the chosen surrogate matrix gives a similar recovery, matrix response and peak signals as the matrix it is mimicking over the entire range of quantification (Cortese et al., 2020; Thakare et al., 2016; Visconti et al., 2023).

- Surrogate analyte methods

The surrogate analyte method uses stable isotope-labeled standards that act as surrogate analyte to construct calibration curves for the target analytes. This method will give direct and sensitive quantification of the analytes and when these surrogate analytes are added before the extraction step, they can correct for all random errors during both the sample extraction and sample analysis leading to improved robustness, precision, and accuracy of the method. However, it should be demonstrated that the extraction recovery and analytical performance are comparable between the surrogate analyte and the original target analyte. The response factor (ratio unlabeled/labeled analyte) has to be similar in the entire calibration range. In addition, these stable-isotope-labeled standards often have limited availability and can be very expensive (Cortese et al., 2020; Thakare et al., 2016; Visconti et al., 2023).

When accurate or precise results cannot be obtained due to matrix effects, it might be necessary to decrease interaction of the target analytes with the matrix components. This can be done by dilution of matrix, reduction of coelution in the chromatographic method or additional clean-up steps in the sample extraction to reduce coanalytes in the matrix. The best approach should be chosen based on the time required for each step and the maximum efficacy of the steps keeping in mind that none of the validation criteria should be compromised due to application of these steps (Cortese et al., 2020; Thakare et al., 2016).

2.6 Interlaboratory comparisons

In this section, we will highlight some general aspects of interlaboratory studies (ILS) and considerations for environmental monitoring programs. Fig. 11.4 visualizes the general design of these ILS exercises. In Section 4, two key

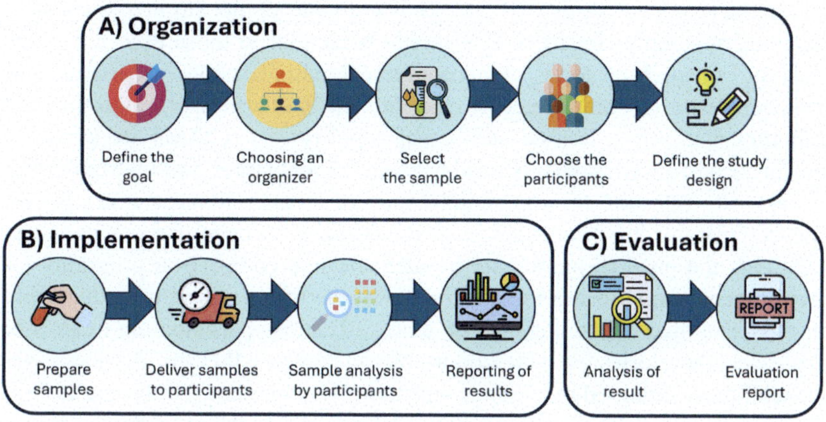

FIG. 11.4 Schematic overview of an interlaboratory study design. *(Modified from Konieczka, P., 2018. Quality Assurance and Quality Control in the Analytical Chemical Laboratory, vol. 294. Taylor & Francis Group, Boca Raton, p. 294. doi:10.1201/9781315295015.)*

examples of ILS will be discussed in greater detail to provide some insights into the organization and structure of these trials. As discussed in Section 2, different international standards (IUPAC, ISO/IEC 17043:2023,...) are available for the organization proficiency testing involving interlaboratory comparisons (International Organization for Standardization, 2023; Thompson et al., 2006). ILS generally serve multiple purposes, including (i) the evaluation of the performance of laboratories for specific analytical measurement, calibration, test, or sampling methods; (ii) the identification of underlying or emerging problems in analytical workflows; (iii) the estimation of measurement uncertainty of conducted analytical measurements; (iv) the assessment of validation parameters; and (v) training and evaluation of new participating laboratories. Regarding the latter, it might be useful to organize mentoring programs to pair new participants with reference laboratories. Interlaboratory exercises could assist in highlighting the limitations associated with in-house separation methods. The majority of international networks for environmental (bio)monitoring (e.g., HBM4EU, NORMAN, PARC) conduct these ILS as an external quality control by distributing test samples to participating members, who analyze them without prior knowledge on the concentrations and/or presence of specific analytes (van Nuijs et al., 2018). In turn, the ILS coordinator transforms the submitted analytical measurements into objective scores that represent the participants and the whole group. These scores can be used to further validate analytical protocols existing in the network and highlight possible problems (van Nuijs et al., 2018).

A major hurdle for QA/QC programs for environmental analysis and human biomonitoring is the preparation of adequate and reliable test samples since only limited reference materials are available that involve the target matrix and biomarker combinations in relevant concentrations for exposure in the general population (Esteban López et al., 2021). If available, composition of the testing materials or samples for these ILS should be similar to those used in routine analysis, especially in terms of environmental matrix, analyte concentrations, and occurrence of cointerferences. For human biomarkers of environmental contaminants, only a limited number of suppliers for (C)RMs are available offering a narrow (though) increasing range of biomarker and matrix combinations with relevant environmental exposure levels (Olivares et al., 2018). This complicates the determination of the accuracy of quantitative results and highlights the need to participate in adequate intercomparison programs for environmental monitoring to obtain data on the trueness and reproducibility for most chemical parameters (Göen et al., 2012). By participating in an ILS, analytical laboratories have the opportunity to compare their results and methods with those obtained by other groups to prove their competence and to continuously monitor the quality of their analyses and highlight potential issues (Konieczka, 2018).

These ILS ensure that the analytical methods of a laboratory are fit-for-purpose by externally validating performance criteria such as intra- and inter-laboratory precision. These ILS should be organized on a multi-year basis to follow up trends in the analytical performance of different laboratories. In any case, the number of test samples, determinants, and participants should be well-defined in the research protocol outlined by the ILS organizer.

For the calculation of z-scores of a participating laboratory (z_i) (Eq. 11.1), the measured values by the participating laboratories (m_i) are compared against an assigned value (M) relative to the standard deviation of the whole group (SD). The assigned value can be determined by one of the following manners: (i) the value obtained by a reference laboratory (i.e., a designated scientific laboratory responsible for developing, validating, and providing guidance on specific analytical methods); (ii) a certified value of a CRM used as test material; (iii) a consensus value by expert laboratories; (iv) a formulation with a well-known analyte content; or (v) the mean of means (Thompson et al., 2006). These z-scores reflect the performance of individual laboratories and the whole group and can alert participating laboratories of unanticipated issues. As van Nuijs et al. (2018) indicated that nominal concentrations might not always be applicable for the calculation of z-scores, and instead group's means can be used as an alternative (either derived from expert laboratories of all participants). In either case, the number of participants should be sufficient to allow reliable statistical evaluation of the outcomes of these trials. This could potentially be an issue for environmental chemicals and, in particular, their metabolites since different laboratories might focus on different subsets of the wide chemical space. The use of nominal values might also be restricted by the unavailability of blank matrix or CRMs. For example, for wastewater samples, the use of spiking levels is impossible because of the presence of unknown native concentrations of the analytes under investigation (van Nuijs et al., 2018). In this case, surrogate matrix (see Section 2.5) can potentially be used to overcome these challenges, but might not fully reflect the actual matrix composition in complex environmental samples (e.g., absence of coeluting interferences and solid particulate matter). The group's mean might also be favored over the nominal concentration when no recognized reference laboratory has been established for the required analysis. More information on the acceptance criteria for the *z*-scores is given in Section 4.1.

$$Z_i = \frac{m_i - M}{SD} \tag{11.1}$$

3 Specific QA/QC systems for chromatography and mass spectrometry

The analysis of organic environmental contaminants commonly involves compound separation using instrumental techniques such as gas chromatography (GC) and liquid chromatography (LC), but also more specialized chromatographic methods including ion chromatography and supercritical fluid chromatography (Schulze et al., 2020; Fisher et al., 2022). While GC is mostly limited to the separation of volatile small molecules or requires derivatization of the target analytes, LC can be applied to a wide range of chemicals using a multitude of mobile and stationary phases including reversed-phase LC (RPLC), mixed-mode LC (MMLC), and hydrophilic liquid interaction chromatography (HILIC). In reality, a combination of separation methods will be needed since no single chromatographic separation system can cover the vast amount of chemicals highly varying in their physicochemical properties present within the sample, commonly referred to as the chemical space. To separate complex mixtures of chemical compounds (e.g., isomeric and isobaric compounds), multidimensional chromatographic methodologies have been proposed. However, this chapter will not delve into two-dimensional chromatography since the number of applications for environmental screening is currently more limited compared to traditional 1-D chromatography and this chapter aims to give a more general overview of QA/QC schemes in environmental chemistry. Besides achieving adequate retention for the analytes of interest, chromatographic methods should also be capable of reducing the degree of ion suppression by isolating remaining matrix interferences from the analytes of interest (Schulze et al., 2020). The optimization of chromatographic parameters (e.g., type, column chemistry, mobile phases, flow rates, gradients) could change the coverage of the analyzable fraction of chemicals and, thus, method selectivity. To address this, analysts should employ IS during method development and validation to assess chromatographic separation.

Different mass spectrometric approaches assist for the qualitative and quantitative screening of environmental contaminants in biological and environmental samples. Chapter 9 provides a more detailed description of the different mass spectrometric approaches. Targeted approaches can only detect a limited number of (already identified and a priori selected) compounds and,

TABLE 11.2 Key differences between targeted and nontargeted analysis based on Fisher et al. (2022).

Performance criteria	Targeted analysis	Nontargeted analysis
Selectivity	• If the confirmation criteria are fulfilled (i.e., retention time, ion ratios, peak shapes), the analyte is known to be truly present in the sample • Targeted analysis can only provide information on a set of a priori defined chemicals	• If an analyst reports the absence of a chemical in the sample, it may not have been correctly identified during data processing even though it is present within the sample (i.e., false negative) • If an analyst reports the presence of a chemical in the sample, it may actually be absent. This could be the case for isomers or incorrect identifications (i.e., false positives)
Sensitivity	• If the analyte response is above the limit of detection (LOD), it is present in a detectable level in the sample	• The higher resolution of nontargeted methods is usually associated with lower sensitivity compared to targeted methods
Accuracy and precision	• Reported concentrations at a defined level of confidence are expected to exist within a defined range	• Concentrations are usually reported without corresponding information on the confidence of the semiquantification method

therefore, do not address the rising issue of environmental contaminants (Schulze et al., 2020; Schulze and Kaserzon, 2024). In contrast, nontargeted screening approaches allow a simultaneous detection of a high number of analytes. Table 11.2 focusses on the key differences between targeted (TA) and nontargeted analysis (NTA) and demonstrates how both approaches can complement each other. In the following sections, specific QC measures for each approach will be discussed in-depth because of the widespread application of these screening methods.

3.1 Targeted analysis

With TA, the focus is on a small fraction of chemicals defined prior to data acquisition for which concentrations of the analytes are measured through the use of isotopically labeled IS and external calibration curves or through the specific approaches enlisted in Section 2.5 (Broadhurst et al., 2018; Fisher et al., 2022). The quality of TA results can be assessed using well-established acceptance criteria for selectivity, sensitivity, accuracy, and precision. These performance criteria are integral parts of analytical method validation that test if the workflow is fit-for-purpose to deliver reliable quantitative information on specific chemicals (see Section 2.5) (Fisher et al., 2022). In TA, standard mixtures and spiked QC samples are frequently employed to assess method performance, especially intra- and interday accuracy and precision. Selectivity of targeted methods is usually guaranteed by compound-specific optimization of the sample preparation and chromatographic separation by spiking reference standards to blank matrix (if available) or surrogates (Fisher et al., 2022). In addition, specific confirmation criteria such as ion ratios, (relative) retention times and peak shapes can further distinguish the presence of specific chemicals from coeluting matrix interferences. Acceptance criteria for these compound identification parameters are described in greater detail in the guidelines from several regulatory bodies such as the Food and Drug Administration, European Union, World Anti-Doping Agency, and other organizations (European Medicines Agency, 2022; Yuan et al., 2015; US Food and Drug Administration, 2024). Fig. 11.5 visualizes the different QC measures for TA during the different analytical steps.

In tandem mass spectrometry, two to four ions should be monitored for compound identification with the most abundant ion in the sample matrix used as the "quantifier" and the remaining transitions serving as "qualifiers" to prevent false positive results. It is important that chosen ion transitions are only minimally influenced by the matrix, to prevent severe interferences

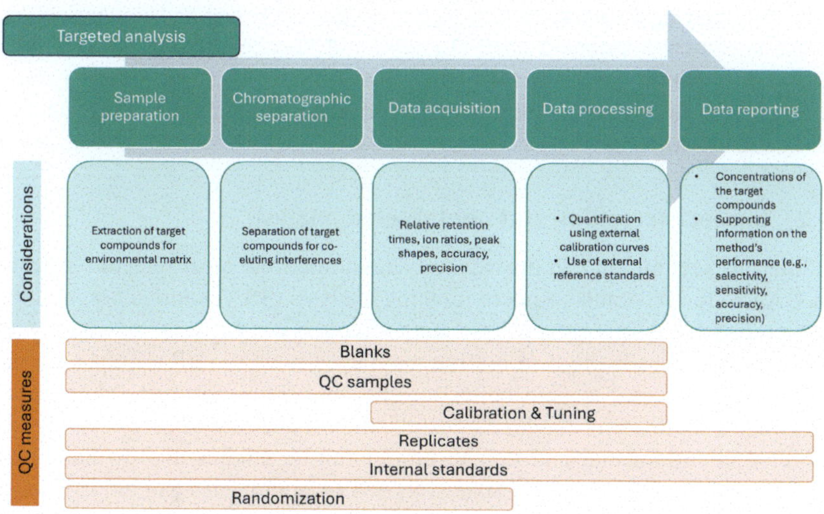

FIG. 11.5 Overview of different analytical steps involved in targeted analysis using chromatography and tandem mass spectrometry with emphasis on the different QC measures during each step. *(Adapted from Schulze, B., Jeon, Y., Kaserzon, S., Heffernan, A.L., Dewapriya, P., O'Brien, J., Gomez Ramos, M.J., Ghorbani Gorji, S., Mueller, J.F., Thomas, K.V., Samanipour, S., 2020. An assessment of quality assurance/quality control efforts in high resolution mass spectrometry non-target workflows for analysis of environmental samples. TrAC Trends Anal. Chem. 1879-3142. 133. Available from: www.elsevier.com/locate/trac. doi:10.1016/j.trac.2020.116063.)*

in a more concentrated matrix. Whenever possible, it is also recommended to select the most compound-specific transitions, however it should be avoided to include ions with a relative abundance <5% (compared to the most abundant ion). Fragments which should preferentially not be included involve: (i) fragments from unspecific neutral losses such as water, (ii) nonspecific fragments with common chemical structures, (iii) solvent adduct ions, and (iv) ions with the same molecular weight of the derivatization reagent (Yuan et al., 2015). However, in specific cases, the precursor ion is fragmented into a limited number of product ions making it necessary to include nonspecific transitions. The ratios between quantifier and qualifiers should fall within specific tolerance ranges that are dependent on the ion intensity, number of qualifiers and methodology (generally 20%–30% RSD) (Yuan et al., 2015; Rivier, 2003).

For the IS, generally one quantifier/qualifier transition should be monitored. Important to note is that large interday variation in ion ratios might exist, which warrants the need to verify them in each analytical batch based on the calibration and QC samples (Yuan et al., 2015; Concheiro et al., 2007). For the relative retention time (i.e., the ratio between the retention time of the analyte and its corresponding IS), a tolerance window of ±0.5% for GC and ±2.5% for LC compared to the calibrants should be employed (Yuan et al., 2015), but variations in these cut-off values exist among different studies (Rivier, 2003).

A solvent blank consisting of reconstitution solvent can be injected at multiple times during the analysis sequence to assess carry-over and contamination in the analytical instrument itself (Lescord and Lepage, 2025; Schulze et al., 2020). In addition, QC samples and calibration curves can be injected multiple times to asses stability of the instrument during the analytical run (Schulze et al., 2020).

3.2 Suspect screening and nontargeted analysis

While this section aims to provide a discussion on the most relevant aspects of QA/QC for NTA and suspect screening analysis (SSA) approaches, a more comprehensive discussion of these methodologies can be found in Chapter 9. Over the last decade, high resolution mass spectrometry (HRMS) has grown into a powerful tool to gain better insights into the exposure to chemicals. The vast complexity of the human exposome cannot be interpreted solely by targeted environmental and human biomonitoring studies due to their limited coverage of the chemical space. NTA/SSA approaches can aid in filling this knowledge gap by providing high specificity and accuracy for compound identification. They allow for the simultaneous detection of all chemicals which can be covered by the applied sample preparation and instrumental methods, allowing for the postacquisition prioritization of compounds of interest depending on the research question. However, they cannot act as a standalone approach since NTA/SSA analysis are unable to provide validated quantitative data which is crucial for a holistic assessment of human exposure (Schulze et al., 2020).

The implementation of QA/QC measures in NTA/SSA workflows is imperative because the generation of data and annotation of molecular formulas or specific molecular structures is influenced by numerous factors such as the sample collection and preparation, chromatographic separation, mass accuracy achieved for both precursors and product ions, data processing, etc. (Knolhoff et al., 2021). Due to the complexity and magnitude of the collected data and acquisition steps, QA/QC approaches conceptualized for TA are not adequate for NTA/SSA. In addition, the choice of the analytical workflow can be different depending on the research question and environmental matrix. The development of comprehensive, accurate, and reproducible NTA and SSA workflows proves to be challenging due to the impact of ion suppression, matrix effects, sample complexity, differences in HRMS instrumentation, etc. on data processing algorithms and the reliability of compound annotations (Knolhoff et al., 2021). Tailored QC strategies for NTA/SSA should therefore be used to identify sources or errors throughout the entire analytical workflow, from sample preparation through data processing.

A major difference between TA and NTA/SSA is that TA-specific QA/QC schemes can rely on the availability of reference analytical standards for all compounds included in the method allowing for a tailored QA/QC approach (Schulze and Kaserzon, 2024). This is not standardized for NTA/SSA due to the large chemical space that can be measured with HRMS and since analytes of interest for NTA are often not selected a priori (Knolhoff et al., 2021). This calls for comprehensive QA/QC guidelines which can be implemented and adapted based on the individual research question and which cover all relevant steps of an NTA/SSA study (see Fig. 11.6).

Relevant QA/QC guidelines introduced for NTA/SSA in literature include the works of Caballero-Casero et al. (2021) and Hollender et al. (2023). The main challenge within NTA and SSA workflows is the inherent presence of false positives (i.e., type I error—compounds are incorrectly annotated) and false negatives (i.e., type II error—compounds present in the sample are not annotated). For false negatives, it is often difficult to distinguish between analyte loss during the different analytical steps (e.g., poor recovery, matrix suppression, poor chromatographic retention) or absence of the chemical in the environmental matrix. Although the majority of existing QA/QC strategies already focus on reporting high confidence levels for compound identification and thereby reducing the occurrence of false positive features, these approaches should also further strive to reduce the number of false negatives as much as possible or discuss in detail what might be omitted (Schulze et al., 2020). These aims should be followed by the implementation of the QA/QC aspects described in the following section.

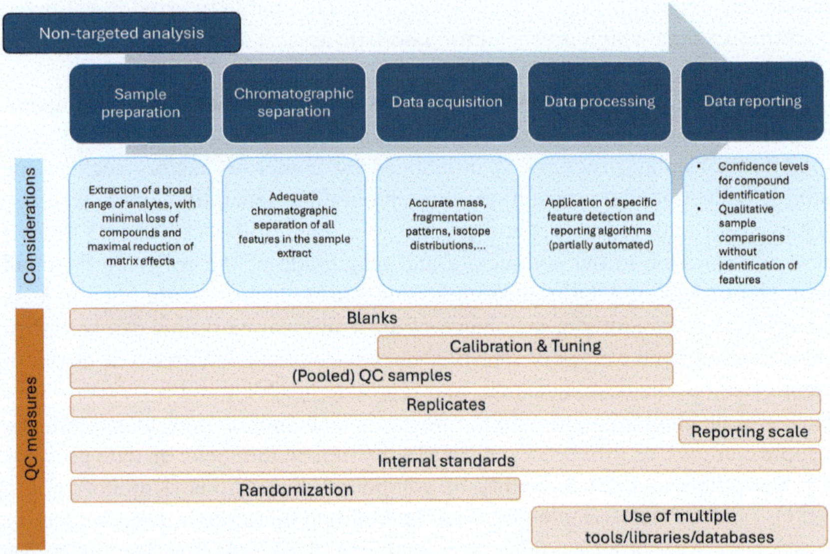

FIG. 11.6 Overview of different analytical steps involved in nontargeted analysis using chromatography and high resolution mass spectrometry with emphasis on the different QC measures during each step. *Adapted from Schulze, B., Jeon, Y., Kaserzon, S., Heffernan, A.L., Dewapriya, P., O'Brien, J., Gomez Ramos, M.J., Ghorbani Gorji, S., Mueller, J.F., Thomas, K.V., Samanipour, S., 2020. An assessment of quality assurance/quality control efforts in high resolution mass spectrometry non-target workflows for analysis of environmental samples. TrAC Trends Anal. Chem. 1879-3142. 133. Available from: www.elsevier.com/locate/trac. doi:10.1016/j.trac.2020.116063.*

Sample extraction and chromatographic methods for NTA/SSA should be able to extract and separate a wide range of chemicals while removing matrix interferences, meanwhile TA workflows focus on the isolation of a predefined selection of target compounds (Schulze et al., 2020). In general, generic extraction methods with limited sample adulteration (including direct injection and "dilute-and-shoot" methods) should be applied to minimize the presence of sample preparation artefacts (e.g., degradation and transformation products, sample contamination) and to avoid the introduction of bias towards (e.g., a compound class best covered by an applied extraction or clean-up method) (Deventer et al., 2014). The inclusion of a selection of IS can aid the evaluation of the extraction recovery, reproducibility, matrix effects and compound stability, and has been commonly accepted as a basic QC measure in most QA/QC guidelines (Knolhoff et al., 2021). However, even a wide selection of IS cannot compensate for the wide variety of chemicals present in the samples and might interfere with other compounds of interest during ionization. Therefore selected IS, usually isotopically labelled, should represent classes of interest and cover a wide range of molecular weights and $\log K_{ow}$ values in both polarities (i.e., positive and negative ionization) to successfully mimic the chemical space present within the samples. However, the use of fortified samples is not always representative for recovery assessment for the wide variety of chemicals present in the sample, since method sensitivity might also be highly impacted by the complex sample matrix background causing high levels of interference. As hypothesized by Schulze et al. (2020) clear guidelines should be constructed for the minimum number of IS for QA/QC schemes applicable to sample matrices with different degrees of complexity. To achieve this, groups of IS can be defined, categorized, and harmonized internationally per matrix type and general research aims.

Even though no standardized method is available for the inclusion of blank samples (Schulze and Kaserzon, 2024), it can aid in vastly reducing the number of false positive compound annotations resulting from contamination, carryover and background noise during sample preparation, chromatography and data acquisition and processing. Therefore the addition of several types of blank samples (e.g., matrix blanks, laboratory blanks, field blanks, instrument blanks) in different steps of the analytical workflow is pivotal to facilitate the identification of relevant features. As stated before, collecting or making a matrix blank might be complicated since it is not possible to guarantee that it will not natively contain any of the broad range of contaminants. The use of calibration or tuning mixtures can be applied during data acquisition and processing to account for false negatives/positives as a result of mass errors in HRMS measurements. Tuning and calibration of the system should be performed periodically to identify potential mass shifts resulting from the electronic hardware, incorrect calibration coefficients, or other aspects. In addition, the use of an online mass calibration through the continuous introduction of ions with known *m/z* values can account for mass shifts occurring during data acquisition.

The use of matrix-matched QC samples, spiked with sets of known reference standards, can help to further evaluate the coverage of the sample preparation and instrumental methods. Often, pooled QC samples are implemented, prepared by pooling aliquots from every study sample included in the analyses. If injected multiple times throughout the sample sequence, these can help to evaluate potential loss in sensitivity during the analytical worklist and to validate the mass accuracy and precision of the HRMS system, including the assessment of chromatographic retention time stability. At the start of the work sequence, the pooled QC sample can be injected ten times to condition the LC-MS system, guaranteeing stabilization of the total ion chromatogram, retention times and peak areas/intensities of IS (Iturrospe et al., 2023). The use of pooled QC samples also enables the monitoring of the reproducibility of data collection and processing for a large amount of molecular features and can be employed to model and compensate for systematic measurement bias (Knolhoff et al., 2021). Pooled biological QCs are widely adopted in metabolomics and food analysis and should be used more routinely for environmental NTA/SSA to assess different parameters associated with sample preparation, data acquisition and data processing. Pooled QC samples and IS with sufficient coverage of the chemical space also permit the discovery and reporting of issues related to ion suppression and other reasons for information loss. Therefore outcomes of the pooled QC analysis should be reported alongside the unknown samples.

Knolhoff et al. (2021) proposed the use of a standard mixture containing a moderate amount of known compounds with different chemical properties (e.g., ionization mode, hydrophilicity, mass and elemental compositions, chemical class) to be spiked to sample matrix as an additional and complementary QC approach and verify the performance of implemented NTA/SSA workflows (e.g., coverage of sample preparation approach, reliability of feature extraction and annotation workflows). However, there is currently no harmonization on the use of specific standard kits since there are only a limited number of commercially available mixtures (e.g., Environmental Protection Agency, European Union Joint Research Centre) that were specifically designed for QA/QC purposes aiming to reflect the compound diversity encountered with environmental screening. Compound detection and annotation in QC samples also depends heavily on the recovery and reproducibility of the sample extraction method and, unfortunately, no single method is available or capable of recovering the entire chemical space present within a sample (Knolhoff et al., 2021). The application of standard mixtures for spiking QCs before sample extraction also allows the assessment of the recovery of sample extraction methods and separation of chromatographic methods for NTA/SSA workflows. Spiking pooled QC samples with larger standard mixtures enables the determination of realistic mass error windows and thresholds for exclusion of false positives for SSA and the estimation of matrix effects on signal suppression or enhancement, resulting in a reduced number of generated molecular features. Harmonized use of these standard mixtures across laboratories could facilitate the comparison of existing and new methodologies during interlaboratory trials.

As evidenced in metabolomics studies (Broadhurst et al., 2018; Knolhoff et al., 2021), a system suitability mixture directly injected in the instrument can be used to monitor instrumental performance based on analyte abundance, mass accuracy, and retention time stability and should be implemented at different stages in the analytical worklist. Another QC measure involves the inclusion of replicates to address reproducibility during all analysis steps (Caballero-Casero et al., 2021). Commonly, at least three technical replicates are included for a subselection of samples to assess, e.g., the reproducibility of feature extraction algorithms. However, the number of repetitions might be limited due to time constraints. Randomization during sample injection could support the minimalization of the occurrence of systematic errors during the workflow, such as carry-over during the chromatographic sequence. Finally, the use of different spectral libraries, in silico protection software and databases could diminish the number of incorrect annotations during data processing and reporting. In this light, the expansion of open access spectral libraries could enable researchers to obtain more transparent, verifiable and comparable results, which is not always the case with specific vendor software. In addition, a wide range of innovative tools and existing approaches (e.g., open access data repositories, retention time prediction models, advanced statistics) are available to evaluate the performance of specific NTA approaches (Schulze et al., 2020). In general, QA/QC schemes for NTA/SSA can benefit from more detailed data reporting guidelines for the identification of unknown features and evaluation of confidence levels, compromising a fixed set of processing and statistical parameters to routinely verify the suitability and reliability of the workflow (Schulze et al., 2020).

The central aim of most NTA/SSA based studies is the reporting of chemicals that could be annotated in the samples of interest to allow for further interpretation of the results, including, e.g., prioritization of compound for targeted analyses, reporting of findings to legislators and policymakers and publication. This raises the question of how to report the confidence in the annotation of a compound without the availability of a reference standard for unequivocal confirmation. Published in 2014 by Schymanski et al. (2014), the 5-level scheme reporting annotation confidence as levels ranging from 1 (highest confidence, confirmed with reference standard) to 5 (lowest confidence, only exact mass available) has been widely established in environmental NTA/SSA studies. Without the availability of a reference standard, confidence level (CL) 2 is the highest annotation level that can be reached within NTA/SSA approaches, provided that experimental fragmentation spectra can be matched with reference library data or provide characteristic evidence allowing to assign one candidate structure, eliminating all other possibilities. This emphasizes the central role of reference spectral repositories, such as MassBank, mzcloud and others, and the need for their constant expansion following FAIR principles (Findable, accessible, interoperable, reusable) (Wilkinson et al., 2016). Compiling and maintaining up-to-date in-house spectral libraries can also complement these repositories to achieve CL1 annotations.

Despite the well-established 5-level system, major discrepancies in the assignment of CLs are often observed between studies indicating no clear agreement on guidelines to be followed for CL annotations. These gaps were recently addressed by *Alygizakis* et al. 2023 introducing an identification scoring system which provides clear guidance on the annotation of CLs. This system is based on assigning a score for each datapoint (*m/z* value, RT, isotopic pattern, fragmentation data etc.) available for the matching (Alygizakis et al., 2023). From the total score of each feature, a confidence level is derived following clearly defined cut-off values. The implementation of this system in NTA/SSA studies alongside a transparent and open communication of the data quality will facilitate a transparent CL annotation and a reliable compound identification.

4 Lessons learned from international collaborations and studies

4.1 HBM4EU—Human biomonitoring for Europe

An example of appropriate QA/QC measures for HBM for internal exposure assessment can be found in the HBM4EU consortium, which founded a specialized Quality Assurance Unit (QAU) to assure reliability and comparability of analytical measurements for HBM studies (Esteban López et al., 2021). Their

QA/QC approach consists of two different schemes that addressed a different number of chemicals and rounds for proficiency tests (Esteban López et al., 2021). The first scheme focused on 73 chemicals that are part of the first priority list (phthalates, DINCH, bisphenols, PFAS, brominated flame retardants, phosphorus flame retardants, polycyclic aromatic hydrocarbons, Cd, Cr, aromatic amines) and encompassed four rounds of proficiency testing with a high number of participants, while the second scheme addressed a reduced selection of 22 biomarkers (for acrylamide, arsenic, mycotoxins, pesticides, and UV-filters) and only involved three rounds of interlaboratory comparisons with a reduced number of expert participants. Depending on the substance group and research aims, these exercises were organized as interlaboratory comparison investigations (ICIs) or the External Quality Assurance Schemes (EQUAS) (Nübler et al., 2022; Esteban López et al., 2021). While the ICI focuses on the comparison of results among participating laboratories for the analysis of HBM4EU samples, the EQUAS provides a comparison of results against a selection of expert laboratories with well-known and documented expertise in previous HBM studies.

In the first scheme, participants were free to decide for which biomarkers they participated, while laboratories should provide data for all biomarkers in the second scheme. The outcomes of these interlaboratory comparisons were used to identify laboratories capable of generating consistent and reliable data for sample analysis in the HBM4EU framework. The SOPs for the interlaboratory comparisons were based on the ISO17043 accreditation guidelines (International Organization for Standardization, 2023). Additional SOPs were designed for the preparation and characterization of control materials and the assessment of the participant's results (Esteban López et al., 2021).

Within the HBM4EU consortium, supporting and participating laboratories were identified by the National Hub Contact points and through an announcement letter disseminated via HBM4EU and the European Environment Agency. Expert laboratories were evaluated through questionnaires aimed to collect information on their experience in the analysis of specific HBM biomarkers in biological samples as well as in organizing interlaboratory trials. Even though capacity building for laboratories is pivotal for sustainable HBM in Europe (since a coherent approach among multiple countries cannot be based on a limited number of specialized expert laboratories), capacity building and comparability of data can become conflicting aims (Schindler et al., 2014). In this light, only analytical laboratories that can demonstrate their analytical capabilities through successful ICI and EQUAS participation should be eligible to perform analytical work (Schindler et al., 2014) (see also Fig. 11.7).

Two control materials with different concentrations were provided to all participants in each proficiency testing round with target concentrations of the HBM biomarkers covering the range commonly observed in the general population with one sample representing a lower exposure while the other sample covered the higher exposure range. These control materials were extensively tested for homogeneity and stability before shipment to the participants (Schindler

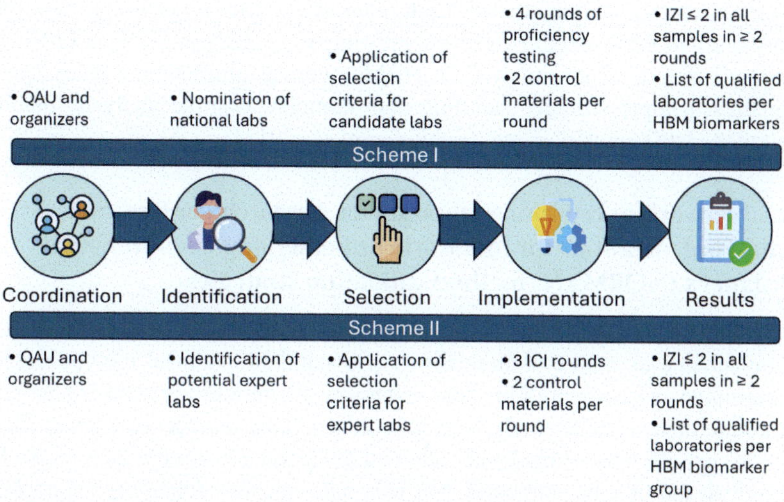

FIG. 11.7 Design of the interlaboratory comparison schemes for the HBM4EU consortium. *Adapted from Esteban López, M., Göen, T., Mol, H., Nübler, S., Haji-Abbas-Zarrabi, K., Koch, H.M., Kasper-Sonnenberg, M., Dvorakova, D., Hajslova, J., Antignac, J.P., Vaccher, V., Elbers, I., Thomsen, C., Vorkamp, K., Pedraza, S., Kolossa-Gehring, D.M., Castaño, A., 2021. The European human biomonitoring platform—design and implementation of a laboratory quality assurance/quality control (QA/QC) programme for selected priority chemicals. Int. J. Hyg. Environ. Health 1618-131X. 234. Available from: http://www.urbanfischer.de/journals/intjhyg. doi:10.1016/j.ijheh.2021.113740, with permission.*

et al., 2014). Homogeneity testing was based on the ISO13528:2015 and the recommendations of Fearn and Thompson (2001). Biomarker/matrix combinations for the control materials in the QA/QC program were selected based on human specificity, biological sensitivity and in-sample stability. For this purpose, scientific literature was scanned to prioritize the most suitable biomarkers and matrices based on technical feasibility, expected exposure levels and policy needs (Vorkamp et al., 2023; Esteban López et al., 2021).

The reliability of HBM data was evaluated by comparing the analytical results for the same control material between multiple laboratories with their own analytical methods. As a means of proficiency, *z*-scores were determined using an assigned value and preset target standard deviation. The assigned value was derived from the participants' results for the ICI, while for the EQUAS the assigned value was based on the analysis of six replicates of control materials by at least three experienced laboratories (Vorkamp et al., 2023). Assigned values are internationally accepted as an approximation of the true value (Esteban López et al., 2021). An absolute *z*-score <2 was considered satisfactory, $2<z<3$ questionable and >3 unsatisfactory (Esteban López et al., 2021).

An important conclusion from these interlaboratory trials is that comparability of analytical results and capacity building of qualified laboratories within Europe for HBM can only be achieved through the harmonization of preanalytical and analytical work (Schindler et al., 2014). For this reason, the Consortium to Perform Human Biomonitoring on a European Scale (COPHES) designed a standardized protocol for HBM covering sampling strategies, questionnaires, sample storage, handling and shipment, sample analysis, quality assurance, data acquisition, etc. (Schindler et al., 2014).

4.2 Network of reference laboratories, research centers, and related organizations for monitoring of emerging environmental substances (NORMAN) method validation strategies

The NORMAN association has developed a common analytical method validation scheme to ensure harmonization and standardization of methodologies to guarantee the comparability and reliability of monitoring data for emerging contaminants (NORMAN Network, 2009). This is necessary since these methods are often developed and validated by different laboratories for specific biological and/or environmental matrices and can therefore only be applied to specific conditions, which complicates data comparisons. The NORMAN method validation schemes address a total of three different method validation strategies based on the level of complexity (Fig. 11.8), including (i) within-laboratory validation at the research laboratory level, (ii) external validation at the level of expert/reference laboratories, and (iii) external validation at the level of routine laboratories (NORMAN Network, 2009). These method validation schemes have been adopted in a wide range of method validation studies for environmental monitoring.

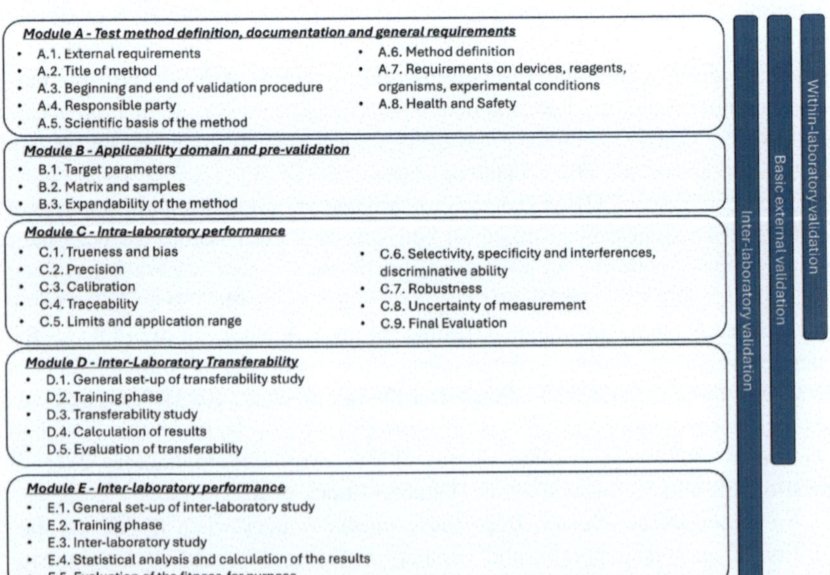

FIG. 11.8 Validation framework of the NORMAN association (NORMAN Network, 2009).

The within-laboratory validation guidelines are applicable in case a method is available for a given (group of) emerging contaminant(s). However, it is not applicable to the specific matrix, compartments or organisms of interest or when its suitability for the intended use regarding the performance criteria has not been sufficiently tested or proven. In this light, the validation guidelines also provide an extensive set of selection criteria to choose the method showing the most potential when multiple analytical methods might be fit for validation at the requested validation level. The within-laboratory performance links back to the concepts of method validation outlined in Section 2.5 (e.g., accuracy, precision, calibration range, traceability, selectivity, specificity).

After successful within-laboratory validation, the NORMAN method validation scheme also provides guidance on the transferability of the analytical method at the expert laboratory level. The analytical method can be regarded as transferable if at least one other laboratory is able to reproduce similar results as observed during internal method validation. The purpose of this transferability study is to minimize the impact of within-laboratory variation on the measures used to characterize and evaluate a test method's performance. For this purpose, a reference standard at a single concentration in a specified matrix may be sufficient to assess transferability, but more concentration levels are recommended. Due to the limited number of participating laboratories in this transferability study, a high number of replicates is required to limit the uncertainty from within-laboratory variability. Among different preset performance criteria, trueness, precision, uncertainty, and application range are evaluated to test the transferability of the analytical method. For each reference material, results of the transferee laboratories should be compared to reference values (i.e., either reference data related to the tested materials or the method) and show similar or better results than the results obtained by the originating laboratory and additionally pass the preset performance requirements. In case of substantial bias, the analytical method might not be fit for its intended purpose.

To be employed in large-scale monitoring programs, an environmental analytical workflow should show sufficient inter-laboratory performance (e.g., robustness and usability) at the level of routine laboratories. These ILS at the routine laboratory level are designed to minimize within-laboratory variation on the measures used to define and assess the performance criteria of the analytical method. These studies not only assess the proficiency or capabilities of participating laboratories, but mainly evaluate the test method's applicability by routine laboratories. The minimum number of participating laboratories is recognized to be eight; however, to account for potential outliers or drop-out of laboratories, 10–12 participating laboratories is regarded as the minimum number of participating laboratories. According to the NORMAN method validation schemes, a total of five test materials should be analyzed in the frame of these interlaboratory trials, containing all compound/matrix/concentration level combinations appropriate for the method's intended use. For multicompound applications, the type and number of analytes in the sample should reflect realistic scenarios in which the analytical method will be used in an environmental context. For each test material, a minimum of three replicates should be considered. One of the main criteria for successful method validation is the availability of a reliable assigned value with additional quantitative information on the degree of uncertainty. For routine analysis, it is important to evaluate the trueness of the analytical method to investigate the potential for systematic bias in the method. If statistical analysis highlights substantial variability between the study mean during validation and the assigned value, it should be evaluated whether fitness-for-purpose of the method is jeopardized. According to the NORMAN documentation, the ratio of outlier values in single laboratories or outlying laboratories should be below 25% for the method to be applicable at the routine laboratory level, including laboratories that were unable to provide results or comply with the validation criteria.

5 Conclusions and recommendations for QA/QC

Robust QA/QC systems are essential for assessing the performance of analytical methods and for generating accurate, comparable, and reproducible data in environmental monitoring. The need for proper QA/QC schemes is further amplified by globalization and need for transferable analytical test methods between different laboratories to minimize interlaboratory variability. This requirement is continuously challenged by the wide range of analytical methods and test matrices to adapt to the increasing range of environmental analytes. The complexity of QA/QC should always be performed in relationship to the fitness of purpose of the analytical approach and level of the laboratory (research versus routine). Even though there is a range of harmonized guidelines (e.g., ISO, IUPAC, EPA, ICH) available for QA/QC systems in analytical chemistry, they are not always fully transferable for environmental monitoring of different analyte/matrix combinations. Therefore further efforts are needed to better adapt these guidelines for environmental monitoring.

The collaborations of HBM4EU, PARC, and NORMAN demonstrate that a wide collaborative network of laboratories (academic, supporting, expert, routine) is essential for continuous external validation of test methods to assure trueness and reproducibility of quantitative measurements. In this sense, capacity building and institutionalization are necessary for sustainable environmental monitoring with increasing levels of complexity (e.g., at the routine laboratory level), although interlaboratory variability also tends to increase simultaneously. The interlaboratory comparisons of HBM4EU and NORMAN also show that accurate and reproducible biomonitoring and environmental data can only be obtained if participating laboratories follow harmonized preanalytical (sampling, storage, transport) and analytical (sample extraction and analysis) workflows including both blanks and spiked samples to evaluate contamination and analyte loss in different steps of the protocol. In this light, both networks present a tiered validation approach, progressing from within-laboratory validation to interlaboratory trials at the routine laboratory level. These strategies ensure that analytical procedures are not only accurate and precise, but also transferable and scalable. Therefore, the recommendations from these collaborative studies can be employed in other international networks for the organization and coordination of interlaboratory trials.

Intra- and interlaboratory method validation should be performed using well-characterized control materials that reflect the realistic application ranges and, preferentially, include the corresponding environmental matrix. However, only a limited number of matrix-specific CRMs are available for this purpose, potentially compromising metrological traceability. In addition, the acquisition of contaminant-free matrix blanks is often challenging because of the widespread occurrence of environmental pollutants in environmental samples. These limitations further highlight the urgent need for additional matrix-specific materials and development of alternative strategies to support the assessment of matrix effects during analytical method validation.

Even though the harmonization of QA/QC strategies is important to assure trueness and reproducibility of analytical workflows, certain analytical approaches require "tailor-made" approaches for QA/QC. For example, TA and NTA demand different QA/QC schemes due to the substantial differences in selectivity, sensitivity, accuracy, and precision of both approaches. Therefore QC measures (e.g., number of replicates, blanks, QC samples) should also be adapted to the specific performance characteristics of the test method and applied continuously to monitor these criteria. Given the widespread application of chromatography and mass spectrometry and targeted and nontargeted screening methods in environmental monitoring, there are possibilities for further standardizing QC measures and maximize the complementarity between both techniques.

References

Alygizakis, N., Lestremau, F., Gago-Ferrero, P., Gil-Solsona, R., Arturi, K., Hollender, J., Schymanski, E.L., Dulio, V., Slobodnik, J., Thomaidis, N.S., 2023. Towards a harmonized identification scoring system in LC-HRMS/MS based non-target screening (NTS) of emerging contaminants. TrAC Trends Anal. Chem. 01659936. 159, 116944. doi:10.1016/j.trac.2023.116944.

Barwick, V., 2018. Eurachem/CITAC Guide: Guide to Quality in Analytical Chemistry: An Aid to Accreditation. Report No.: ISBN978-0-948926-32–7.

Bopp, S.K., Barouki, R., Brack, W., Dalla Costa, S., Dorne, J.L.C.M., Drakvik, P.E., Faust, M., Karjalainen, T.K., Kephalopoulos, S., van Klaveren, J., Kolossa-Gehring, M., Kortenkamp, A., Lebret, E., Lettieri, T., Nørager, S., Rüegg, J., Tarazona, J.V., Trier, X., van de Water, B., van Gils, J., Bergman, Å., 2018. Current EU research activities on combined exposure to multiple chemicals. Environ. Int. 18736750. 120, 544–562. Available from: www.elsevier.com/locate/envint. doi:10.1016/j.envint.2018.07.037.

Broadhurst, D., Goodacre, R., Reinke, S.N., Kuligowski, J., Wilson, I.D., Lewis, M.R., Dunn, W.B., 2018. Guidelines and considerations for the use of system suitability and quality control samples in mass spectrometry assays applied in untargeted clinical metabolomic studies. Metabolomics. 15733890. 14 (6). doi:10.1007/s11306-018-1367-3.

Caballero-Casero, N., Belova, L., Vervliet, P., Antignac, J.P., Castaño, A., Debrauwer, L., López, M.E., Huber, C., Klanova, J., Krauss, M., Lommen, A., Mol, H.G.J., Oberacher, H., Pardo, O., Price, E.J., Reinstadler, V., Vitale, C.M., van Nuijs, A.L.N., Covaci, A., 2021. Towards harmonised criteria in quality assurance and quality control of suspect and non-target LC-HRMS analytical workflows for screening of emerging contaminants in human biomonitoring. TrAC Trends Anal. Chem. 18793142. 136. Available from: http://www.elsevier.com/locate/trac. doi:10.1016/j.trac.2021.116201.

Comero, S., Dalla Costa, S., Cusinato, A., Korytar, P., Kephalopoulos, S., Bopp, S., Gawlik, B.M., 2020. A conceptual data quality framework for IPCHEM—the European Commission Information Platform for chemical monitoring. TrAC Trends Anal. Chem. 01659936. 127, 115879. doi:10.1016/j.trac.2020.115879.

Concheiro, M., De Castro, A., Quintela, O., Cruz, A., López-Rivadulla, M., 2007. Determination of illicit drugs and their metabolites in human urine by liquid chromatography tandem mass spectrometry including relative ion intensity criterion. J. Anal. Toxicol. 1945-2403. 31 (9), 573–580. doi:10.1093/jat/31.9.573.

Cortese, M., Gigliobianco, M.R., Magnoni, F., Censi, R., Di Martino, P., 2020. Compensate for or minimize matrix effects? strategies for overcoming matrix effects in liquid chromatography-mass spectrometry technique: a tutorial review. Molecules. 14203049. 25 (13). Available from: https://www.mdpi.com/1420-3049/25/13/3047/pdf. doi:10.3390/molecules25133047.

Deventer, K., Pozo, O.J., Verstraete, A.G., Van Eenoo, P., 2014. Dilute-and-shoot-liquid chromatography-mass spectrometry for urine analysis in doping control and analytical toxicology. TrAC Trends Anal. Chem. 01659936. 55, 1–13. doi:10.1016/j.trac.2013.10.012.

Ellison, S.L.R., Williams, A., 2019. Eurachem/CITAC guide: metrological traceability. Anal. Meas. 2. Available from: http://www.eurachem.org. ISBN: 978-0-948926-34-1n.

Environmental Protection Agency, 2017. Quality control tools factsheet for blanks. Available from: https://19january2017snapshot.epa.gov/quality/quality-control-tools-factsheet-blanks_.html.

Environmental Protection Agency, 2025. Method validation and peer review policies and guidelines. Available from: https://www.epa.gov/measurements-modeling/method-validation-and-peer-review-policies-and-guidelines.

Esteban López, M., Göen, T., Mol, H., Nübler, S., Haji-Abbas-Zarrabi, K., Koch, H.M., Kasper-Sonnenberg, M., Dvorakova, D., Hajslova, J., Antignac, J.P., Vaccher, V., Elbers, I., Thomsen, C., Vorkamp, K., Pedraza, S., Kolossa-Gehring, D.M., Castaño, A., 2021. The European human biomonitoring platform—design and implementation of a laboratory quality assurance/quality control (QA/QC) programme for selected priority chemicals. Int. J. Hyg. Environ. Health 1618131X. 234. Available from: http://www.urbanfischer.de/journals/intjhyg. doi:10.1016/j.ijheh.2021.113740.

European Medicines Agency, 2022. ICH M10 on bioanalytical method validation—scientific guideline. Available from: https://www.ema.europa.eu/en/ich-m10-bioanalytical-method-validation-scientific-guideline.

Fearn, T., Thompson, M., 2001. A new test for 'sufficient homogeneity'. Analyst. 00032654. 126 (8), 1414–1417. doi:10.1039/b103812p.

Fisher, C.M., Peter, K.T., Newton, S.R., Schaub, A.J., Sobus, J.R., 2022. Approaches for assessing performance of high-resolution mass spectrometry–based non-targeted analysis methods. Anal. Bioanal. Chem. 1618-2642. 414 (22), 6455–6471. doi:10.1007/s00216-022-04203-3.

Göen, T., Schaller, K.H., Drexler, H., 2012. External quality assessment of human biomonitoring in the range of environmental exposure levels. Int. J. Hyg. Environ. Health. 1618131X. 215 (2), 229–232. doi:10.1016/j.ijheh.2011.08.012.

Hollender, J., Schymanski, E.L., Ahrens, L., Alygizakis, N., Béen, F., Bijlsma, L., Brunner, A.M., Celma, A., Fildier, A., Fu, Q., Gago-Ferrero, P., Gil-Solsona, R., Haglund, P., Hansen, M., Kaserzon, S., Kruve, A., Lamoree, M., Margoum, C., Meijer, J., Merel, S., Rauert, C., Rostkowski, P., Samanipour, S., Schulze, B., Schulze, T., Singh, R.R., Slobodnik, J., Steininger-Mairinger, T., Thomaidis, N.S., Togola, A., Vorkamp, K., Vulliet, E., Zhu, L., Krauss, M., 2023. Norman guidance on suspect and non-target screening in environmental monitoring. Environ. Sci. Eur. 21904715. 35 (1). Available from: https://enveurope.springeropen.com. doi:10.1186/s12302-023-00779-4.

International Organization for Standardization, 2007. ISO/IEC Guide 99:2007: International Vocabulary of Metrology—Basic and General Concepts and Associated Terms (VIM), International Organization for Standardization, Geneva.

International Organization for Standardization, 2015. ISO 9001:2015: Quality Management Systems—Requirements, International Organization for Standardization, Geneva.

International Organization for Standardization, 2016. ISO 17034:2016: General Requirements for the Competence of Reference Material Producers, International Organization for Standardization, Geneva.

International Organization for Standardization, 2017. ISO/IEC 17025:2017: General Requirements for the Competence of Testing and Calibration Laboratories, International Organization for Standardization, Geneva.

International Organization for Standardization, 2022. ISO 13528: Statistical Methods for Use in Proficiency Testing by Interlaboratory Comparison, International Organization for Standardization, Geneva.

International Organization for Standardization, 2023. ISO/IEC 17043:2023: Conformity Assessment—General Requirements for the Competence of Proficiency Testing Providers, International Organization for Standardization, Geneva.

Iturrospe, E., da Silva, K.M., van de Lavoir, M., Robeyns, R., Cuykx, M., Vanhaecke, T., van Nuijs, A.L.N., Covaci, A., 2023. Mass spectrometry-based untargeted metabolomics and lipidomics platforms to analyze cell culture extracts. Methods Mol. Biol. 2571, 189–206. Available from: http://www.springer.com/series/7651. doi:10.1007/978-1-0716-2699-3_19.

Knolhoff, A.M., Premo, J.H., Fisher, C.M., 2021. A proposed quality control standard mixture and its uses for evaluating nontargeted and suspect screening LC/HR-MS method performance. Anal. Chem. 15206882. 93 (3), 1596–1603. Available from: http://pubs.acs.org/journal/ancham. doi:10.1021/acs.analchem.0c04036.

Konieczka, P., 2018. Quality Assurance and Quality Control in the Analytical Chemical Laboratory, vol. 294. Taylor & Francis Group, Boca Raton, p. 294. doi:10.1201/9781315295015.

Landrigan, P.J., Fuller, R., Acosta, N.J.R., Adeyi, O., Arnold, R., Basu, N., Baldé, A.B., Bertollini, R., Bose-O'Reilly, S., Boufford, J.I., Breysse, P.N., Chiles, T., Mahidol, C., Coll-Seck, A.M., Cropper, M.L., Fobil, J., Fuster, V., Greenstone, M., Haines, A., Hanrahan, D., Hunter, D., Khare, M., Krupnick, A., Lanphear, B., Lohani, B., Martin, K., Mathiasen, K.V., McTeer, M.A., Murray, C.J.L., Ndahimananjara, J.D., Perera, F., Potočnik, J., Preker, A.S., Ramesh, J., Rockström, J., Salinas, C., Samson, L.D., Sandilya, K., Sly, P.D., Smith, K.R., Steiner, A., Stewart, R.B., Suk, W.A., van Schayck, O.C.P., Yadama, G.N., Yumkella, K., Zhong, M., 2018. The Lancet Commission on pollution and health. Lancet. 1474547X. 391 (10119), 462–512. Available from: http://www.journals.elsevier.com/the-lancet/. doi:10.1016/S0140-6736(17)32345-0.

Lescord, G.L., Lepage, A.T., 2025. Making waves: report your quality assurance and control (QA/QC) information when publishing analytical data. Water Res. 18792448. 278. Available from: https://www.sciencedirect.com/science/journal/00431354. doi:10.1016/j.watres.2025.123167.

Lippa, K.A., Aristizabal-Henao, J.J., Beger, R.D., Bowden, J.A., Broeckling, C., Beecher, C., Clay Davis, W., Dunn, W.B., Flores, R., Goodacre, R., Gouveia, G.J., Harms, A.C., Hartung, T., Jones, C.M., Lewis, M.R., Ntai, I., Percy, A.J., Raftery, D., Schock, T.B., Sun, J., Theodoridis, G., Tayyari, F., Torta, F., Ulmer, C.Z., Wilson, I., Ubhi, B.K., 2022. Reference materials for MS-based untargeted metabolomics and lipidomics: a review by the metabolomics quality assurance and quality control consortium (mQACC). Metabolomics. 15733890. 18 (4). Available from: https://link.springer.com/journal/11306. doi:10.1007/s11306-021-01848-6.

Matuszewski, B.K., Constanzer, M.L., Chavez-Eng, C.M., 2003. Strategies for the assessment of matrix effect in quantitative bioanalytical methods based on HPLC-MS/MS. Anal. Chem. 00032700. 75 (13), 3019–3030. doi:10.1021/ac020361s.

NORMAN Network, 2009. Protocol for the validation of chemical and biological monitoring methods. Network of reference laboratories and related organisations for monitoring and biomonitoring of emerging environmental pollutants. Available from: https://www.norman-network.com/.

Nübler, S., Schäfer, M., Haji-Abbas-Zarrabi, K., Marković, S., Marković, K., Esteban López, M., Castaño, A., Mol, H., Koch, H.M., Antignac, J.-P., Hajslova, J., Thomsen, C., Vorkamp, K., Göen, T., 2022. Interlaboratory comparison investigations (ICIs) for human biomonitoring of chromium as part of the quality assurance programme under HBM4EU. J. Trace Elem. Med. Biol. 0946672X. 70, 126912. doi:10.1016/j.jtemb.2021.126912.

Olivares, I.R.B., Souza, G.B., Nogueira, A.R.A., Toledo, G.T.K., Marcki, D.C., 2018. Trends in developments of certified reference materials for chemical analysis—focus on food, water, soil, and sediment matrices. TrAC Trends Anal. Chem. 18793142. 100, 53–64. Available from: www.elsevier.com/locate/trac. doi:10.1016/j.trac.2017.12.013.

Organisation for Economic Co-operation and Development, OECD, 1998. Series on principles of good laboratory practice and compliance monitoring number 1. Paris: Organisation for Economic Co-operation and Development, Environmental Health and Safety Publications. Available from: https://www.oecd.org/content/dam/oecd/en/publications/reports/1998/01/oecd-principles-on-good-laboratory-practice_g1gh32e8/9789264078536-en.pdf.

Queensland Government, 2009. Quality Control for Water and Sediment Sampling. Environmental Protection (Water) Policy.

Quevauviller, P., Maier, E.A., 2007. Quality assurance and quality control for environmental monitoring. In: Quevauviller, P. (Ed.), Quality Assurance in Environmental Monitoring: Sampling and Sample Pretreatment. Wiley Blackwell, Belgium, pp. 1–24. Available from: http://onlinelibrary.wiley.com/book/10.1002/9783527615216. doi:10.1002/9783527615216.ch01.

Rivier, L., 2003. Criteria for the identification of compounds by liquid chromatography-mass spectrometry and liquid chromatography-multiple mass spectrometry in forensic toxicology and doping analysis. Anal. Chim. Acta 00032670. 492 (1–2), 69–82. Available from: http://www.journals.elsevier.com/analytica-chimica-acta/. doi:10.1016/S0003-2670(03)00889-4.

Sabbioni, G., Castaño, A., Esteban López, M., Göen, T., Mol, H., Riou, M., Tagne-Fotso, R., 2022. Literature review and evaluation of biomarkers, matrices and analytical methods for chemicals selected in the research program Human Biomonitoring for the European Union (HBM4EU). Environ. Int. 01604120. 169, 107458. doi:10.1016/j.envint.2022.107458.

Sandstrom, M., Lewis, J., 2009. Instructions for field use of spike solutions for organic-analyte samples. In: Field Spike Solutions. Version 1.0. US Geological Survey, Field spike solutions. USGS Publications Warehouse, pp. 1–8.

Sauvé, S., Desrosiers, M., 2014. A review of what is an emerging contaminant. Chem. Cent. J. 1752153X. 8 (1). Available from: http://journal.chemistrycentral.com/content/8/1/15. doi:10.1186/1752-153X-8-15.

Schindler, B.K., Esteban, M., Koch, H.M., Castano, A., Koslitz, S., Cañas, A., Casteleyn, L., Kolossa-Gehring, M., Schwedler, G., Schoeters, G., Hond, E.D., Sepai, O., Exley, K., Bloemen, L., Horvat, M., Knudsen, L.E., Joas, A., Joas, R., Biot, P., Aerts, D., Lopez, A., Huetos, O., Katsonouri, A., Maurer-Chronakis, K., Kasparova, L., Vrbík, K., Rudnai, P., Naray, M., Guignard, C., Fischer, M.E., Ligocka, D., Janasik, B., Reis, M.F., Namorado, S., Pop, C., Dumitrascu, I., Halzlova, K., Fabianova, E., Mazej, D., Tratnik, J.S., Berglund, M., Jönsson, B., Lehmann, A., Crettaz, P., Frederiksen, H., Nielsen, F., McGrath, H., Nesbitt, I., De Cremer, K., Vanermen, G., Koppen, G., Wilhelm, M., Becker, K., Angerer, J., 2014. The European COPHES/DEMOCOPHES project: towards transnational comparability and reliability of human biomonitoring results. Int. J. Hyg. Environ. Health. 1618131X. 217 (6), 653–661. Available from: www.urbanfischer.de/journals/intjhyg. doi:10.1016/j.ijheh.2013.12.002.

Schulze, B., Jeon, Y., Kaserzon, S., Heffernan, A.L., Dewapriya, P., O'Brien, J., Gomez Ramos, M.J., Ghorbani Gorji, S., Mueller, J.F., Thomas, K.V., Samanipour, S., 2020. An assessment of quality assurance/quality control efforts in high resolution mass spectrometry non-target workflows for analysis of environmental samples. TrAC Trends Anal. Chem. 18793142. 133. Available from: www.elsevier.com/locate/trac. doi:10.1016/j.trac.2020.116063.

Schulze, B., Kaserzon, S.L., 2024. Quality assurance and quality control (QA/QC) for high-resolution mass spectrometry (HRMS) non-target screening methods. Handb. Environ. Chem. 138, 23–40. Available from: https://www.springer.com/series/698. doi:10.1007/698_2024_1090.

Schymanski, E.L., Jeon, J., Gulde, R., Fenner, K., Ruff, M., Singer, H.P., Hollender, J., 2014. Identifying small molecules via high resolution mass spectrometry: communicating confidence. Environ. Sci. Technol. 15205851. 48 (4), 2097–2098. doi:10.1021/es5002105.

Tadić, Đ., Pires de Lima, A., Ricci, M., 2025. Quality assurance and quality control for human biomonitoring data—focus on matrix reference materials. Anal. Bioanal. Chem. 16182650. 417 (16), 3513–3528. Available from: https://www.springer.com/journal/216. doi:10.1007/s00216-025-05859-3.

Thakare, R., Chhonker, Y.S., Gautam, N., Alamoudi, J.A., Alnouti, Y., 2016. Quantitative analysis of endogenous compounds. J. Pharm. Biomed. Anal. 1873264X. 128, 426–437. Available from: www.elsevier.com/locate/jpba. doi:10.1016/j.jpba.2016.06.017.

Thompson, M., Ellison, S.L.R., Wood, R., 2006. The International Harmonized Protocol for the proficiency testing of analytical chemistry laboratories: (IUPAC technical report). Pure Appl. Chem. 00334545. 78 (1), 145–196. doi:10.1351/pac200678010145.

Thompson, M., Wood, R., 1995. Harmonized guidelines for internal quality control in analytical chemistry laboratories (Technical Report). Pure Appl. Chem. 1365-3075. 67 (4), 649–666. doi:10.1351/pac199567040649.

US Food and Drug Administration, 2021. Q2(R2) Validation of Analytical Procedures. Available from: https://www.fda.gov/regulatory-information/search-fda-guidance-documents/q2r2-validation-analytical-procedures.

US Food and Drug Administration, 2024. Q2(R2) validation of analytical procedures—guidance for industry. Available from: https://www.fda.gov/regulatory-information/search-fda-guidance-documents/q2r2-validation-analytical-procedures.

Valcárcel, M., Lucena, R., 2013. Synergistic relationships between analytical chemistry and written standards. Anal. Chim. Acta. 18734324. 788, 1–7. doi:10.1016/j.aca.2013.04.008.

van Nuijs, A.L.N., Lai, F.Y., Been, F., Andres-Costa, M.J., Barron, L., Baz-Lomba, J.A., Berset, J.D., Benaglia, L., Bijlsma, L., Burgard, D., Castiglioni, S., Christophoridis, C., Covaci, A., de Voogt, P., Emke, E., Fatta-Kassinos, D., Fick, J., Hernandez, F., Gerber, C., González-Mariño, I., Grabic, R., Gunnar, T., Kannan, K., Karolak, S., Kasprzyk-Hordern, B., Kokot, Z., Krizman-Matasic, I., Li, A., Li, X., Löve, A.S.C., Lopez de Alda, M., McCall, A.K., Meyer, M.R.,

Oberacher, H., O'Brien, J., Quintana, J.B., Reid, M., Schneider, S., Simoes, S.S., Thomaidis, N.S., Thomas, K., Yargeau, V., Ort, C., 2018. Multi-year inter-laboratory exercises for the analysis of illicit drugs and metabolites in wastewater: development of a quality control system. TrAC Trends Anal. Chem. 18793142. 103, 34–43. Available from: www.elsevier.com/locate/trac. doi:10.1016/j.trac.2018.03.009.

Visconti, G., Boccard, J., Feinberg, M., Rudaz, S., 2023. From fundamentals in calibration to modern methodologies: a tutorial for small molecules quantification in liquid chromatography–mass spectrometry bioanalysis. Anal. Chim. Acta. 00032670. 1240, 340711. doi:10.1016/j.aca.2022.340711.

Vorkamp, K., Esteban López, M., Gilles, L., Göen, T., Govarts, E., Hajeb, P., Katsonouri, A., Knudsen, L.E., Kolossa-Gehring, M., Lindh, C., Nübler, S., Pedraza-Díaz, S., Santonen, T., Castaño, A., 2023. Coordination of chemical analyses under the European Human Biomonitoring Initiative (HBM4EU): concepts, procedures and lessons learnt. Int. J. Hyg. Environ. Health. 14384639. 251, 114183. doi:10.1016/j.ijheh.2023.114183.

Wilkinson, M.D., Dumontier, M., Aalbersberg, I.J., Appleton, G., Axton, M., Baak, A., Blomberg, N., Boiten, J.-W., Bonino da Silva Santos, L., Bourne, P.E., Bouwman, J., Brookes, A.J., Clark, T., Crosas, M., Dillo, I., Dumon, O., Edmunds, S., Evelo, C.T., Finkers, R., Gonzalez-Beltran, A., Gray, A.J.G., Groth, P., Goble, C., Grethe, J.S., Heringa, J., 't Hoen, P.A.C., Hooft, R., Kuhn, T., Kok, R., Kok, J., Lusher, S.J., Martone, M.E., Mons, A., Packer, A.L., Persson, B., Rocca-Serra, P., Roos, M., van Schaik, R., Sansone, S.-A., Schultes, E., Sengstag, T., Slater, T., Strawn, G., Swertz, M.A., Thompson, M., van der Lei, J., van Mulligen, E., Velterop, J., Waagmeester, A., Wittenburg, P., Wolstencroft, K., Zhao, J., Mons, B., 2016. The FAIR guiding principles for scientific data management and stewardship. Sci. Data. 2052-4463. 3 (1). doi:10.1038/sdata.2016.18.

Wise, S.A., Phillips, M.M., 2019. Evolution of reference materials for the determination of organic nutrients in food and dietary supplements—a critical review. Anal. Bioanal. Chem. 16182650. 411 (1), 97–127. Available from: http://link.springer.de/link/service/journals/00216/index.htm. doi:10.1007/s00216-018-1473-0.

Yuan, C., Chen, D., Wang, S., 2015. Drug confirmation by mass spectrometry: identification criteria and complicating factors. Clin. Chim. Acta 18733492. 438, 119–125. Available from: www.elsevier.com/locate/clinchim. doi:10.1016/j.cca.2014.08.021.

Index

Note: Page numbers followed by "*f*" indicate figures, "*t*" indicate tables.

L

M

N

T

U

V

W

Printed and bound by CPI Group (UK) Ltd, Croydon, CR0 4YY
25/09/2026
02222749-0001